T0336919

D-WAVE SUPERCONDUCTIVITY

This volume provides a comprehensive introduction to the theory of d-wave superconductivity, focused on d-wave pairing symmetry and its physical consequences in the superconducting state. It discusses the basic concepts and methodologies related to high-temperature superconductivity and compares experimental phenomena with theoretical predictions. After a brief introduction to the basic theory of superconductivity and several models for high-temperature superconductivity, this book presents detailed derivations and explanations for various single-particle and collective properties of d-wave superconductors that can be monitored experimentally, including thermodynamics, angular-resolved photo-emission, single-particle and Josephson tunneling, impurity scattering, magnetic and superfluid responses, transport and optical properties, and mixed states. Various universal behaviors of d-wave superconductors are highlighted. Aimed primarily at graduate students and research scientists in condensed matter and materials physics, this text enables readers to understand systematically the physical properties of high-temperature superconductors.

TAO XIANG is a Professor at the Institute of Physics, Chinese Academy of Sciences (CAS), working on condensed matter physics. He is an elected CAS member and a fellow of the World Academy of Sciences. He has received the Ho Leung Ho Lee Prize for Scientific and Technological Progress and several other awards.

CONGJUN WU is a Professor at Westlake University, working on exploring new states of matter in condensed matter and cold atom systems, including superconductivity, magnetism, orbital physics, topological states, and quantum Monte-Carlo simulations. He was elected to be a fellow of the American Physical Society in 2018, and awarded the Sloan Research Fellowship in 2008.

D-WAVE SUPERCONDUCTIVITY

TAO XIANG

Chinese Academy of Sciences, Beijing

CONGJUN WU

Westlake University, Hangzhou

CAMBRIDGE
UNIVERSITY PRESS

University Printing House, Cambridge CB2 8BS, United Kingdom

One Liberty Plaza, 20th Floor, New York, NY 10006, USA

477 Williamstown Road, Port Melbourne, VIC 3207, Australia

314–321, 3rd Floor, Plot 3, Splendor Forum, Jasola District Centre,
New Delhi – 110025, India

103 Penang Road, #05–06/07, Visioncrest Commercial, Singapore 238467

Cambridge University Press is part of the University of Cambridge.

It furthers the University's mission by disseminating knowledge in the pursuit of
education, learning, and research at the highest international levels of excellence.

www.cambridge.org
Information on this title: www.cambridge.org/9781009218597
DOI: 10.1017/9781009218566

First published 2022

A catalogue record for this publication is available from the British Library.

ISBN 978-1-009-21859-7 Hardback

Contents

Preface

Superconductivity, a state wherein a conductor exhibits perfect diamagnetism and conducts electricity without resistance below a critical temperature T_c, is believed to result from the formation and condensation of so-called Cooper pairs. A Cooper pair behaves like a boson. It is formed when two electrons couple together to become a bound state through a weak attractive interaction induced by the electron–lattice or other interactions. The order parameter that governs the physical properties of a superconductor is the binding energy of Cooper pairs, which is also called the gap function. The gap function, on the other hand, is determined by the internal wave function of a Cooper pair and classified according to its rotational symmetry, known as the pairing symmetry. The pairing symmetry dictates the momentum dependence of the superconducting order parameter. It carries the birthmark of interactions that glue electrons together to form Cooper pairs, being one of the key parameters that need to be unambiguously determined in order to unveil the mechanism of superconductivity.

The establishment of d-wave superconductivity in cuprate superconductors is one of the most important progresses achieved in the investigation of the high-T_c mechanism after the discovery of these systems by Bednorz and Müller in 1986 [1]. Supported by numerous experimental as well as theoretical studies, it demonstrates that, unlike most conventional metallic superconductors, the superconducting electrons in high-T_c copper oxides possess the so-called d-wave pairing symmetry, in which two electrons inside each pair carry a relative angular momentum of two. For comparison, most metallic superconductors exhibit s-wave pairing symmetry and each Cooper pair carries a relative angular momentum of zero. The gap parameter of an s-wave superconductor is finite over the entire Fermi surface, and the corresponding low energy excitations are thermally activated. In contrast, the energy gap of a d-wave superconductor has nodes on the Fermi surface at which the activation energy is zero. These two types of superconductor exhibit qualitatively different behaviors in their linear and nonlinear responses to thermal, physical, or chemical pressure, and to electromagnetic excitations. To quantify and reveal these differences, through comparative and comprehensive theoretical and experimental investigations, has been one of the central themes in the study of high-T_c superconductivity.

The significant development of d-wave or other unconventional superconductivity with nontrivial pairing symmetry in the past three decades has greatly broadened our knowledge beyond conventional s-wave superconductivity. However, a systematic introduction to the theory of d-wave superconductivity is still not available. Traditional textbooks on superconductivity [2–4] mainly deal with metallic superconductors, focusing on the phenomenological and microscopic description of s-wave superconductors. A number of review articles published in recent years have discussed the physical properties of d-wave superconductors but focused more on experimental observations. Many theoretical works related to the topics discussed in this book are scattered in journals, so it is rather difficult even for an experienced research scientist to gain a global picture of d-wave superconductivity just by skimming through the literature.

This book is intended to fill this gap by giving a systematic introduction to the theory of d-wave superconductivity, concentrating on d-wave pairing symmetry and its physical consequences in the superconducting state. It is written based on the progress achieved in the study of cuprate superconductors. For comparison with theoretical predictions, some experimental results obtained in high-T_c copper oxides are analyzed and summarized. This book does not aim at providing an introduction to a topic to which a coherent and unified theoretical picture has not yet been established. Neither the microscopic origin of d-wave superconductivity nor the puzzling normal state properties of cuprate superconductors, such as the pseudogap and the linear resistivity, will be discussed.

The d-wave superconductor is one kind of non-s-wave superconductor. It is also one of the representatives of the entire family of unconventional superconductors. The theory and formulas introduced in this book hold more generally. They can be applied directly or with slight modification to other unconventional superconductors with or without gap nodes, including, for example, $s + id$ and p-wave superconductors. Iron-based superconductors, first discovered by Hosono and coworkers in 2008 [5], belong to another family of high-T_c superconductors. Similar as in cuprate superconductors, antiferromagnetic fluctuations and their interplay with orbital ordering play an important role in pairing electrons in these materials. However, iron-based superconductors are multi-band systems whose pairing symmetry has not been indisputably determined. It is commonly believed that the pairing gap on each band has simple s-wave symmetry, but there are sign changes on different bands. Loosely speaking, an iron-based superconductor could be regarded as a sign-change superconductor. As the sign change in the gap function is also a characteristic feature of a d-wave energy gap, some physical pictures drawn from d-wave superconductors without particularly considering the contribution of the gap nodes can also be used to understand physical behaviors of iron-based superconductors.

The book is aimed primarily at graduate students and researchers in condensed matter and materials physics. We attempt to provide an encyclopedic introduction to d-wave superconductivity with an overarching perspective on unifying concepts and methodologies. It is assumed that the reader is familiar with the elements of quantum mechanics, second quantization, and Green's functions [6]. We start from a brief introduction to the fundamental phenomena and the Bardeen–Cooper–Schrieffer (BCS) theory of superconductivity in Chapter 1. This chapter also includes some concepts that are important to the understanding of superconductivity but not well introduced in the standard textbooks. Chapter 2 introduces the microscopic models of high-temperature superconductivity. This chapter is relatively independent and provides a background that may facilitate the readers toward a more in-depth study of the microscopic mechanism of d-wave superconductivity. The later chapters systematically introduce the physical properties of different thermodynamic and electrodynamic response functions of d-wave superconductors. These chapters highlight various universal behaviors of d-wave superconductors. Detailed derivations of theoretical formulas, together with detailed comparisons with the experimental results of cuprate superconductors, are presented. In addition, there are seven appendices that provide a detailed introduction to some of the theorems and mathematical formulas and methods used in the main text.

This book is a revised and updated version of a Chinese book [7] of the same title published by one of the authors, Tao Xiang, in 2007. Both authors have participated in the revision of this book. Besides the changes that are made to reflect more comprehensively the progress achieved in theoretical and experimental studies of d-wave superconductivity in recent years, and to provide more background information on the general properties of superconductivity with more detailed deviations, we have added a new chapter, §13, and a number of new sections, including §1.10, §1.16, §7.8, §14.1, and two appendices (F and G).

In writing this book, we received warm encouragements and kind support from Changde Gong, Zhao-bin Su, Lu Yu, and Zhongxian Zhao. We have also benefited from discussions with Jun Chang, Xianhui Chen, Shiping Feng, Rushan Han, W. N. Hardy, Jianxin Li, Chengshi Liu, Honggang Luo, Jianlin Luo, Bruce Normand, C. Panagopoulos, Yuehua Su, Nanlin Wang, Qianghua Wang, Haihu Wen, Zhengyu Weng, J. M. Wheatley, Liping Yang, Guangming Zhang, Qingming Zhang, and many other colleagues. We would like to express our heartfelt thanks to them.

Abbreviations

- AC — alternating current
- ARPES — angle-resolved photo–emission spectroscopy
- BCS theory — Bardeen–Cooper–Schrieffer theory
- BdG equation — Bogoliubov–de Gennes equation
- BTK — Blonder–Tinkham–Klapwijk
- DC — direct current
- EDC — energy distribution curve
- GL — Ginzburg–Landau
- H_{c1} — lower critical field
- H_{c2} — upper critical field
- MDC — momentum distribution curve
- NMR — nuclear magnetic resonance
- ODLRO — off-diagonal long-range order
- QPI — quasiparticle interference
- SQUID — superconducting quantum interference device
- STM — scanning tunneling microscopy
- T_c — superconducting transition temperature

1

Introduction to Superconductivity

1.1 Basic Properties of Superconductivity

Superconductivity, as an emergent macroscopic quantum phenomenon, is one of the most important subjects of contemporary condensed matter physics. It was first discovered by Dutch physicist Heike Kamerlingh Onnes on April 8, 1911 [8–10]. In 1908, Onnes and his assistants successfully liquefied helium and for the first time reached low temperatures below 4.25K. This was a historic breakthrough for low temperature physics. When they applied this technique and measured the resistance of mercury, they found that its resistance dropped abruptly from $0.1\,\Omega$ to below $10^{-6}\,\Omega$ within a narrow temperature range of 0.01 K around 4.2 K. This important discovery opened up the field of superconductivity and related applications. It also greatly stimulated the study of quantum emergent phenomena in condensed matter physics.

Understanding the phenomena and exploring the mechanism of superconductivity are historically important in the development of condensed matter physics. In the early days, condensed matter physics was not considered as fundamental as quantum field theory by the mainstream of physics. Various classical and quantum mechanical theories were developed to study solid state phenomena, such as the Drude theory of transport, the Sommerfeld theory of electrons, the Debye theory of phonons, and the Bloch theory of energy band structures. However, there were few original fundamental principles arising from this field. This situation was changed when the mechanism of superconductivity as well as that of superfluidity was revealed.

A superconductor has two characteristic electromagnetic features, namely zero direct current resistance and perfect diamagnetism. Zero resistance means that superconductors are ideal conductors, and there is no energy loss during electric energy transport using superconducting transmission lines. Moreover, superconductors are more than just ideal conductors. More fundamentally, superconductors exhibit perfect diamagnetism which expels magnetic flux lines from the interior of superconductor. The external magnetic field can only penetrate into superconductors within a short length scale near the surface called the penetration length. The perfect diamagnetism of superconductivity was discovered by W. Meissner and R. Ochsenfeld in 1933. It is also called the Meissner effect [11]. The Meissner effect is *not* a consequence of zero resistance but an independent fundamental property resulting from the phase coherence of superconductivity.

The Meissner effect distinguishes a superconductor from an ideal normal conductor in their responses to an applied magnetic field. If a magnetic field is applied to a normal metal, Faraday's law, or Lentz's law, says that a screening eddy current is induced to expel the magnetic flux. However, due to the existence of resistance, the induced eddy current dissipates and eventually decays to zero, allowing the magnetic field to penetrate into the interior of the conductor. On the other hand, if the magnetic field is applied to an ideal conductor or a superconductor at low temperatures, as there is no resistance in either case, a persistent eddy current exists which expels the magnetic field from within the bulk. Now if the temperature is raised so that both systems return back to their normal metallic states, the magnetic field penetrates to the bulks again. So far we have not seen any difference between a superconductor and an ideal conductor.

A sharp contrast between an ideal conductor and a superconductor appears when both systems are cooled down. In an ideal conductor, the magnetic field remains inside the system, while in a superconductor, the magnetic field is expelled to the outside. Thus, for an ideal conductor, it matters if it is field cooled or zero field cooled, whereas for a superconductor, regardless of the external field and its history, the magnetic field becomes zero inside the bulk.

The zero resistance and the Meissner effect are two defining properties of superconductors that cannot be understood in the framework of the single-electron theory, or, the band theory. In the macroscopic world, dissipation and friction are nearly unavoidable. How can electric currents be free of dissipation? Diamagnetism is found in nearly all materials, but it is generally very weak and can only be observed in materials that do not exhibit other forms of magnetism. The perfect diamagnetism exhibited in superconductors is even more puzzling than the appearance of zero resistance. Quite a number of noble metals, such as gold, silver, and cooper, are in fact not superconducting at all at ambient pressure. Thus superconductivity is not a consequence of weak dissipation. Instead, it is a macroscopic phenomenon, resulting from the collective interplay of electrons.

The superconducting state is a distinct thermodynamic phase. It occurs when the temperature is reduced below a critical temperature, denoted as T_c, through a second order phase transition in the absence of an external magnetic field. The superconducting transition temperatures are generally below 25 K. High temperature superconductors are ideally defined as materials that superconduct at temperatures above the boiling point of liquid nitrogen, i.e. 77 K. However, in the literature, materials with T_c close to or larger than 40 K are all referred to as high-T_c superconductors. The high-T_c superconductors that have been discovered include: (1) hole or electron doped perovskite copper oxides, first discovered by Bednorz and Müller in 1986 [1]; (2) electron or hole doped iron pnictides [5] or chalcogenides [12], first discovered by Hosono and coworkers in 2008; (3) superhydride compounds under ultrahigh pressure, anticipated by Ashcroft for metallic hydrogen [13] and hydrogen enriched materials [14], and first confirmed experimentally in H_3S by Drozdov et al. in 2015 [15]; (4) Magnesium diboride with $T_c \sim 39$ K, discovered by Nagamatsu et al. in 2001 [16]. The current highest T_c record holder is $HgBa_2Ca_2Cu_3O_{8+\delta}$ (133 K) at ambient pressure [17], and carbonaceous sulfur hydride (288 K) under 267 GPa [18].

The phase transition from a normal metallic or insulating state to a superconducting state corresponds to the formation of superconducting long-range order. Different from ferromagnetism, the superconducting order is an off-diagonal long-range order which does not have a classical correspondence [19]. By lowering temperatures, there exists a critical temperature range within which the resistance drops to zero. The width of this critical region is determined by the fluctuation of superconducting order parameter. In conventional metal-based superconductors, this critical temperature range is very narrow, and the resistance drops to zero abruptly. However, in high-T_c copper oxides or iron-based superconductors, or in dirty superconductors of metals and alloys, fluctuations are strong. The corresponding critical regions are broad and the resistance drops are relatively slow.

A superconductor has exactly zero direct-current resistance and is able to maintain an electric current without generating an external voltage in the superconducting state. It loses the superconducting phase coherence and exhibits a small but finite resistance in the presence of an alternative current. One can also turn a superconductor into a normal conductor by applying a strong magnetic field or a direct electric current. For a given temperature, the highest applied magnetic field or electric current under which a material remains superconducting are called the upper critical field or the critical current.

1.2 Two-Fluid Model and London Equations

Historically, an important phenomenological theory of superconductivity is the two-fluid model first proposed by Groter and Casimir [20]. The key assumption of this model is the existence of two different types of electrons in superconductors, namely normal and superconducting electrons. The density of normal electrons is called the normal fluid density and that of superconducting electrons is called the superfluid density. The sum of these two kinds of densities gives the total density of electrons. Normal electrons carry entropy and behave similarly as in ordinary metals. Their states are changed by scattering with phonons and impurities. In contrast, super-conducting electrons are resistance free. They do not carry entropy and have no contribution to thermodynamic quantities such as the specific heat. A static electric field cannot exist in an equilibrium superconducting state. Otherwise, supercon-ducting electrons would be accelerated without attenuation, leading to a divergent electric current. The existence of superconducting electrons with zero electric field explains why the resistance is zero. However, the two-fluid model does not answer the question of how superconducting electrons are formed, neither can it explain the Meissner effect.

In order to explain the Meissner effect, Fritz and Hentz London brothers pro-posed an electromagnetic equation [21] to describe the superconducting current. This equation connects the superconducting current density $\mathbf{J}_s$ with the electromag-netic vector potential $\mathbf{A}$. Under the Coulomb gauge (also known as the transverse gauge) where $\nabla \cdot \mathbf{A} = 0$, it can be expressed as

$$\mathbf{J}_s = -\frac{n_s e^2}{m} \mathbf{A}, \tag{1.1}$$

where n_s is the superfluid density of electrons. This equation is called the London equation. It cannot be deduced from the Maxwell equations and should be viewed as an independent electromagnetic equation by treating superconductors as a special class of electromagnetic media.

The London equation could be rigorously derived only after the microscopic the-ory of superconductivity has been established. For better understanding its physical meaning, a heuristic argument is commonly given to formally "derive" this equation within theory of classical electromagnetism. A basic assumption is that electrons are moving in a frictionless state, so that

$$m\dot{\mathbf{v}}_s = -e\mathbf{E}, \tag{1.2}$$

where $\mathbf{v}_s$ is the velocity of superconducting electrons and $\mathbf{E}$ is the electric field. The supercurrent $\mathbf{J}_s = -en_s\mathbf{v}_s$ is then governed by the equation

$$\frac{\partial \mathbf{J}_s}{\partial t} = \frac{e^2 n_s}{m} \mathbf{E}, \tag{1.3}$$

which is referred to as the first London equation. Then, using the Maxwell equation,

$$\nabla \times \mathbf{E} = -\frac{\partial \mathbf{B}}{\partial t}, \tag{1.4}$$

we immediately arrive at

$$\frac{\partial}{\partial t} \left(\nabla \times \mathbf{J}_s + \frac{e^2 n_s}{m} \mathbf{B} \right) = 0. \tag{1.5}$$

This describes the behavior of an ideal conductor. To describe the Meissner effect, the constant of integration must be chosen to zero so that

$$\nabla \times \mathbf{J}_s + \frac{e^2 n_s}{m} \mathbf{B} = 0. \tag{1.6}$$

This is the second London equation.

The two London equations can be combined into a single one, i.e. Eq. (1.1), in terms of the vector potential in the Coulomb gauge. One can also write the London equation in an arbitrarily chosen gauge. In that case, the London equation becomes

$$\mathbf{J}_s = \frac{e^2 n_s}{m} \left(-\mathbf{A} + \nabla \varphi\right), \tag{1.7}$$

which differs from Eq. (1.1) by a gradient of a scalar field φ. Later on, we will see that this scalar field is just the condensation phase field and the corresponding term reflects the nonlocal effect of electromagnetic responses. $\nabla \varphi$ is to shift the vector potential from an arbitrary gauge to the Coulomb gauge.

If the second London equation is manipulated by applying Ampere's law,

$$\nabla \times \mathbf{B} = \mu_0 \mathbf{J}_s, \tag{1.8}$$

it turns into the Helmholtz equation for the magnetic field:

$$\nabla^2 \mathbf{B} = \frac{\mu_0 n_s e^2}{m} \mathbf{B}. \tag{1.9}$$

In a semi-infinite plate of superconductor with its surface perpendicular to the x-direction, the solution of Eq. (1.9) is simply given by

$$B(x) = B(x_0) e^{-(x-x_0)/\lambda}, \tag{1.10}$$

where

$$\lambda = \sqrt{\frac{m}{\mu_0 n_s e^2}} \tag{1.11}$$

is the London penetration depth describing the decay length of an external magnetic field and x_0 is the x-coordinate of the superconductor-vacuum interface. In the limit $x - x_0 \gg \lambda$, the magnetic field decays to zero. This gives a phenomenological explanation to the Meissner effect.

In spite of its simplicity, the two-fluid model captures the key features of superconductors. The key concepts – the normal and superconducting electrons – were broadly used in the construction of the microscopic theory of superconductivity. The normal and superconducting electrons correspond to the quasiparticle excitations and the superconducting paired electrons, respectively. The two-fluid model has played an important role in the study of superconductivity, although it does not explain the microscopic mechanism of superconductivity. Even after the establishment of the microscopic theory of superconductivity, it is still useful to apply the two-fluid model to understand qualitatively experimental results of superconductors.

1.3 Cooper Pairing

Superconductivity is a quantum many-body effect and cannot be understood based on the single-electron theory and its perturbative expansion. In 1956, Cooper considered a two-electron problem which turned out to be one of the most crucial steps toward a microscopic understanding of superconductivity [22]. He showed that if there exists an effective attraction interaction, no matter how weak it is, between two electrons in a background of the Fermi sea, the Fermi surface is no longer stable. Electrons on the Fermi surface will pair each other to form bound states, so that the ground state energy is reduced. The bound state of paired electrons is called a Cooper pair.

The Cooper instability results from the interplay between the weak attractive interaction and the Fermi sea. The appearance of the Fermi sea is crucial. Otherwise, the Cooper pairing instability would not happen in an arbitrarily weak attractive potential. In free space, two electrons can form a bound state only if the attractive interaction between them is sufficiently strong (above a finite threshold) in three dimensions.

The proof given by Cooper is based on a simple variational calculation. He considered how the ground state energy is changed by adding two extra electrons with opposite momenta and spins to a filled Fermi sea at zero temperature. Due to the Pauli exclusion principle, these two electrons can only be put outside the Fermi sea. For simplicity in the calculation, he assumed that the attractive potential is nonzero only when both electrons lie between the Fermi energy E_F and $E_F + \omega_D$, and the amplitude of the potential V_0 is momentum independent. Here the cutoff ω_D is a characteristic energy scale determined by the mechanism or resource from which the attraction is induced. If the effective attraction is induced by the electron–phonon interaction, ω_D is just the characteristic frequency of phonons, namely the Debye frequency. After a simple variational calculation, Cooper found that the two electrons form a bound state with the binding energy

$$\Delta E = 2\Delta = -2\hbar\omega_D e^{-2/N_F g}, \qquad (1.12)$$

where N_F is the electron density of states on the Fermi surface, and g is the coupling strength. This is also the energy needed to break a Cooper pair. This result shows that the Fermi surface is unstable against a small attractive interaction. It also reveals two important parameters in describing a superconducting state. One is the characteristic attraction energy scale ω_D, and the other is the dimensionless coupling constant defined by the product of the density of states at the Fermi level and the depth of the attractive interaction. As discussed later, these two parameters also determine the superconducting transition temperature T_c. The calculation made by Cooper is simple, but it captured the main character of superconductivity.

Equation (1.12) shows that the dependence of the binding energy on the interaction strength g is singular. It implies that the microscopic theory of superconductivity cannot be established through perturbative calculations based on normal conducting states. This is actually the major difficulty in the study of the superconducting mechanism, which obstructed the development of a microscopic theory of superconductivity for nearly fifty years after its discovery.

To see more clearly how the Cooper pairing energy comes about, let us follow Cooper to solve a simple model of two electrons added to a rigid Fermi sea at zero temperature. It is assumed that the two electrons interact with each other but not with those in the Fermi sea. To reduce the repulsive interaction applied by the exclusion principle, the two electrons should form a spin singlet so that their spin wave function is antisymmetric and their spatial wave function is symmetric. Moreover, the lowest energy state should have zero total momentum so that the electrons must have opposite momenta. Therefore, the wave function has the form

$$|\Psi\rangle = \sum_{\mathbf{k}} \alpha(\mathbf{k}) c_{\mathbf{k}\uparrow}^{\dagger} c_{-\mathbf{k}\downarrow}^{\dagger} |0\rangle, \tag{1.13}$$

where $|0\rangle$ is the vacuum composed of the rigid Fermi sea.

This interacting system of two electrons is governed by the Hamiltonian

$$H = \sum_{\mathbf{k}} (\varepsilon_{\mathbf{k}} - \mu) c_{\mathbf{k}\sigma}^{\dagger} c_{\mathbf{k}\sigma} - \sum_{\mathbf{k},\mathbf{k}'} V_{\mathbf{k}\mathbf{k}'} c_{\mathbf{k}\uparrow}^{\dagger} c_{-\mathbf{k},\downarrow}^{\dagger} c_{-\mathbf{k}',\downarrow} c_{\mathbf{k}',\uparrow}, \tag{1.14}$$

where $\varepsilon_{\mathbf{k}}$ is the energy dispersion of electrons and μ is the chemical potential. $V_{\mathbf{k},\mathbf{k}'}$ is the scattering potential between two Cooper pairs with momenta $(\mathbf{k}\uparrow, -\mathbf{k}\downarrow)$ and $(\mathbf{k}'\uparrow, -\mathbf{k}'\downarrow)$. For simplicity, the attractive interaction between the two electrons is assumed to be momentum independent and to take a simple form

$$V_{\mathbf{k},\mathbf{k}'} = \frac{g}{V}, \tag{1.15}$$

with V the system volume. From the Schrödinger equation

$$H|\Psi\rangle = E|\Psi\rangle, \tag{1.16}$$

we find the equation that $\alpha(\mathbf{k})$ satisfies

$$2\xi_{\mathbf{k}}\alpha(\mathbf{k}) - \frac{g}{V} \sum_{\mathbf{k}'} \alpha(\mathbf{k}') = (E - E_0)\alpha(\mathbf{k}), \tag{1.17}$$

where E_0 is the energy of the filled Fermi sea and

$$\xi_{\mathbf{k}} = \varepsilon_{\mathbf{k}} - \mu. \tag{1.18}$$

Equation (1.17) can be rewritten as

$$\alpha(\mathbf{k}) = \frac{g}{2\xi_{\mathbf{k}} - \Delta E} \frac{1}{V} \sum_{\mathbf{k}'} \alpha(\mathbf{k}'), \tag{1.19}$$

where $\Delta E = E - E_0$ is the energy gap of the system with respect to the vacuum. Summing over all momentum points allows $\alpha(\mathbf{k})$ to be cancelled out from both sides. This leads to the gap equation

$$\frac{1}{g} = \frac{1}{V} \sum_{\mathbf{k}} \frac{1}{2\xi_{\mathbf{k}} - \Delta E} = N_F \int_0^{\hbar\omega_D} d\xi \frac{1}{2\xi - \Delta E}. \tag{1.20}$$

By solving this equation, we find that

$$\frac{1}{g} = \frac{N_F}{2} \ln \frac{2\hbar\omega_D - \Delta E}{-\Delta E} \approx \frac{N_F}{2} \ln \frac{2\hbar\omega_D}{|\Delta E|}. \tag{1.21}$$

This yields the result shown in Eq. (1.12).

1.4 BCS Mean Field Theory

In 1957, John Bardeen, Leon Cooper, and John Robert Schrieffer (BCS) proposed the microscopic theory of superconductivity based on the concept of Cooper pairing [23]. Their work established a fundamental theory of superconductivity. It also provided tremendous progress toward the understanding of microscopic quantum world.

In the BCS framework, there are two preconditions for the formation of super-conducting condensation. The first is the formation of Cooper pairs through an attraction interaction. The second is the development of phase coherence among Cooper pairs. Cooper pairing refers to the process that electrons near the Fermi surface form bound states. It is a prerequisite of superconductivity because Cooper pairs carry the feature of bosons that eliminates the effective repulsion induced by the Fermi statistics of electrons, and can condense into a superfluid state by forming phase coherence. Cooper pairs are found to exist in all superconductors discovered so far. This gives strong support to the BCS theory.

The BCS work is a variational theory. It is based on the BCS variational wave-function first proposed by Schrieffer. This wavefunction generalizes the solution of Cooper pair to a many-body system. It captures the main picture of Cooper for the superconducting condensation of paired electrons. The BCS theory is equivalent to the mean-field theory later developed based on the Bogoliubov transformation.

This mean-field theory is to take the Gaussian or saddle-point approximation in the framework of quantum field theory. It handles the thermal average of operators, rather than the variational wavefunction of the ground state. Fluctuations of Cooper pairs around the saddle point can be included, for example, by taking the one-loop expansion in the path-integral formulism.

The BCS mean field theory starts by considering the reduced pairing Hamiltonian defined by Eq. (1.14). This Hamiltonian is a simplification to the complex interactions of electrons. It highlights the interaction in the pairing channel and neglects interactions in other channels.

Equation (1.14) is applicable to superconductors with spin singlet pairing. It can be extended to describe spin triplet superconductors with slight modifications. This Hamiltonian considers the Cooper pairs with zero center-of-mass momentum, and neglects the pairing with finite center-of-mass momentum. The zero momentum pairing is physically reasonable because the phase space for the finite momentum pairing is strongly constrained by the Fermi surface geometry and by the momentum conservation [6]. In an external magnetic field, the Fermi surfaces of up- and down-spin electrons are split, and the pairing with finite center-of-mass momentum is favored. Cooper pairs in a current-carrying superconducting state have finite pairing momenta. But the pairing energy is suppressed and becomes zero when the current exceeds a critical current.

To define

$$A = \sum_{\mathbf{k}} c_{-\mathbf{k}\downarrow} c_{\mathbf{k}\uparrow}, \tag{1.22}$$

we can rewrite the BCS reduced Hamiltonian as

$$H = \sum_{\mathbf{k}\sigma} \xi_{\mathbf{k}} c_{\mathbf{k}\sigma}^{\dagger} c_{\mathbf{k}\sigma} - \frac{g}{V} A^{\dagger} A. \tag{1.23}$$

Taking the mean-field approximation for the interaction term,

$$-A^{\dagger} A = -\langle A^{\dagger} \rangle A - \langle A \rangle A^{\dagger} + \langle A^{\dagger} \rangle \langle A \rangle, \tag{1.24}$$

we obtain the BCS mean-field Hamiltonian

$$H_{MF} = \sum_{\mathbf{k}} \left(\sum_{\sigma} \xi_{\mathbf{k}} c_{\mathbf{k}\sigma}^{\dagger} c_{\mathbf{k}\sigma} + \Delta c_{\mathbf{k}\uparrow}^{\dagger} c_{-\mathbf{k}\downarrow}^{\dagger} + \Delta^{*} c_{-\mathbf{k}\downarrow} c_{\mathbf{k}\uparrow} \right) + \frac{V}{g} |\Delta|^{2}. \tag{1.25}$$

$\langle A \rangle$ represents the expectation value of operator A. Δ is the superconducting order parameter determined by the equation

$$\Delta = -\frac{g}{V} \langle A \rangle = -\frac{g}{V} \sum_{\mathbf{k}} \langle c_{-\mathbf{k}\downarrow} c_{\mathbf{k}\uparrow} \rangle. \tag{1.26}$$

$\langle c_{-\mathbf{k}\downarrow} c_{\mathbf{k}\uparrow} \rangle$ depends on the value of Δ.

Equation (1.26) is just the celebrated BCS gap equation. It determines completely the low energy quasiparticle excitation spectra in the superconducting state. By solving this equation self-consistently, one can determine all the thermodynamic quantities.

H_{MF} does not conserve the particle number, but the total spin, $\sum_k \sigma c_{k\sigma}^\dagger c_{k\sigma}$, and the total momentum of the Cooper pairs remain conserved. H_{MF} can be diagonalized by a unitary matrix using the Bogoliubov transformation introduced in Appendix A

$$
\begin{pmatrix} c_{k\uparrow} \\ c_{-k\downarrow}^\dagger \end{pmatrix} = \begin{pmatrix} u_k & v_k \\ -v_k^* & u_k^* \end{pmatrix} \begin{pmatrix} \alpha_k \\ \beta_k^\dagger \end{pmatrix}.
\tag{1.27}
$$

After the diagonalization, the Hamiltonian becomes

$$
H_{MF} = \sum_k E_k \left(\alpha_k^\dagger \alpha_k + \beta_k^\dagger \beta_k \right) + \sum_k (\xi_k - E_k) + \frac{V}{g}\Delta^2.
\tag{1.28}
$$

$\alpha_k^\dagger$ and $\beta_k^\dagger$ are the creation operators of the Bogoliubov quasiparticles. They describe the single-particle excitations above the superconducting gap, corresponding to the normal electrons in the two-fluid model. The quasiparticle excitation energy is given by

$$
E_k = \sqrt{\xi_k^2 + \Delta^2}.
\tag{1.29}
$$

On the Fermi surface, $\xi_k = 0$ and $E_k = |\Delta|$. Thus Δ_k is the gap function of quasiparticles in momentum space. The matrix elements u_k and v_k satisfy the normalization condition, $u_k^2 + v_k^2 = 1$, and are determined by

$$
u_k = \sqrt{\frac{1}{2} + \frac{\xi_k}{2E_k}},
\tag{1.30}
$$

$$
v_k = -\mathrm{sgn}(\Delta)\sqrt{\frac{1}{2} - \frac{\xi_k}{2E_k}}.
\tag{1.31}
$$

By calculating the pairing correlation function using the above solution, we can express explicitly the gap equation as

$$
1 = \frac{g}{V} \sum_k \frac{1}{2E_k} \tanh \frac{\beta E_k}{2}.
\tag{1.32}
$$

The temperature dependence of the energy gap can be determined by self-consistently solving this equation. Moreover, the superconducting transition temperature T_c can be solved from this equation by setting $\Delta = 0$.

At zero temperature, there are no quasiparticle excitations, and both $\langle \alpha_k^\dagger \alpha_k \rangle$ and $\langle \beta_k^\dagger \beta_k \rangle$ are zero. The ground state wavefunction can be obtained by projecting out both α- and β-types of quasiparticles from an arbitrary initial state $|\Psi_0\rangle$ not orthogonal to the ground state

$$|\Psi\rangle = \prod_{\mathbf{k}} \left(1 - \alpha_{\mathbf{k}}^{\dagger} \alpha_{\mathbf{k}}\right) \left(1 - \beta_{\mathbf{k}}^{\dagger} \beta_{\mathbf{k}}\right) |\Psi_0\rangle . \tag{1.33}$$

To set $|\Psi_0\rangle$ as the vacuum state $|0\rangle$, the above wavefunction after renormalization then becomes

$$|\Psi\rangle = \prod_{\mathbf{k}} \left(u_{\mathbf{k}} + v_{\mathbf{k}} c_{\mathbf{k}\uparrow}^{\dagger} c_{-\mathbf{k}\downarrow}^{\dagger}\right) |0\rangle = \prod_{\mathbf{k}} u_{\mathbf{k}} \exp\left(\frac{v_{\mathbf{k}}}{u_{\mathbf{k}}} c_{\mathbf{k}\uparrow}^{\dagger} c_{-\mathbf{k}\downarrow}^{\dagger}\right) |0\rangle , \tag{1.34}$$

which is just the BCS variational wavefunction with $v_{\mathbf{k}}^2$ the pairing probability. In the above expression, the states inside and outside the Fermi surface can be separated

$$|\Psi\rangle = \prod_{|\mathbf{k}|>k_F} \left(u_{\mathbf{k}} + v_{\mathbf{k}} c_{\mathbf{k}\uparrow}^{\dagger} c_{-\mathbf{k}\downarrow}^{\dagger}\right) \prod_{|\mathbf{k}|<k_F} \left(u_{\mathbf{k}} c_{-\mathbf{k}\downarrow} c_{\mathbf{k}\uparrow} + v_{\mathbf{k}}\right) |\text{Fermi Sea}\rangle . \tag{1.35}$$

Based on this expression, it is clear that quasiparticle excitations with momenta outside and inside the Fermi surface are electron- and hole-like, respectively. Here the definition of electrons and holes is the same as in normal conductors.

The quasiparticle operators $\alpha_{\mathbf{k}}$ and $\beta_{\mathbf{k}}$ contain both the creation and annihilation operators of electrons. Clearly, they are not particle-number eigen-operators. However, the real physical process should preserve the total electric charge. Is it proper or improper to use these operators to describe physical observables? The answer is affirmative. To gain an intuitive understanding, let us introduce the creation and annihilation operators of Cooper pairs, $\hat{B}^{\dagger}$ and $\hat{B}$, and redefine the quasiparticle operators $\alpha_{\mathbf{k}}$ and $\beta_{\mathbf{k}}$ as

$$\alpha_{\mathbf{k}} = u_{\mathbf{k}} c_{\mathbf{k}\uparrow} - v_{\mathbf{k}} \hat{B} c_{-\mathbf{k}\downarrow}^{\dagger}, \tag{1.36}$$

$$\beta_{\mathbf{k}}^{\dagger} = v_{\mathbf{k}} \hat{B}^{\dagger} c_{\mathbf{k}\uparrow} + u_{\mathbf{k}} c_{-\mathbf{k}\downarrow}^{\dagger}. \tag{1.37}$$

The pair operators $\hat{B}^{\dagger}$ and $\hat{B}$ create and annihilate two electrons, respectively. Operators $\alpha_{\mathbf{k}}$ and $\beta_{\mathbf{k}}$ so defined maintain the charge conservation. They change the particle number by -1 and 1, respectively. This gives a more rigorous definition for the creation and annihilation operators of Bogoliubov quasiparticles. In the superconducting state, Cooper pairs condense, and $\hat{B}$ and $\hat{B}^{\dagger}$ are replaced by their expectation values, $\hat{B} = \hat{B}^{\dagger} \approx \langle B \rangle$. Thus we can set $\hat{B}$ and $\hat{B}^{\dagger}$ as constants and eliminate them from the above expressions by absorbing them into the redefinition of $v_{\mathbf{k}}$. (However, it should be emphasized that $\hat{B}$ and $\hat{B}^{\dagger}$ are not dimensionless quantities. They carry $-2e$ and $2e$ charges, respectively.) Equations (1.36, 1.37) then return to the original expressions of $\alpha_{\mathbf{k}}$ and $\beta_{\mathbf{k}}$. This implies that charge conservation is still preserved in the BCS theory, although it is formally broken in the definition of quasiparticle operators $\alpha_{\mathbf{k}}$ and $\beta_{\mathbf{k}}$. Thus quasiparticle excitations and related physical quantities described by these operators are physically observable. This is in fact confirmed by a great many experimental measurements.

Using the mean-field solution of superconducting quasiparticles, we can express the gap equation (1.32) as an integral equation

$$\Delta = gN_F \int_{-\hbar\omega_D}^{\hbar\omega_D} d\xi \frac{\Delta}{2\sqrt{\xi^2 + \Delta^2}} \tanh \frac{\beta\sqrt{\xi^2 + \Delta^2}}{2}. \tag{1.38}$$

At zero temperature, $T = 0$, it becomes

$$\frac{1}{gN_F} = \int_0^{\hbar\omega_D} d\xi \frac{1}{\sqrt{\xi^2 + \Delta^2}} = \sinh^{-1} \frac{\hbar\omega_D}{\Delta}. \tag{1.39}$$

In the limit $\omega_D \gg \Delta$ and $gN_F \ll 1$, it further reduces to

$$\Delta \approx 2\hbar\omega_D e^{-1/gN_F}. \tag{1.40}$$

This result differs from Eq. (1.12) by a factor of 2 in the exponents. This is because, in the two-body Cooper's problem, only the excitations above the Fermi level are considered, whereas in the BCS mean-field theory, the excitations both above and below the Fermi level are included.

On the other hand, around the critical temperature T_c, Δ approaches 0, the gap equation becomes

$$\frac{1}{gN_F} = \int_0^{\hbar\omega_D} d\xi \frac{1}{\xi} \tanh \frac{\beta_c \xi}{2} = \int_0^{\frac{1}{2}\beta_c \hbar\omega_D} dx \frac{\tanh x}{x}. \tag{1.41}$$

The integral on the right-hand side is estimated to be

$$\int_0^a dx \frac{\tanh x}{x} \approx \ln 2.28a \tag{1.42}$$

in the large a limit. Thus in the limit $k_B T_c \ll \hbar\omega_D$, we have

$$k_B T_c = 1.14\hbar\omega_D e^{-1/gN_F}. \tag{1.43}$$

If the superconducting pairing is driven by the electron–phonon interaction, ω_D is the Debye frequency, which is inversely proportional to the square root of the mass of atom M. This implies that

$$T_c \propto M^{-\frac{1}{2}}. \tag{1.44}$$

This isotope effect of superconducting transition temperature has been observed in a variety of superconductors, including Hg, Pb, Mg, Sn, and Tl, lending support to the pairing mechanism driven by electron–phonon interactions.

Both the zero-temperature energy gap Δ and the transition temperature T_c depend sensitively on gN_F. This implies that neither Eq. (1.40) nor Eq. (1.43) is of much predictive power. However, their ratio is a universal number in the weak coupling limit $gN_F \ll 1$

$$\frac{2\Delta}{k_B T_c} \approx 3.53, \tag{1.45}$$

independent of microscopic details. This is a remarkable result of the BCS theory. It is widely used in literature for distinguishing weak from strong coupling superconductors. For strong coupling superconductors this ratio is generally higher than 3.53. For example, $2\Delta/(k_B T_c) \approx 4.6$ and 5.2 for Hg and Pb superconductors, respectively.

1.5 Bogoliubov–de Gennes Self-consistent Equation

The BCS gap equation and other formulas introduced in the preceding section are derived based on translation invariance. They need to be modified in a system with impurities or magnetic vortices where the translation symmetry is broken. In order to describe the spatial variations of superconducting order parameters and other physical quantities, it is more convenient to work directly in coordinate space rather than in momentum space.

In a spatially inhomogeneous system, if there are no magnetic impurities or other sources of interactions that break time-reversal symmetry, the BCS mean-field Hamiltonian can be generally expressed as

$$H_{MF} = \int d\mathbf{r} d\mathbf{r}' (c_{\mathbf{r}\uparrow}^{\dagger}, c_{\mathbf{r}\downarrow}) \begin{pmatrix} H_0(\mathbf{r})\delta(\mathbf{r}-\mathbf{r}') & \Delta(\mathbf{r},\mathbf{r}') \\ \Delta^*(\mathbf{r},\mathbf{r}') & -H_0(\mathbf{r})\delta(\mathbf{r}-\mathbf{r}') \end{pmatrix} \begin{pmatrix} c_{\mathbf{r}'\uparrow} \\ c_{\mathbf{r}'\downarrow}^{\dagger} \end{pmatrix}, \tag{1.46}$$

where

$$H_0 = -\frac{\hbar^2}{2m}\nabla^2 + U(\mathbf{r}) - \mu, \tag{1.47}$$

and $U(\mathbf{r})$ is a scalar scattering potential. In real space, the gap function $\Delta(\mathbf{r},\mathbf{r}')$ is defined as the pairing order parameter for the two electrons at $\mathbf{r}$ and $\mathbf{r}'$

$$\Delta(\mathbf{r},\mathbf{r}') = -g\langle c_{\mathbf{r}\uparrow} c_{\mathbf{r}'\downarrow}\rangle, \tag{1.48}$$

which should be determined self-consistently.

The Hamiltonian defined in Eq. (1.46) is quadratic. Its trace is zero, i.e. $\mathrm{Tr}H_{MF} = 0$. From the particle–hole symmetry, it can be shown that if E_n is an eigenvalue of H_{MF}, so is $-E_n$. H_{MF} can be diagonalized using an unitary matrix through the Bogoliubov transformation

$$\begin{pmatrix} c_{\mathbf{r}\uparrow} \\ c_{\mathbf{r}\downarrow}^{\dagger} \end{pmatrix} = \sum_n \begin{pmatrix} u_n(\mathbf{r}) & -v_n^*(\mathbf{r}) \\ v_n(\mathbf{r}) & u_n^*(\mathbf{r}) \end{pmatrix} \begin{pmatrix} \alpha_n \\ \beta_n^{\dagger} \end{pmatrix}. \tag{1.49}$$

For superconducting systems without time-reversal symmetry, for example, in the presence of an external magnetic field where up- and down-spin electrons are mixed by the Zeeman interaction, a similar Bogoliubov transformation can be defined. But the above 2×2 transformation matrix needs to be generalized and replaced by a 4×4 matrix.

$u_n(\mathbf{r})$ and $v_n(\mathbf{r})$ define the wavefunction of Bogoliubov quasiparticles. They are determined by the eigenequation of H_{MF}

$$\int d\mathbf{r'} \left(\begin{array}{cc} H_0(\mathbf{r})\delta(\mathbf{r} - \mathbf{r'}) & \Delta(\mathbf{r},\mathbf{r'}) \\ \Delta^*(\mathbf{r},\mathbf{r'}) & -H_0(\mathbf{r})\delta(\mathbf{r} - \mathbf{r'}) \end{array} \right) \left(\begin{array}{c} u_n(\mathbf{r'}) \\ v_n(\mathbf{r'}) \end{array} \right) = E_n \left(\begin{array}{c} u_n(\mathbf{r}) \\ v_n(\mathbf{r}) \end{array} \right).$$

(1.50)

In the quasiparticle representation, the gap function $\Delta(\mathbf{r},\mathbf{r'})$ can be expressed using $u_n(\mathbf{r})$ and $v_n(\mathbf{r})$ as

$$\Delta(\mathbf{r},\mathbf{r'}) = -\frac{g}{2} \sum_n \left[u_n(\mathbf{r}) v_n^*(\mathbf{r'}) + u_n(\mathbf{r'}) v_n^*(\mathbf{r}) \right] \tanh \frac{\beta E_n}{2}.$$

(1.51)

Equations (1.50) and (1.51) are the Bogoliubov–de Gennes (BdG) self-consistent equations [24]. They have been widely used to solve the problems related to impurity scattering, elementary excitations around vortex lines, surface states, and Andreev reflection.

The BdG self-consistent equations are equivalent to the Green's function theory of superconductivity at the mean-field level. In a spatially inhomogeneous system, the Green's function $G(\mathbf{r},\mathbf{r'})$ depends on both $\mathbf{r}$ and $\mathbf{r'}$, not just on their difference $\mathbf{r} - \mathbf{r'}$. In this case, it is usually more convenient to solve the BdG equation than the Green's function because the BdG wavefunction $(u_n(\mathbf{r}), v_n(\mathbf{r}))$ depends only on coordinate $\mathbf{r}$.

1.6 Charge and Probability Current Density Operators

As mentioned previously, the Bogoliubov quasiparticles determined by Eq. (1.50) are not eigenstates of the electron number operator, and the total electron number is not conserved. This can be more clearly understood from the Bogoliubov transformation of quasiparticle operators given in Eq. (1.49). The breaking of the electron number conservation implies that the probability of quasiparticles, $\rho_P(\mathbf{r})$, is not proportional to the density of electrons, $\rho_Q(\mathbf{r})$. Correspondingly, the current density of quasiparticles, $J_P(\mathbf{r})$, is also not proportional to the electric current density, $J_Q(\mathbf{r})$. This is markedly different from the situation in a normal metal where the electron probability (current) density and the corresponding charge (current) density are essentially equivalent and satisfy the simple equations, $\rho_P(\mathbf{r}) = e\rho_Q(\mathbf{r})$ and $J_Q(\mathbf{r}) = eJ_P(\mathbf{r})$. The difference results from the Cooper pair condensation in

the superconducting state. A thorough understanding to it is important to the understanding of the gauge invariance and the scattering problem of electrons in superconductors.

The definition of the Bogoliubov quasiparticle current density and the electric
current density depends on the symmetry of the gap function $\Delta(\mathbf{r}, \mathbf{r}')$. In an isotropic
s-wave superconductor, the pairing interaction is entirely local, these quantities are
relatively simple to define, and the Bogoliubov wavefunctions u and v are governed
by the equation

$$
i\hbar \frac{\partial}{\partial t} \begin{pmatrix} u(\mathbf{r}) \\ v(\mathbf{r}) \end{pmatrix} = \begin{pmatrix} H_0(\mathbf{r}) & \Delta(\mathbf{r}) \\ \Delta^*(\mathbf{r}) & -H_0(\mathbf{r}) \end{pmatrix} \begin{pmatrix} u(\mathbf{r}) \\ v(\mathbf{r}) \end{pmatrix}, \tag{1.52}
$$

where $\Delta(\mathbf{r}) = \Delta(\mathbf{r}, \mathbf{r}')\delta(\mathbf{r} - \mathbf{r}')$ is the superconducting order parameter. In a d-wave
or other unconventional superconductor, the gap function becomes non-local, and
a few off-diagonal nonlocal terms needs to be added to the definitions of these
quantities.

In the isotropic s-wave superconductor, the gap function is independent of
momentum, i.e. $\Delta_{\mathbf{k}} = \Delta$, the quasiparticle density contains the contribution from
both particles (u) and holes (v),

$$
\rho_P(\mathbf{r}) = |u(\mathbf{r})|^2 + |v(\mathbf{r})|^2. \tag{1.53}
$$

Its time-derivative, $\partial \rho_P / \partial t$, can be obtained using Eq. (1.52). The conservation law
of probability is described by the equation

$$
\frac{\partial}{\partial t} \rho_P + \nabla \cdot \mathbf{J}_P = 0. \tag{1.54}
$$

Based on this equation, we find that the quasiparticle probability current is defined as

$$
\mathbf{J}_P = \frac{\hbar}{m} \mathrm{Im} \left(u^* \nabla u - v^* \nabla v \right). \tag{1.55}
$$

As expected, particles and holes have opposite contributions to the probability current density.

Particles and holes carry opposite charges. The charge density of superconducting
quasiparticles is therefore defined by

$$
\rho_Q(\mathbf{r}) = e \left(|u(\mathbf{r})|^2 - |v(\mathbf{r})|^2 \right). \tag{1.56}
$$

From the time-evolution equation of u and v, Eq. (1.52), we find that the charge
density satisfies the equation

$$
\frac{\partial}{\partial t} \rho_Q(\mathbf{r}) + \nabla \cdot \mathbf{J}_Q(\mathbf{r}) = \frac{4e}{\hbar} \mathrm{Im} \left(\Delta u^*(\mathbf{r}) v(\mathbf{r}) \right), \tag{1.57}
$$

where

$$\mathbf{J}_Q(\mathbf{r}) = \frac{e\hbar}{m}\text{Im}\left(u^*\nabla u + v^*\nabla v\right) \tag{1.58}$$

is the electric charge current density of quasiparticles. In comparison with the probability conservation equation, the electric charge conservation equation contains an extra term, contributed by the superconducting paired electrons. This term is proportional to the product of both particle and hole wavefunctions of Bogoliubov eigenstates. If we define a supercurrent density operator $\mathbf{J}_S$ by the equation

$$\nabla \cdot \mathbf{J}_S(\mathbf{r}) = -\frac{4e}{\hbar}\text{Im}\left[\Delta u^*(\mathbf{r})v(\mathbf{r})\right], \tag{1.59}$$

the charge conservation law becomes

$$\frac{\partial}{\partial t}\rho_Q(\mathbf{r}) + \nabla \cdot \left[\mathbf{J}_Q(\mathbf{r}) + \mathbf{J}_S(\mathbf{r})\right] = 0. \tag{1.60}$$

In a translation invariant system, momentum $\mathbf{k}$ is a good quantum number. The Bogoliubov quasiparticle wavefunctions are determined by the BCS mean-field equations, Eqs. (1.30) and (1.31). In real space, they are given by

$$u(\mathbf{r}) = \frac{1}{\sqrt{V}}e^{i\mathbf{k}\cdot\mathbf{r}}\sqrt{\frac{1}{2} + \frac{\xi_\mathbf{k}}{2E_\mathbf{k}}}, \tag{1.61}$$

$$v(\mathbf{r}) = -\frac{1}{\sqrt{V}}e^{i\mathbf{k}\cdot\mathbf{r}}\sqrt{\frac{1}{2} - \frac{\xi_\mathbf{k}}{2E_\mathbf{k}}}. \tag{1.62}$$

When $\mathbf{k}$ is real, the probability current density is equal to

$$\mathbf{J}_P = \frac{\hbar\xi_\mathbf{k}\mathbf{k}}{mVE_\mathbf{k}}. \tag{1.63}$$

In contrast, the normal charge current $\mathbf{J}_Q$ and the supercurrent $\mathbf{J}_S$ are given by

$$\mathbf{J}_Q = \frac{e\hbar\mathbf{k}}{mV}, \tag{1.64}$$

$$\mathbf{J}_S = 0. \tag{1.65}$$

This indicates that a Bogoliubov quasiparticle with a real momentum $\mathbf{k}$ will not decay to generate a supercurrent by forming a Cooper pair with another quasiparticle. Thus the supercurrent vanishes, $\mathbf{J}_S = 0$. Both $\mathbf{J}_P$ and $\mathbf{J}_Q$ are proportional to the momentum $\hbar\mathbf{k}$, but $\mathbf{J}_P$ contains the factor of $\xi_\mathbf{k}/E_\mathbf{k}$. The charge current density is in the same direction as $\mathbf{k}$. For a particle-like quasiparticle with $\xi_\mathbf{k} > 0$, its probability current density is also parallel to $\mathbf{k}$. But for a hole-like quasiparticle with $\xi_\mathbf{k} < 0$, its probability current density is antiparallel to $\mathbf{k}$.

On the other hand, if **k** contains a small imaginary part, say $\mathbf{k} = \mathbf{k}_0 + i\eta\hat{x}$ with $\eta > 0$, the wavefunction of quasiparticle decays exponentially along the x-direction. In this case, the charge current of quasiparticles becomes

$$\mathbf{J}_Q = \frac{e\hbar\mathbf{k}_0}{mV}e^{-2\eta x}. \qquad (1.66)$$

This also decays along the x-direction. The supercurrent is still zero along the y- and z-directions, $J_S^y = J_S^z = 0$. However, it is finite along the x-direction

$$J_S^x = -\frac{e\Delta^2}{\eta\hbar V}\left(1 - e^{-2\eta x}\right)\operatorname{Im}\frac{1}{E_\mathbf{k}}. \qquad (1.67)$$

In the limit $\eta \ll |\mathbf{k}_0|$, J_S^x is approximately given by

$$J_S^x = \frac{e\Delta^2\hbar\xi_{k_0}k_{0,x}}{mVE_{k_0}^3}\left(1 - e^{-2\eta x}\right). \qquad (1.68)$$

This indicates that the charge current of quasiparticles is transformed into the supercurrent of Cooper pairs. The inverse of the imaginary part of the quasiparticle momentum η^{-1} is a characteristic length scale of quasiparticles to form condensed superconducting Cooper pairs.

1.7 Off-Diagonal Long-Range Order

The superconducting transition is a continuous transition from a high temperature normal conducting phase to a low temperature macroscopic long-range ordered phase. In 1962, C. N. Yang pointed out [19] that the long-range order of superconductivity is an *off-diagonal long-range order* (ODLRO), which is fundamentally different from a diagonal long-range order, such as the crystalline order of crystals. This kind of order is induced purely by quantum effects and there is no correspondence in classical systems.

The concept of ODLRO provides a mathematical foundation for the microscopic theory of superconductivity as well as the corresponding theory of macroscopic quantum phase transition. The variational wavefunction proposed by Bardeen–Cooper–Schrieffer, Eq. (1.34), actually possesses ODLRO. This is in fact the reason that Bardeen, Cooper, and Schrieffer could achieve great success in establishing the microscopic theory of superconductivity. ODLRO plays a similar role to the diagonal long-range crystalline order in solids. It is impossible to establish correctly the theory of superconductivity if the superconducting ODLRO is not properly included in the wavefunction.

ODLRO exists only in quantum fluids, including quantum gases and liquids, such as the Fermi liquid state of conducting electrons in metals. In insulators, charge fluctuations are short-ranged and the ODLRO is suppressed. Nevertheless, ODLRO could coexist with diagonal long-range orders. For example, a superfluid ODLRO, can coexist with a diagonal density-wave order in a supersolid.

In order to understand ODLRO, let us consider the following two-particle reduced density matrix

$$\rho_2(i\sigma_i, j\sigma_j; k\sigma_k, l\sigma_l) = \text{Tr}\left(c_{i\sigma_i} c_{j\sigma_j} \rho c^{\dagger}_{k\sigma_k} c^{\dagger}_{l\sigma_l}\right), \tag{1.69}$$

where

$$\rho = \frac{e^{-\beta H}}{\text{Tr} e^{-\beta H}} \tag{1.70}$$

is the density matrix. It is simple to show that ρ and ρ_2 are semi-positive definite, namely all their eigenvalues are always larger than or equal to zero.

In a normal metallic state of N electrons, the eigenvalues of ρ_2 are typically of order 1, much smaller than N. Hence there is not a state which can be occupied by macroscopically many pairs of electrons with the same quantum numbers. In this case, an infinitesimally small energy, in comparison with the total energy which is proportional to N, is able to change the microscopic distribution of ρ_2, and thus the system is dissipative. On the contrary, if ρ_2 has an eigenvalue of order N (assuming it to be αN with α a number of order 1), it is no longer easy to change the behavior of electrons in this eigenstate by applying a macroscopically small perturbation, which implies that the system is macroscopically coherent and dissipationless. This is just the most prominent feature of superconductivity arising from the macroscopic pair condensation. That eigenstate is just a superconducting condensed state. In this case, one can separate that eigenstate from ρ_2 and rewrite ρ_2 as

$$\rho_2(i\sigma_i, j\sigma_j; k\sigma_k, l\sigma_l) = \alpha N \phi(i\sigma_i, j\sigma_j)\phi^*(k\sigma_k, l\sigma_l)$$
$$+ \rho_2'(i\sigma_i, j\sigma_j; k\sigma_k, l\sigma_l), \tag{1.71}$$

where $\phi(i\sigma_i, j\sigma_j)$ is the normalized eigenfunction corresponding to the eigenvalue of αN. The normalization requires $\phi(i\sigma_i, j\sigma_j)$ to be inversely proportional to the system volume V. Thus $N\phi(i\sigma_i, j\sigma_j)\phi^*(k\sigma_k, l\sigma_l)$ is proportional to the electron density. ρ_2' is a regular reduced density matrix whose eigenvalues are all macroscopically small compared to N.

Equation (1.71) suggests that electron pairs are long-range correlated. This is because no matter how far a pair of electrons at sites i and j is from another pair of electrons at sites k and l, their correlation function, $\langle c_{i\sigma_i} c_{j\sigma_j} c^{\dagger}_{k\sigma_k} c^{\dagger}_{l\sigma_l} \rangle$, remains finite. A superconducting state possesses ODLRO because it has a finite probability of annihilating a local pair of electrons and simultaneously creating another local pair of electrons separated in an arbitrary long distance. On the other hand, if a system possesses ODLRO, it can be also shown that its two-particle reduced density matrix has at least one eigenvalue of order N. A superconducting transition emerges when the maximal eigenvalue ρ_2 changes from a number of order 1 to a number of order N. The coefficient α can be defined as the superconducting order parameter. It is finite in the superconducting state and becomes 0 at and above the superconducting transition temperature T_c.

In Eq. (1.71), if we take an approximation by neglecting the regular ρ_2' term, then ρ_2 becomes

$$\rho_2(i\sigma_i, j\sigma_j; k\sigma_k, l\sigma_l) \approx \alpha N \phi(i\sigma_i, j\sigma_j)\phi^*(k\sigma_k, l\sigma_l). \qquad (1.72)$$

In this case, the electron pair correlation function is factorized. This is just the basic assumption made in the BCS mean-field theory.

In textbooks and literature, the quasiparticle excitation gap is generally defined as the superconducting order parameter. Rigorously speaking, this definition is not that accurate. A system is superconducting as long as it possesses an ODLRO, no matter whether it has an energy gap or not. In fact, there are gapless superconductors. For example, there is no gap in the quasiparticle excitation spectra in a superconductor with magnetic impurities. Conceptually, the superconducting energy gap and the superconducting order parameter are different. However, in most superconductors, it is not necessary to distinguish these two concepts because the superconducting energy gap is proportional to the superconducting order parameter at least at the mean-field level.

In both one and two dimensions, it was proven by Hohenberg (see Appendix B) that there is no ODLRO at any nonzero temperatures if the f-sum rule is valid [25]. For the BCS reduced Hamiltonian, defined by Eq. (1.14), the pairing potential is long-ranged and the f-sum rule is violated. In that case, a superconducting ODLRO with the corresponding phase transition is allowed to exist in finite temperatures even in one or two dimensions [26].

1.8 Ginzburg–Landau Free Energy

The BCS theory provides a microscopic framework to describe superconducting properties. However, it is not that convenient to use in the study of the dynamical properties of magnetic fluxes, and in the quantitative characterization of superconducting phase transition as well as many other macroscopic phenomena of superconductors. In this case, it is technically simpler and conceptually more transparent to describe superconducting properties by adopting the phenomenological theory first proposed by Ginzburg and Landau (GL) [27].

The phenomenological GL theory was introduced before the establishment of the BCS microscopic theory. It relies on the assumption that the superconducting state is a macroscopic quantum state that can be described by an order parameter.

As discussed in the preceding section, §1.7, the pairing condensation results from the formation of superconducting ODLRO. The corresponding order parameter is determined by the pairing correlation function, which is off-diagonal and difficult to visualize. On the contrary, a classical phase transition, such as the Ising transition, is induced by the ordering in the diagonal channel and the corresponding order parameter is simply the magnetization density. It was a miracle that Ginzburg and Landau introduced the concept of superconducting order parameter just from the general argument of spontaneous symmetry breaking even without knowing its true physical origin. This is similar to the discovery of the periodical table of elements by Mendeleev without any knowledge of quantum mechanics.

As the superconducting electrons can couple directly with electromagnetic fields, it is natural to assume that the superconducting order parameter is a complex field $\psi(\mathbf{r})$ whose dynamics is governed by the minimal coupling. Furthermore, it was assumed that the macroscopic properties of superconductors are completely described by the free energy, independent of their microscopic details. Thus if the spatial variation of the order parameter is slow in comparison with the scale of coherence length, the free energy can be expanded as a functional of the order parameter ψ and its spatial gradient $\nabla \psi$ as

$$f = f_n + \frac{1}{2m^*} \left| (-i\hbar\nabla - e^*\mathbf{A}) \, \psi \right|^2 + \alpha \, |\psi|^2 + \frac{\beta}{2} \, |\psi|^4 + \frac{\mathbf{H}^2}{2\mu_0}, \tag{1.73}$$

where f_n is the free energy in the normal state, $\mathbf{H} = \nabla \times \mathbf{A}$ is the external magnetic field, and $\mathbf{A}$ is the associated vector potential. In obtaining this expression, the variance of the order parameter with time is assumed small and negligible.

In a homogeneous system without external magnetic fields, the GL free energy becomes

$$f = f_n + \alpha \, |\psi|^2 + \frac{\beta}{2} \, |\psi|^4. \tag{1.74}$$

In the normal state, $\alpha > 0$, and the order parameter is zero, $\psi = 0$, and $f = f_n$ is the free energy. In the superconducting state, α becomes negative ($\alpha < 0$) and the system is in an ordered state and the value of the order parameter ψ_0 is determined by the minimum of the free energy and given by

$$|\psi_0|^2 = -\frac{\alpha}{\beta}. \tag{1.75}$$

Substituting this into Eq. (1.74), we find the difference in the free energy between the superconducting and normal phases to be

$$f_s - f_n \equiv -\frac{H_c^2}{2\mu_0} = -\frac{\alpha^2}{\beta}, \tag{1.76}$$

where H_c is the thermodynamic critical field of the superconducting state.

In the GL theory, parameters α and β are unknown but can be determined from the BCS theory or alternatively from the measurement values of the critical field H_c and the magnetic penetration length λ using the formulas

$$\alpha(T) = -\frac{2e^{*2}}{m^*} H_c^2(T)\lambda^2(T), \qquad (1.77)$$

$$\beta(T) = \frac{4\mu_0 e^{*4}}{m^{*2}} H_c^2(T)\lambda^4(T). \qquad (1.78)$$

In 1959, Gor'kov [28] showed that the GL free energy could be derived from the BCS theory around the transition temperature T_c under the condition that both ψ and $\mathbf{A}$ do not vary too fast over the coherence length scale ξ. He found that, as expected, the order parameter ψ is proportional to the quasiparticle energy gap Δ. Furthermore, he showed that the effective charge that ψ carries is $e^* = 2e$, and the corresponding effective mass $m^* \approx 2m$ under the free electron approximation, as a clear indication of the pairing nature of ψ. Substituting $e^* = 2e$ into Eq. (1.73), the GL free energy becomes

$$f = f_n + \frac{1}{2m^*} |(-i\hbar\nabla - 2e\mathbf{A})\,\psi|^2 + \alpha\,|\psi|^2 + \frac{\beta}{2}\,|\psi|^4 + \frac{\mathbf{H}^2}{2\mu_0}. \qquad (1.79)$$

Gor'kov's work established a microscopic foundation for the GL theory. It clarifies the condition of validity and the limitation of the GL theory, and provides a clear guidance to its application in real materials.

In the equilibrium, the free energy is minimized. From the variance of the free energy f with respect to ψ^*, we obtain the first GL equation

$$\frac{1}{4m}(-i\hbar\nabla - 2e\mathbf{A})^2\,\psi + \left(\alpha + \beta\psi^*\psi\right)\psi = 0. \qquad (1.80)$$

Furthermore, by taking the variation with respect to $\mathbf{A}$ and using Ampere's law

$$\mathbf{J}_s = \frac{1}{\mu_0}\nabla \times \mathbf{H}, \qquad (1.81)$$

we obtain the expression of the supercurrent,

$$\mathbf{J}_s = \frac{ie\hbar}{2m}\left(\psi^*\nabla\psi - \psi\nabla\psi^*\right) + \frac{2e^2}{m}\mathbf{A}\psi^*\psi, \qquad (1.82)$$

which is also called the second GL equation.

In the polar representation of the order parameter

$$\psi(\mathbf{r}) = |\psi(\mathbf{r})|e^{i\phi(\mathbf{r})}, \qquad (1.83)$$

the supercurrent becomes

$$\mathbf{J}_s = \frac{2e^2\,|\psi|^2}{m}\left(\mathbf{A} - \frac{\hbar}{2e}\nabla\phi\right). \qquad (1.84)$$

This is nothing but the London equation. By comparison with Eq. (1.7), we find that the square of the order parameter is proportional to the superfluid density.

$$n_s = 2|\psi|^2, \tag{1.85}$$

and $\hbar\phi/(2e)$ serves as the role of scalar potential. At the saddle point,

$$n_s = 2|\psi_0|^2 = -\frac{2\alpha}{\beta}. \tag{1.86}$$

The corresponding penetration depth is

$$\lambda = \sqrt{\frac{m\beta}{2\mu_0|\alpha|e^2}}. \tag{1.87}$$

Around the critical point

$$\alpha \propto T - T_c, \tag{1.88}$$

and β is roughly a constant. Therefore, in the limit $T \to T_c$,

$$n_s \propto T_c - T \tag{1.89}$$

in the superconducting phase.

From Eq. (1.84), we can find the magnetic flux penetrating a superconducting area enclosed by a closed loop C far from the boundary on which the supercurrent is zero

$$\mathbf{J}_s \propto \mathbf{A} - \frac{\hbar}{2e}\nabla\phi = 0, \qquad (\mathbf{r} \in C). \tag{1.90}$$

From the integral of the vector gauge field, we find the magnetic flux threading such a loop is quantized

$$\Phi = \oint d\mathbf{r} \cdot \mathbf{A} = \frac{\hbar}{2e}\oint d\mathbf{r} \cdot \nabla\phi(\mathbf{r}) = \frac{hn}{2e} = n\Phi_0, \tag{1.91}$$

where n is an integer. The basic flux quantum is

$$\Phi_0 = \frac{h}{2e} = 2.07 \times 10^{-15}\text{Wb}. \tag{1.92}$$

The quantization of magnetic flux was first predicted by Fritz London. It was confirmed experimentally by B. S. Deaver and W. M. Fairbank [29] and, independently, by R. Doll and M. Näbauer [30].

1.9 Two Length Scales

There exist two characteristic length scales in superconductors. The first is the penetration depth, λ, used for characterizing the Meissner effect. As discussed in §1.2 and §1.8, the penetration depth is inversely proportional to the square root of the superfluid density. It measures the half decay length of an external magnetic field penetrating into the superconductor.

The second length scale is the coherence length ξ. This is a concept that was first introduced by Pippard in 1953 to explain the numerous experimental results that deviate from the predictions of the London theory [31]. He pointed out that the response of the supercurrent to an external magnetic field is nonlocal, in the sense that the value of $\mathbf{J}_s$ measured at a point $\mathbf{r}$ depends on the value of $\mathbf{A}$ throughout a volume of radius ξ, which is also referred to as the Pippard coherence length, surrounding the point $\mathbf{r}$. A nonlocal generalization of the London equation was first established by Pippard [31]. In the Coulomb gauge, it is written as

$$\mathbf{J}_s(\mathbf{r}) = C \int d\mathbf{r}' \frac{[\mathbf{A}(\mathbf{r}') \cdot \mathbf{R}]\mathbf{R}}{R^4} e^{-R/\xi}, \tag{1.93}$$

where $\mathbf{R} = \mathbf{r} - \mathbf{r}'$. The coefficient C in the above equation can be determined by considering the limit where $\mathbf{A}$ varies slowly in space. In that limit, the above equation goes back to the London equation. By setting $\mathbf{A} \parallel \hat{z}$, we have

$$\mathbf{J}_s(\mathbf{r}) = C\mathbf{A} \int \frac{\cos^2 \theta}{R^2} e^{-R/\xi} R^2 dR \sin\theta d\theta d\phi = \frac{4\pi}{3} C\xi \mathbf{A}. \tag{1.94}$$

Compared with the London equation, we find that

$$C = -\frac{3}{4\pi} \frac{n_s e^2}{m\xi}. \tag{1.95}$$

The Ginzburg-Landau coherence length ξ is the correlation length associated with the phase transition from superconducting to the normal state. As with any other phase transition, this length describes how far fluctuations of the order parameter propagate. In other words, it is the length scale that characterizes the spatial variation of the superconducting order parameter under a local perturbation. Particularly, the value of ξ can be determined by considering a superconductor in zero magnetic field below its transition temperature and subject to a small perturbation which forces the order parameter ψ to deviate from its equilibrium value ϕ_0 by $\delta\psi$ at some point. To the leading order approximation in $\delta\psi$, the first Ginzburg–Landau equation (1.80) becomes

$$\nabla^2 \psi(\mathbf{r}) = \frac{4m}{\hbar^2} \left(\alpha + 3\beta|\psi_0|^2\right) \delta\psi(\mathbf{r}) = \frac{2}{\xi^2} \delta\psi(\mathbf{r}). \tag{1.96}$$

It shows that $\delta\psi$ decays exponentially to its equilibrium value with length $(\xi/\sqrt{2})$ as the coordinate moves away from the point of perturbation, and

$$\xi = \frac{\hbar}{\sqrt{4m\left(\alpha + 3\beta|\psi_0|^2\right)}} = \frac{\hbar}{\sqrt{4m|\alpha|}}. \tag{1.97}$$

As this expression is obtained from the Ginzburg–Landau theory, ξ so obtained is also called the Ginzburg–Landau coherence length.

The nonlocal response of the order parameter to a local disturbance results from the fact that a Cooper pair has a finite size. As a matter of fact, the Pippard or Ginzburg–Landau coherence length is of the same order as the Cooper pair size (i.e. the distance between two electrons in a Cooper pair) at zero temperature. For the isotropic s-wave superconductor, it turns out that the Cooper pair size equals the BCS coherence length in the weak-coupling limit

$$\xi_{\mathrm{BCS}} = \frac{\hbar v_f}{\pi \Delta}, \tag{1.98}$$

where v_f is the Fermi velocity.

As you may have already noticed, three kinds of coherence lengths have been introduced from three different perspectives in the analysis of different superconducting properties. They are of the same order of magnitude and differ from each other by at most a constant factor at low temperatures.

1.10 Two Types of Superconductor

Both the coherent length and the penetration depth are temperature dependent. They diverge at the superconducting transition temperature. The competition between these two length scales has important consequences for the physical properties of superconductors. Particularly, superconductors are classified into two types according to the value of the GL parameter

$$\kappa = \frac{\lambda(T)}{\xi(T)} = \frac{m}{\hbar e}\sqrt{\frac{2\beta}{\mu_0}}, \tag{1.99}$$

which turns out to be weakly temperature dependent. They are called type-I and type-II superconductors if ξ/λ is larger or smaller than $\sqrt{2}$, respectively. Magnetic fluxes are expelled from the interior of type-I superconductors, but can penetrate into the interior of type-II superconductors, forming quantized vortex lines.

For each vortex line, the magnetic field is maximum at the center of the line. Going outwards from the core center, the magnetic field drops exponentially due to the electromagnetic screening by the supercurrent in a characterizing radius of the penetration depth λ. On the other hand, the superconducting order parameter is reduced only in a small core regime of radius of the coherence length ξ.

1.10.1 Upper Critical Field

The upper critical field, abbreviated as H_{c2}, is the minimum magnetic flux density that completely suppresses superconductivity in a type-II superconductor. Near H_{c2}, the order parameter is small and governed by the linearized GL equation

$$\frac{1}{4m} (-i\hbar\nabla - 2e\mathbf{A})^2 \psi = -\alpha\psi. \tag{1.100}$$

The β-term is dropped because it is negligible for small ψ. Moreover, in the small ψ limit, the supercurrent provides negligible screening and the gauge field $\mathbf{A}$ corresponds to a uniformly applied magnetic field. In this case, the above GL equation is identical to the Schödinger equation for a free particle of mass $2m$ and charge $2e$ in a uniform magnetic field, and $|\alpha|$ serves the role of eigenvalue. The solutions are the Landau orbitals in the layer perpendicular to the applied field. The corresponding cyclotron frequency is

$$\omega_c = \frac{eH}{m} \tag{1.101}$$

and the energy gap between two Landau levels is $\hbar\omega_c$. This equation has a solution only when $|\alpha|$ is larger than or equal to the zero-point motion energy $\hbar\omega_c/2 = |\alpha|$. This implies that the upper critical field is simply determined by the equation

$$|\alpha| = \frac{\hbar e H_{c2}}{2m}, \tag{1.102}$$

hence

$$H_{c2} = \frac{2m|\alpha|}{\hbar e} = \frac{\Phi_0}{2\pi\xi^2}. \tag{1.103}$$

Since the size of a vortex core is of the order of $\xi(T)$, it means that H_{c2} actually happens when the vortex cores touch one another.

The upper critical field is clearly not the thermodynamic critical field H_c, defined by Eq. (1.76), which is in fact determined by both $\lambda(T)$ and $\xi(T)$

$$H_c(T) = \frac{\Phi_0}{2\pi\xi\lambda}. \tag{1.104}$$

The ratio between these two critical fields is simply determined by the GL parameter

$$\frac{H_{c2}}{H_c} = \frac{\lambda(T)}{\xi(T)} = \kappa. \tag{1.105}$$

Hence by comparing H_{c2} with H_c, one can also determine which type a superconductor is.

1.10.2 Lower Critical Field

In addition to the upper critical field, there exists another critical field which is called the lower critical field H_{c1}. It corresponds to the critical field at which an applied magnetic field starts to penetrate the interior of superconductor.

To understand the physics governing the lower critical field, we first solve a single vortex line problem in the extreme type-II limit $\lambda \gg \xi$. As the vortex core radius ξ is very small in comparison with λ, we may take the amplitude of the order parameter as a constant in space and neglect the contribution of the vortex hard core due to the suppression of the superconducting order parameter. The vortex line energy is then given by the formula

$$F = \int_{r>\xi} d\mathbf{r} \left[\frac{|\psi|^2}{4m} (\hbar \nabla \phi - 2e\mathbf{A})^2 + \frac{1}{2\mu_0} \mathbf{H}^2 \right]. \tag{1.106}$$

Using the expression of the supercurrent (1.84) and Ampere's law (1.81), F can be written as

$$F = \int_{r>\xi} d\mathbf{r} \frac{1}{2\mu_0} \left[\lambda^2 (\nabla \times \mathbf{H})^2 + \mathbf{H}^2 \right]. \tag{1.107}$$

Minimizing the free energy with respect to $\mathbf{H}$, we obtained the usual London equation

$$\mathbf{H} + \lambda^2 \nabla \times (\nabla \times \mathbf{H}) = 0, \qquad (r > \xi). \tag{1.108}$$

In the interior of the vortex core, this equation should be modified. Since the core size is very small, we may replace the corresponding singularity simply by a two-dimensional delta function

$$\mathbf{H} + \lambda^2 \nabla \times (\nabla \times \mathbf{H}) = \Phi \hat{z} \delta^2(\mathbf{r}), \tag{1.109}$$

where Φ is a parameter of the dimension of the magnetic flux.

To determine the value of Φ, let us consider a two-dimensional integral for the above equation around an area bounded by a loop with $r \gg \lambda$ around the core, which reads

$$\int d\boldsymbol{\sigma} \cdot \mathbf{H} + \lambda^2 \oint \nabla \times \mathbf{H} \cdot d\mathbf{l} = \Phi. \tag{1.110}$$

The integral of the first term gives the total flux of the vortex, i.e. Φ_0. The second term is a loop integral for the supercurrent $\mathbf{J}_s \propto \nabla \times \mathbf{H}$. As the supercurrent is very small in the limit $r \gg \lambda$, this loop integral is negligible. The above equation thus indicates that Φ is just the fundamental flux quanta, i.e. $\Phi = \Phi_0$.

When r is much smaller than λ but still outside the vortex core, $\xi < r \ll \lambda$, the contribution from the first term scales as r^2/λ^2 and can be neglected in the extreme type-II case. We then have $\mathbf{H}(\mathbf{r}) = H(r)\hat{z}$ and

$$\lambda^2 \oint \nabla \times \mathbf{H} \cdot d\mathbf{l} = -2\pi r \lambda^2 \frac{dH}{dr} = \Phi_0. \tag{1.111}$$

The solution is

$$H(r) = \frac{\Phi_0}{2\pi \lambda^2} \ln \frac{\lambda}{r}. \tag{1.112}$$

More generally, the solution is

$$H(r) = \frac{\Phi_0}{2\pi \lambda^2} K_0 \left(\frac{r}{\lambda} \right), \tag{1.113}$$

where $K_0(x)$ is the zeroth order imaginary argument Bessel function. It gives rise to short distance behavior as in Eq. (1.112), and the long distance behavior at $r \gg \lambda$ as

$$H(r) = \frac{\Phi_0}{2\pi \lambda^2} \sqrt{\frac{\pi \lambda}{2r}} e^{-r/\lambda}. \tag{1.114}$$

Inserting the above solution into Eq. (1.107) and using Eq. (1.109), the vortex energy is found to be

$$F = \frac{\lambda^2}{2\mu_0} \int_{r \geqslant \xi} d\mathbf{r} \left[-\mathbf{H} \cdot \nabla \times (\nabla \times \mathbf{H}) + (\nabla \times \mathbf{H})^2 \right]$$

$$= \frac{\pi \xi L \lambda^2}{\mu_0} H(\xi) |\nabla \times H(\xi)|, \tag{1.115}$$

where L is the length of the vortex line. Plugging in

$$H(\xi) = \frac{\Phi_0}{2\pi \lambda^2} \ln \frac{\lambda}{\xi}, \tag{1.116}$$

we have

$$\frac{F}{L} = \frac{\Phi_0^2}{4\pi \mu_0 \lambda^2} \ln \frac{\lambda}{\xi}. \tag{1.117}$$

We now deduce the Gibbs free energy to determine the lower critical field. In the dilute limit, the distance between any two vortices is significantly larger than the penetration depth so that the interaction among vortex lines is negligible. In that limit, the Gibbs function per unit area and per unit length of the vortex line is simply given by

$$g = \frac{n_v F}{L} - MH. \tag{1.118}$$

The first term is the energy sum of each of the individual vortices. The second term represents the effect of the field $\mathbf{H}$, which favors large magnetization M. n_v is the number of vortices per unit area, which is related to M by

$$M = \frac{n_v \Phi_0}{\mu_0}. \tag{1.119}$$

The Gibbs density can also be expressed as

$$g = \frac{n_v \Phi_0}{\mu_0} \left(\frac{\Phi_0}{4\pi\lambda^2} \ln \frac{\lambda}{\xi} - H \right). \tag{1.120}$$

It indicates that the Gibbs energy density becomes negative when the field H is larger than a critical field

$$H_{c1} = \frac{\Phi_0}{4\pi\lambda^2} \ln \frac{\lambda}{\xi}. \tag{1.121}$$

This is just the lower critical field above which the quantized magnetic fluxes begin to emerge. Particularly, two superconducting phases exist in type-II superconductors:

(i) The Meissner phase: $0 < H < H_{c1}$. This is the phase without any vortices. g is an increasing function of n_v. To minimize g, n_v remains zero.
(ii) The mixed phase: $H_{c2} > H > H_{c1}$. g drops with n_v and the quantized vortices coexist with the superconducting domains.

1.11 Spontaneous Symmetry Breaking and Meissner Effect

Superconductors are not only ideal conductors but also perfect diamagnets. This is due to the formation of ODLRO with spontaneous breaking of U(1) electromagnetic gauge symmetry. It is simple to show that the GL free energy is invariant under the following $U(1)$ gauge transformation

$$\psi \rightarrow \psi' = e^{i\varphi}\psi, \tag{1.122}$$

$$\mathbf{A} \rightarrow \mathbf{A}' = \mathbf{A} + \frac{\hbar}{2e}\nabla\varphi, \tag{1.123}$$

where φ is an arbitrary single-valued scalar function. This U(1) gauge invariance of the GL free energy is a consequence of electric charge conservation. It is valid independent of the detailed formulism of the free energy. The free energy is invariant under the above gauge transformation because

$$\left(-i\hbar\nabla - 2e\mathbf{A}'\right)\psi' = e^{i\varphi}\left(-i\hbar\nabla - 2e\mathbf{A}\right)\psi. \tag{1.124}$$

The gauge invariance implies that the electromagnetic field and the phase field of superconducting order parameter are interchangeable. They can be transformed into each other by the gauge transformation. The phase ϕ of the order parameter $\psi = |\psi|\exp(i\phi)$ is a Goldstone boson field. If we take the gauge in which φ in Eq. (1.122) equals the phase field ϕ (dubbed as the unitary gauge in literature), then the free energy defined by Eq. (1.79) becomes

$$f = f_n + \frac{\hbar^2}{2m^*}(\nabla|\psi|)^2 + \frac{m_A^2}{2}\mathbf{A}^2 + \alpha|\psi|^2 + \frac{\beta}{2}|\psi|^4 + \frac{\mathbf{H}^2}{2\mu_0}, \qquad (1.125)$$

where

$$m_A^2 = \frac{4e^2}{m^*}|\psi|^2. \qquad (1.126)$$

Under this gauge, the phase field ϕ of the order parameter is completely absorbed by the gauge field $\mathbf{A}$ and does not appear explicitly in the expression of the GL free energy. However, the gauge field $\mathbf{A}$ now acquires a mass m_A due to the spontaneous symmetry breaking, i.e. $|\psi| \neq 0$, in the superconducting phase. Hence, the onset of superconductivity generates a mass for the vector potential, so that the electromagnetic field become massive. This is the celebrated Anderson–Higgs mechanism associated with the spontaneous breaking of the $U(1)$ gauge symmetry in the context of superconductivity [26].

The above results show that under the unitary gauge, the phase field (or the Goldstone boson field) is completely absorbed by the gauge field and has no contribution to the free energy. It seems that the total degrees of freedom are reduced. This is in fact not the case. The massless U(1) gauge field, i.e. the electromagnetic field, is a transverse field with only two degrees of freedom. It has not the longitudinal component. After it has acquired mass by absorbing the Goldstone boson, the longitudinal component of the gauge field emerges, which maintains the total degrees of freedom of the system.

The Meissner effect is therefore a consequence of spontaneous symmetry breaking. It is a manifestation of the Anderson–Higgs mechanism, resulting from the interplay between the phase field and the gauge field. Under an external magnetic field, the spatial variance of the phase field in ψ generates a persistent supercurrent to screen the applied field. Therefore, the applied field becomes massive and decays inside the superconductor in a length scale characterized by the penetration depth.

1.12 Two Characteristic Energy Scales

There are two important energy scales in superconductors. One is the quasiparticle excitation gap Δ, which is also the binding energy of Cooper pairs. The other is the phase coherence energy, T_θ, which is determined by the phase fluctuation of Cooper pairs. Cooper pairs can develop global phase coherence only when their phase fluctuation energy is lower than T_θ. Superconducting properties, in particular the superconducting transition temperature T_c, are strongly influenced by the competition of these two energy scales. Both depairing (i.e. breaking Cooper pairs) and dephasing (i.e. disrupting the phase coherence of Cooper pairs) effects can suppress superconductivity.

In order to determine the energy scale of phase coherence, let us consider a superconducting system in the absence of an external magnetic field. If we ignore the amplitude fluctuation and keep the phase fluctuation of the order parameter, the free energy, according to Eq. (1.79), is then given by

$$f = \frac{\hbar^2}{2m^*} |\psi|^2 (\nabla\phi)^2. \qquad (1.127)$$

The long-range correlation of Cooper pairs is completely suppressed if the phase fluctuation over the coherence length ξ reaches the order of 2π. The energy scale corresponding to this critical fluctuation is just the energy of phase coherence, T_θ. Its value, which measures the phase stiffness of superconducting electrons, is estimated to be

$$T_\theta \approx \frac{\hbar^2}{m^*}|\psi|^2 \xi^3 \left(\frac{2\pi}{\xi}\right)^2 = \frac{4\pi^2\hbar^2\xi}{m^*}|\psi|^2 = \frac{\pi^2\hbar^2\xi}{2\mu_0 e^2 \lambda^2}. \qquad (1.128)$$

This expression agrees with the result given in Ref. [32] up to a constant factor of order 1. Thus the phase coherence energy is proportional to the superfluid density, $T_\theta \propto |\psi|^2$, which measures the capacity of paired electrons carrying superconducting currents.

In a highly anisotropic system, for example, a cuprate superconductor, λ in Eq. (1.128) and the corresponding superfluid density n_s should be their values along the c-axis. This is because the phase fluctuation is the strongest and hence n_s is the smallest along this direction. Similarly, ξ is the correlation length along the c-axis. If the correlation length is shorter than the interlayer distance d, then ξ should be set to d.

If T_θ is much larger than Δ, electrons immediately become phase coherent once they form Cooper pairs. In this case, the pair breaking is the main destructor of superconductivity, and T_c is entirely determined by the superconducting gap Δ. Thus the superconducting transition temperature is approximately proportional to the energy gap, $T_c \sim \Delta$. This is just the result of the BCS mean-field theory by neglecting phase fluctuations. In almost all conventional superconductors, made of metals or alloys, T_c is indeed found to scale approximately with the energy gap Δ. For the isotropic s-wave superconductor, the BCS mean-field theory predicts that

$$\Delta = 1.76T_c. \qquad (1.129)$$

On the other hand, if $\Delta \gg T_\theta$, electrons form Cooper pairs well before they develop phase coherence, and the BCS mean-field approximation is no longer valid. In this case, superconductivity can be eliminated by destroying the phase coherence, but without breaking Cooper pairs. Hence, it is dephasing, rather than depairing, that

Table 1.1. *Phase fluctuation energy scale T_θ versus T_c for conventional metal-based, organic, and high-T_c superconductors (from Ref. [32])*

materials	$T_c(K)$	T_θ/T_c
Pb	7	2×10^5
Nb_3Sn	18	2×10^3
UBe_{13}	0.9	3×10^2
$LaMO_6S_8$	5	2×10^2
$B_{0.6}K_{0.4}BiO_3$	20	50
K_3C_{60}	19	17
$(BEDT)_2Cu(NCS)_2$	8	1.7
$Nd_{2-x}Ce_xCu_2O_{4+\delta}$	21	16
$Tl_2Ba_2CuO_{6+\delta}$	80	2
	55	3.6
$Bi_2Sr_2CaCu_2O_8$	84	1.5
$Bi_2Pb_xSr_2Ca_2Cu_3O_{10}$	106	$0.8 \sim 1.4$
$La_{2-x}Sr_xCuO_{4+\delta}$	28	1
	38	2
$YBa_2Cu_3O_{7-\delta}$	92	1.4
$YBa_2Cu_4O_8$	80	0.7

becomes the main destructor of superconductivity. Consequently, the superconducting transition temperature is controlled by the phase coherence energy, T_θ, instead of the pairing energy gap Δ. Hence T_c is roughly proportional to T_θ,

$$T_c \sim T_\theta. \tag{1.130}$$

This is a result predicted by the theory of preformed pairs, applicable to systems with strong phase fluctuations. It could be observed in systems with small superfluid densities, where electrons could form pairs but without developing long-range phase coherence at relatively high temperatures. In underdoped high-T_c cuprates, it was found by experimental measurements that T_c is proportional to the superfluid density n_s, not the energy gap Δ [33]. This linear relationship between T_c and n_s is believed to be a consequence of strong phase fluctuations. It is a key experimental fact that should be seriously taken into account in the study of the phase diagram of high-T_c superconductors.

Table 1.1 shows the ratio T_θ/T_c estimated from experimental results for a number of superconductors. The smaller is T_θ/T_c, the stronger is the phase fluctuation. In conventional three-dimensional metal-based superconductors, T_θ/T_c is typically larger than 10^2, and T_c is hardly affected by phase fluctuations. By contrast, in organic or underdoped high-T_c superconductors, T_θ/T_c is close to 1, and phase fluctuations are very strong.

Phase fluctuations can suppress the long-range phase coherence of superconducting order parameters. It can also induce particle number fluctuation to enhance the charge fluctuation, since the particle number is conjugate to the phase of Cooper pairs. This is purely a quantum effect, which is difficult to observe in conventional metal-based superconductors.

1.13 Pairing Mechanism

As explained before, there are two steps for electrons to become superconducting: to form Cooper pairs and to develop phase coherence. Correspondingly, there are two questions that need to be resolved in the study of pairing mechanisms:

(1) What is the main interaction that glues electrons to form Cooper pairs?

(2) How do the Cooper pairs form phase coherence and condense?

The first question has been thoroughly discussed in textbooks and literature, although no consensus has been reached for high-T_c superconductors. Discussions of the second one are rather limited. In fact, our understanding of the dynamics of phase coherence is inadequate. This is not a serious problem in the study of metal- or alloy-based superconductors because phase fluctuations in these materials are weak and electrons become condensed almost immediately after they form Cooper pairs. However, in underdoped high-T_c cuprates, phase fluctuations are very strong. A thorough understanding of phase coherence is indispensable for the understanding of microscopic mechanism of high-T_c superconductivity.

Investigation of pairing mechanism plays a central role in the establishment of a microscopic theory of superconductivity. Once the pairing mechanism, especially the main interaction that drives electrons to superconduct, is determined, one can control and synthesize materials with certain targeted structures and chemical stoichiometries to enhance pairing interactions so that both the superconducting transition temperature and the critical current density can be enhanced.

In conventional superconductors, the pairing arises from the electron–phonon interaction. This has been verified by numerous experimental measurements. A frequently mentioned experimental evidence is the isotope effect. If one type of atoms is partially or completely replaced by its isotope in a superconductor, then the characteristic phonon frequency is changed due to the change of the atomic mass. Under the BCS mean-field approximation, it is predicted that the superconducting transition temperature induced by electron–phonon interactions is inversely proportional to the square root of the atomic mass. This prediction has been confirmed in a number of superconductors of simple metals.

However, the transition temperature induced by electron–phonon interactions is generally not very high. It is estimated to be less than 40K at ambient pressure according to the McMillan formula [34] because the energy scale of the Debye frequency is just of the order of room temperature, and there is not much room to greatly enhance it in laboratory. The electron–phonon coupling cannot be significantly enhanced either. Otherwise, it may cause instability in crystal structures. On the other hand, if the pairing arises from the interaction of electrons with optical phonons, the superconducting transition temperature is not constrained by the Debye frequency. As the characteristic frequency of optical phonons can be significantly higher than the Debye frequency, T_c can in principle exceed 40K.

The theory of high-T_c superconductivity remains one of the most fundamental and challenging problems. The simple electron–phonon mechanism fails to explain why the superconducting transition temperatures for both cuprate and iron-based superconductors can be much higher than the so-called McMillan limit ($\sim$ 40K) [34]. On the contrary, the electron–electron interaction, in particular the antiferromagnetic fluctuation, is very strong in both cuprate [35] and iron-based [36] superconductors. It might be the driving force for high-T_c superconductivity [37].

It should be pointed out that once electrons form Cooper pairs with macroscopic phase coherence, their physical behaviors become universal, no matter whether the superconducting pairing arises from the electron–phonon interaction or from the electron–electron interaction. As long as we know the characteristic energy scale of pairing interaction and the quasiparticle spectra function, we can accurately predict all dynamic and thermodynamic properties of superconducting states. This is the reason why we can still discuss and successfully predict physical properties of a high-T_c superconductor without knowing clearly its microscopic pairing mechanism.

1.14 Classification of Pairing Symmetry

Superconductors can be classified according to the internal symmetry of Cooper pairs. The wavefunction of a Cooper pair depends on both the spatial coordinates and the spin configurations of two electrons. In the absence of spin-orbit coupling or other interactions that break the spin rotational symmetry, the total spin is conserved and the pairing wavefunction can be factorized as a product of the spatial and spin wavefunctions

$$\Psi(\sigma_1, \mathbf{r}_1; \sigma_2, \mathbf{r}_2) = \chi(\sigma_1, \sigma_2) \Delta(\mathbf{R}, \mathbf{r}), \tag{1.131}$$

where $(\sigma_1, \mathbf{r}_1)$ and $(\sigma_2, \mathbf{r}_2)$ are the spin and spatial coordinates of the first and second electrons, respectively. $\mathbf{R} = (\mathbf{r}_1 + \mathbf{r}_2)/2$ is the coordinate of the center of mass and $\mathbf{r} = \mathbf{r}_1 - \mathbf{r}_2$ is the relative coordinate of the two electrons.

A Cooper pair can be either in a spin singlet or in a spin triplet state depending on whether the total spin is 0 or 1. The spin wavefunction is antisymmetric, $\chi(\sigma_1, \sigma_2) = -\chi(\sigma_2, \sigma_1)$, for the spin singlet state, and symmetric, $\chi(\sigma_1, \sigma_2) = \chi(\sigma_2, \sigma_1)$, for the spin triplet state. Since the full pairing wavefunction, $\Psi(\sigma_1, \mathbf{r}_1; \sigma_2, \mathbf{r}_2)$, is always antisymmetric under the exchange of two electrons, the spatial wavefunction corresponding to the spin singlet and triplet pairing states should be symmetric and antisymmetric, respectively.

Under the exchange of two electrons, the coordinate of the center of mass $\mathbf{R}$ is invariant, but the relative coordinate $\mathbf{r}$ changes sign. The pairing symmetry is classified by the symmetry of the spatial wavefunction under the change of the relative coordinate $\mathbf{r}$. If the Hamiltonian is rotationally invariant, the orbital angular momentum is conserved and $\Delta(\mathbf{R}, \mathbf{r})$ is an eigenfunction of the orbital angular momentum $\mathbf{L} = -i\hbar \mathbf{r} \times \nabla$. Thus the spatial wavefunction can be classified according to the eigenvalues of L^2

$$\mathbf{L}^2 \Delta(\mathbf{R}, \mathbf{r}) = l(l+1)\hbar^2 \Delta(\mathbf{R}, \mathbf{r}), \qquad (1.132)$$

where l is an integer. The eigenstate of the orbital angular momentum is symmetric if l is even, or antisymmetric if l is odd. Thus the orbital angular momentum is even for the spin singlet pairing state, and odd for the spin triplet one. In a translation invariant system, $\Delta(\mathbf{R}, \mathbf{r})$ can be further factorized as a product of the wavefunction for the coordinate of the center of mass, $\Delta_0(\mathbf{R})$, and that for the relative coordinate, $\psi(\mathbf{r})$

$$\Delta(\mathbf{R}, \mathbf{r}) = \Delta_0(\mathbf{R})\psi(\mathbf{r}), \qquad (1.133)$$

where $\psi(\mathbf{r})$ is the eigenfunction of orbital angular momentum. The pairing symmetry is determined by $\psi(\mathbf{r})$.

Cuprate superconductors have d-wave symmetry whose orbital angular momentum equals 2, i.e. $l = 2$, by adopting the convention of atomic physics. Superconductors with pairing orbital angular momenta $l = 0, 1, 2, 3, 4$ are called s, p, d, f, and g-wave superconductors, respectively. Among them, the s, d, and g-wave superconductors have spin singlet pairing, and the p- and f-wave superconductors have spin triplet pairing. The $l = 0$ state of the orbital angular momentum is isotropic and nondegenerate. The corresponding s-wave pairing state is also nondegenerate and spatially isotropic. The $l = 2$ states are five-fold degenerate, and the corresponding d-wave superconductors possess five different representations or pairing symmetries. They are generally denoted as d_{xy}, $d_{x^2-y^2}$, d_{xz}, d_{yz}, and $d_{3z^2-r^2}$ according to the eigenvalue of the third component of the angular momentum, respectively.

Physical properties of superconductors with different pairing symmetries are markedly different. The gap functions of d-wave superconductors (or any other superconductors with $l \neq 0$) can have gap nodes at which $\Delta(\mathbf{R}, \mathbf{r}) = 0$. In contrast, the gap function of s-wave superconductors is nodeless, namely $\Delta(\mathbf{R}, \mathbf{r})$ is always finite. This is the major difference between the s- and d-wave superconductors, which can significantly affect their low energy properties. In s-wave superconductors, the density of states of Bogoliubov quasiparticles vanishes inside the gap, and thermodynamic quantities decay exponentially with temperature and energy. However, in d-wave superconductors with gap nodes, the low energy density of states is linear, and thermodynamic quantities exhibit power-law behaviors at low temperatures, qualitatively different from s-wave superconductors.

In solids, the continuous rotational symmetry is broken into the discrete lattice point group symmetries. The definition of orbital angular momentum must be consistent with the point group symmetries. If the superconducting pairing is local in real space and the size of Cooper pairs is comparable to the lattice constant, such as in a high-T_c superconductor, the lattice symmetry needs to be considered in order to determine the pairing symmetry. The pairing symmetry should be classified according to the eigenstates of the point group.

For quasi-two-dimensional materials with tetragonal symmetry, the gap function $\psi(\mathbf{r})$ is invariant under the rotation around the c-axis in an s-wave superconductor. However, in a p- or d-wave superconductor, the gap function changes sign under the rotation of $180°$ or $90°$ around the c-axis. The p-wave pairing has two degenerate representations, p_x or p_y. A p-wave superconductor can be in either one of these states, or in a combined $p_x \pm i p_y$ pairing state with spontaneous breaking of time-reversal symmetry. There are also two possible representations for a d-wave superconductor, namely d_{xy} and $d_{x^2-y^2}$, but they are generally not degenerate even if the lattice is tetragonal.

In momentum space, the gap function is defined by Eq. (1.26), i.e. $\Delta_{\mathbf{k}} = \Delta_0 \psi_{\mathbf{k}}$, where $\psi_{\mathbf{k}}$ is the Fourier transformation of the gap function, $\psi(\mathbf{r})$. $\psi_{\mathbf{k}}$ is also the form factor of the pairing interaction. For the $d_{x^2-y^2}$ superconductor,

$$\psi_{\mathbf{k}} = c_1(\cos k_y - \cos k_x), \qquad (1.134)$$

where c_1 is a normalization factor. For the d_{xy}-wave superconductor, $\psi_{\mathbf{k}}$ is defined by

$$\psi_{\mathbf{k}} = c_2 \sin k_x \sin k_y, \qquad (1.135)$$

where c_2 is the inverse of the maximal value of $\sin k_x \sin k_y$ on the Fermi surface.

The nodal points of the $d_{x^2-y^2}$- and d_{xy}-wave superconductors are different in the Brillouin zone. The gap nodes lie along the diagonal lines of the Brillouin zone, i.e. $k_x = \pm k_y$, in the former case, and along the two axes of the Brillouin zone, i.e. $k_x = 0$ or $k_y = 0$, in the latter case. Apart from this, physical properties of these two kinds of d-wave pairing states are similar. The conclusion drawn from a $d_{x^2-y^2}$-wave superconductor is applicable to a d_{xy}-wave superconductor simply by rotating the axes by $\pi/4$, and vice versa. However, it should be emphasized that the microscopic origins leading to these two kinds of state could be different.

In the study of low energy physics, only the quasiparticle excitations around the Fermi surface are physically important. In this case, the pairing function can be simplified. For the $d_{x^2-y^2}$-wave superconductor, $\psi_{\mathbf{k}}$ can be approximately represented using the azimuthal angle of the wavevector, φ, and the gap function becomes

$$\Delta_\varphi = \Delta_0 \cos(2\varphi), \qquad (1.136)$$

where $\varphi = \tan^{-1} k_y/k_x$. This simplified expression is convenient to use in analytic calculations. Around the nodal points, Δ_k can also be written as

$$\Delta_k = \Delta_0 \frac{k_x^2 - k_y^2}{k_F^2}. \qquad (1.137)$$

For the $d_{x^2-y^2}$-wave superconductors, the above three expressions of ψ are physically equivalent. One can use the most convenient one in dealing with a concrete problem.

The relative wavefunction, $\psi(\mathbf{r})$, which discloses the internal structure of Cooper pairs, is determined by pairing interaction. Usually the gap energy decreases with the increase of the orbital angular momentum of paired electrons. Thus for all spin singlet superconductors, the s-wave pairing is generally more favored in energy and has the highest probability of being observed. Indeed, most superconductors discovered are s-wave ones.

However, in strongly correlated electronic systems, a non-s-wave pairing might be energetically more favored. This is because in strongly correlated systems, the local Coulomb repulsion is generally strong, which tends to reduce the probability of two electrons approaching each other. In a non-s-wave superconductor, the gap function vanishes at $\mathbf{r} = 0$, i.e. $\psi(\mathbf{r} = 0) = 0$. This releases the energy raised by the Coulomb repulsion between two electrons. On the contrary, in an s-wave pairing state, the gap function is finite at $\mathbf{r} = 0$, which is not favored by strong Coulomb repulsion.

Energetically, it is difficult to find Cooper pairs in the g- or even high angular momentum channels. However, it is not completely impossible. Evidence supporting g-wave pairing was reported in heavy fermion superconductors.

Triplet pairing breaks the time reversal symmetry and is rarer to discover. This kind of pairing is energetically favored in materials with strong ferromagnetic fluctuations. A possible candidate is Sr_2RuO_4, although no consensus on the pairing symmetry of this material has been reached [38]. Sr_2RuO_4 has a similar lattice structure to the Mott insulator La_2CuO_4, but it is a metal with strong ferromagnetic correlation.

Pairing symmetry is determined not just by pairing interactions, but also by the lattice symmetry. It can be classified according to the value of orbital angular momentum as the s, d, or other pairing state only if the system possesses perfect tetragonal or other lattice symmetries in two dimensions. Otherwise, different pairing channels are mixed. The level of mixing is determined by the lattice anisotropy between the two principal axes. The mixing may result from an explicit breaking of lattice symmetry induced, for example, by an uniaxial stress. It may also arise from spontaneous breaking of lattice symmetry generated, for example, by some nonlinear interactions.

It should be emphasized that there is not a one-to-one correspondence between a pairing symmetry and a pairing interaction. In fact, different interactions may lead to the same pairing symmetry. One should not infer the origin of pairing interaction just from the gap symmetry.

1.15 Pairing Symmetry of Cuprate Superconductors

To identify and verify the pairing symmetry of high-T_c Cooper pairs has been a great challenge and also one of the major achievements in the study of high-T_c superconductivity. It is an indispensable and key step toward the understanding of fundamental pairing mechanism and the establishment of a microscopic theory of high-T_c superconductivity.

Different from conventional superconductors whose normal states are Landau Fermi liquids, the normal states of high-T_c copper oxides are much more complicated and believed to be non-Fermi liquid-like. However, in the superconducting phase, the difference between these two kinds of superconductors is small, except that phase fluctuations are weaker and coherence lengths are longer in conventional superconductors. It is generally believed that the BCS theory of superconductivity is applicable to high-T_c superconductors, no matter whether the normal state is a Landau Fermi liquid or not. This is a basic assumption made in the analysis of experimental data of cuprate as well as iron-based high-T_c superconductors. It implies that we can identify the pairing symmetry of high-T_c superconductivity by comparing experimental measurements with theoretical predictions from the BCS theory, without knowing its pairing mechanism.

Both the pairing mechanism and the symmetry of Cooper pairs are determined by low-energy electronic structures and electron–electron interactions. In conventional superconductors of metals, the pairing results from electron–phonon interactions, and Cooper pairs have isotropic energy gaps with s-wave symmetry. Low-energy physics of high-T_c cuprates is determined by the conducting electrons in the two-dimensional CuO_2 planes on which Cooper pairing is expected to arise. Without doping, high-T_c cuprates are antiferromagnetic insulators with strong antiferromagnetic exchange interactions. The pairing in high-T_c cuprates is likely to arise from antiferromagnetic fluctuations, rather than from electron–phonon interactions. Based on the scenario of antiferromagnetic fluctuations, it was predicted that high-T_c superconductivity would have $d_{x^2-y^2}$-wave pairing symmetry [39–42].

To apply the BCS theory, one needs to first verify experimentally whether there exists Cooper pairs in high-T_c superconductors and whether the superconducting phase transitions therein are due to pair condensation. On the condition that Cooper pairing does exist, the next step is to determine the spin structure and pairing symmetry of Cooper pairs.

To determine whether electrons are paired and condensed at low temperatures, one needs to examine their characteristic effects and make comparison with theoretical predictions. The main physical phenomena or effects that have been utilized to judge the existence of Cooper pairs in high-T_c cuprates include:

(1) The direct-current (DC) and alternating-current (AC) Josephson effects: In addition to the single electron tunneling, there is also the Josephson pair tunneling in a junction between two superconductors. The response of pair tunneling to an applied electric or magnetic field behaves differently from that of normal single electron tunneling. It exhibits a number of characteristic coherent effects which can be used to determine the pairing state and its phase coherence.

(2) Andreev reflection: When a beam of electrons is incident onto the surface of a metal, part of the beam will be reflected. However, when a beam of electrons is incident on the surface of a superconductor, in addition to the reflection of normal electrons, there is also the reflection of holes due to pair condensation, which enhances the reflection current. At zero bias, the reflection current can be twice that of the incident electric current. Thus we can determine the pairing and phase coherence through the measurement of the Andreev reflection current.

(3) The Little–Parks magnetic flux quantization: The magnetic flux enclosed by a superconducting ring is quantized due to the phase coherence of superconducting order parameter, since the phase variable is gauge equivalent to a vector potential. The minimal quantized value of flux is $h/2e$ instead of h/e, determined by the total charge of a Cooper pair, $2e$, instead of the charge of a single electron. This experiment can be used to test if there is a flux quantization and if the minimal quantized flux is $h/2e$.

(4) The electron–hole mixing: In the superconducting states, the number of electrons is not conserved due to pair condensation. Electron and hole are mixed and manifested as Bogoliubov quasiparticle excitations. This mixing is also strong evidence of superconducting pairing. It can be probed by angular resolved photoemission spectroscopy (ARPES).

For high-T_c superconductors, there were a great deal of experimental investigations on the above four effects. All the experimental measurements on the Josephson effect [43–45], the Andreev reflection [46, 47], flux quantization [48–50], and electron–hole mixing [51] agree with the predictions of BCS theory. In addition, a large amount of measurement of thermodynamic and dynamic properties is also qualitatively consistent with the Cooper pairing picture. It all convincingly shows that high-T_c superconducting transitions are still due to condensation of electron pairs, just as in conventional superconductors.

The spin structure of Cooper pairs can be determined from the Knight shift of nuclear magnetic resonance (NMR). The Knight shift measures the magnetic susceptibility of electrons. In a spin singlet superconductor, the spin excitation is gapped and the Knight shift is suppressed at low temperatures, exhibiting a thermally activated exponential behavior. By contrast, in a triplet superconductor, the spin excitation is gapless, and the spin susceptibility in the superconducting states is comparable to or the same as in the normal state, hence the Knight shift is nearly temperature independent across the transition temperature. The Knight shift experiments on high-T_c superconductors are consistent with the prediction based on the spin singlet pairing picture [52, 53]. They show that high-T_c pairing happens in the spin singlet channel and the gap function is spatially symmetric under the exchange of two electrons.

A variety of experimental techniques have been used to measure the orbital angular momentum or to probe the pairing symmetry of high-T_c superconductors. This also generates many interesting problems for theoretical studies. Useful information on the pairing symmetry, properties of quasiparticle excitations and their interactions were drawn from nearly all kinds of thermodynamic and dynamical measurements. This is important not just for identifying the pairing symmetry, but also for exploring the origin of many anomalous behaviors of high-T_c cuprates in the normal state.

The experimental results depend strongly on the quality of the samples measured. If the sample quality is not that good then measurement errors are large, experimental results might not reflect the intrinsic properties of superconductors, making the judgement on the pairing symmetry difficult or even wrong in some cases. For example, in the early years of high-T_c study, particularly before 1993, most experimental measurements on thermodynamic as well as transport properties suggested that the high-T_c pairing had the *s*-wave symmetry, similar to conventional phonon-mediated superconductors. However, the conclusion was completely changed after high quality single crystals became available. In the meanwhile, theoretical studies also made great progress, providing important guidance toward a thorough understanding of experimental results. Now more and more experimental and theoretical studies have overwhelmingly shown that the high-T_c pairing has *d*-wave rather than *s*-wave symmetry. To learn more about the early history in this respect, please refer to Ref. [54].

A *d*-wave superconductor differs from an *s*-wave one in two respects. First, the *d*-wave gap function changes sign under a rotation of 90° around the *c*-axis. In contrast, the *s*-wave gap function does not change sign under rotation. Second, there are nodes in the *d*-wave gap function and the low-energy density of states of superconducting quasiparticles scales linearly with energy. Consequently, all thermodynamic quantities of d-wave superconductors exhibit power-law behavior as functions of energy or temperature at low temperatures. In contrast, the isotropic s-wave gap function is fully gapped over the entire Fermi surface, and all thermodynamic quantities exhibit activated behaviors at low temperatures. These qualitative differences set up criteria for identifying pairing symmetry in high-T_c cuprates. Correspondingly, experimental measurements can be divided into two categories.

The first category contains all the experiments that are sensitive to the phase of the gap function, by detecting the phase variation over the Fermi surface through the measurement of quantum interference effects induced in various Josephson junctions. This kind of experiment is not sensitive to the gap amplitude, but can be used to detect the positions of gap nodes and the sign change of the phase variable. It provides an indisputable way to differentiate a $d_{x^2-y^2}$-wave from a strongly anisotropic s-wave pairing state.

The second category includes experimental measurements of ARPES, magnetic penetration depth, NMR, optical conductivity, thermal conductivity, specific heat, and so on. This category does not include any experiment that is phase sensitive. It intends to identify the pairing symmetry by directly detecting the nodal positions and the gap anisotropy through measurements of response functions of low energy excitations to various applied perturbations, like heat, light, electromagnetic fields, and so on. In particular, ARPES can directly measure the momentum dependence of the gap function on the Fermi surface, from which the pairing symmetry is inferred. Raman scattering can selectively probe the gap function along different directions on the Fermi surface by changing the directions of incident and scattered lights. From the temperature dependence of the penetration depth, NMR, or specific heat, one can determine the low-energy density of states of quasiparticles. Measurement results of magnetic penetration depth and NMR are relatively simple to interpret because these probes measure directly physical properties of superconducting electrons, without worrying about the contribution of phonons and other effects. At low temperatures, physical properties of d-wave superconductors are governed by low energy quasiparticle excitations around the gap nodes, not by the size and the shape of the Fermi surface. But disorder effects induced by sample inhomogeneities, impurities, and dislocations can strongly affect low-energy behaviors of d-wave superconductors. These extrinsic effects should be considered in the analysis of measurement data.

Many physical properties of superconductors in the vicinity of T_c are also sensitive to pairing symmetry. But experimental results are difficult to analyze because both superconducting phase fluctuations and antiferromagnetic fluctuations become strong around T_c. A collective resonance may emerge in neutron scattering spectroscopy when the momentum transfer equals the momentum difference between two gap nodes. This can also be used to identify the locations of gap nodes.

However, as the momentum difference between the two gap nodes is close to the characteristic wave vector of antiferromagnetic fluctuations in high-T_c cuprates, it is not that simple to distinguish a neutron resonance peak from a peak induced purely by antiferromagnetic fluctuations.

It should be emphasized that different experimental techniques have their own limitations. It is impossible to draw a decisive conclusion simply based on a single experimental measurement. Instead, unified and unbiased explanations of all experimental results are important in the analysis of high-T_c superconductivity.

To summarize, we have achieved significant progress in the study of pairing symmetry in high-T_c copper oxides. In hole-doped cuprate superconductors, most experimental and theoretical studies have suggested that the gap function is strongly anisotropic and possesses the $d_{x^2-y^2}$-symmetry. However, there are still debates on whether the pairing symmetry has the same symmetry in electron-doped materials. In this book, we give a general introduction to the theory of d-wave superconductors by taking high-T_c cuprates as a prototype system. We hope this may deepen our understanding of this class of novel quantum phenomena, and provide useful guidance for further exploration and analysis of novel superconductors.

1.16 Pairing Symmetry of Iron-Based Superconductors

In conventional BCS theory of superconductivity, magnetic moments, like Fe^{2+} ions, are believed to be detrimental to superconductivity. The discovery of iron-based superconductors, however, has overturned this viewpoint. Iron-based superconductors, including iron pnictides and iron chalcogenites, are quasi-two-dimensional materials with strong antiferromagnetic fluctuations, similar to cuprate superconductors. The parent compounds of iron-based materials exhibit various antiferromagnetic orders [36, 55, 56]. These orders are driven predominantly by the magnetic interactions between Fe spins; among them the most important is the superexchange interaction between Fe spins mediated by As or Se 4p electrons [57, 58]. At low temperatures, most of the parent compounds, including LaFeAsO, $BaFe_2As_2$, and FeTe, are in the antiferromagnetic metallic phase. They become superconducting upon electron or hole doping. Some parent compounds, including LaFePO, LiFeAs, and FeSe, are superconducting even without doping at low temperatures.

Iron-based superconducting materials are multiband systems. The Fermi surface are centered around either $M = (\pi, 0)$ and its equivalent points or the zone center $\Gamma = (0, 0)$ [57, 59]. The low energy excitations contribute mainly by Fe $3d$ electrons, particularly by d_{xz}, d_{yz}, and d_{xy} orbitals. These orbitals couple strongly with each other and with the other two Fe $3d$ orbitals, $d_{x^2-y^2}$ and d_{z^2}, by the Hund's rule interactions. These $3d$ orbitals are partially itinerary and partially localized, behaving as in an orbital selective Mott system [60]. In other words, the low energy charge dynamics is governed by itinerant $3d$ electrons and behaves more like in a conventional metal with weak correlation, whereas the spin dynamics is essentially governed by localized moments and behaves more like in a strong coupling system.

To determine whether the superconducting pairing happens in the spin singlet or triplet channel, the Knight shift was measured on several Fe-based superconductors, including $LaO_{1-x}F_xFeAs$ [61] and $Ba(Fe_{1-x}Co_x)_2As_2$ [62]. The measurement results show that the Knight shift decreases for all angles of the applied magnetic field with respect to the crystallographic axes, which effectively excludes the possibility of triplet pairing. It is commonly believed that all Fe-based superconducting electrons are spin singlet paired. A possible exception is FeSe in its nematic phase where a p-wave-like gap anisotropy is observed [63].

From mean-field calculations, it was predicted that the Fe-based superconducting energy gap possesses conventional s-wave symmetry [64–67], namely in the identity representation of point group. This prediction was confirmed by spectroscopy as well as transport measurements on most of iron-based superconductors. However, Fe-based superconductors are multiband systems. The relative phases of gap functions can be different on the two bands located around the M and Γ points, respectively. The pairing is said to have s_{++} symmetry if the gap function has the same phase on these two Fermi surfaces. On the other hand, if the gap function changes sign on the two Fermi surfaces, the pairing is said to have s_{+-} symmetry.

The relative phase is determined by the interaction between Cooper pairs on different bands. If the pairing is induced by the antiferromagnetic fluctuations, interaction between Cooper pairs on the two bands whose centers are linked by the characteristic wave vector of the antiferromagnetic interaction are generally repulsive. In this case, the superconducting phases are opposite on the two Fermi surfaces, and the gap function has s_{+-} symmetry [64–67]. On the other hand, if the pairing is induced by the orbital fluctuation in the A_{1g} channel, the interaction between Cooper pairs on the band is attractive, and the gap function generally has s_{++} symmetry [68]. Thus, the relative phases of the gap function could reveal important information on the pairing interaction.

In literature, there are many discussions of the phase structure of the gap function. However, as most of experiments are not phase sensitive, it is actually very difficult to resolve unambiguously this seemly simple phase problem. The situation becomes more complicated and exotic when the Fermi surface pocket around either the Γ or M point completely disappears. The phase sensitive experiment that provided strong evidence for d-wave pairing in cuprate superconductors has proven extremely difficult to implement for the following two reasons: (1) it is difficult to fabricate clean junctions; (2) more importantly, the s_{++} and s_{+-} states have the same angular symmetry, and there is no way to distinguish them by tunneling into different faces of a crystal through the corner-junction experiment.

For an s_{+-} superconductor, it is expected that a strong neutron resonance peak exists around the momentum linking hole and electron Fermi surfaces, i.e. at $M = (\pi, 0)$ and equivalent points [69, 70]. This resonance peak has indeed been observed in nearly all iron-based superconductors [71–73], in agreement with the theory that predicts the pairing to have s_{+-} symmetry. However, it should be noted that in situations where the quasiparticle relaxation rate exhibits a strong energy dependence, an s_{++}-state may also show a sharp peak, similar to the resonance peak observed, in the superconducting state.

From the experimental observation of quantum interference of quasiparticles with magnetic or nonmagnetic impurities, it was also found that an s_{+-} pairing is more likely [74]. On the other hand, from the Anderson theorem, it is well known that nonmagnetic impurity scattering does not have much effect on the transition temperature for s_{++} superconductors, but it may reduce strongly the transition temperature for s_{+-} superconductors [68]. In particular, the transition temperature of an s_{+-} superconductor should decrease with increasing impurity concentration. However, for iron-based superconductors, the critical transition temperature does not depend much on the quality of the samples. This seems to suggest that the s_{++} pairing is more favored.

In a sign-changing energy gap system, a localized resonance may emerge near a nonmagnetic impurity. This kind of in-gap nonmagnetic impurity resonance state has indeed been observed in several iron-based superconductors, including LiFeAs [75], FeSe [76], and $Na(Fe_{0.96}Co_{0.03}Cu_{0.01})As$ [77]. These observations provided evidence in favor of s_{+-}-pairing, but more studies are needed to rule out magnetic characters of impurities. An unusual impurity state was also observed at an interstitial Fe site in Fe(Se,Te) [78]. This state is located at zero energy, to within experimental resolution, and unusually stable against an applied magnetic field. This was interpreted as a Majorana fermion, which is topologically protected by an energy barrier.

The superconducting and antiferromagnetic orders are two competing orders. Generally they repel each other. On the other hand, if the pairing has s_{+-} symmetry, theoretical calculations have suggested that these two kinds of order can coexist [79]. Experimentally, this kind of coexistence has indeed been observed in $BaFe_2As_2$, $Ba_{1-x}K_xFe_2As_2$, and $SmFeAsO_{1-x}F_x$ with Co substituting Fe or with P substituting As [80, 81], and in $K_xFe_2Se_2$. However, in these systems in which the coexistence was observed, the superconducting gap was also found to have line nodes. It is unknown whether this coexistence is caused by the s_{+-} pairing symmetry or by the line nodes.

For some Fe-based materials, a large gap anisotropy is observed on individual Fermi surface sheets. For KFe_2As_2 [82], $BaFe_{2-x}Ru_xAs_2$, and nearly all phosphorus-based superconductors, including LaFePO, LiFeP, and $BaFe_2As_{2-x}P_x$, it has even been found that the gap nodes exist. A p-wave-like state was also observed in the nematic phase of FeSe superconductor with twofold rotational symmetry [63, 83]. A number of scenarios, including d-wave or other exotic pairing symmetry [84], were proposed to interpret the experimental data. The presence of gap nodes generally implies that the pairing symmetry is unconventional, although an extended s-wave pairing may have accidental nodes on one or more Fermi surfaces.

It is still unclear why the pairing function can show such a large difference on different iron-based superconductors. Nevertheless, it is not surprising that the superconducting state shows such nonuniversal behavior in the Fe-based materials. As mentioned before, both the orbital and antiferromagnetic fluctuations are very strong. This, combined with the multiband feature, implies that this kind of material is sensitive to small changes in pressure, doping, or disorder, and that the superconducting state may exhibit a full gap in some compounds and clear evidence of nodal quasiparticle excitations in others.

2

Microscopic Models for High Temperature Superconductors

2.1 Phase Diagram of Cuprate Superconductors

Perovskite copper oxides, or cuprates, are the first class of high-T_c superconductors to have been discovered [1]. The parent compounds of these superconductors are antiferromagnetic Mott insulators, exhibiting an antiferromagnetic long-range order at low temperatures. Chemical doping, by element substitutions or by changing oxygen or other atomic contents, introduces conducting electrons to the parent compounds. This suppresses antiferromagnetic fluctuations and drives copper oxides into the superconducting phase. For example, La_2CuO_4 and $YBa_2Cu_3O_6$ are two typical parent compounds. They are antiferromagnetic Mott insulators and become high-T_c superconductors upon hole-doping. In fact, they were the first two families of high-T_c superconductors to be discovered. Cuprate superconductors are obtained from the parent compounds by either hole or electron doping. The resultants are called hole- and electron-doped high-T_c superconductors, respectively.

So far more than 10 families of high-T_c superconductors with different lattice and chemical structures have been discovered. They all have layered structures. The layers are composed of CuO_2 planes (see Fig. 2.1), whose crystalline axes are denoted as a and b, respectively. The c-axis is perpendicular to the ab-plane. Strong anisotropy exists between the ab-plane and the c-axis. In the presence of free charge carriers, the conductivity in the CuO_2 plane is usually two to four orders of magnitude higher than that along the c-axis. As verified by both band structure calculations and numerous experimental measurements, transport properties and low energy thermal excitations are governed by electrons in the CuO_2 planes. This is a basic property of cuprates that should be considered in the analysis of high-T_c superconductivity.

As mentioned, cuprate superconductors are quasi-two-dimensional materials, which possess two characteristic features: first, quantum and thermal fluctuations are very strong; and second, the Coulomb screening is poor so that electron–electron

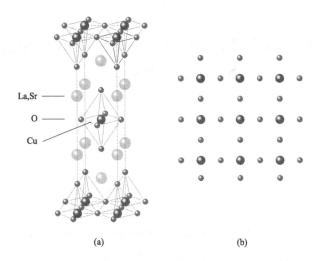

La,Sr

O

Cu

(a)   (b)

Figure 2.1 (a) Crystalline structure of high-T_c superconductor $La_{2-x}Sr_xCuO_4$. (b) Lattice structure of a CuO_2 plane.

interactions are strong. These features are responsible for various strongly correlated and anomalous behaviors observed in high-T_c cuprates. High-T_c cuprates are prototype systems of strongly correlated electrons. Investigation into physical properties of cuprates is important not only for the understanding of the microscopic mechanism of high-T_c superconductivity but also for a comprehensive understanding of general low-dimensional strongly correlated systems.

The microscopic models introduced in this chapter serve as a starting point for understanding the effective low energy physics of high-T_c cuprates. At the current stage, it remains unclear whether the high-T_c problem can be solved simply based on these models. It is even unknown whether these models possess the superconducting ODLRO. Each of these models has its own limitations and conditions of validity. Before clarifying these subtleties, we first give a brief overview of the physical properties, especially the phase diagram, of high-T_c superconductivity.

Physical properties of high-T_c cuprates depend crucially on temperature as well as doping level. Applying pressure and strong electromagnetic fields may also significantly alter the physical properties of high-T_c cuprates. Figure 2.2 shows a typical phase diagram of high-T_c cuprates. The high-T_c cuprates are antiferromagnetic insulators at low doping. Superconductivity emerges when the doping exceeds a critical level. The superconducting transition temperature increases with the doping at the beginning, then drops after passing a maximal value. The doping level at which the transition temperature reaches the maximum is called the optimal doping. The doping above and below the optimal doping is called overdoping and underdoping, respectively.

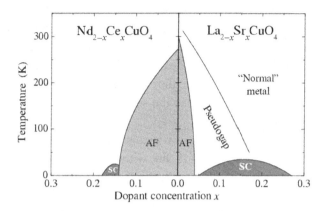

Figure 2.2 Phase diagram of $La_{2-x}Sr_xCuO_4$ and $Nd_{2-x}Ce_xCuO_4$ superconductors (from Ref. [85]). The left- and right-hand sides represent electron- and hole-doped cases, respectively. AF represents the antiferromagnetic long-range ordered phase, and SC represents the superconducting phase.

The phase diagram of high-T_c superconductivity is asymmetric with respect to electron- and hole-dopings. In the hole-doped case, the antiferromagnetic insulating state disappears above 3%, and superconductivity emerges when the doping level exceeds 5%. The optimal doping takes place around 15%. On the other hand, in the electron-doped case, the antiferromagnetic insulating state disappears at a doping level higher than 13%. The superconductivity appears in a much narrower range than in the hole-doped case.

Cuprate superconductors behave very differently from conventional metal-based ones. Some of the phenomena discovered in cuprates can be reasonably understood, but many of them lack a unified and comprehensive explanation. This includes the spin-charge separation [86, 87], the pseudogap phenomenon observed in the underdoped cuprates [88], and the intrinsic charge inhomogeneity [89]. Understanding these anomalous properties is not only important to the understanding of measurement data, but also crucial to the construction of high-T_c theory.

Among various anomalous properties, the pseudogap is one of the most important effects observed in the normal state. The pseudogap is a manifestation of the suppression in the density of states of low-lying electronic excitations of underdoped high-T_c cuprates. It shares many similarities with the superconducting gap of quasiparticle excitations in the superconducting state. For example, the pseudogap suppresses various physical quantities at low temperatures such as the specific heat, the magnetic susceptibility, the optical conductivity, and the spectra weight of electrons. It is also conceivable that the pseudogap has the same symmetry as the superconducting energy gap. However, the pseudogap is not a superconducting order parameter. There is no phase transition associated with a pseudogap. The transition from the normal metallic phase to the pseudogap phase is continuous without any singularities in the specific heat and other thermodynamic quantities. Therefore, it is very difficult to accurately determine the crossover temperature of the pseudogap phase. In the heavily underdoped regime, the onset temperature of the pseudogap is about one order of magnitude higher than the superconducting transition temperature, but drops with increasing doping.

The physical origin of the pseudogap remains unclear. One possibility is that it results from "preformed" Cooper pairs, but without developing collective phase coherence. This scenario is consistent with the fact that the superfluid density is low and the phase fluctuation is strong in underdoped high-T_c superconductors, and supported by the experimental measurement of the Nernst effect of transverse thermal conductivity [90]. However, we still lack a quantitative understanding of phase fluctuations. It is difficult to make a conclusive judgment on the validity of this scenario. In addition, the pseudogap appears in the vicinity of the antiferromagnetic phase, where strong antiferromagnetic fluctuations obscure the picture of this puzzling phenomenon.

The stripe phase, or the intrinsic charge inhomogeneity, is another important effect observed in underdoped high-T_c cuprates [91]. The key experimental evidence comes from the incommensurate peaks of spin structure factors measured by neutron scattering spectroscopy. These peaks appear near the characteristic wave vector (π, π) of antiferromagnetic fluctuations. The stripe phase is not observed in all underdoped cuprate superconductors. In most high-T_c cuprates, the static stripe phase is not observed, and there is also no direct or strong evidence for the existence of dynamic stripes. Similar to the pseudogap, theoretical study on the stripe phase is immature, and a quantitative description of it is still not available.

In the overdoped regime, the pseudogap effect and antiferromagnetic fluctuations are weakened. The temperature and energy dependencies of various thermodynamic quantities and transport coefficients behave similarly to conventional metals, as predicted by the Landau Fermi liquid theory.

It is unclear if there is a quantum phase transition between underdoped and overdoped high-T_c materials. This is an important question that needs to be resolved by experiments in the future. Experimentally, it was found that there might exist a critical regime that separates the underdoped pseudogap phase and the overdoped doped Fermi liquid phase, which implies the existence of a quantum critical point at zero temperature and that the phases on the two sides of this critical point are different [92]. This quantum critical point lies in the slightly overdoped regime [93], but the scaling behavior in the vicinity of this putative critical point and the associated discontinuity of thermodynamic quantities were not observed. It remains an open question whether this quantum critical point really exists.

2.2 Antiferromagnetic Insulating States

In the undoped insulating parent compounds, the copper and oxygen ions in the CuO_2 plane are in the Cu^{2+} and O^{2-} valence states, respectively. The outer shell electron configuration of O^{2-} is $2p^6$, whose three $2p$ orbitals are fully occupied. The outer shell electron configuration of Cu^{2+} is $3d^9$. Among the five $3d$ orbitals of Cu^{2+}, four of them are fully filled, and the one with the highest energy, $3d_{x^2-y^2}$, is singly occupied, namely in the half-filled state (Fig. 2.3). In such a configuration, Cu^{2+} carries a spin of $S = \frac{1}{2}$.

Based on the standard Bloch band theory, solid state materials with half-filled bands are metallic in the absence of Peierls-type lattice structure transitions. However, experimentally La_2CuO_4 and other parent compounds of cuprate superconductors are actually antiferromagnetic insulators at low temperatures, indicating that the $3d_{x^2-y^2}$ electrons of Cu^{2+} are localized around the copper sites and have no contribution to the charge current. This class of insulators are called Mott insulators, which are fundamentally different from the band insulators with either empty or

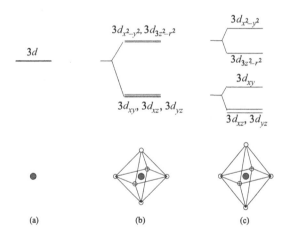

Figure 2.3 Crystal field splitting of the five $3d$ orbitals of Cu^{2+}. (a) In an isolated Cu^{2+} (solid circle), the five $3d$ orbitals are degenerate. (b) In a CuO_6 octahedron with cubic symmetry, the five d orbitals split into three t_{2g} orbitals, $3d_{xy}$, $3d_{xz}$, and $3d_{yz}$, and two e_g orbitals, $3d_{x^2-y^2}$ and $3d_{3z^2-r^2}$. The energy level of e_g electrons is higher because the charge clouds of e_g orbitals point toward the oxygen anions at the vertexes with stronger Coulomb repulsion. (c) Energy level splitting in a CuO_6 octahedron elongated along the c-axis. The energy level of $d_{3z^2-r^2}$ becomes lower because the wavefunction overlap between this orbital and the $2p$ orbitals of the two apical oxygens becomes smaller. The three t_{2g} orbitals also become non-degenerate. d_{xy} has higher energy because the Coulomb repulsions between the other two t_{2g} orbitals and the two apical oxygens are reduced. The energy levels of d_{xz} and d_{yz} remain degenerate if the octahedron has $\pi/2$ rotational symmetry along the c-axis, but can be split by the Jahn–Teller effect if the occupation numbers in these orbitals are different.

fully filled bands. The Mott insulator results from the Coulomb interaction and is an effect of many-body strong correlations. In comparison, band insulators are purely a consequence of Pauli's exclusion principle of Fermi statistics.

The strongest interaction in the CuO_2 plane is the Coulomb interaction between two electrons at the outmost $3d$ orbital of Cu^{2+} cation. The Coulomb repulsion between different Cu^{2+} cations is relatively weaker. Removing one electron from a Cu^{2+} site to one of its neighboring sites creates a doubly occupied site and an empty site which is energetically unfavored. This effective Coulomb interaction is modeled by the Hubbard interaction whose Hamiltonian is defined by $H_I = U \sum_i n_{i\uparrow} n_{i\downarrow}$, with $n_{i\uparrow}$ and $n_{i\downarrow}$ the up- and down-spin electron number operators in the $3d_{x^2-y^2}$ orbital at site i, respectively. In cuprate superconductors, the effective Coulomb repulsion energy U is about a few electron volts, larger than the bandwidth of conducting electrons.

At half-filling, if the Hubbard interaction is strong enough, electrons are localized on lattice sites and do not conduct. In the meanwhile, the antiferromagnetic exchange interaction between two spins on the neighboring sites is unscreened and becomes the most important interaction that governs low energy excitations. It leads to the antiferromagnetic long-range order at low temperatures.

The Mott insulators are intimately connected with the antiferromagnetic orders. In fact, the antiferromagnetic orders are discovered in nearly all the Mott insulating materials. The antiferromagnetic order is absent in the one dimensional Hubbard model at half-filling, at which the antiferromagnetic correlations exhibit an algebraic decay. In two or three dimensions, the Mott insulating states without long-range antiferromagnetic ordering, namely the spin-liquid states, have not been found without doubt experimentally.

The antiferromagnetic Heisenberg model is the fundamental model describing low energy antiferromagnetic exchange interactions. The corresponding Hamiltonian reads

$$H = J \sum_{\langle ij \rangle a} S_{i,a} S_{j,a}, \tag{2.1}$$

where $S_{i,a}$ with $a = x, y, z$ are the spin operators of the Cu^{2+} site; $\langle ij \rangle$ represents the summation over the nearest neighboring sites i and j. According to the Wagner–Mermin theorem, there is no long-range magnetic order at any finite temperature for this SU(2) invariant spin model in two dimensions. Nevertheless, high-T_c cuprates are not exactly two-dimensional materials. They exhibit a quasi-two-dimensional layered structure with weak coupling along the c-axis. The antiferromagnetic long-range order may appear at low temperatures.

The spin operator $S_{i,a}$ can be expressed in terms of electron operators as

$$S_{i,a} = d_i^\dagger \frac{\sigma_a}{2} d_i, \tag{2.2}$$

where σ's are Pauli matrices; $d_i = (d_{i\uparrow}, d_{i\downarrow})$ are the annihilation operators of Cu $3d_{x^2-y^2}$ electrons. At half-filling, every site is singly occupied, and d_i satisfies the constraint

$$d_i^\dagger d_i = 1. \tag{2.3}$$

In this case, the spin operator $S_{i,a}$ is invariant under the following local SU(2) transformation

$$\begin{pmatrix} d_{i\uparrow} & d_{i\downarrow} \\ d_{i\downarrow}^\dagger & -d_{i\uparrow}^\dagger \end{pmatrix} \rightarrow g_i \begin{pmatrix} d_{i\uparrow} & d_{i\downarrow} \\ d_{i\downarrow}^\dagger & -d_{i\uparrow}^\dagger \end{pmatrix}, \tag{2.4}$$

where g_i is a local SU(2) transformation matrix. This local SU(2) symmetry is equivalent to the particle–hole symmetry, and is a consequence of the particle–hole invariance of the Hubbard model at half-filling. It plays an important role in the mean-field study of high-T_c superconductors.

For the two local fermion operators $d_{i\sigma}$, the largest algebra that they may generate is SO(4) [94]. SO(4) could be decomposed as a product of two SU(2) algebras: the usual spin SU(2) algebra whose generators are defined in Eq. (2.2), and the pseudospin SU(2) algebra spanned by the generators

$$\eta_{i,a} = \frac{1}{2}(d_{i\uparrow}^\dagger, d_{i\downarrow})\tau_a \begin{pmatrix} d_{i\uparrow} \\ d_{i\downarrow}^\dagger \end{pmatrix}, \tag{2.5}$$

where τ_a are the Pauli matrices in the Nambu channel. They generate the local SU(2) transformation matrix g_i.

2.3 Three-Band Model

Undoped cuprates are antiferromagnetic Mott insulators. The antiferromagnetic coupling between two neighboring Cu^{2+} spins results mainly from the superexchange interaction induced by the strong Coulomb repulsion U_d between two electrons occupying the same Cu $3d_{x^2-y^2}$ orbital. However, as the highest O $2p$ levels fall between the single and double occupied Cu $3d_{x^2-y^2}$ states (corresponding to the Cu^{2+} and Cu^+ states, respectively), the low-energy charge dynamics is governed predominantly by the difference between U_d and the charge-transfer gap Δ between O $2p$ and Cu $3d_{x^2-y^2}$ states. Thus undoped cuprates belong to a particular class of Mott insulators – charge transfer insulators [95].

For cuprates, the Coulomb repulsion of two electrons (or holes) in the Cu $3d_{x^2-y^2}$ orbit is about $U_d \approx 5$ eV. On the other hand, the energy difference between the Cu $3d_{x^2-y^2}$ orbit and the oxygen $2p$-orbit is $\Delta = \varepsilon_p - \varepsilon_d \approx 2$ eV. Furthermore, the hopping integral between these two orbitals are $t_{pd} \approx 1.3$ eV. Roughly speaking, the following conditions are satisfied in cuprates

$$U_d \gg \Delta = \varepsilon_p - \varepsilon_d > |t_{pd}|. \tag{2.6}$$

The electronic states of high-T_c cuprates are significantly changed after doping. In the hole-doped cuprates, holes are mainly doped onto the oxygen sites, and the valence configuration of the doped oxygen site changes from O^{2-} to O^-. Owing to the hybridization between the oxygen p orbitals and the copper d orbitals, a small but finite portion of holes occupies the copper sites, which changes the valence configuration of Cu from Cu^{2+} to Cu^{3+}. In contrast, in the electron-doped case, most electrons are doped onto the copper sites, which changes the valence configuration of the copper cations from Cu^{2+} to Cu^+.

Doping opens new conducting channels which allow electrons to move without encountering the penalty of on-site Coulomb repulsion. In the hole-doped cuprates, a hole can hop from one site to another without creating double occupancy. Similarly, in electron-doped materials, the hopping of electrons between a doubly occupied site and a singly occupied one does not cost extra Coulomb repulsion. Therefore, doping, whether hole-doping or electron-doping, can always destabilize the Mott insulating state and enhance conductivity. The high-T_c superconductivity emerges when the doping reaches a critical level.

The high-T_c physics is determined by the orbitals in the CuO_2 plane, particularly the Cu $3d_{x^2-y^2}$ orbitals with their hybridized O $2p$ orbitals. The oxygen orbital that couples to the $3d_{x^2-y^2}$ orbitals on the neighboring copper sites is determined by the p orbital orientation. Based on the symmetry analysis, only the $2p_{x(y)}$ orbital can form a σ-bond along the $x(y)$-direction with a Cu $3d_{x^2-y^2}$ orbital (Fig. 2.4). Other oxygen $2p$ orbitals do not couple to the copper $3d_{x^2-y^2}$ orbitals because the overlap integrals between these orbitals vanish.

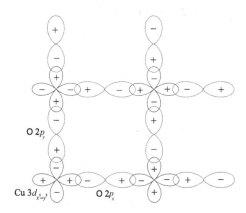

Figure 2.4 Three most important orbitals that govern the low-energy physics of high temperature superconductors: copper $3d_{x^2-y^2}$, and oxygen $2p_x$ and $2p_y$-orbitals. The relative phases of these Wannier orbitals are not uniquely defined. The convention used here is determined by requiring that the overlap integrals between Cu $3d_{x^2-y^2}$ and O $2p$ orbitals are negative.

Low energy excitations of high-T_c cuprates are governed by the electrons in the copper $3d_{x^2-y^2}$ orbitals, the oxygen $2p_x$ orbitals along the a-axis, and the oxygen $2p_y$ orbitals along the b-axis. In the hole-doped materials, the interactions among these orbitals are described by the Hamiltonian [96]:

$$H = -\sum_{\langle il\rangle} t_{pd}\left(p_l^\dagger d_i + d_i^\dagger p_l\right) + \sum_l \varepsilon_p p_l^\dagger p_l + \sum_i \varepsilon_d d_i^\dagger d_i$$
$$+ \sum_l U_p p_{l\uparrow}^\dagger p_{l\uparrow} p_{l\downarrow}^\dagger p_{l\downarrow} + \sum_i U_d d_{i\uparrow}^\dagger d_{i\uparrow} d_{i\downarrow}^\dagger d_{i\downarrow}, \qquad (2.7)$$

where i and l represent the coordinates of copper and oxygen sites, respectively. The summation $\langle il\rangle$ runs over the nearest neighboring copper cation and oxygen anion sites. $p_l = (p_{l\uparrow}, p_{l\downarrow})$ is the annihilation operator of holes in the oxygen $2p$ orbital. The first term describes the hybridization between a copper $3d_{x^2-y^2}$ orbital and an oxygen $2p$ orbital. The second and third terms are the on-site Coulomb potentials of holes in the oxygen $2p$ and copper $3d_{x^2-y^2}$ orbitals, respectively. U_p and U_d are the Coulomb repulsions on the oxygen and copper sites, respectively. A convenient phase convention for the Wannier wavefunctions of these orbitals is shown in Figure 2.4.

The above Hamiltonian Eq. (2.7) is called the *three-band model* of high-T_c superconductors. It offers a starting point for the study of the high-T_c mechanism. However, this model includes too many degrees of freedom and parameters, and is difficult to handle.

2.4 *dp* Model of Interacting Spins and Holes

As discussed previously, in the hole-doped cuprates, the $3d_{x^2-y^2}$ orbitals on the copper sites are singly occupied, and doped holes are predominantly located on the oxygen sites. In this case, one can take the first term in Eq. (2.7) as a perturbation and the other terms as the zeroth-order Hamiltonian, and use the degenerate perturbation theory introduced in Appendix C to simplify the three-band model as an effective low energy Hamiltonian. For this purpose, we define the zeroth-order Hamiltonian H_0 and the perturbation H_1 as

$$H_0 = \varepsilon_d \sum_i d_i^\dagger d_i + \varepsilon_p \sum_l p_l^\dagger p_l + U_d \sum_i d_{i\uparrow}^\dagger d_{i\uparrow} d_{i\downarrow}^\dagger d_{i\downarrow}, \qquad (2.8)$$

$$H_1 = -t_{pd} \sum_{\langle il\rangle}\left(p_l^\dagger d_i + d_i^\dagger p_l\right). \qquad (2.9)$$

In high-T_c cuprates, the hole density is low and the chance two holes of occupying the same oxygen site is very low. Thus we can neglect the oxygen Coulomb repulsion term, i.e. the U_p-term, in Eq. (2.7).

In the CuO_2 plane, every unit cell contains one copper atom and two oxygen atoms. For convenience, we treat these two oxygen atoms separately. If we use $p_{x,\mathbf{k}}$ and $p_{y,\mathbf{k}}$ to represent respectively these oxygen $2p$ orbitals in the momentum space, the Fourier transform of the oxygen hole operators p_l is then defined as

$$p_{i+\hat{x}/2,\sigma} = \frac{1}{\sqrt{N}} \sum_{\mathbf{k}} p_{x,\mathbf{k},\sigma} \exp\left[i\mathbf{k} \cdot \left(\mathbf{R}_i + \frac{\hat{x}}{2}\right)\right], \qquad (2.10)$$

$$p_{i+\hat{y}/2,\sigma} = \frac{1}{\sqrt{N}} \sum_{\mathbf{k}} p_{y,\mathbf{k},\sigma} \exp\left[i\mathbf{k} \cdot \left(\mathbf{R}_i + \frac{\hat{y}}{2}\right)\right]. \qquad (2.11)$$

Substituting them into Eqs. (2.8) and (2.9) and after simplification, we obtain

$$H_0 = \varepsilon_d \sum_i d_i^\dagger d_i + \varepsilon_p \sum_i \left(a_i^\dagger a_i + b_i^\dagger b_i\right) + U_d \sum_i d_{i\uparrow}^\dagger d_{i\uparrow} d_{i\downarrow}^\dagger d_{i\downarrow}, \quad (2.12)$$

$$H_1 = -t_{pd} \sum_{ij} u\,(i - j) \left(a_j^\dagger d_i + d_i^\dagger a_j^\dagger\right), \qquad (2.13)$$

where

$$a_{i\sigma} = \frac{1}{\sqrt{N}} \sum_{\mathbf{k}} \frac{c_x p_{x,\mathbf{k},\sigma} + c_y p_{y,\mathbf{k},\sigma}}{\sqrt{c_x^2 + c_y^2}} e^{i\mathbf{k}\cdot\mathbf{R}_i}, \qquad (2.14)$$

$$b_{i\sigma} = \frac{1}{\sqrt{N}} \sum_{\mathbf{k}} \frac{c_y p_{x,\mathbf{k},\sigma} - c_x p_{y,\mathbf{k},\sigma}}{\sqrt{c_x^2 + c_y^2}} e^{i\mathbf{k}\cdot\mathbf{R}_i}, \qquad (2.15)$$

$$u\,(\mathbf{r}) = \frac{2}{N} \sum_{\mathbf{k}} \sqrt{c_x^2 + c_y^2}\, e^{i\mathbf{k}\cdot\mathbf{r}}, \qquad (2.16)$$

and $c_x = \cos(k_x/2)$; $c_y = \cos(k_y/2)$; a_i and b_i are independent fermion operators, $\{a_i, b_i\} = \{a_i, b_i^\dagger\} = 0$; $u(\mathbf{r})$ satisfies the equation

$$\sum_i u\,(i - j)\, u\,(i - j') = 4\delta_{j,j'} + \delta_{\langle j, j'\rangle}. \qquad (2.17)$$

$|u(\mathbf{r})|$ is a fast-decay function of r. When $r \gg 1$, $u(r)$ approximately decays as $1/r^3$. The first three largest values of $u(\mathbf{r})$ are $u(0,0) = 1.91618$, $u(1,0) = 0.280186$, and $u(1,1) = -0.0470135$.

The above equations show that the interactions only exist between a- and d electrons, and there is no interaction between b and d electrons. Thus b_i is a nonbonding orbital and a_i represents a bonding orbital. The energy of b electrons (holes more precisely) lies above the Fermi energy, which has no contribution to dynamics and can be neglected. This leads to an effective two-band model which contains only a and d electrons. This equivalence between the two-band model and the three-band one is reached based on the assumption that the Coulomb interaction on the oxygen orbitals is negligible. If this term is included, a and b electrons are mixed. In this case, the three-band model cannot be reduced to a two-band model.

The ground states of H_0 are highly degenerate. All the states in which the copper $3d$-orbitals are singly occupied are the ground states of H_0. Below we use the degenerate perturbation theory to project the Hamiltonian into this degenerate ground state subspace and derive the low energy effective model. We use P to denote the projection operator for the ground state of H_0. Its effect is to project the Hamiltonian into the physical subspace in which all copper 3d orbitals are singly occupied, i.e. $d_i^\dagger d_i = 1$.

The hopping terms in H_1 change the occupation number of copper 3d orbitals. Thus the first-order correction of H_1 to the ground state is 0,

$$H_{eff}^{(1)} = P H_1 P = 0. \tag{2.18}$$

Similarly, it can be shown that all odd perturbation terms of H_1 vanish.

The second-order perturbation contribution from H_1 is given by

$$H_{eff}^{(2)} = P H_1 (1 - P) \frac{1}{E_0 - H_0} (1 - P) H_1 P. \tag{2.19}$$

After neglecting an irrelevant constant term, we find that

$$H_{eff}^{(2)} = -t_P \sum_{\langle ij \rangle} P a_i^\dagger a_j P + J_P \sum_i P d_{i\sigma}^\dagger d_{i\sigma'} \tilde{a}_{i\sigma'}^\dagger \tilde{a}_{i\sigma} P, \tag{2.20}$$

where

$$t_P = \frac{t_{pd}^2}{\varepsilon_p - \varepsilon_d}, \tag{2.21}$$

$$J_P = \frac{t_{pd}^2}{\varepsilon_p - \varepsilon_d} + \frac{t_{pd}^2}{U_d - \varepsilon_p + \varepsilon_d}, \tag{2.22}$$

$$\tilde{a}_i = \sum_j u(i - j) a_j. \tag{2.23}$$

$H_{eff}^{(2)}$ contains both the hopping and interaction terms of oxygen holes. In the undoped system, $H_{eff}^{(2)} = 0$. In order to study the interaction between Cu spins in the low doping limit, we need to calculate the contribution from the fourth-order perturbation in H_1.

$H_{eff}^{(4)}$ contains more terms than $H_{eff}^{(2)}$. Some of them are just to renormalize the coupling constants in $H_{eff}^{(2)}$. These terms can be absorbed into $H_{eff}^{(2)}$ just by modifying the coefficients. The terms, which are new and important at low doping, include the Heisenberg exchange interactions among copper spins, and the hopping terms of oxygen a electrons between the next-nearest and next-next-nearest neighboring sites:

$$H_{eff}^{(4)} = J \sum_{\langle ij \rangle} P \mathbf{S}_i \cdot \mathbf{S}_j P + t_P' \sum_{\langle ij \rangle'} P(a_i^\dagger a_j + h.c.) P + t_P'' \sum_{\langle ij \rangle''} P(a_i^\dagger a_j + h.c.) P, \quad (2.24)$$

where $\langle \rangle'$ and $\langle \rangle''$ represent summations over the next-nearest and next-next-nearest neighbor sites, respectively. The Heisenberg exchange constant is given by

$$J = \frac{t_{pd}^4}{(\varepsilon_p - \varepsilon_d)^2} \left(\frac{1}{U_d} + \frac{1}{\varepsilon_p - \varepsilon_d} \right). \quad (2.25)$$

The sum of $H_{eff}^{(2)}$ and $H_{eff}^{(4)}$ defines the low energy effective Hamiltonian for cuprates. It describes the interaction among electrons in the copper $3d_{x^2-y^2}$ and oxygen 2p orbitals. It is correct up to the 4th order of H_1, represented as

$$H_{dp} = H_{eff}^{(2)} + H_{eff}^{(4)}. \quad (2.26)$$

2.5 Zhang–Rice Singlet

In H_{dp}, J_P is a relatively large energy scale. It describes the interaction between the local spins in the copper $3d$ orbitals and the holes in the oxygen $2p$ orbitals. Because the off-site interaction between a d electron and an a hole is much smaller than the on-site interaction, we can take the approximation $u(r) \approx u(0)\delta_{r,0}$. In this case, the J_p-term becomes

$$H_{J_P} = J_P u^2(0) \sum_i P \left(a_i^\dagger a_i - 2e_i^\dagger e_i \right) P, \quad (2.27)$$

where e_i is a spin singlet operator formed by a and d electrons

$$e_i = \frac{1}{\sqrt{2}} \left(d_{i\uparrow} a_{i\downarrow} - d_{i\downarrow} a_{i\uparrow} \right). \quad (2.28)$$

In the limit J_P is much larger than t_P and other parameters in H_{dp}, the above equation shows that it is energetically favored for a copper $3d$ electron (d_i) and an oxygen $2p$-hole (a_i) to form a spin singlet bound state. It has an energy lower than both a unbounded state and a spin triplet one. In the low energy limit, e_i should be treated as a composite operator, and the two-band model can be further simplified as a single-band model. Based on this observation, Zhang and Rice derived an effective single-band model for high-T_c superconductors in 1988 [37, 97]. We call a localized spin singlet formed by the copper $3d$ spin and the oxygen hole a Zhang–Rice singlet. The energy difference between a Zhang–Rice singlet and the corresponding triplet is given by

$$E_{ZR} = 2J_P u^2(0). \tag{2.29}$$

The single-band model is obtained by projecting the dp model H_{dp} onto the subspace spanned by the ground states in which each lattice site is either in a state with singly occupied d orbitals, or in a Zhang–Rice singlet, limited by the constraint,

$$e_i^\dagger e_i + d_i^\dagger d_i = 1. \tag{2.30}$$

If P_{ZR} is the corresponding projection operator, the effective single-band model is determined by

$$H = P_{ZR} H_{dp} P_{ZR}. \tag{2.31}$$

The rule of projection is simple. A Zhang–Rice singlet exists at site i if and only if there is an oxygen hole at that site. d_i is invariant after the projection, i.e. $P_{ZR} d_i P_{ZR} = d_i$. The hole operator a_i, after projection, becomes

$$P_{ZR} a_{i\sigma} P_{ZR} = -\frac{1}{\sqrt{2}} \sigma d_{i\bar\sigma}^\dagger e_i. \tag{2.32}$$

It simply means that annihilating an oxygen hole with spin σ is equivalent to annihilating a Zhang–Rice singlet and at the same time creating an electron with opposite spin. The coefficient $1/\sqrt{2}$ is due to the fact that in the Zhang–Rice singlet state, the spin of the oxygen hole a_i only has half probability in the state of σ. Applying these results to H_{dp}, and after neglecting some dynamically irrelevant constant terms, we find the effective single-band Hamiltonian to be

$$H = -\sum_{ij} t_{ij} d_i^\dagger d_j e_j^\dagger e_i + J \sum_{\langle ij \rangle} \left(\mathbf{S}_i \cdot \mathbf{S}_j - \frac{1}{4} n_i n_j \right), \tag{2.33}$$

where

$$t_{ij} = t\delta_{\langle ij \rangle} + t'\delta_{\langle ij \rangle'} + t''\delta_{\langle ij \rangle''} \tag{2.34}$$

and $t = t_P/2$, $t' = t'_P/2$, $t'' = t''_P/2$.

2.6 Hubbard Model

The one-band Hubbard model, or, simply, the Hubbard model, is a fundamental model of interacting electrons on a lattice. It was first introduced to understand the microscopic origin of itinerant ferromagnetism. Now it has been widely used to investigate antiferromagnetism and the metal–insulator transitions. Soon after the discovery of high-T_c superconductors, P. W. Anderson first proposed to use the one-band Hubbard model to study the mechanism of high-T_c superconductivity [37].

The one-band Hubbard model is an effective low-energy model. For sufficiently strong Coulomb repulsion, its ground state is a Mott insulator at half-filling, exhibiting an antiferromagnetic long-range order with strong antiferromagnetic fluctuations, similar as in the undoped high-T_c cuprates. The one-band Hubbard model is much simpler to analyze than the three-band one. In the strong coupling limit, the one-band Hubbard model is equivalent to the t–J model at low doping.

The Hubbard model is defined by the Hamiltonian

$$H = -t \sum_{\langle ij \rangle} \left(c_i^\dagger c_j + c_j^\dagger c_i \right) + U \sum_i c_{i\uparrow}^\dagger c_{i\uparrow} c_{i\downarrow}^\dagger c_{i\downarrow}, \qquad (2.35)$$

where c_i is the annihilation operation of an electron. The t-term describes the hopping of electrons between two neighboring sites and the U-term represents the on-site Coulomb repulsion. In high-T_c cuprates, $U > t$, the Coulomb repulsive energy is higher than the kinetic energy.

In spite of its seeming simplicity, the Hubbard model is notoriously difficult to solve. In one dimension, it is integrable and can be solved by employing the Bethe-ansatz method [98]. The celebrated Lieb–Wu solution [98] shows that a charge gap is open at half-filling at an infinitesimally small U, while the spin excitation remains gapless at any value of U. There is no rigorous solution for this model in higher dimensions.

In the limit $U \gg t$, we can treat the U-term as the zeroth order Hamiltonian and the hopping term as the perturbation. By applying the degenerate perturbation theory, this model can be simplified by projecting out all high energy states with doubly occupied states. To the leading order approximation, an electron can hop from an occupied site to an empty neighbor. The hopping term is described by the t-term in Eq. (2.35), but in the constrained basis space. To the second order perturbation, a virtual hopping between two singly occupied sites connected by the hopping term introduces an antiferromagnetic exchange interaction between the two spins through an intermediate doubly occupied state. The exchange coupling constant is

$$J = \frac{4t^2}{U}. \qquad (2.36)$$

This effective Heisenberg interaction lowers the energy of a total spin singlet state.

Thus up to the second order in t/U, the low-energy physics of the Hubbard model is governed by the t–J model

$$H = -t \sum_{\langle ij \rangle} \left(c_i^\dagger c_j + c_j^\dagger c_i \right) + J \sum_{\langle ij \rangle} \mathbf{S}_i \cdot \mathbf{S}_j. \qquad (2.37)$$

In this model, the Hilbert space of each site contains three states and the double occupation state is excluded. Thus the Hilbert space of the t–J model is constrained.

This is an advantage in the numerical study of the t–J model. But it is difficult to treat this constraint analytically.

In the constrained Hilbert space, the electron operator $c_{i\sigma}$ does not satisfy the usual anticommutation relation of fermions, and the standard method of quantum field theory does not apply. Of course, we can force $c_{i\sigma}$ to satisfy the fermion anticommutation relation. In this case, the constraint becomes an inequality

$$c_i^\dagger c_i \leqslant 1, \qquad (2.38)$$

which is difficult to implement in the analytical calculations.

To remove the complication introduced by the above constraint, a commonly used approach is to introduce the slave-particle representation to convert the inequality into an equality. For doing this, one has to factorize the electron operator $c_{i\sigma}$ as a product of a holon operator, e_i, and a spinon operator, $d_{i\sigma}$

$$c_{i\sigma} = e_i^\dagger d_{i\sigma}. \qquad (2.39)$$

e_i carries charge but without spin. On the other hand, $d_{i\sigma}$, is a pure spin operator but charge neutral. This slave-particle representation enlarges the Hilbert space at each lattice site. To get rid of unphysical degrees of freedom in this representation, the following constraint is imposed

$$e_i^\dagger e_i + \sum_\sigma d^\dagger d_{i\sigma} = 1. \qquad (2.40)$$

In the slave-particle representation, the t–J model takes the same form as that defined by Eq. (2.33) with $t_{ij} = t\delta_{i,j}$. Thus the low-energy physics of the three-band model is also described by the single-band t–J model in the strong coupling limit.

To maintain the fermion nature of $c_{i\sigma}$, we may assign e_i as a boson operator and $s_{i\sigma}$ as a fermion operator, or vice versa. According to the statistics of e_i, Eq. (2.39) is called the slave-boson or slave-fermion representation of electrons. These two kinds of representation are physically equivalent and should yield the same results if the local constraint (2.40) is rigorously implemented. In case the constraint is only approximately treated, the results obtained with these two representations could be different. Typically, the slave-fermion approach emphasizes more the antiferromagnetic correlations, especially under doping, while the slave-boson approach favors more the superconducting long-range order.

Thus the analysis of the t–J model may not be easier than the Hubbard model. Does this mean we should abandon this model to study directly the Hubbard model? The answer is no because the t–J model only contains the low-energy degrees of freedom that are most relevant to the high-T_c physics. It is easier to catch the key physical properties of cuprate superconductors by taking some approximations for the t–J model than for the Hubbard model. Moreover, as shown previously, the t–J model is also a low-energy effective model of the three-band Hubbard model.

2.7 Interlayer Electronic Structures

Dynamics of electrons along the c-axis is dramatically different from that along the ab-plane in high-T_c cuprates. The difference is not only quantitative but also qualitative in many aspects. For example, in underdoped cuprates, the in-plane resistivity is metal-like, while the c-axis resistivity is semiconductor-like. Various theories were proposed to explain the difference between the in-plane and c-axis charge dynamics. It was conjectured that the interlayer hopping of electrons is incoherent, i.e. electron momentum is not conserved. It was also proposed that in analogy to the quark confinement, electrons could be dynamically confined to the CuO_2 plane. These phenomenological hypotheses are simplified interpretations of experimental results. It is not a genuinely microscopic description of the electron motion along the c-axis. With the progress of high-T_c study, it has gradually been realized that, in order to correctly describe electron dynamics along the c-axis, a comprehensive understanding of the microscopic picture of electron hopping along the c-axis is desired.

As mentioned, there are three key orbitals that are responsible for the low energy physics in high-T_c cuprates: the copper $3d_{x^2-y^2}$ orbital, and the oxygen $2p_x$ and $2p_y$ orbitals. These orbitals couple to each other and determine the low-energy physics of each CuO_2 plane. However, these orbitals have strong two-dimensional characters. Their charge clouds extend mainly along the CuO_2 plane. The characteristic length scale of these orbitals along the c-axis is less than $1\mathring{A}$. The overlap between these orbitals on different CuO_2 planes is almost zero. Thus electrons can hardly hop along the c-axis. This is the reason why the c-axis conductivity is so small in the layered cuprates. However, in real high-T_c materials, the hopping along the c-axis is not exactly zero. Electrons in the copper $3d_{x^2-y^2}$ and oxygen $2p_x, 2p_y$ orbitals can hop between CuO_2 layers via other orbitals. Among them, the most important one is the Cu $4s$ orbital, which is rotationally symmetric with respect to the c-axis.

On the same site, the Wannier wave functions of Cu $3d_{x^2-y^2}$ and $4s$ orbitals are orthogonal to each other. Therefore, the copper $4s$ orbital cannot assist electrons in the copper $3d_{x^2-y^2}$ orbital to hop along the c-axis. Rather, it can facilitate the interlayer hopping of electrons between two oxygen $2p$ orbitals. In fact, this is the main channel through which electrons hop along the c-axis. The microscopic hopping process [99, 100] is

$$(O\,2p)_1 \rightarrow (Cu\,4s)_1 \rightarrow (*)_{12} \rightarrow (Cu\,4s)_2 \rightarrow (O\,2p)_2, \qquad (2.41)$$

where the subscripts denote the indices of CuO_2 planes, $(*)_{12}$ represents the orbitals assisting electrons hopping between two neighboring CuO_2 planes. This is a virtual hopping process because the energy of Cu $4s$ orbital is above the Fermi energy.

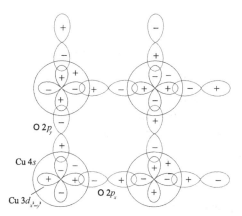

Figure 2.5 Wavefunctions of Cu $3d_{x^2-y^2}$, Cu $4s$, and the bonding O $2p_x$ and $2p_y$ orbitals. The overlap integrals between Cu $4s$ and O $2p$ orbitals possess the $d_{x^2-y^2}$ symmetry.

The effective interlayer hopping integral, t_c, between the oxygen $2p$ orbitals of the first and the second layers, is proportional to the product of the matrix elements of all virtual hopping steps, that is,

$$t_c \sim \langle (O\,2p)_2 | (Cu\,4s)_2 \rangle \langle (Cu\,4s)_2 | (*)_{12} \rangle \langle (*)_{12} | (Cu\,4s)_1 \rangle \langle (Cu\,4s)_1 | (O\,2p)_1 \rangle, \quad (2.42)$$

where $\langle a|b \rangle$ represents the hopping integral between the Wannier orbitals $|a\rangle$ and $|b\rangle$. The values of these integrals depend on the crystal and electronic structures. Nevertheless, these overlaps possess certain symmetry, which holds generally, independent of detailed properties of materials. In particular, the overlap between the copper $4s$ orbital and the oxygen $2p$ orbitals within the same CuO_2 plane, i.e. $\langle O\,2p | Cu\,4s \rangle$, possesses the $d_{x^2-y^2}$ symmetry under the rotation around the c-axis. This symmetry can be identified from the phase structure of the overlap between Cu $4s$ and O $2p_x$ or $2p_y$ orbitals shown in Fig. 2.5: the overlap between Cu $4s$ and O $2p_x$ orbitals is positive, while that between Cu $4s$ and O $2p_y$ orbitals is negative. This wavefunction overlap has precisely the $d_{x^2-y^2}$ symmetry. In momentum space, it implies that the corresponding overlap can be represented as

$$\langle (Cu\,4s)_n | (O\,2p)_n \rangle \propto \cos k_a - \cos k_b, \quad (2.43)$$

where $n = 1$ or 2. The right-hand side of the equation is just the wavefunction of $d_{x^2-y^2}$ orbital in momentum space.

The other two overlap integrals, $\langle (Cu\,4s)_2 | (*)_{12} \rangle$ and $\langle (*)_{12} | (Cu\,4s)_1 \rangle$, are related to the crystal and chemical structures between two neighboring CuO_2 planes. Generally, they do not possess specific symmetry. Here we treat them as constants.

Therefore, we have

$$t_c \sim t_\perp (\cos k_a - \cos k_b)^2 . \tag{2.44}$$

This shows that the interlayer hopping of electrons strongly depends on the momentum direction in the CuO_2 plane. For the in-plane momentum along the diagonal lines, i.e. $|k_a| = |k_b|$, the c-axis hopping integral equals zero. In other words, when $|k_a| = |k_b|$, electrons are dispersionless along the c-axis, which is a peculiar and important property of high-T_c cuprates. The coincidence of the zeros of t_c and the nodal line of the $d_{x^2-y^2}$-wave pairing leads to many anomalous effects observed in experiments. It is still unclear whether this coincidence is related to the pairing symmetry in high-T_c cuprates.

Equation (2.44) is a general property of high-T_c cuprates, independent of specific crystalline structures and chemical ingredients. It is valid for all the monolayer, bilayer, trilayer, and even infinite-layer compounds, because the $d_{x^2-y^2}$ symmetry of the overlap integral between Cu $4s$ and O $2p$ orbitals results simply from a symmetry property of wavefunctions within each CuO_2-plane, independent of the interlayer coupling. For $Bi_2Sr_2CaCu_2O_8$, or other high-T_c cuprates whose unit cell contains two CuO_2 planes, the coupling between two CuO_2 plane leads to a bilayer splitting of the energy bands with a splitting energy scale of the order of $2t_c$. This splitting, as confirmed by the angle-resolved photoemission spectroscopy (ARPES) experimental observation, is highly anisotropic. It vanishes along the nodal line of the $d_{x^2-y^2}$-wave pairing gap, but takes a maximal value along the antinodal direction. The value of t_c, measured by experiment agrees quantitatively with Eq. (2.44) within experimental errors [101].

For $La_{2-x}Sr_xCuO_4$ and other cuprates with body-centered lattice symmetry, the coefficient of $(\cos k_a - \cos k_b)^2$, i.e. $t_\perp$, also depends on k_a and k_b. t_c generally has the form,

$$t_c \propto \cos \frac{k_a}{2} \cos \frac{k_b}{2} (\cos k_a - \cos k_b)^2 . \tag{2.45}$$

It also vanishes when $k_a = \pi$ or $k_b = \pi$. This is a general property of cuprates with body-centered lattice symmetry in the tight-binding approximation. It has also been verified experimentally [102].

2.8 Systems with Zn or Ni Impurities

Doping magnetic or nonmagnetic impurities is an important approach to perturb and probe high-T_c superconductors, in addition to measuring the response of a system to a perturbation generated by an external electric, magnetic, or thermal field. Both theoretical and experimental studies on the impurity effects have greatly deepened our understanding on the mechanism of high-T_c superconductivity.

There are various ways to dope impurities into high-T_c materials. The most common one is the element substitution. Depending on different types of dopants and elements substituted, the responses of superconductors to impurities are different. Zinc (Zn) and nickel (Ni) are the two elements closest to Cu in the periodical table. They are also the impurity elements that have been systematically investigated in cuprate superconductors.

The zinc and nickel substitutions affect strongly physical properties of high-T_c cuprates because they replace mainly the copper elements in the CuO$_2$-plane. The zinc impurity is a strong scattering center, and is known to be the strongest pair-breaker. Experimentally it was found that the scattering phase shift induced by the zinc impurity potential approaches the limit of resonant scattering $\pi/2$. For YBa$_2$Cu$_3$O$_{7-\delta}$ superconductors, around 7% zinc impurity concentration can completely suppress the superconducting long-range order and reduces the transition temperature T_c to zero. In contrast, the nickel impurity has a weaker influence on the high-T_c properties. It suppresses T_c three times weaker than Zn, indicating that the nickel impurity is a weak scattering center.

Just as for Cu, both Zn and Ni are divalent elements. Substituting Cu^{2+} by Zn^{2+} or Ni^{2+} neither increases or reduces the carrier number in the system. Figure 2.6 shows the $3d$ electron configurations of these cations. The $3d$ shell of Zn^{2+} is fully occupied, hence Zn^{2+} is nonmagnetic. Ni^{2+} is different in that both the $3d_{x-y^2}$ and $3d_{3z^2-r^2}$ orbitals are singly occupied. According to the Hund's rule, the spin states in these two orbitals are parallelized. Thus Ni^{2+} carries a magnetic moment and serves as a magnetic impurity.

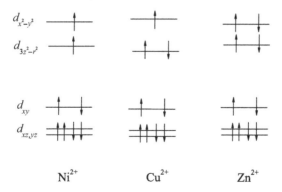

Figure 2.6 Configurations of the $3d$-electrons in Ni^{2+}, Cu^{2+}, and Zn^{2+} cations in an octahedral crystal field.

For conventional s-wave superconductors, the pair-breaking effect from a magnetic impurity is much stronger than that from a nonmagnetic one. However, in high-T_c superconductors, the nonmagnetic zinc impurity suppresses superconductivity much more strongly than the magnetic nickel impurity. This implies that the impurity scattering potential of zinc is much stronger than that of nickel. The underlying physics can be understood only by correctly constructing the microscopic model for these impurities.

In a background with strong antiferromagnetic fluctuations, nonmagnetic impurities can induce magnetic moments around them. Therefore, they exhibit many features of magnetic impurities. Based on this reasoning, some theoretical and experimental works tend to attribute the strong scattering effect of zinc to the induced magnetic moments. This sounds to be a correct picture. However, it can not explain why the scattering from zinc is stronger than from nickel. First, the magnetic moment induced by a zinc impurity is a secondary effect in comparison with the intrinsic magnetic moment of nickel. It is unlikely that a zinc impurity will exhibit stronger pair-breaking effect than a nickel impurity. Second, in overdoped high-T_c superconductors the antiferromagnetic correlations are significantly weakened. The argument of induced magnetic moments by the zinc impurity should not work in this regime. A weak pair-breaking effect of the zinc impurity is expected, but it is not consistent with experimental observations. Therefore, the pair-breaking effect of the zinc impurity comes predominantly from the nonmagnetic potential scattering. In fact, the difference in the suppression of the transition temperature T_c by both zinc and nickel impurities results mainly from the potential scattering effect.

As already mentioned, the scattering potential of the zinc impurity is in the limit of the resonant scattering and the scattering phase shift η_0 approaches $\pi/2$, and the substitution of Cu^{2+} by Zn^{2+} does not change the total carrier number. These two seemingly unrelated facts are actually inconsistent with each other. They violate the Friedel sum rule in quasi-two-dimensional systems [6]:

$$\Delta Z = \frac{2}{\pi} \left(\eta_0 + 2 \sum_{l \neq 0} \eta_l \right), \qquad (2.46)$$

where η_l is the phase shift of the scattering at the Fermi level in the channel of angular momentum l, and ΔZ is the number of electrons added to the system by the impurity. In high-T_c materials, $\Delta Z = 0$ because the zinc substitution does not change the number of charge carriers. The right-hand side equals 1 because $\eta_l \approx \pi/2$ for $l = 0$ and almost zero for $l \geqslant 1$. This discrepancy actually results from the correlated effect of high-T_c superconductors. It can be resolved by considering the correction to the Zn scattering potential by the correlation effect of electrons on the CuO_2 planes.

Below we introduce briefly the Friedel sum rule and derive the effective one-band model for a zinc or nickel impurity, starting from the corresponding three-band model [103]. The idea guiding the derivation holds generally, and can be extended to other impurities similar to zinc or nickel.

2.8.1 Friedel Sum Rule

Now we provide a simple derivation for the Friedel sum rule using the standard partial-wave method. Let us consider a scattering problem of an electron in a short-range potential $V(r)$ in two dimensions,

$$\left[-\frac{\hbar^2}{2m} \left(\frac{\partial^2}{\partial x^2} + \frac{\partial^2}{\partial y^2} \right) + V(r) \right] \psi(\mathbf{r}) = E \psi(\mathbf{r}). \tag{2.47}$$

In the limit $r \to +\infty$, $V(r)$ becomes zero and the scattering wave function is asymptotically given by

$$\psi(\mathbf{r}) \to e^{ik_x x} + f(\phi) \frac{e^{i\mathbf{k} \cdot \mathbf{r}}}{\sqrt{r}}, \tag{2.48}$$

in the polar coordinate system $\mathbf{r} = (r, \phi)$ with ϕ the azimuthal angle. $\mathbf{k}$ is the wave vector, which is related to the eigenenergy by the formula

$$E = \frac{\hbar^2 \mathbf{k}^2}{2m}. \tag{2.49}$$

The scattering amplitude $f(\phi)$ can be expanded using the angular momentum basis states as

$$f(\phi) = \sum_{l=0}^{\infty} f_l \cos(l\phi). \tag{2.50}$$

Similarly, the incident plane-wave can be expanded as

$$e^{ik_x x} = e^{ikr \cos\phi} = \sum_{l=0}^{\infty} i^l J_l(kr) \cos(l\phi), \tag{2.51}$$

where $J_l(kr)$ is the lth order Bessel function of the first kind. In the limit $r \to \infty$,

$$J_l(kr) \to \sqrt{\frac{2}{\pi kr}} \cos\left[kr - \frac{\pi}{2} \left(l + \frac{1}{2} \right) \right]. \tag{2.52}$$

On the other hand, the wavefunction $\psi(\mathbf{r})$ can be expanded as

$$\psi(\mathbf{r}) = \sum_{l=0}^{\infty} A_l R_l(kr) \cos(l\phi), \tag{2.53}$$

where A_l is the coefficient, and

$$R_l(kr) = J_l(kr) + \frac{1}{2}H_l(kr)\left(e^{2i\eta_l} - 1\right),$$ (2.54)

$$H_l(kr) = J_l(kr) + iN_l(kr),$$ (2.55)

with η_l the phase shift. H_l is the Hankel function of the first kind and N_l is the Neumann function. In the limit $r \to \infty$,

$$R_l(kr) \to \sqrt{\frac{2}{\pi kr}}e^{-i\eta_l}\cos\left[kr - \frac{\pi}{2}\left(l + \frac{1}{2}\right) + \eta_l\right].$$ (2.56)

Now let us consider a circular disc with a radius R. In the absence of the impurity scattering potential, the wave vector k is determined by the zeros of the Bessel function $J_l(kR) = 0$. From the asymptotic expressions of J_l in the limit $kR \to \infty$, we find that k is quantized

$$kR = \left(n + \frac{l}{2}\right)\pi - \frac{\pi}{4},$$ (2.57)

where n is an integer. Hence,

$$\frac{dn}{dk} = \frac{R}{\pi}.$$ (2.58)

In the presence of the scattering potential, the quantization condition changes to

$$kR = \left(n + \frac{l}{2}\right)\pi - \frac{\pi}{4} - \eta_l(k),$$ (2.59)

and

$$\frac{dn}{dk} = \frac{R}{\pi} + \frac{1}{\pi}\frac{d\eta_l(k)}{k}.$$ (2.60)

Compared to the case of the absence of the impurity, the change of the number of states in each partial-wave channel is

$$\frac{d\Delta n_l}{dk} = \frac{1}{\pi}\frac{d\eta_l(k)}{dk}.$$ (2.61)

After integrating k from 0 to the Fermi wave vector k_F and counting the spin and orbital angular momentum degeneracies, we immediately obtain the Friedel sum rule (2.46) for the change of the charge number in the presence of an impurity in two dimensions.

In three dimensions, the phase shift should be multiplied by the degeneracy number in each partial-wave channel l, and the corresponding Friedel sum rule becomes

$$\Delta Z = \frac{2}{\pi}\sum_{l=0}^{\infty}(2l + 1)\eta_l.$$ (2.62)

2.8.2 Zn Impurity

Physically it is interesting to consider a system with low Zn concentration in which the interaction among these impurities is small and negligible. In this case, we just need to solve a single impurity problem. The result can be readily extended to a many-impurity system in the dilute limit.

As the five $3d$ electron orbitals of Zn^{2+} are fully occupied, the total spin of Zn^{2+} is zero. This cation is very stable against valence fluctuations. It is difficult to change Zn^{2+} to Zn^{3+} by removing one electron, or to Zn^+ by adding one more electron. Therefore, Zn^{2+} is an inert nonmagnetic impurity and has no charge transfer with the surrounding O^{2-} anions and Cu^{2+} cations.

The three-band Hamiltonian, corresponding to Eqs. (2.8) and (2.9), for the system including a zinc impurity is defined by

$$H^{Zn} = H_0^{Zn} + H_1^{Zn}, \tag{2.63}$$

$$H_0^{Zn} = \varepsilon_p \sum_l p_l^\dagger p_l + \sum_{i \neq i_0} \left(\varepsilon_d d_i^\dagger d_i + U_d d_{i\uparrow}^\dagger d_{i\uparrow} d_{i\downarrow}^\dagger d_{i\downarrow} \right), \tag{2.64}$$

$$H_1^{Zn} = - \sum_{\langle il \rangle i \neq i_0} t_{pd} \left(p_l^\dagger d_i + d_i^\dagger p_l \right), \tag{2.65}$$

where i_0 is the position of the zinc impurity. Using the bonding and non-bonding operators of the oxygen holes, the above equations can be expressed as

$$H_0^{Zn} = \varepsilon_p \sum_i \left(a_i^\dagger a_i + b_i^\dagger b_i \right) + \sum_{i \neq i_0} \left(\varepsilon_d d_i^\dagger d_i + U_d d_{i\uparrow}^\dagger d_{i\uparrow} d_{i\downarrow}^\dagger d_{i\downarrow} \right), \tag{2.66}$$

$$H_1^{Zn} = -t_{pd} \sum_{i \neq i_0, j} u(i - j) \left(a_j^\dagger d_i + d_i^\dagger a_j \right). \tag{2.67}$$

In the limit of $U_d \gg \varepsilon_p - \varepsilon_d \gg t_{pd}$, we can take H_1^{Zn} as a perturbation to project the above Hamiltonians onto the ground state subspace spanned by the Zhang–Rice singlets and the unpaired copper spins using the method introduced in §2.4 and §2.5. Following the derivation steps previously introduced, the effective low energy one-band Hamiltonian is found to be

$$H_{Zn} = \sum_i V_{Zn}(i) d_i^\dagger d_i - \sum_{i \neq j} t_{ij}^{Zn} d_j^\dagger d_i + \sum_{\langle ij \rangle \neq i_0} J S_i \cdot S_j. \tag{2.68}$$

Similarly to the standard t–J model, the d electrons at site $i \neq i_0$ satisfy the constraint

$$d_i^\dagger d_i \leqslant 1. \tag{2.69}$$

At the impurity site, the operator d_{i_0} is not the annihilation operator of $3d$ electrons. Instead it is defined by the bonding operator a_{i_0} of the oxygen hole at that site

$$d_{i_0\sigma} = -\sigma a_{i_0\bar{\sigma}}^\dagger. \tag{2.70}$$

Unlike the d electron operators at the other lattice sites, there is no constraint on d_{i_0}. Therefore, two electrons with opposite spins can occupy the impurity site. Formally, this is consistent with the fact that there is only one electron at the $3d_{x^2-y^2}$ orbital in Cu^{2+} and two electrons at that orbital in Zn^{2+}. However, it should be emphasized that there is no unoccupied electrons at the zinc site in the original three-band Hamiltonian Eq. (2.63).

The electron hopping integral is

$$t_{ij}^{Zn} = \frac{\tilde{t}_{ij}}{2}\delta_{i\neq i_0, j\neq i_0} + \frac{\tilde{t}_{i_0 j}}{\sqrt{2}}\delta_{i, i_0} + \frac{\tilde{t}_{i_0 i}}{\sqrt{2}}\delta_{j, i_0} - t'\delta_{\langle ij\rangle'\neq i_0} - t''\delta_{\langle ij\rangle''\neq i_0}, \qquad (2.71)$$

where

$$\tilde{t}_{ij} = t_P \delta_{\langle i, j\rangle} - t_P u(i_0 - i)u(i_0 - j). \qquad (2.72)$$

The first term is the hopping integral in the absence of the impurity. The second term is the correction introduced by the zinc impurity, which is small but non-local.

The effective impurity potential is defined by

$$V_{Zn}(i) = -t_P u^2(i_0 - i)\delta_{i\neq i_0} - (t_P + J_P)u^2(0)\delta_{i, i_0}, \qquad (2.73)$$

which is an attractive interaction for electrons. This is not an on-site potential. It decays approximately as $|V_{Zn}(i)| \propto |i - i_0|^{-6}$ in the large $|i - i_0|$ limit. This potential at the impurity site, $|V_{Zn}(i_0)| \approx (t_P + J_P)u^2(0)$, is about two orders of magnitude larger than that on the nearest neighboring sites, which is about $0.0785 t_P$. It is also more than one order of magnitude larger than the effective hopping integral $t_P/2$. Thus the zinc impurity is a strong scattering center. The effective attractive potential arises from the strong repulsion of the Zn^{2+} cation to the bonding oxygen holes on the impurity site, for the following two reasons. First, Zn^{2+} is spinless and cannot form a Zhang–Rice singlet with an oxygen hole. This leads to a relative increase of the oxygen hole energy, $J_P u^2(0)$, at the impurity site. Second, electrons cannot hop between O $2p$ and Zn $3d$ orbitals, resulting in a loss of kinetic energy, $t_P u^2(0)$. Therefore, the total energy loss is $(t_P + J_P)u^2(0)$, which is equivalent to having a repulsive potential for oxygen holes, or an attractive potential for electrons, at the impurity site.

The above discussion reveals an intimate connection between the scattering potential of the zinc impurity and the Zhang–Rice singlet. It shows that the binding energy of the Zhang–Rice singlet can be determined by measuring the Zn impurity scattering potential.

Assuming the total number of doped holes is N_h, then,

$$\sum_i a_i^\dagger a_i = N_h. \qquad (2.74)$$

Because the impurity potential is strongly repulsive to holes, no oxygen hole in low energy can exist on this site, thus we have

$$\sum_i d_i^\dagger d_i = (N - N_h) + 1, \qquad (2.75)$$

where N is the total number of lattice sites. This expression shows that, in the effective low-energy one-band model, each Zn impurity contributes an extra electron to the system. It also shows that although both Cu^{2+} and Zn^{2+} ions are divalent, Zn^{2+} should be treated as having one more electron than Cu^{2+} in the effective single-band model.

Therefore, in the strongly correlated CuO_2 plane, the phase space of electrons is enlarged by the substitution of Zn impurities. Effectively, each Zn introduces an extra electron to the system so that

$$\Delta Z = 1, \qquad (2.76)$$

rather than $\Delta Z = 0$, which modifies the Friedel sum rule Eq. (2.46). This is a consequence of strongly correlated effect. It shows that the resonant scattering phase shift induced by the zinc impurity scattering, $\eta_0 = \pi/2$, is consistent with the Friedel sum rule, resolving the aforementioned puzzle about the Friedel sum rule.

2.8.3 Ni Impurity

A Ni^{2+} cation has eight electrons in its $3d$ shell. Due to the strong Hund's rule coupling between $3d_{3z^2-r^2}$ and $3d_{x^2-y^2}$ electrons, the total spin is 1. Similarly to Zn^{2+}, Ni^{2+} is very stable. However, as the $3d_{3z^2-r^2}$ and $3d_{x^2-y^2}$ orbitals are not fully filled, the hybridizations between these two orbitals and the surrounding oxygen $(2p_x, 2p_y)$ orbitals are strong.

Similarly to the zinc impurity, the three-band model for the system with one nickel impurity is defined by the Hamiltonian

$$H^{Ni} = H^{Zn} + \sum_\alpha \varepsilon_\alpha^{Ni} c_\alpha^\dagger c_\alpha - \sum_{\langle li_0\rangle\alpha} t_\alpha^{Ni} \left(p_l^\dagger c_\alpha + h.c. \right)$$
$$- J_H c_1^\dagger \frac{\sigma}{2} c_1 \cdot c_2^\dagger \frac{\sigma}{2} c_2 + \sum_\alpha U_\alpha^{Ni} c_{\alpha\uparrow}^\dagger c_{\alpha\uparrow} c_{\alpha\downarrow}^\dagger c_{\alpha\downarrow}, \qquad (2.77)$$

where $\alpha = 1$ and 2 represent the $3d_{x^2-y^2}$ and $3d_{3z^2-r^2}$ orbitals of Ni^{2+}, respectively. $c_\alpha = (c_{\alpha\uparrow}, c_{\alpha\downarrow})$ are the electron annihilation operators of these orbitals. The corresponding onsite energy and Coulomb repulsion are denoted as ε_α^{Ni} and U_α^{Ni}, respectively. t_α^{Ni} is the hybridization between the Ni $3d$ and the surrounding oxygen $2p$ electrons. J_H is the Hund's coupling constant between $3d_{x^2-y^2}$ and $3d_{3z^2-r^2}$ spins.

Compared to the Hund's rule coupling, the hopping term t_α^{Ni} is a relatively small quantity and can be treated as a perturbation. The effective one-band model of the nickel impurity can be derived similarly to the zinc impurity. However, as the total spin of Ni^{2+} is one, the Ni spin cannot form a Zhang–Rice singlet with an oxygen hole. Instead, they will form a Zhang–Rice-like spin doublet, which reduces strongly the scattering potential so that Ni^{2+} behaves like a weak scattering center. This is a subtle difference between Zn and Ni impurities.

Using the degenerate perturbation theory, the effective one-band Hamiltonian for describing a system with one nickel impurity is found to be

$$H_{Ni} = \sum_{i \neq j \neq i_0} t_{ij}^{Ni} d_j^\dagger d_i + \sum_i V_{Ni}(i) d_i^\dagger d_i + \sum_{\langle ij \rangle} J_{ij} S_i \cdot S_j, \qquad (2.78)$$

where $d_{i_0} = (d_{i_0\uparrow}, d_{i_0\downarrow})$ are the annihilation operators of the spin doublet formed by the Ni^{2+} spin and the oxygen hole at the impurity site. Similarly, if $i \neq i_0$, the d electrons satisfy the constraint $d_i^\dagger d_i \leqslant 1$. However, on the impurity site, the nickel spin and the oxygen hole spin form a spin doublet. In this effective single-band model, the Ni^{2+} spin is partially screened, and can be effectively identified as a spin-$\frac{1}{2}$ magnetic impurity.

The hopping integrals in the first term of H_{Ni} are given by

$$t_{ij}^{Ni} = t_{ij}^{Zn} + t_p' u(i_0 - i) u(i_0 - j), \qquad (2.79)$$

$$t_p' = \sum_\alpha \frac{\left(t_\alpha^{Ni}\right)^2}{U_\alpha^{Ni} - \varepsilon_p + \varepsilon_\alpha^{Ni} + J_H/4}. \qquad (2.80)$$

The exchange energy $J_{ij} = J$ when neither i nor j equals i_0. When either i or j equals i_0, $J_{ij} \neq J$ but remains at the same order.

The scattering potential of the nickel impurity is given by

$$V_{Ni}(i) = \left(t_P + J_P - \frac{1}{2}J_P' - \frac{3}{2}t_P'\right) u^2(0)\delta_{i, i_0} - \left(t_P - t_P'\right) u^2(i_0 - i)\delta_{i \neq i_0}, \quad (2.81)$$

and

$$J_P' = \sum_\alpha \frac{\left(t_\alpha^{Ni}\right)^2}{\varepsilon_p - \varepsilon_\alpha^{Ni} + J_H/4}. \qquad (2.82)$$

The J_P' term arises from the hybridization between the nickel $3d$ orbitals and the oxygen $2p$ orbitals. The t_P' term is the binding energy of the local spin doublet formed by the Ni^{2+} spin and the oxygen hole. These terms suppress the scattering potential generated by the t_P and J_P terms. On the impurity site, $V_{Ni}(i) = \left(t_P + J_P - \frac{1}{2}J_P' - \frac{3}{2}t_P'\right) u^2(0)$, which is much smaller than the corresponding Zn scattering potential. Thus the nickel impurity is a weaker scattering center compared to the zinc impurity, consistent with the experimental result.

The above discussions demonstrate the complexity of strongly correlated electronic systems. The difference between the zinc and nickel impurities does not arise from the distribution of electrons in their $3d$ orbitals, but from their correlation effects with the surrounding oxygen holes. It indicates that the influence of zinc and nickel impurities in high-T_c superconductors results predominantly from the potential scattering, rather than from the magnetic interaction generated by the effective spin of the nickel impurity or the induced magnetic moment around the zinc impurity.

3

Basic Properties of d-wave Superconductors

3.1 d-Wave Gap Equation

In §1.4, we derived the gap equation (1.32) for conventional s-wave superconductors under the BCS mean-field approximation. Here we use the two-dimensional $t-J$ model defined by Eq. (2.37) on the square lattice as a toy model to derive the gap equation for d-wave superconductors. The purpose is to demonstrate how the gap equation is modified by the pairing symmetry of superconducting electrons, rather than on the strong correlation nature of cuprates. More specifically, we treat the constraint of no double occupancy just on average. This is equivalent to add a chemical potential (or a spatial independent Lagrangian multiplier) to the system so that the $t-J$ model is reduced to a conventional model of interacting electrons without constraint. The Hamiltonian now becomes

$$H = -t \sum_{\langle ij \rangle \sigma} \left(c^\dagger_{i\sigma} c_{j\sigma} + h.c. \right) + J \sum_{\langle ij \rangle} \mathbf{S}_i \cdot \mathbf{S}_j - \mu \sum_i n_i. \tag{3.1}$$

n_i is the number of electrons at site i.

The Heisenberg interaction term (i.e. the J-term) favors the formation of a spin singlet on the two neighboring sites. It can be expressed using the nearest neighbor pairing operator

$$\Delta(i, i + \delta) = c_{i\uparrow} c_{i+\delta,\downarrow} - c_{i\downarrow} c_{i+\delta,\uparrow}, \tag{3.2}$$

as

$$H_J = -\frac{J}{2} \sum_{\langle ij \rangle} \Delta^\dagger_{ij} \Delta_{ij} + \frac{J}{4} \sum_{\langle ij \rangle} n_i n_j. \tag{3.3}$$

Near the half-filling, the electron number per site is close to 1 and with very small fluctuation if J is not significantly smaller than the hopping constant t. In that case, the second term on the right-hand side of the above equation is nearly a constant and can be dropped out. Moreover, the probability of pairing at the same site is completely suppressed by the constraint of no double occupancy.

Now we take a mean-field approximation to reduce the many-body interactions in H_J to an effective one-body problem. The t–J model then becomes

$$H \approx -t \sum_{\langle ij \rangle \sigma} \left(c_{i\sigma}^{\dagger} c_{j\sigma} + h.c. \right) + \sum_{\langle ij \rangle} \left(\phi_{ij}^* \Delta_{ij} + \phi_{ij} \Delta_{ij}^{\dagger} + \frac{2}{J} \phi_{ij}^* \phi_{ij} \right) - \mu \sum_i n_i \tag{3.4}$$

up to a constant. ϕ_{ij} is an effective field acting on the bond (ij), which is determined by the mean-field equation

$$\phi_{ij} = -\frac{J}{2} \langle \Delta_{ij} \rangle. \tag{3.5}$$

On a square lattice with both rotational and translational symmetries, there are two possible solutions for ϕ_{ij}

$$\phi_{ij} = \Delta \left(\delta_{j,i+\hat{x}} + \delta_{j,i-\hat{x}} \right) \pm \Delta \left(\delta_{j,i+\hat{y}} + \delta_{j,i-\hat{y}} \right), \tag{3.6}$$

corresponding to the extended *s*- and *d*-wave pairing symmetry, respectively. By performing the Fourier transformation for the electron operator

$$c_{i\sigma} = \frac{1}{\sqrt{N}} \sum_{\mathbf{k}} c_{\mathbf{k}\sigma} e^{i\mathbf{k}\cdot\mathbf{r}_i}, \tag{3.7}$$

we obtain the expression of the above mean-field Hamiltonian in the momentum space

$$H = \sum_{\mathbf{k}} \begin{pmatrix} c_{\mathbf{k}\uparrow}^{\dagger} & c_{-\mathbf{k}\downarrow} \end{pmatrix} \begin{pmatrix} \xi_{\mathbf{k}} & \Delta_{\mathbf{k}} \\ \Delta_{\mathbf{k}} & -\xi_{\mathbf{k}} \end{pmatrix} \begin{pmatrix} c_{\mathbf{k}\uparrow} \\ c_{-\mathbf{k}\downarrow}^{\dagger} \end{pmatrix} + \sum_{\mathbf{k}} \xi_{\mathbf{k}} + \frac{4N}{J} \Delta^2. \tag{3.8}$$

N is the lattice size and

$$\xi_{\mathbf{k}} = -2t(\cos k_x + \cos k_y) - \mu. \tag{3.9}$$

$\Delta_{\mathbf{k}}$ is the gap function

$$\Delta_{\mathbf{k}} = \Delta \gamma_{\mathbf{k}} \tag{3.10}$$

and $\gamma_{\mathbf{k}}$ is the form factor that characterizes the pairing symmetry

$$\gamma_{\mathbf{k}} = \begin{cases} \cos k_x - \cos k_y, & d\text{-wave} \\ \cos k_x + \cos k_y, & \text{extended } s\text{-wave} \end{cases}. \tag{3.11}$$

The corresponding gap equation becomes

$$\Delta = -\frac{J}{4N} \sum_{\mathbf{k}} \gamma_{\mathbf{k}} \langle c_{-\mathbf{k}\downarrow} c_{\mathbf{k}\uparrow} \rangle. \tag{3.12}$$

By diagonalizing the mean-field Hamiltonian with the Bogoliubov transformation (1.27), we obtain the energy eigenspectrum

$$E_{\mathbf{k}} = \sqrt{\xi_{\mathbf{k}}^2 + \Delta_{\mathbf{k}}^2}, \tag{3.13}$$

and the corresponding eigenvectors of superconducting quasiparticles. From the expectation value of the pairing operator, the gap equation is found to be

$$1 = \frac{J}{4N} \sum_{\mathbf{k}} \frac{\gamma_{\mathbf{k}}^2}{2E_k} \tanh \frac{\beta E_{\mathbf{k}}}{2}. \tag{3.14}$$

The chemical potential μ, on the other hand, is determined by the equation

$$n = \frac{1}{N} \sum_{\mathbf{k}\sigma} \langle c_{\mathbf{k}\sigma}^{\dagger} c_{\mathbf{k}\sigma} \rangle = 1 - \frac{1}{N} \sum_{\mathbf{k}} \frac{\xi_{\mathbf{k}}}{E_{\mathbf{k}}} \tanh \frac{\beta E_{\mathbf{k}}}{2}. \tag{3.15}$$

Near half-filling, the Fermi surface is close to the antiferromagnetic Brillouin zone boundary $\cos k_x + \cos k_y = 0$. In this case, the extended s-wave gap symmetry function $\gamma_{\mathbf{k}}$ is very small on the Fermi surface. To fulfill the gap equation, Δ has to be small so that the ratio $\gamma_{\mathbf{k}}^2/E_{\mathbf{k}}$ is large on the Fermi surface. This suggests that the extended s-wave pairing is dramatically suppressed and not energetically favored in comparison with the d-wave pairing near half-filling [104]. On the other hand, the value $\gamma_{\mathbf{k}}$ on the Fermi surface increases very quickly when the filling factor n moves away from the half-filling for the extended s-wave pairing state. In comparison, the d-wave gap function always intersects with the Fermi surface and has gapless excitations. Thus in the limit with $n \ll 1$, it is expected that the extended s-wave pairing would become the dominant. This implies that the pairing symmetry is determined more by the topology of the Fermi surface, rather than the pairing interaction itself [104]. In other words, simply from the pairing symmetry, it is difficult to reveal the origin of the pairing interaction.

3.2 Temperature Dependence of the *d*-Wave Energy Gap

We now determine the superconducting transition temperature and the temperature dependence of the gap parameter by solving the gap equation (3.14) for d-wave superconductors. For convenience in the calculation, we use $\gamma_\varphi = \cos 2\varphi$ to replace the form factor $\gamma_{\mathbf{k}} = \cos k_x - \cos k_y$. This replacement is appropriate because in the long wavelength limit, $\gamma_{\mathbf{k}} = \cos k_x - \cos k_y \propto \cos 2\varphi$, where φ is the polar angle of $\mathbf{k}$. Furthermore, we neglect the angular dependence of $\xi_{\mathbf{k}}$ so that the wave vector summation can be replaced by an integral

$$\frac{1}{N} \sum_{\mathbf{k}} = \int \frac{d^2\mathbf{k}}{(2\pi)^2} \to \int \frac{d\varphi}{2\pi} \int_{-\omega_0}^{\omega_0} d\xi \rho(\xi), \tag{3.16}$$

where $\rho(\xi)$ is the density of states of normal electrons. ω_0 is the characteristic energy scale of the pairing interaction, which is generally assumed to be larger than the superconducting transition temperature T_c. This allow us to write the gap equation as

$$g \int \frac{d\varphi}{2\pi} \int_0^{\omega_0} d\xi \rho(\xi) \frac{\gamma_\varphi^2}{\sqrt{\xi^2 + \Delta^2 \gamma_\varphi^2}} \tanh \frac{\beta \sqrt{\xi^2 + \Delta^2 \gamma_\varphi^2}}{2} = 1, \qquad (3.17)$$

where $g = J/4$.

In a metal-based or other conventional superconductor induced by the electron–phonon interaction, ω_0 approximately equals the Debye frequency ω_D. For high-T_c superconductors, it is unclear what the energy scale ω_0 represents. The density of states $\rho(\xi)$ is determined by the bandwidth which is usually much larger than ω_0. In this case, $\rho(\xi)$ is approximately determined by its value at the Fermi energy, $\rho(\xi) \approx \rho(\xi_F) = N_F$. This simplifies Eq. (3.17) to

$$g N_F \int \frac{d\varphi}{2\pi} \int_0^{\omega_0} d\xi \frac{\gamma_\varphi^2}{\sqrt{\xi^2 + \Delta^2 \gamma_\varphi^2}} \tanh \frac{\beta \sqrt{\xi^2 + \Delta^2 \gamma_\varphi^2}}{2} = 1. \qquad (3.18)$$

At the critical transition temperature, $\Delta = 0$, the above equation reduces to

$$\int_0^{\omega_0/2k_B T_c} dx \frac{\tanh x}{x} = \frac{2}{g N_F}. \qquad (3.19)$$

After integration by parts, it becomes

$$\ln \frac{\omega_0}{2k_B T_c} \tanh \frac{\omega_0}{2k_B T_c} - \int_0^{\omega_0/2k_B T_c} dx \ln x \, \text{sech}^2 x = \frac{2}{g N_F}. \qquad (3.20)$$

The integral on the left-hand side of the equation converges as $x \to \infty$, thus the upper limit of the integral can be safely set to $+\infty$ in the limit $\omega_0 \gg T_c$. The transition temperature T_c is then found to be

$$k_B T_c \approx c_0 \omega_0 e^{-\frac{2}{g N_F}}, \qquad (3.21)$$

where

$$c_0 = \frac{1}{2} \exp\left(-\int_0^\infty dx \ln x \, \text{sech}^2 x\right) \approx 1.134. \qquad (3.22)$$

At zero temperature, after integrating out ξ, Eq. (3.18) becomes

$$\frac{g N_F}{\pi} \int_0^\pi d\varphi \gamma_\varphi^2 \ln \frac{\omega_0 + \sqrt{\omega_0^2 + \Delta_0^2 \gamma_\varphi^2}}{\Delta_0 |\gamma_\varphi|} = 1, \qquad (3.23)$$

where $\Delta_0 = \Delta(T = 0)$ is the zero-temperature energy gap amplitude. In the limit $\omega_0 \gg \Delta_0$, this equation becomes

$$\frac{gN_F}{\pi} \int_0^\pi d\varphi \gamma_\varphi^2 \ln \frac{2\omega_0}{\Delta_0 |\gamma_\varphi|} \approx 1. \tag{3.24}$$

The solution is

$$\Delta_0 = c_1 \omega_0 \exp(-2/gN_F), \tag{3.25}$$

where

$$c_1 = 2\exp\left(-\frac{4}{\pi} \int_0^{\pi/2} d\varphi \cos^2 \varphi \ln \cos \varphi\right) = 4e^{-0.5} \approx 2.426. \tag{3.26}$$

The superconducting transition temperature T_c and the zero temperature gap Δ_0 are the two fundamental parameters of superconductors. Δ_0^2 is proportional to the condensation energy of the superconducting state. If the corrections from the Coulomb repulsion among electrons are included, the d-wave pairing may win over the s-wave one. The d-wave pairing can reduce the on-site Coulomb repulsion energy because the probability for two paired electrons occupying the same lattice site in a d-wave superconductor vanishes.

The above results indicate that the ratio between Δ_0 and T_c depends on the d-wave pairing function γ_φ, but not on the detailed band structures in the limit $\omega_0 \gg k_B T_c$,

$$\frac{2\Delta_0}{k_B T_c} \simeq 4.28, \tag{3.27}$$

which is larger than the corresponding ratio, $2\Delta_0/(k_B T_c) \simeq 3.53$, for the isotropic s-wave superconductor shown in Eq. (1.45).

In a conventional superconductor, the deviation of $2\Delta_0/k_B T_c$ from 3.53 is a characteristic parameter in quantifying the coupling strength of Cooper pairs. In high-T_c cuprates, however, the ratio $2\Delta_0/k_B T_c$ exhibits strong sample and doping dependence. In the optimal or overdoped regime, this ratio is around 4.28, close to the value predicted by the mean-field theory. In the underdoped regime, however, $2\Delta_0/k_B T_c$ increases rapidly with decreasing doping and is generally much larger than the mean-field value. The suppression of T_c is likely to be induced by strong phase fluctuations.

Rigorously speaking, $\Delta_k = \cos 2\varphi$ is a good approximation only in the vicinity of the d-wave gap nodes. The value of $2\Delta_0/k_B T_c$ obtained from this approximation is valid in the low energy limit. However, the value of $2\Delta_0/k_B T_c$ obtained by experiments is often an average of $2\Delta(\varphi)/k_B T_c$ over the entire Fermi surface. It could also be a value of $2\Delta(\varphi)/k_B T_c$ along a particular momentum direction. This is the reason why the experimental values of $2\Delta_0/k_B T_c$ exhibit large fluctuations.

Strong correlation effects, such as antiferromagnetic fluctuations, may also change this ratio significantly.

In order to determine the temperature dependence of Δ, we take the difference between Eq. (3.18) and the corresponding equation in the low-temperature limit $T \to 0$. By utilizing Eqs. (3.18) and (3.23) in the limit $\omega_0 \gg \Delta$, we find that the gap function is determined by the equation

$$
\int_{-\pi}^{\pi} \frac{d\varphi}{2\pi} \int_{0}^{\omega_0} d\xi \, \frac{\gamma_\varphi^2}{\sqrt{\xi^2 + \Delta^2 \gamma_\varphi^2}} \left(1 - \tanh \frac{\beta \sqrt{\xi^2 + \Delta^2 \gamma_\varphi^2}}{2} \right)
$$

$$
= \int_{-\pi}^{\pi} \frac{d\varphi}{2\pi} \int_{0}^{\omega_0} d\xi \left(\frac{\gamma_\varphi^2}{\sqrt{\xi^2 + \Delta^2 \gamma_\varphi^2}} - \frac{\gamma_\varphi^2}{\sqrt{\xi^2 + \Delta_0^2 \gamma_\varphi^2}} \right)
$$

$$
\approx \langle \gamma_\varphi^2 \rangle_{FS} \ln \frac{\Delta_0}{\Delta}, \tag{3.28}
$$

where $\langle \gamma_\varphi^2 \rangle_{FS}$ is the average of γ_φ^2 on the Fermi surface. It is equal to 1 and $\frac{1}{2}$ for the isotropic s-wave ($\gamma_\varphi = 1$) and d-wave superconductors, respectively.

In an s-wave superconductor, there are very few low-lying excitations and $\Delta(T)$ approaches $\Delta(0)$ exponentially at low temperatures

$$
\Delta(T) = \Delta_0 - \sqrt{2\pi k_B T \Delta_0} \exp \left(-\frac{\Delta_0}{k_B T} \right), \qquad (T \ll T_c). \tag{3.29}
$$

For a d-wave superconductor, Eq. (3.28) can be rewritten as

$$
\frac{\pi \beta^2 \Delta^2}{2} \ln \frac{\Delta_0}{\Delta}
$$

$$
= \int_{0}^{\beta\omega_0} dy \int_{0}^{\beta\Delta} \frac{dx}{\sqrt{\beta^2 \Delta^2 - x^2}} \frac{x^2}{\sqrt{x^2 + y^2}} \left(1 - \tanh \frac{\sqrt{x^2 + y^2}}{2} \right). \tag{3.30}
$$

At low temperatures, the integral contributes mainly from the domain of $x \ll \beta\Delta$ where $\sqrt{\beta^2\Delta^2 - x^2} \approx \beta\Delta$. This allows us to simplify the above equation to

$$
\frac{\pi \beta^3 \Delta^3}{2} \ln \frac{\Delta_0}{\Delta} = \int_{0}^{\beta\omega_0} dy \int_{0}^{\beta\Delta} dx \frac{x^2}{\sqrt{x^2 + y^2}} \left(1 - \tanh \frac{\sqrt{x^2 + y^2}}{2} \right). \tag{3.31}
$$

In the limit $\beta\omega_0 \gg 1$ and $\beta\Delta \gg 1$, it is safe to set the upper limits of the above two integrals to infinity,

$$
\frac{\pi \beta^3 \Delta^3}{2} \ln \frac{\Delta_0}{\Delta} = \int_{0}^{\infty} dy \int_{0}^{\infty} dx \frac{x^2}{\sqrt{x^2 + y^2}} \left(1 - \tanh \frac{\sqrt{x^2 + y^2}}{2} \right). \tag{3.32}
$$

Solving this equation yields

$$\Delta(T) = \Delta_0 \exp\left(-\alpha_0 \frac{k_B^3 T^3}{\Delta^3}\right) \approx \Delta_0\left(1 - \alpha_0 \frac{k_B^3 T^3}{\Delta_0^3}\right), \tag{3.33}$$

where

$$\alpha_0 = \int_0^\infty r^2\left(1 - \tanh\frac{r}{2}\right) \approx 3.606. \tag{3.34}$$

Thus Δ approaches $\Delta(0)$ cubically with temperature, unlike the s-wave case.

In the vicinity of the superconducting transition temperature, the gap value Δ is small and we can expand Eq (3.18) in terms of Δ. Taking the approximation up to the second order terms in Δ, we have

$$g N_F\left[\langle\gamma_\varphi^2\rangle_{FS}\int_{-\omega_0}^{\omega_0} d\xi \frac{\tanh(\beta\xi/2)}{\xi} + \langle\gamma_\varphi^4\rangle_{FS}\Delta^2 p\right] \approx 1, \tag{3.35}$$

where

$$p = \int_0^{\omega_0} d\xi\left[\frac{\beta\text{sech}^2(\beta\xi/2)}{4\xi^2} - \frac{\tanh(\beta\xi/2)}{2\xi^3}\right]. \tag{3.36}$$

Using Eq. (3.19), Eq. (3.35) can be further simplified as

$$\langle\gamma_\varphi^2\rangle_{FS}\ln\frac{T_c}{T} + \langle\gamma_\varphi^4\rangle_{FS}\Delta^2 p \approx 0. \tag{3.37}$$

Thus the solution of Δ is given by

$$\Delta = \sqrt{\frac{\langle\gamma_\varphi^2\rangle_{FS}}{\langle\gamma_\varphi^4\rangle_{FS} p}\ln\frac{T}{T_c}} = \sqrt{\frac{\langle\gamma_\varphi^2\rangle_{FS}k_B^2 T^2}{\langle\gamma_\varphi^4\rangle_{FS}g_0}\ln\frac{T_c}{T}}, \tag{3.38}$$

where

$$g_0 = -\int_0^\infty dx\left[\frac{\text{sech}^2(x/2)}{4x^2} - \frac{\tanh(x/2)}{2x^3}\right] \approx 0.107. \tag{3.39}$$

In the limit $T_c - T \ll T_c$,

$$\ln\frac{T}{T_c} \approx -\left(1 - \frac{T}{T_c}\right), \tag{3.40}$$

and Δ is approximately given by

$$\Delta(T) \approx c_2 k_B T_c\sqrt{1 - \frac{T}{T_c}}, \tag{3.41}$$

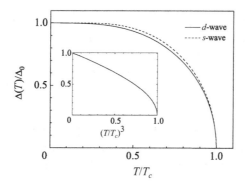

Figure 3.1 Temperature dependence of the gap parameter $\Delta(T)/\Delta_0$. The inset shows $\Delta(T)/\Delta_0$ as a function of $(T/T_c)^3$ for the d-wave superconductor.

where

$$c_2 = \left[\frac{\langle \gamma_\varphi^2 \rangle_{FS}}{\langle \gamma_\varphi^4 \rangle_{FS} g_0} \right]^{1/2}. \tag{3.42}$$

c_2 equals 3.063 and 3.537 for the isotropic s- and d-wave superconductor, respectively.

The gap equation in the whole superconducting phase can be solved numerically. Fig. 3.1 shows the temperature dependence of Δ for both s- and d-wave superconductors. The difference between these two kinds of superconductors is small. The main difference occurs at low temperatures, where Δ scales as T^3 in the d-wave state but varies exponentially with temperature in the s-wave state.

3.3 Density of States

The density of states $\rho(\omega)$ of superconducting quasiparticle excitations is an important quantity of superconductors. Given the quasiparticle energy dispersion, $E_{\mathbf{k}} = \sqrt{\xi_{\mathbf{k}}^2 + \Delta_{\mathbf{k}}^2}$, the density of states is defined as

$$\rho(\omega) = \frac{1}{N} \sum_{\mathbf{k}} \delta(\omega - E_{\mathbf{k}}) = \int_0^{2\pi} \frac{d\varphi}{2\pi} \int d\xi \rho_0(\xi) \delta(\omega - \sqrt{\xi^2 + \Delta^2 \gamma_\varphi^2}). \tag{3.43}$$

ρ_0 is the density of states in the normal state. In the low-energy limit, $\rho_0(\xi) \approx N_F$, the above expression can be written as

$$\rho(\omega) = N_F \int_0^{2\pi} \frac{d\varphi}{2\pi} \frac{\omega}{\sqrt{\omega^2 - \Delta^2 \gamma_\varphi^2}} \theta(\omega - \Delta |\gamma_\varphi^2|), \tag{3.44}$$

where $\theta(x)$ is the Heaviside step function.

In the isotropic s-wave superconductor, $\gamma_\varphi = 1$, Eq. (3.44) shows that the density of states is finite only when $\omega > \Delta$ as a result of the opening of an isotropic energy gap on the whole Fermi surface

$$\rho(\omega) = \frac{N_F \omega \theta(\omega - \Delta)}{\sqrt{\omega^2 - \Delta^2}}. \tag{3.45}$$

As expected, $\rho(\omega)$ approaches the normal state density of states N_F in the limit $\omega \gg \Delta$. However, right at the gap edge, $\omega = \Delta$, $\rho(\omega)$ diverges as $(\omega - \Delta)^{-1/2}$. This divergence strongly affects the physical properties of superconductors. For example, it yields a coherence peak in the spin-lattice relaxation rate of nuclear magnetic resonances (NMR) at the critical transition temperature, and a divergence in the optical conductivity at $\omega = 2\Delta$. The divergence happens at $\omega = 2\Delta$ not at $\omega = \Delta$ because in the optical conductivity measurement it is always a pair of quasiparticles are excited due to the momentum conservation.

In the d-wave superconductor, the density of states is determined by

$$\rho(\omega) = N_F \int \frac{d\varphi}{2\pi} \frac{\omega \, \theta(\omega - \Delta|\cos 2\phi|)}{\sqrt{\omega^2 - \Delta^2 \cos^2 2\varphi}}. \tag{3.46}$$

When $\omega > \Delta$, $\rho(\omega)$ can be further expressed as

$$\rho(\omega) = \frac{2N_F}{\pi} \int_0^1 dx \frac{1}{\sqrt{1 - x^2}\sqrt{1 - (\Delta/\omega)^2 x^2}}, \tag{3.47}$$

the right-hand side is an elliptical integral. On the other hand, when $\omega < \Delta$, $\rho(\omega)$ is given by

$$\rho(\omega) = \frac{2N_F \omega}{\pi \Delta} \int_0^1 dx \frac{1}{\sqrt{1 - x^2}\sqrt{1 - (\omega/\Delta)^2 x^2}}. \tag{3.48}$$

As the quasiparticle dispersion is linear around the gap nodes in a d-wave superconductor, it is simple to show from this equation that the low-energy density of states varies linearly with ω

$$\rho(\omega) \simeq \frac{N_F \omega}{\Delta}, \qquad \omega \ll \Delta. \tag{3.49}$$

This is a characteristic property of the d-wave or other two-dimensional superconductors with gap nodes. It has a strong impact on the low temperature properties of superconductors.

As ω approaches Δ from high frequency, the elliptical integral in Eq. (3.47) can be approximately integrated out. This yields

$$\rho(\omega \to \Delta^+) \approx \frac{N_F}{\pi} \ln \frac{8}{1 - \Delta/\omega}. \tag{3.50}$$

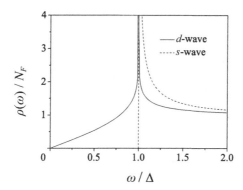

Figure 3.2 The quasiparticle density of states of s- and d-wave superconductors.

As ω approaches Δ from low frequency, the density of states is given by

$$\rho(\omega \to \Delta^-) \approx \frac{N_F \omega}{\pi \Delta} \ln \frac{8}{1 - \omega/\Delta}. \tag{3.51}$$

Thus from either direction $\rho(\omega)$ diverges logarithmically at $\omega = \Delta$, weaker than the square root divergence in the s-wave superconductor. This divergence of $\rho(\omega)$ can also induce a coherence peak in the NMR spin-lattice relaxation rate at the superconducting transition temperature and a divergent optical conductivity at $\omega = 2\Delta$. However, in real d-wave superconductors, this divergence is often smeared out by strong coupling effects or by impurity scattering, and is difficult to observe experimentally.

3.4 Entropy

Thermodynamic properties of d-wave superconductors differ significantly from those of conventional s-wave superconductors. In particular, the d-wave symmetry renders the low-energy Bogoliubov excitations gapless and dramatically changes the low temperature behaviors of thermodynamic quantities.

The entropy is an important quantity characterizing thermodynamic fluctuations. At the mean-field level, it can be expressed using the Fermi distribution function of superconducting quasiparticles

$$f_{\mathbf{k}} = \frac{1}{e^{\beta E_{\mathbf{k}}} + 1} \tag{3.52}$$

as

$$S = -\frac{2k_B}{N} \sum_{\mathbf{k}} [(1 - f_{\mathbf{k}}) \ln (1 - f_{\mathbf{k}}) + f_{\mathbf{k}} \ln f_{\mathbf{k}}]. \tag{3.53}$$

Expressed using the quasiparticle density of states $\rho(\omega)$, it becomes

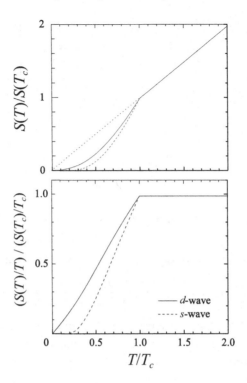

Figure 3.3 Temperature dependence of the normalized entropy for *d*- and *s*-wave superconductors.

$$S = 2k_B \int d\omega \rho(\omega) \left[\beta \omega f(\omega) + \ln\left(1 + e^{-\beta \omega}\right) \right]$$

$$= 2k_B^2 T \int dx \rho\, (k_B T x) \left[\frac{x}{1 + e^x} + \ln\left(1 + e^{-x}\right) \right]. \qquad (3.54)$$

The temperature dependence of the entropy can be evaluated numerically based on the above formula. Figure 3.3 compares the entropy as a function of temperature for the *s*- and *d*-wave superconductors. The difference lies mainly in the low temperature region where the entropy of the *s*-wave superconductor drops much faster than the *d*-wave one. As the low-energy density of states is linear for the *d*-wave superconductor

$$\rho\, (k_B T x) \simeq \frac{N_F k_B T x}{\Delta(T)}, \qquad (3.55)$$

the low temperature entropy varies quadratically with temperature

$$S \simeq \frac{\alpha_1 N_F k_B^3 T^2}{\Delta(T)}, \qquad T \ll T_c. \qquad (3.56)$$

This is different from the activated temperature dependence of the entropy in the s-wave superconductor. In Eq. (3.56)

$$\alpha_1 = 2 \int_0^\infty dx \left[\frac{x^2}{1 + e^x} + x \ln\left(1 + e^{-x}\right) \right] \simeq 5.41. \tag{3.57}$$

Below but close to the critical temperature T_c, $\Delta_{\mathbf{k}}$ is small and the entropy S can be expanded in terms of $\Delta_{\mathbf{k}}$. To the second order approximation in $\Delta_{\mathbf{k}}$, the entropy of the d-wave superconductor is

$$
\begin{aligned}
S(T) &\simeq S_N(T) - \frac{1}{k_B T^2 N} \sum_{\mathbf{k}} \frac{e^{\beta \xi_{\mathbf{k}}} \Delta_{\mathbf{k}}^2}{\left(1 + e^{\beta \xi_{\mathbf{k}}}\right)^2} \\
&\simeq S_N(T) - c_1^2 k_B^2 N_F \langle \gamma_\varphi^2 \rangle (T_c - T),
\end{aligned} \tag{3.58}
$$

where $S_N(T)$ is the normal state entropy obtained by linear extrapolation from the entropy above T_c. The superconducting state is more ordered than the normal state. Thus its entropy is smaller than the extrapolated normal state value $S_N(T)$. As shown in Fig. 3.3, $S(T)/T$ varies linearly with temperature in both s- and d-wave superconductors, but with different slopes, when T approaches T_c in the superconducting state.

Above the critical temperature, the superconductivity disappears, and $S(T)$ equals S_N. Assuming $\rho_0(\omega) \approx N_F$, independent of the energy, it is simple to show that $S_N(T)$ varies linearly with temperature

$$S_N \simeq \frac{2\pi^2}{3} N_F k_B^2 T, \qquad T > T_c. \tag{3.59}$$

It reaches zero entropy if it is extrapolated to zero temperature.

In the superconducting state, the entropy is lowered due to Cooper pair condensation. It should recover its normal state value at T_c. This is a consequence of the conservation of total degrees of freedom. If above T_c, the entropy does not reach the extrapolated value based on the high temperature data, it simply means that there is an entropy loss below T_c. This entropy loss does not exist in the conventional BCS superconductors. It must result from other physical effects, such as competing orders.

Figure 3.4 shows the temperature dependence of the entropy for $YBa_2Cu_3O_{6+x}$ at several different doping levels [105]. The entropy is obtained from the temperature integral of the specific heat measured by experiments. In the overdoped regime, for example the curve of $x = 0.97$, the entropy behaves similarly to an ideal BCS superconductor, as shown in Fig. 3.3. Above T_c, the entropy varies linearly with temperature and is extended to the origin if it is extrapolated to low temperatures.

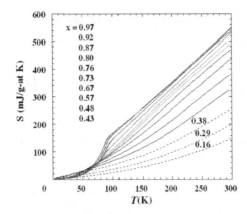

Figure 3.4 Temperature dependence of the entropy of YBa$_2$Cu$_3$O$_{6+x}$ at different doping levels. The larger is x, the higher is the doping level. (The experimental data are from Ref. [105])

In the underdoped regime, for example the curve of $x = 0.38$, the entropy behaves differently. First, there is no cusp on the entropy curve at T_c. It is impossible to determine T_c based on the singularity in the entropy. Second, if we linearly extrapolate the entropy curve around $T \sim 300$K to low temperatures, the extrapolated line will have a negative intercept at zero temperature. This means that there is an entropy loss even above the superconducting transition temperature. This entropy loss is clearly not due to the superconducting condensation.

The suppression of the low energy density of states in the normal state in high-T_c cuprates suggests that an energy gap similar to the superconducting gap exists in the normal state. It is this normal state gap that is responsible for the suppression of low-energy density of states of electrons. However, unlike the superconducting gap, there is no coherent condensation associated with this normal state energy gap. In literature, this gap is called a pseudogap. The pseudogap strongly affects the physical properties of high-T_c cuprates. It yields various anomalous behaviors in the specific heat, magnetic susceptibility, resistance, and many other thermodynamic and transport coefficients. The entropy loss in the normal state is one of them.

3.5 Specific Heat

The specific heat is determined by the temperature derivative of the entropy. From Eq. (3.53), we have

$$C = T \frac{\partial S}{\partial T} = \frac{2}{N} \sum_{\mathbf{k}} E_{\mathbf{k}} \frac{\partial f(E_{\mathbf{k}})}{\partial T}. \tag{3.60}$$

It should be noted that the energy dispersion of the superconducting quasi-particle $E_\mathbf{k}$ is temperature dependent, i.e. $\partial E_\mathbf{k}/\partial T \neq 0$. The above expression can be also written as

$$C = \frac{2}{k_B N} \sum_\mathbf{k} \frac{e^{\beta E_\mathbf{k}}}{\left(1 + e^{\beta E_\mathbf{k}}\right)^2} \left(\frac{E_\mathbf{k}^2}{T^2} - \frac{E_\mathbf{k}}{T} \frac{\partial E_\mathbf{k}}{\partial T} \right). \tag{3.61}$$

At low temperatures, $\Delta_\mathbf{k}$ depends weakly on temperature, and the specific heat is simply proportional to the number of quasiparticles within the energy scale of $k_B T$. In this case, the specific heat is mainly contributed by the first term in Eq. (3.61). Since the low-energy density of states is linear in a d-wave superconductor, the low temperature specific heat varies quadratically with temperature

$$C \simeq \frac{2\alpha_1 N_F k_B^3 T^2}{\Delta(T)}. \tag{3.62}$$

This can be compared with the low temperature specific heat of s-wave superconductors. In an s-wave superconductor, since the low energy density is zero, the specific heat decays exponentially with temperature at low temperatures

$$C \simeq \frac{2\sqrt{2\pi} N_F \Delta_0^{5/2}}{k_B T^{3/2}} e^{-\Delta_0/k_B T}. \tag{3.63}$$

Around T_c, $\Delta(T)$ is given by Eq. (3.41). The derivative of $E_\mathbf{k}$ with respect to temperature is finite, and the specific heat has a finite jump at T_c. Above T_c, C is just the specific heat of electrons in the normal state

$$C_N(T) = \frac{2}{k_B T^2 N} \sum_\mathbf{k} \frac{\xi_\mathbf{k}^2 e^{\beta \xi_\mathbf{k}}}{\left(1 + e^{\beta \xi_\mathbf{k}}\right)^2} \simeq \frac{2\pi^2}{3} k_B^2 N_F T. \tag{3.64}$$

The jump of the specific heat at T_c,

$$\Delta C(T_c) = C(T_c^-) - C(T_c^+) = C(T_c^-) - C_N(T_c), \tag{3.65}$$

is determined by

$$\Delta C(T_c) = \frac{c_1^2 k_B}{N} \sum_\mathbf{k} \frac{e^{\beta \xi_\mathbf{k}}}{\left(1 + e^{\beta \xi_\mathbf{k}}\right)^2} \gamma_\varphi^2 \simeq c_1^2 k_B^2 N_F T_c \langle \gamma_\varphi^2 \rangle_{FS}. \tag{3.66}$$

Using the result of $C_N(T)$, we find the ratio between ΔC and C_N at T_c to be

$$\frac{\Delta C(T_c)}{C_N(T_c)} = \frac{3 c_1^2 \langle \gamma_\varphi^2 \rangle_{FS}}{2\pi^2}. \tag{3.67}$$

This ratio equals 1.43 and 0.95 for s- and d-wave superconductors, respectively. The specific heat jump of the d-wave superconductor is smaller than the s-wave one.

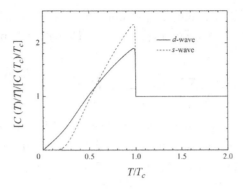

Figure 3.5 Temperature dependence of C/T for d- and s-wave superconductors.

The above discussion indicates that there are two major differences in the specific heats between s- and d-wave superconductors. First, the specific heat of the s-wave superconductor decays exponentially at low temperatures. It drops only quadratically with temperature in the d-wave superconductor. Second, the specific heat jump in the s-wave superconductor is larger than that in the d-wave one. These differences can be clearly seen from Fig. 3.5, in which the temperature dependence of the specific heat coefficient, $\gamma = C/T$, is depicted.

Figure 3.6 shows the specific heat coefficient γ as a function of temperature for $YBa_2Cu_3O_{6+x}$ at different doping obtained from the differential heat capacity measurements [105]. Similar to the entropy, the specific heat behaves differently in the overdoped and underdoped regimes. In fact, the entropy curves shown in Fig. 3.4 were obtained by the temperature integration of the specific heat data shown in Fig. 3.6. In the overdoped regime, for example, the case $x = 0.97$, the experimental result agrees qualitatively with the theoretical prediction for the d-wave superconductor shown in Fig. 3.5. However, there are two differences. First the specific heat jump at T_c is larger than the theoretical value for the d-wave superconductor, but closer to the value for the s-wave superconductor. It is unknown whether this relatively large jump is an intrinsic property of high-T_c superconductivity or is simply due to measurement errors. Second, $\gamma(T)$ at low temperatures in the d-wave superconductor should be a linear function of T, but the experimental data scale as T^2. This difference may result from the disorder effect. In the presence of impurity scattering, the temperature dependence of the low-energy density of states changing from linear to quadratic, which can alter the low temperature specific heat coefficient γ changing from T to T^2. A detailed discussion on the correction of impurity scattering to the specific heat is given in Chapter 8.

In the underdoped regime, the specific heat jump at T_c is dramatically suppressed. Unlike in the overdoped samples, the specific heat coefficient begins to drop above

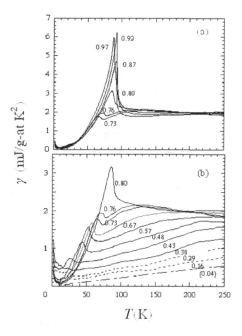

Figure 3.6 Temperature dependence of the specific heat coefficient $\gamma = C(T)/T$ for $YBa_2Cu_3O_{6+x}$ at several different doping levels. (Reproduced using the data published in Ref. [105])

T_c. Moreover, the temperature at which the specific heat coefficient begins to drop increases with decreasing doping. In the superconducting state, the decrease of γ is due to the opening of the quasiparticle energy gap. The drop of γ in the normal state implies a suppression in the normal state density of states, which is a manifestation of the pseudogap effect. As the microscopic origin of pseudogap is unknown, it is difficult to perform a quantitative analysis for the temperature dependence of the specific heat in the underdoped regime.

At low temperatures, the specific heat of $YBa_2Cu_3O_{6+x}$ shows an upturn. This is probably caused by magnetic impurities. In $La_{2-x}Sr_xCuO_{4+\delta}$, the contribution from magnetic impurities is significantly weakened, and the specific heat curve does not show any upturn. Instead, it follows the T^2-law at low temperatures [106] as predicted, in support of theory of d-wave superconductivity.

3.6 Gap Operators in the Continuum Limit

In a homogeneous system, the gap function of d-wave superconductors is diagonal in momentum space. However, if the translational symmetry is broken by, for example, vortex lines or disorders, it is more convenient to study the superconducting state using the BdG equation in real space. For high-T_c cuprates, the coherence length is short, the BdG equation can be discretized according to the lattice symmetry and solved numerically. However, the physical properties of low-energy quasiparticles are not sensitive to the lattice structure because their de Broglie wavelengths are very long. In this case, the BdG equation can be simplified by linearizing the Hamiltonian in the continuum limit. This is a commonly used approach in the study of the low-energy electromagnetic response functions and scaling behaviors of d-wave superconductors. To do this, we define the energy gap operator $\hat{\Delta}$ through its action on the wavefunction $\psi(\mathbf{r})$ as

$$\hat{\Delta}\psi(\mathbf{r}_1) = \int d\mathbf{r}_2 \Delta(\frac{\mathbf{r}_1 + \mathbf{r}_2}{2}, \mathbf{r}_1 - \mathbf{r}_2)\psi(\mathbf{r}_2). \tag{3.68}$$

The order parameter Δ is determined by the gap equation

$$\Delta(\mathbf{R}, \mathbf{r}) = -g(\mathbf{r})\langle\psi_\uparrow(\mathbf{r}_1)\psi_\downarrow(\mathbf{r}_2)\rangle, \tag{3.69}$$

where $\mathbf{R} = (\mathbf{r}_1 - \mathbf{r}_2)/2$ and $\mathbf{r} = \mathbf{r}_1 - \mathbf{r}_2$. $g(\mathbf{r})$ is the pairing interaction between electrons.

Using the gap operator, the BdG equation can be expressed as

$$i\hbar\partial_t\psi = H\psi, \tag{3.70}$$

where $\psi = (u, v)^T$ and ε_F is the Fermi energy. H is the BCS mean-field Hamiltonian defined by

$$H = \begin{pmatrix} \hat{h} + U(\mathbf{r}) - \varepsilon_F & \hat{\Delta} \\ \hat{\Delta}^\dagger & -\hat{h}^* - U(\mathbf{r}) + \varepsilon_F \end{pmatrix}. \tag{3.71}$$

$U(\mathbf{r})$ is a scattering potential and $\hat{h}$ is the kinetic energy operator

$$\hat{h} = \frac{1}{2m}\left(-i\hbar\nabla - \frac{e}{c}\mathbf{A}\right)^2. \tag{3.72}$$

In order to determine the expression of the gap operator of d-wave superconductors, we take the Fourier transformation for the relative coordinate $\mathbf{r}$ and rewrite the gap parameter Δ as

$$\Delta(\mathbf{R}, \mathbf{k}) = \int d\mathbf{r}\Delta(\mathbf{R}, \mathbf{r})e^{i\mathbf{k}\cdot\mathbf{r}}. \tag{3.73}$$

The pairing symmetry is determined by the relative momentum dependence of $\Delta(\mathbf{R}, \mathbf{k})$. For the $d_{x^2-y^2}$-wave superconductor, $\Delta(\mathbf{R}, \mathbf{k})$ in the continuum limit is defined by

$$\Delta(\mathbf{R}, \mathbf{k}) = \Delta_0(\mathbf{R})\frac{k_x^2 - k_y^2}{k_F^2}, \tag{3.74}$$

where k_F is the Fermi momentum. $\Delta_0(\mathbf{R})$ measures the center of mass distribution of the order parameter, independent of pairing symmetry.

Taking an inverse transformation to convert $\Delta(\mathbf{R}, \mathbf{k})$ back to the coordinate space, we find that the gap operator can be expressed as

$$k_F^2 \hat{\Delta} \psi(\mathbf{r}_1)$$
$$= -\int d\mathbf{r}_2 \Delta_0(\frac{\mathbf{r}_1 + \mathbf{r}_2}{2}) \left(\partial_x^2 - \partial_y^2 \right) \delta(\mathbf{r}_1 - \mathbf{r}_2) \psi(\mathbf{r}_2),$$
$$= -\partial_x^2 \Delta_0(\mathbf{r}_1) \psi(\mathbf{r}_1) + \partial_x \frac{\partial \Delta_0(\mathbf{r}_1)}{\partial x} \psi(\mathbf{r}_1) - \frac{1}{4} \frac{\partial^2 \Delta_0(\mathbf{r}_1)}{\partial x^2} \psi(\mathbf{r}_1) - (\partial_x \rightarrow \partial_y).$$

Using the identity,

$$\{\partial_x, f(\mathbf{r})\} = 2\partial_x f(\mathbf{r}) - \frac{\partial f(\mathbf{r})}{\partial x}, \tag{3.75}$$

we can further express $\hat{\Delta}$ as

$$\hat{\Delta} = \frac{1}{4p_F^2} \{p_x, \{p_x, \Delta_0(\mathbf{r})\}\} - \frac{1}{4p_F^2} \{p_y, \{p_y, \Delta_0(\mathbf{r})\}\}, \tag{3.76}$$

where $p_F = \hbar k_F$. p_x and p_y are momentum operators. $\{a, b\} = ab + ba$ is the anticommutator.

In the study of low-energy properties, only low-energy quasiparticles around the gap nodes are important. Therefore, we can expand the Hamiltonian in the nodal region, by just keeping the terms linear in momentum and neglecting all other high order terms. The linearization needs to be performed around each of the four nodes in the $d_{x^2-y^2}$-wave superconductor. Here, as an example, we consider the expansion around the node at $\mathbf{k}_1 = (k_F/\sqrt{2}, k_F/\sqrt{2})$. It is straightforward to generalize the derivation to the other three nodal points.

The linearization is to perform a Galilean transformation to change the origin of coordinates to the frame that moves with the wave vector $\mathbf{k}_1$. The Hamiltonian is projected onto this reference frame by expressing the wavefunction $\psi(\mathbf{r})$ as a product of the plane-wave of momentum $\mathbf{k}_1$ and a wavefunction $\tilde{\psi}(\mathbf{r})$ in the new reference frame

$$\psi(\mathbf{r}) = e^{ik_F(x+y)/\sqrt{2}} \tilde{\psi}(\mathbf{r}). \tag{3.77}$$

The linearization is to find the equation that $\tilde{\psi}(\mathbf{r})$ satisfies.

Substituting this expression into Eq. (3.70) and keeping the leading order terms in momentum, we then obtain the following linearized BdG equation:

$$i\hbar \partial_t \tilde{\psi} \approx \hat{H}_0 \tilde{\psi}, \tag{3.78}$$

where $\hat{H}_0$ is given by

$$\hat{H}_0 = \begin{pmatrix} \frac{v_F}{\sqrt{2}}(\hat{x} + \hat{y}) \cdot (\mathbf{p} - \frac{e}{c}\mathbf{A}) + U & \frac{1}{\sqrt{2}p_F}\{p_x - p_y, \Delta_0(\mathbf{r})\} \\ \frac{1}{\sqrt{2}p_F}\{p_x - p_y, \Delta_0^*(\mathbf{r})\} & \frac{v_F}{\sqrt{2}}(\hat{x} + \hat{y}) \cdot (\mathbf{p} + \frac{e}{c}\mathbf{A}) - U \end{pmatrix}. \tag{3.79}$$

The Hamiltonian that is neglected,

$$\hat{H}_1 = \begin{pmatrix} \hat{h} & \hat{\Delta} \\ \hat{\Delta}^\dagger & -\hat{h}^* \end{pmatrix}, \tag{3.80}$$

contains only the higher order terms in momentum.

Around the gap nodes, the energy dispersion of the quasiarticle excitations is approximately given by the eigenvalues of $\hat{H}_0$

$$\varepsilon_p = \sqrt{(v_F \hbar k_\perp)^2 + \left(\frac{\Delta_0 k_\parallel}{k_F}\right)^2}, \tag{3.81}$$

where $k_\perp$ and $k_\parallel$ are the momenta parallel and perpendicular to the tangent direction of the Fermi surface at the nodal point, respectively. At a given temperature, T, the thermal energy scales linearly with T

$$\langle \hat{H}_0 \rangle \sim \varepsilon_p \sim k_B T. \tag{3.82}$$

This implies that the corresponding wave vectors have the scaling properties

$$k_\perp \sim \frac{T}{\sqrt{2}v_F}, \tag{3.83}$$

$$k_\parallel \sim \frac{T \hbar k_F}{\sqrt{2}\Delta_0} \gg k_\perp. \tag{3.84}$$

The inequality holds under the condition $\varepsilon_F \gg \Delta_0$. The energy scale of $\hat{H}_1$ can be estimated using the above expressions. As $k_\parallel \gg k_\perp$, it can be shown that the leading term in $\hat{H}_1$ is given by

$$\langle \hat{H}_1 \rangle \sim \frac{1}{2}\left(\frac{k_B T}{\Delta_0}\right)^2 \varepsilon_F. \tag{3.85}$$

Thus the ratio between $\langle \hat{H}_1 \rangle$ and $\langle \hat{H}_0 \rangle$ scales linearly with temperature

$$\frac{\langle \hat{H}_1 \rangle}{\langle \hat{H}_0 \rangle} \sim \frac{k_B T \varepsilon_F}{2\Delta_0^2}. \tag{3.86}$$

By requesting $\langle \hat{H}_1 \rangle \ll \langle \hat{H}_0 \rangle$, we then obtain the condition at which the linear approximation is valid

$$T \ll \frac{2\Delta_0^2}{k_B \varepsilon_F}. \tag{3.87}$$

The above derivation can be readily extended to the d_{xy}-wave pairing state. In that case, $\Delta(\mathbf{R}, \mathbf{k})$ is defined by

$$\Delta(\mathbf{R}, \mathbf{k}) = \Delta_0(\mathbf{R}) \frac{k_x k_y}{k_F^2}, \tag{3.88}$$

and the corresponding gap operator is

$$\hat{\Delta} = -\frac{1}{4k_F^2} \left\{ \frac{\partial}{\partial y}, \left\{ \frac{\partial}{\partial x}, \Delta_0(\mathbf{r}) \right\} \right\} = \frac{1}{4p_F^2} \left\{ p_y, \{p_x, \Delta_0(\mathbf{r})\} \right\}. \tag{3.89}$$

This expression of the gap operator can also be obtained from Eq. (3.76) by taking $45°$-rotation for the coordinates. The linearized Hamiltonian now becomes

$$\hat{H}_0 = \begin{pmatrix} v_F(p_x - \frac{e}{c}A_x) + U & \frac{1}{2p_F}\{p_y, \Delta_0(\mathbf{r})\} \\ \frac{1}{2p_F}\{p_y, \Delta_0^*(\mathbf{r})\} & -v_F(p_x + \frac{e}{c}A_x) - U \end{pmatrix}. \tag{3.90}$$

3.7 Current Operators

The probability density and charge density of superconducting quasiparticles for d-wave superconductors are similarly defined as for s-wave superconductors. From the probability or charge conservations, the continuity equation for the probability or charge conservation can be derived from the time-dependent BdG equation. From these equations, we can define the expressions of the probability current density, $\mathbf{J}_P$, and the electric current density, $\mathbf{J}_Q$, for d-wave superconductors. A major difference between s- and d-wave superconductors is that the gap function is nonlocal in the latter case, and this nonlocal gap function has also contribution to the probability current operator.

The derivation of $\mathbf{J}_P$ and $\mathbf{J}_Q$ for the d-wave superconductor is similar to the s-wave one. Here we skip the detail of the derivation. For the $d_{x^2-y^2}$-wave superconductor, the probability current density $\mathbf{J}_P$ includes both the diagonal term from the kinetic energy and the off-diagonal term from the pairing energy, and is given by

$$\mathbf{J}_P = \mathbf{J}_P^{(1)} + \mathbf{J}_P^{(2)}, \tag{3.91}$$

$$\mathbf{J}_P^{(1)} = \frac{\hbar}{m} \text{Im}(u^* \nabla u - v^* \nabla v), \tag{3.92}$$

$$\mathbf{J}_P^{(2)} = \frac{2}{\hbar k_F^2} \text{Im} \left[u^*(\hat{x}\partial_x - \hat{y}\partial_y)\Delta_0(\mathbf{r})v + v^*(\hat{x}\partial_x - \hat{y}\partial_y)\Delta_0^*(\mathbf{r})u \right], \tag{3.93}$$

where $\mathbf{J}_P^{(1)}$ is independent of the pairing symmetry. $\mathbf{J}_P^{(2)}$ is absent in the s-wave superconductor.

The charge current density of quasiparticles $\mathbf{J}_Q$ does not depend on the pairing symmetry. It has the same form as in the s-wave superconductor

$$\mathbf{J}_Q = \frac{e\hbar}{m}\mathrm{Im}\left(u^*\nabla u + v^*\nabla v\right). \tag{3.94}$$

However, the supercurrent density $\mathbf{J}_S$ is determined by the equation

$$\nabla \cdot \mathbf{J}_S = \frac{e}{2\hbar k_F^2}\mathrm{Im}\left(u^*\{\partial_x, \{\partial_x, \Delta_0(\mathbf{r})\}\}v - v^*\{\partial_x, \{\partial_x, \Delta_0^*(\mathbf{r})\}\}u\right) - (\partial_x \to \partial_y). \tag{3.95}$$

For a translation invariant d-wave superconductor, $\Delta_0(\mathbf{r}) = \Delta_0$, the quasiparticle wavefunctions $u(\mathbf{r})$ and $u(\mathbf{r})$ are given by

$$u(\mathbf{r}) = \frac{1}{\sqrt{V}}e^{i\mathbf{k}\cdot\mathbf{r}}\sqrt{\frac{1}{2} + \frac{\xi_\mathbf{k}}{2E_\mathbf{k}}}, \tag{3.96}$$

$$v(\mathbf{r}) = -\frac{\mathrm{sgn}(k_x^2 - k_y^2)}{\sqrt{V}}e^{i\mathbf{k}\cdot\mathbf{r}}\sqrt{\frac{1}{2} - \frac{\xi_\mathbf{k}}{2E_\mathbf{k}}}. \tag{3.97}$$

If $\mathbf{k}$ is real, the wavefunctions do not decay, and the probability current and the electric current vectors become

$$\mathbf{J}_P^{(1)} = \frac{\hbar\xi_\mathbf{k}\mathbf{k}}{mVE_\mathbf{k}}, \tag{3.98}$$

$$\mathbf{J}_P^{(2)} = -\frac{2\Delta_0\Delta_\mathbf{k}(k_x\hat{x} - k_y\hat{y})}{\hbar k_F^2 VE_\mathbf{k}}, \tag{3.99}$$

$$\mathbf{J}_Q = \frac{e\hbar\mathbf{k}}{mV}, \tag{3.100}$$

$$\mathbf{J}_S = 0. \tag{3.101}$$

Compared with the results for s-wave superconductors, only the definition of the probability current vector is changed. It contains an extra term $\mathbf{J}_P^{(2)}$. All other terms are unchanged.

4

Quasiparticle Excitation Spectra

4.1 Single-Particle Spectral Function

The single-particle spectral function $A(\mathbf{k}, \omega)$ is an important quantity characterizing the physical properties of interacting electrons. It measures the weight of a system after adding an electron with a given momentum $\mathbf{k}$ and energy ω, or removing a hole with opposite momentum and energy. The spectral function can be measured through angle-resolved photoemission spectroscopy (ARPES).

$A(\mathbf{k}, \omega)$ in the superconducting state can be used to extract the momentum dependence of the energy gap, and the scattering lifetime of superconducting quasiparticles. It is also a basic parameter for describing interaction effects and electromagnetic response functions. In order to study the behavior of single-particle excitations in d-wave superconductors, let us define the Matsubara Green's function of superconducting quasiparticles at finite temperatures as

$$G(\mathbf{k}, \tau - \tau') = -\left\langle T_\tau \begin{pmatrix} c_{\mathbf{k}\uparrow}(\tau) \\ c_{-\mathbf{k}\downarrow}^\dagger(\tau) \end{pmatrix} \begin{pmatrix} c_{\mathbf{k}\uparrow}^\dagger(\tau') & c_{-\mathbf{k}\downarrow}(\tau') \end{pmatrix} \right\rangle. \qquad (4.1)$$

$G(\mathbf{k}, \tau)$ is a 2×2 matrix. Its diagonal terms are the normal propagators of electrons. The off-diagonal terms of $G(\mathbf{k}, \tau)$ are the anomalous propagators of electrons which contain the information of superconducting pairing. Here the Nambu spinor representation of electrons is invoked, which treats electron creation and annihilation operators on an equal footing. It provides a convenient representation to study superconducting states in which the fermion number is not conserved. The Green's function in the Nambu representation can be similarly treated as in a system where the total electron number is conserved. A more detailed introduction to the Green's functions, including its analytic properties, can be found in Appendix F.

In the BCS weak coupling limit, the single-electron Green's function without considering the correction from disorder or other physical effects can be derived by diagonalizing the mean-field Hamiltonian defined by Eq. (1.25). The Green's function satisfies

$$[\partial_\tau - H_0(\mathbf{k})]\, G^{(0)}(\mathbf{k}, \tau) = \delta(\tau), \tag{4.2}$$

where H_0 is the mean-field Hamiltonian in the Nambu representation

$$H_0(\mathbf{k}) = \xi_{\mathbf{k}}\sigma_3 + \Delta_{\mathbf{k}}\sigma_1, \tag{4.3}$$

σ_1 and σ_3 are Pauli matrices. In the Matsubara frequency space, the solution is

$$G^{(0)}(\mathbf{k}, i\omega_n) = \frac{1}{i\omega_n - \xi_{\mathbf{k}}\sigma_3 - \Delta_{\mathbf{k}}\sigma_1} = \frac{i\omega_n + \xi_{\mathbf{k}}\sigma_3 + \Delta_{\mathbf{k}}\sigma_1}{(i\omega_n)^2 - E_{\mathbf{k}}^2}, \tag{4.4}$$

with

$$E_{\mathbf{k}} = \sqrt{\xi_{\mathbf{k}}^2 + \Delta_{\mathbf{k}}^2}. \tag{4.5}$$

This Green's function, as indicated by the superscript "0," is taken as the unperturbed Green's function. This will be modified by the electron–electron interaction as well as the impurity or other scattering effects. The renormalized Green's function is determined by the Dyson equation

$$G^{-1}(\mathbf{k}, i\omega_n) = \left[G^{(0)}(\mathbf{k}, i\omega_n)\right]^{-1} - \Sigma(\mathbf{k}, i\omega_n). \tag{4.6}$$

The self-energy $\Sigma(\mathbf{k}, i\omega_n)$ contains the correction of the perturbed Hamiltonian, and is generally difficult to rigorously evaluate.

Physical measurement quantities are related to retarded Green's functions. They can be obtained from the Matsubara Green's function through analytical continuation by taking the Wick rotation

$$G^R(\mathbf{k}, \omega) = G(\mathbf{k}, i\omega_n \to \omega + i0^+). \tag{4.7}$$

The imaginary part of the diagonal component $G^R(\mathbf{k}, \omega)$ is proportional to the spectral function of electrons,

$$A(\mathbf{k}, \omega) = -\frac{1}{\pi}\mathrm{Im}G_{11}^R(\mathbf{k}, \omega). \tag{4.8}$$

Its momentum integral equals the density of states of electrons

$$\rho(\omega) = \frac{1}{N}\sum_{\mathbf{k}} A(\mathbf{k}, \omega). \tag{4.9}$$

On the other hand, the Matsubara frequency summation of $A(\mathbf{k}, \omega)$ satisfies the following sum rules

$$\int_{-\infty}^{\infty} d\omega A(\mathbf{k}, \omega) = 1, \tag{4.10}$$

$$\int_{-\infty}^{\infty} d\omega f(\omega) A(\mathbf{k}, \omega) = n_{\mathbf{k}} = \sum_{\sigma} \langle c_{\mathbf{k}\sigma}^{\dagger} c_{\mathbf{k}\omega} \rangle, \tag{4.11}$$

where $n_{\mathbf{k}}$ is the momentum distribution function of electrons and $f(\omega)$ is the Fermi distribution function.

For an ideal BCS superconductor, the self-energy vanishes and the spectral function reads

$$A^{(0)}(\mathbf{k}, \omega) = u_{\mathbf{k}}^2 \delta(\omega - E_{\mathbf{k}}) + v_{\mathbf{k}}^2 \delta(\omega + E_{\mathbf{k}}), \tag{4.12}$$

where $u_{\mathbf{k}}$ and $v_{\mathbf{k}}$ are defined by Eqs. (1.30) and (1.31), respectively. The two δ-functions represent the contribution from electron and hole excitations with the corresponding spectra weight given by $u_{\mathbf{k}}^2$ and $v_{\mathbf{k}}^2$, respectively.

4.2 ARPES

A light, shining on a solid, could emit electrons from the surface if its frequency exceeds a threshold determined by the work function of that material. This is nothing but the photoelectric effect first discovered by Hertz in 1887. The measurement on the cross-section, or the electric current density of photoelectrons with specific momentum and energy, gives rise to the angle-resolved photoemission spectra. This measurement reveals the property of electron spectral functions. ARPES is an important method to analyze properties of superconducting electrons, and has played an important role in the study of high-T_c superconductivity.

4.2.1 Photoelectric Current

In an ARPES measurement, the photoelectric current is proportional to the flux of incoming light, or the square of the vector potential $|\mathbf{A}(\mathbf{r}, t)|^2$ of the light. The photoelectric spectra measure the nonlinear response to an external electromagnetic field. We can treat the interaction between incident photons and electrons as a perturbation if the light intensity is not too strong, and evaluate the photoelectric current intensity using the perturbation theory. For this purpose, we divide the Hamiltonian into two parts

$$H = H_0 + H_{\text{int}}, \tag{4.13}$$

where H_0 is the Hamiltonian of electrons, and H_{int} is the minimal electron–photon interaction defined by

$$H_{int} = \int d\mathbf{r} \mathbf{A}(\mathbf{r}, t) \cdot \mathbf{J}(\mathbf{r}). \qquad (4.14)$$

$\mathbf{A}(\mathbf{r}, t)$ is the vector potential of the external electromagnetic field. $\mathbf{J}(\mathbf{r})$ is the electric current density operator. $\mathbf{r}$ is the coordinate of electron. We neglect the interaction among electrons already escaping the surface and their interactions with the external electromagnetic field.

We consider the perturbative correction of H_{int} to the photoelectric current density $\langle \mathbf{J}(\mathbf{R}, t) \rangle$ at the location of detector $\mathbf{R}$. At the zeroth order, the photoelectric current is proportional to the expectation value $\langle \mathbf{J}(\mathbf{R}, t) \rangle$ of the current operator under the ensemble average with respect to H_0. This contribution is zero since there is no current in the detector in the absence of applied electromagnetic field. The first order perturbation is proportional to $\langle J_\alpha(\mathbf{R}, t) J_\beta(\mathbf{r}', t') \rangle$ or $\langle J_\beta(\mathbf{r}', t') J_\alpha(\mathbf{R}, t) \rangle$. Since the matrix elements of $J_\alpha(\mathbf{R}, \mathbf{t})$ is nonzero if and only if there is an electron at $\mathbf{R}$, we have $\langle J_\alpha(\mathbf{R}, t) J_\beta(\mathbf{r}', t') \rangle = \langle J_\beta(\mathbf{r}', t') J_\alpha(\mathbf{R}, t) \rangle = 0$. Hence the first order perturbation of H_{int} has no contribution either. The finite contribution starts from the second order perturbation.

Using the theory of second order perturbation, it can be shown that the photoelectric current is determined by the correlation function of three current operators [107, 108]

$$\langle J_\alpha(\mathbf{R}, t) \rangle \propto \sum_{\mu, \nu} \int d\mathbf{r}' dt' d\mathbf{r}'' dt'' A_\mu(\mathbf{r}', t') A_\nu(\mathbf{r}'', t'')$$
$$\left\langle J_\mu(\mathbf{r}', t') J_\alpha(\mathbf{R}, t) J_\nu(\mathbf{r}'', t'') \right\rangle. \qquad (4.15)$$

As there is no photoelectric current at $\mathbf{R}$ in both the initial and final states, $J_\alpha(\mathbf{R}, t)$ should be sandwiched between $J_\mu(\mathbf{r}', t')$ and $J_\nu(\mathbf{r}'', t'')$, otherwise the above current-current-current correlation function should be zero

$$\left\langle J_\mu(\mathbf{r}', t') J_\nu(\mathbf{r}'', t'') J_\alpha(\mathbf{R}, t) \right\rangle = \left\langle J_\alpha(\mathbf{R}, t) J_\mu(\mathbf{r}', t') J_\nu(\mathbf{r}'', t'') \right\rangle = 0. \qquad (4.16)$$

Equation (4.15) is the basic formula for evaluating photoelectric current intensity. However, in order to accurately calculate the correlation function of three current operators, we need to know accurately the band structure of electrons. At the same time, we also need to comprehensively consider the corrections to this correlation function from electron–electron, electron–phonon, electron–impurity, and electron–surface interactions. This type of calculation is formidable for strongly correlated electronic systems. Moreover, the physical picture governing this formula is not transparent and in some case even counterintuitive.

4.2.2 The Three-Step Model

The most commonly used model for the interpretation of photoemission spectra is the so-called three-step model [85, 109, 110]. It is a phenomenological approach, which has nonetheless proven to be quite successful. It breaks up the complicated photoemission process into three steps: (1) excitation of electrons by the incident light; (2) propagation of excited electrons to the surface; (3) excited electrons are detected after escaping from the surface. The intensity of the photoelectric current is determined by the product of the probabilities for these three processes, namely the probability of an electron being excited by the light, the scattering probability of the excited electrons during the propagation from the bulk to the surface, and the probability of tunneling through the surface barrier to the vacuum. Compared to the complicated perturbative result of Eq. (4.15), this simplified model neglects the interference effect between the bulk electrons and the excited surface electrons, as well as the quantum interference effect of electrons during the relaxation within the bulk. Nevertheless, the physical picture revealed by this simplified model is clear. It contains all the essential points of photoelectric scattering and is a valuable empirical model for analyzing the ARPES experiments.

In the three-step model, the angle-resolved photoelectric spectra are obtained by calculating the scattering probability of photoelectrons. The scattering process of photoelectrons is determined by the energy and momentum of electrons in both the initial and final states, and the energy and direction of the incident photon. From the Fermi golden rule, the transition probability is approximately found to be

$$w_{fi} = \frac{2\pi}{\hbar} \left| \langle \Psi_f^N | H_{\text{int}} | \Psi_i^N \rangle \right|^2 \delta \left(E_f^N - E_i^N - h\nu \right), \tag{4.17}$$

where ν is the frequency of the incident photon. The energy of the final state is the sum of the kinetic energy of the outgoing electron E_{kin}, the surface work function ϕ, and the total energy after one electron has escaped from the system

$$E_f^N = E_{\text{kin}} + \phi + E_f^{N-1}. \tag{4.18}$$

According to energy conservation, the kinetic energy of the photoelectron and the frequency of the incident photon satisfy

$$h\nu = E_{\text{kin}} + \phi + E_B, \tag{4.19}$$

where $E_B = -\xi_{\mathbf{k}}$ is the band energy of the electron with respect to the Fermi level. Using these equations, w_{fi} can be further expressed as

$$w_{fi} = \frac{2\pi}{\hbar} \left| \langle \Psi_f^N | H_{\text{int}} | \Psi_i^N \rangle \right|^2 \delta \left(\xi_{\mathbf{k}} + E_f^{N-1} - E_i^N \right). \tag{4.20}$$

Both E_{kin} and ϕ can be determined experimentally. From the measurement of the frequency of incident photon and the energy of the outgoing electron, we can determine the energy dispersion of electrons $\xi_{\mathbf{k}}$.

In order to further analyze the angle-resolved photoelectron spectra, we need to simplify Eq. (4.20) using the "sudden" approximation. This approximation assumes that the time scale for the photon excited electrons to escape from the location of excitation to the vacuum is much shorter than the typical scattering time in the bulk, so that the excited electron is not scattered by other particles, including electrons, phonons, and photons. It implies that the interaction between the excited electron and other electrons is negligible and the wavefunction of the final state is a direct product of the wavefunction of the excited photoelectron and that of other electrons

$$\Psi_f^N = \mathcal{A}\phi_f^{\mathbf{k}}\Psi_f^{N-1}, \tag{4.21}$$

where $\mathcal{A}$ is the antisymmetrization operator to ensure that the wavefunction is completely antisymmetric under the exchange of any two electrons. $\phi_f^{\mathbf{k}}$ is the wavefunction of the photoelectron. Ψ_f^{N-1} is the wavefunction of the $N-1$ electrons not being excited by photons. It can be at any one of the eigenstates of electrons Ψ_m^{N-1}. To obtain the total transition probability, we need to sum over all possible eigenstates Ψ_m^{N-1} of these $N-1$ electrons.

The "sudden" approximation can be examined by comparing the escape-time with the average scattering time of electrons in the bulk. The kinetic energy of photoelectron is typically of the order of 20 eV and the corresponding velocity is about $v \approx 3 \times 10^8$ cm/s. The escape depth of photoelectrons is typically of the order of 10 Å. Thus escape-time is approximately of the order $t_e \sim 3 \times 10^{-16}$ sec. In solids, the characteristic scattering time induced by electron–electron interactions can be estimated from the plasma frequency ω_p. For high-T_c cuprate, $\omega_p \sim 1$ eV, and the corresponding interaction time scale is approximately $t_s = 2\pi/\omega_p \sim 4 \times 10^{-15}$ sec. As t_e is about one order smaller than t_s, the "sudden" approximation is justified.

In the treatment of the initial state, the independent particle approximation is generally used. The initial wavefunction is a direct product of the wavefunction of the excited electron and that of other electrons

$$\Psi_i^N = \mathcal{A}\phi_i^{\mathbf{k}}\Psi_i^{N-1}, \tag{4.22}$$

and

$$\Psi_i^{N-1} = c_{\mathbf{k}\sigma}\Psi_i^N. \tag{4.23}$$

Unlike Ψ_f^{N-1}, Ψ_i^{N-1} is generally not an eigenstate of electrons.

Under the above approximation, the scattering matrix element between the initial and final states becomes

$$\langle \Psi_f^N | H_{\text{int}} | \Psi_i^N \rangle = \langle \phi_f^{\mathbf{k}} | H_{\text{int}} | \phi_i^{\mathbf{k}} \rangle \langle \Psi_f^{N-1} | \Psi_i^{N-1} \rangle. \tag{4.24}$$

Substituting this into Eq. (4.20), we obtain the expression of the photoelectric current density

$$w_{f,i} = I_0(\mathbf{k}, \nu) \left| \langle \Psi_f^{N-1} | c_{\mathbf{k}\sigma} | \Psi_i^N \rangle \right|^2 \delta \left(\varepsilon_{\mathbf{k}} + E_f^{N-1} - E_i^N \right), \tag{4.25}$$

where $I_0(\mathbf{k}, \nu)$ is proportional to the single-electron dipole matrix element

$$I_0(\mathbf{k}, \nu) \propto \langle \phi_f^{\mathbf{k}} | H_{\text{int}} | \phi_i^{\mathbf{k}} \rangle, \tag{4.26}$$

which is assumed to be independent of Ψ_i^{N-1} and Ψ_f^{N-1}.

The right-hand side of Eq. (4.25) is related to the electron spectral function. From the theory of Green's functions, it can be shown that

$$\frac{1}{Z} \sum_{fi} e^{-\beta E_i^N} \left| \langle \Psi_f^{N-1} | c_{\mathbf{k}\sigma} | \Psi_i^N \rangle \right|^2 \delta \left(\varepsilon_{\mathbf{k}} + E_f^{N-1} - E_i^N \right) = f(\varepsilon_{\mathbf{k}}) A(\mathbf{k}, \varepsilon_{\mathbf{k}}), \tag{4.27}$$

where $f(\varepsilon)$ is the Fermi distribution function. Substituting (4.27) into (4.25), we have

$$I(\mathbf{k}, \omega) = I_0(\mathbf{k}, \omega) f(\omega) A(\mathbf{k}, \omega). \tag{4.28}$$

This is just the formula that is used in the analysis of ARPES measurement data. $A(\mathbf{k}, \omega)$ contains the information of electron–phonon and electron–electron interactions. It should be borne in mind that $A(\mathbf{k}, \omega)$ measured by ARPES, mainly contributes from the surface states, which might be different from the bulk ones. Here we assume that there is no difference between the surface and bulk states for simplicity. Otherwise, a correction should be included. In the analysis of ARPRS data using Eq. (4.28), one should also consider the contribution of the background arising from the inelastic scattering of photoelectrons before they escape to the vacuum.

Equation (4.28) shows that the photoelectron current density is proportional to the single-electron spectral function. Thus the single-electron spectra can be deduced from ARPES. This is an advantage of ARPES in comparison with other experimental measurements which probe physical quantities related to two- or multiparticle correlation functions. Theoretical analysis of measurement data in the latter case is clearly much more involved.

In Eq. (4.28), $I_0(\mathbf{k}, \omega)$ depends on $\mathbf{k}$, the polarization as well as the frequency of the incident photon. It is less sensitive to the photoelectron energy ω and temperature T. Thus the ω dependence of the ARPES lineshape is determined purely by the electron spectral function and the Fermi distribution function. In this case, it is convenient to analyze the momentum distribution function of electrons from the measurement data using the sum rule presented in Eq. (F.16).

4.2.3 Energy and Momentum Distribution Curves

The ARPES results are usually presented in terms of the energy distribution curve (EDC), which shows $I(\mathbf{k}, \omega)$ as a function of ω for a given $\mathbf{k}$. It is intuitive and convenient to use EDC to analyze properties of superconducting gaps and pseudogaps in high-T_c cuprates. However, EDC does not exhibit a Lorentzian lineshape. It is difficult to use to quantitatively analyze experimental data unless we have a good understanding on the self-energy of electrons. There are two reasons that may cause the EDC lineshape to be non-Lorentzian. First, the Fermi distribution function $f(\omega)$ suppresses the spectral density at $\omega > 0$, which introduces an asymmetry in the lineshape. Second, the energy dependence of the spectral function $A(\mathbf{k}, \omega)$ is generally complicated, and it may not be Lorentzian itself. Moreover, the contribution of the background to the spectra may also show a complicated ω-dependence, making a quantitative analysis of experimental data even more difficult.

An intriguing feature revealed by ARPES for $Ba_2Sr_2CaCu_2O_{8+x}$ is that there is a peak-dip-hump structure in the EDC near the $(\pi, 0)$ point of the Brillouin zone in the superconducting state [111]. This structure consists of a sharp low-energy peak, followed by a dip, and then a hump at higher energy. This feature is widely observed in the spectrum of several high-T_c superconductors [85]. Similar structures are also observed in the tunneling spectra of high-T_c copper oxides [112]. The main feature of the peak-dip-hump structure could be understood from the bilayer splitting, where the peak and hump correspond to the antibonding and bonding bands, respectively [113]. The coupling of quasiparticles with a collective boson mode [114], particularly the magnetic resonance mode observed in inelastic neutron scattering experiments [115], is also believed to play an important role in generating this peculiar structure.

In recent years, with the improvement in momentum and angular resolution, it becomes feasible to analyze the ARPES data using the momentum distribution curve (MDC). The MDC is generally symmetric and has approximately a Lorentzian lineshape, which is more convenient to analyze than EDC, especially in the case where the excitation becomes gapless. In particular, it is more convenient to use MDC to analyze the spectral functions of gapless excitations near the Fermi surface. If $I_0(\mathbf{k}, \omega)$ is momentum independent and $\Sigma(\mathbf{k}, \omega)$ depends weakly on the momentum component perpendicular to the Fermi surface, $k_\perp$, then the MDC lineshape should be a Lorentzian according to Eq. (F.30). The center of this Lorentzian peak is located at

$$k = k_F + \frac{\omega - \mathrm{Re}\Sigma(\omega)}{v_F^0} \tag{4.29}$$

and the half-width is given by $\mathrm{Im}\Sigma(\omega)/v_F^0$. The peak position measures the renormalized energy–momentum dispersion of electrons. The half-width is proportional to the scattering rate or the inverse lifetime of electrons. Moreover, the contribution from the background scattering to MDC depends weakly on momentum. Therefore, MDC provides a useful approach for extracting information on the self-energy from ARPES.

ARPES is a unique technique that can directly detect the single-particle spectral function with both momentum and energy resolutions. It has played an important role in determining the Fermi surface structures, the characteristic energy scale of low-energy excitations, the momentum dependence of the superconducting gap and the pseudogap, and the pairing symmetry for cuprate superconductors. In recent years, the momentum and energy resolutions of ARPES have been significantly improved, making ARPES an indispensable experimental tool for studying various anomalous physical phenomena in high-T_c cuprates and other low-dimensional properties of electrons. For the experimental progress of ARPES, please refer to the three review articles [85, 116, 117].

4.3 Fermi Surface and Luttinger Sum Rule

The Fermi wave vector and the Fermi surface are the two basic physical quantities characterizing interacting electron systems. As only the electrons around the Fermi level can be excited at low energy, the Fermi surface structure is important to the understanding of both electromagnetic and thermodynamic properties. In a normal metallic state, the volume enclosed by the Fermi surface is fixed regardless of the geometry of the Fermi surface. This volume is proportional to the electron density and is referred to as the Luttinger volume. The formula that reveals the relationship between the volume of the Fermi surface and the density of electrons n is called the Luttinger sum rule, or the Luttinger theorem, which reads

$$n = 2 \int_{G(\mathbf{k},\omega=0)>0} \frac{\mathrm{d}^d\mathbf{k}}{(2\pi)^d}, \tag{4.30}$$

where d is the spatial dimension. The integral is performed over the region in momentum space surrounded by the surface defined by positive $G(\mathbf{k}, \omega = 0)$. In a superconducting state, $G(\mathbf{k}, \omega)$ in the above equation is the diagonal compound of the Green's function $G_{11}(\mathbf{k}, \omega)$ in the Nambu representation.

This equation shows that in a many-electron system, the momentum space volume in which the zero-energy single-particle Green's function takes positive values, is proportional to the electron density N/V, independent of interaction. This formula is applicable not only to normal electrons but also to superconducting electrons. A proof of this theorem can be found in Refs. [118, 119].

The domains with $G(\mathbf{k}, 0) > 0$ might be connected or disconnected. In either case, the domains between $G(\mathbf{k}, 0) > 0$ and $G(\mathbf{k}, 0) < 0$ are separated by one or a few $(d-1)$-dimensional surfaces. There are two possibilities for this surface:

(i) G diverges on this surface, approaching $+\infty$ and $-\infty$ from the two sides of the surface. In a normal metal, this surface is just the Fermi surface usually defined. On the Fermi surface, the lifetime of quasiparticles is infinite and the imaginary part of the Green's function vanishes. When the momentum $\mathbf{k}$ crosses through the Fermi surface, the divergence in the real part of the zero-energy Green's function changes sign. In this case, the right-hand side of Eq. (4.30) is just the phase space volume enclosed by the Fermi surface, which equals the density of electrons. This is the statement of the Luttinger theorem in the normal metallic state. When $G(\mathbf{k}, 0)$ diverges, its residue is finite and proportional to the spectral weight of the electrons. In this case, the zero-energy spectral function of electrons diverges and manifests as a sharp resonance peak in the ARPES data. The Fermi wave vector can be determined from the measurement of this sharp resonance peak.

(ii) $G(\mathbf{k}, 0) = 0$ on this surface and changes sign on the two sides of this surface. This kind of surface does not exist in conventional metals. Nevertheless, one can still regard it as a Fermi surface. It is actually a remnant of the usual Fermi surface in a gapped state. Such a situation occurs in an insulating state or a superconducting state. In a superconducting state, under the BCS mean-field approximation, the diagonal component of the Green's function is given by

$$G_{11}(\mathbf{k}, \omega) = \frac{\omega + \xi_{\mathbf{k}}}{\omega^2 - \xi_{\mathbf{k}}^2 - \Delta_{\mathbf{k}}^2}. \tag{4.31}$$

On the Fermi surface, $\xi_{\mathbf{k}} = 0$ but $\Delta_{\mathbf{k}} \neq 0$. Hence, $G_{11}(\mathbf{k}, 0) = 0$ and there is no divergence on the Fermi surface. Similar situations also exist in other gapped systems. In this case, the spectra weights of electrons are completely suppressed on the Fermi surface, and no low-energy resonance peaks appear in the ARPES measurement. The divergence in G_{11} happens at $\omega = \pm\sqrt{\xi_{\mathbf{k}}^2 + \Delta_{\mathbf{k}}^2}$, not at zero energy. Hence, the Fermi surface cannot be determined simply from low-energy peaks of ARPES.

In the overdoped high-T_c cuprates, the ARPES measurement shows that there exists a closed Fermi surface, similarly to the conventional metallic state. The normalized area enclosed by the Fermi surface equals the electron density $1-x$ (x is the doped hole concentration) within experimental errors. Thus the Luttinger theorem

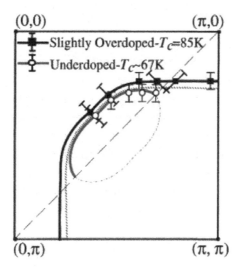

Figure 4.1 Fermi surfaces of slightly overdoped and underdoped high-T_c cuprates Bi$_2$Sr$_2$CaCu$_2$O$_{8+\delta}$ determined by ARPES [120]. The Fermi surface of the underdoped cuprate is an open arc, which is markedly different from the Fermi surface in a normal metal.

is obeyed. On the other hand, in the underdoped regime, the Fermi surface is not closed at low temperatures. Instead, it exhibits four disconnected arcs near the momenta $(\pm\pi/2, \pm\pi/2)$ [120], and the Luttinger theorem seems to be violated. Nevertheless, at high temperatures, these Fermi arcs become connected [121] and the Luttinger theorem seems to be restored. Figure 4.1 shows the Fermi surface structure for the overdoped and underdoped high-T_c cuprates Bi$_2$Sr$_2$CaCu$_2$O$_{8+\delta}$. The anomalous behavior of the Fermi surface in the underdoped cuprate is clearly related to the pseudogap phenomenon. It is a consequence of the strong correlation effect. It leads to a variety of anomalous phenomena. The microscopic origin of the Fermi arc and the pseudogap remains unclear and needs to be further explored both experimentally and theoretically.

4.4 Particle–Hole Mixing and Superconducting Energy Gap

In the normal metallic state, the spectral function of electrons $A(\mathbf{k}, \omega)$ exhibits an approximately Lorentzian peak centered at $\omega = \xi_{\mathbf{k}}$. When the momentum of an electron passes through the Fermi surface from a point below the Fermi level to a point above, the peak energy changes from negative to zero, and then to positive. Hence from ARPES, we can see how the spectral peak moves toward the Fermi energy from an energy below. This corresponds to the shift of momentum $\mathbf{k}$ from a point below the Fermi level to the Fermi momentum. However, when $\mathbf{k}$ is above the

Fermi surface, the spectral intensity is suppressed by the Fermi function $f(\omega)$ and the corresponding peak is difficult to probe by ARPES.

The evolution of the ARPES lineshape with momentum $\mathbf{k}$ is markedly different in the superconducting state due to particle–hole mixing. In an ideal BCS superconductor, the single-particle spectral function is given by Eq. (4.12). However, in real materials, the δ-function-like spectral peak in $A^{(0)}(\mathbf{k}, \omega)$ is broadened or even completely suppressed by the self-energy correction induced by the scattering of electrons. In this case, in order to describe correctly the behavior of spectral function, an accurate evaluation of the Green's function of electrons is desired but difficult to fulfill. To take a quantitative analysis of experimental results, a commonly adopted approximation is to assume that the broadened peak has a Lorentzian lineshape and the spectral function in the superconducting state is approximately given by

$$A(\mathbf{k}, \omega) \approx \frac{1}{\pi} \left[\frac{u_{\mathbf{k}}^2 \Gamma}{(\omega - E_{\mathbf{k}})^2 + \Gamma^2} + \frac{v_{\mathbf{k}}^2 \Gamma}{(\omega + E_{\mathbf{k}})^2 + \Gamma^2} \right], \tag{4.32}$$

where $u_{\mathbf{k}}^2$ and $v_{\mathbf{k}}^2$ measure the spectral weights of the particle and hole excitations, respectively, and Γ is the scattering rate. The particle–hole mixing implies that in the superconducting phase, the ARPES measurement can detect not only the spectral peaks of normal electrons below the Fermi surface, but also those above. According to Eq. (4.28), the photoelectric current is proportional to $f(\omega)A(\mathbf{k}, \omega)$. The Fermi distribution function $f(\omega)$ effectively suppresses the contribution of the $u_{\mathbf{k}}$-term, protruding the contribution of the $v_{\mathbf{k}}$-term. When the momentum $\mathbf{k}$ moves toward the Fermi surface from a point below the Fermi level, the peak energy grows with the increase of energy. The peak stops moving upward when the peak energy reaches a maximum at $|\omega| = |\Delta_{\mathbf{k}}|$ and the momentum $\mathbf{k}$ touches the normal state Fermi surface. Unlike in the normal state, the spectral peak does not disappear immediately when the momentum is further increased. Instead, the peak moves toward a lower energy. The peak disappears gradually as $\mathbf{k}$ moves further away from the Fermi surface.

Figure 4.2 shows the evolution of the energy distribution curves with momentum in both the normal and superconducting states of $Bi_2Sr_2CaCu_2O_{8+\delta}$ [51]. The momentum dependence of the spectra in the superconducting phase is quite different from that in the normal state. In the normal state, the spectral peak is invisible above the Fermi surface. However, in the superconducting phase, the spectral peaks exist on both sides of the Fermi momentum. The experimental results agree qualitatively with the theoretical description based on the picture of BCS quasiparticles. It implies that the particle–hole mixing induced by the superconducting pairing exists in high-T_c superconductors, similarly to conventional metal-based superconductors.

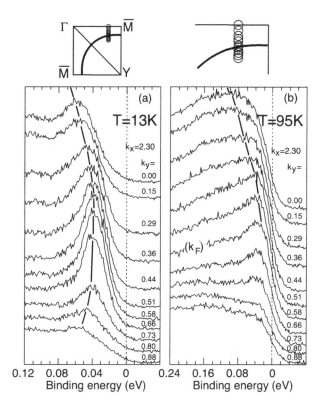

Figure 4.2 The ARPES spectral lines of the high-T_c cuprates $Bi_2Sr_2CaCu_2O_{8+\delta}$ at the momentum k marked by the circles in the figures above. (a) $T = 13$ K in the superconducting state, (b) $T = 95$ K in the normal state, (From Ref. [51])

The single-particle excitation spectra and the quasiparticle energy gap can be measured through ARPES. The pairing symmetry can be also determined by analyzing the momentum dependence of the superconducting gap on the Fermi surface. This is an advantage of ARPES. It plays an important role in the study of high-T_c cuprates.

There are two approaches to extract the values of superconducting gaps from ARPES. The simplest and most intuitive is to determine the gap directly from the relative shift in the leading edges of the ARPES spectra between the superconducting state and the normal state. This approach does not depend on the detailed formulism of quasiparticle spectra, and can be used even in the case where a comprehensive understanding of the superconducting quasiparticles is not available. It is based on this approach that Z. X. Shen et al. discovered the anisotropy in the high-T_c gap function [123]. In particular, they found that the ARPES spectra do not change much in the middle point of the leading edge in both the normal and superconducting states for $Bi_2Sr_2CaCu_2O_{8+\delta}$, along the diagonal direction of the Brillouin zone, indicating

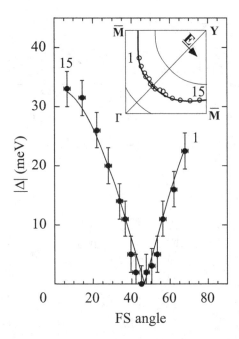

Figure 4.3 Angular dependence of superconducting energy gap on the Fermi surface of $Bi_2Sr_2CaCu_2O_{8+\delta}$ ($T_c = 87$ K) (From Ref. [122]). The bold solid line in the inset is the Fermi surface. The open circles represent the momenta measured by ARPES.

that there is no gap along that direction. Around $(0, \pi)$, on the other hand, the ARPES spectra in the normal and superconducting states behave very differently. In comparison with the normal state, the leading edge in the superconducting state clearly shifts down to lower energy, exhibiting a large energy gap [123]. Their results showed unambiguously that the gap function of high-T_c superconductors is strongly anisotropic, consistent with the $d_{x^2-y^2}$-wave symmetry.

A more quantitative approach is to directly extract the energy gaps by fitting the low-energy data using a spectral function which is obtained under proper assumptions and with a full consideration of the energy and momentum resolutions. Assuming the spectral function is phenomenologically determined by Eq. (4.32), Ding et al. analyzed the variation of the gap function with the momentum on the Fermi surface. They found that the momentum dependence of the gap function agrees well with the $d_{x^2-y^2}$-wave formulae $\Delta_{\mathbf{k}} = \Delta_0(\cos k_x - \cos k_y)$, as shown in Fig. 4.3 [122].

However, it should be emphasized that ARPES is not a phase sensitive measurement. It can only measure the magnitude but not the sign change of the $d_{x^2-y^2}$ gap function. In other words, it cannot distinguish a $d_{x^2-y^2}$ gap function from an anisotropic s-wave gap function which shows the same anisotropy as the d-wave gap but without changing sign on the whole Fermi surface.

4.5 Scattering Between Quasiparticles

The interplay between superconducting quasiparticles can lower the lifetime and change the behavior of quasiparticles, altering the thermodynamic and transport properties of superconductors. The quasiparticle lifetime can be probed by ARPES or through measurements of electric conductivity, thermal conductivity, or other transport coefficients. In general, the quasiparticle lifetime probed by electric conductivity is different from that probed by ARPES. The lifetime probed by ARPES is determined by the scattering between electrons. However, the electric conductivity, or resistivity, is determined by the current–current correlation function of electrons, which is described by a two-electron Green's function. As the total momentum of two electrons is conserved, the normal momentum-conserving scattering process between electrons has no contribution to the resistivity. Only the Umklapp scattering between electrons, which breaks the momentum conservation, contributes to the resistivity. The Umklapp scattering happens when the total momenta of two electrons before and after scattering differ by an integer multiple of the reciprocal lattice vector. In a d-wave superconductor, this condition of Umklapp scattering is generally difficult to fulfill and the Umklapp scattering is strongly suppressed. The quasiparticle transport lifetime is dominated by the impurity scatterings and thus is different from the single particle lifetime. It is important to understand this difference in the analysis of different experimental results for d-wave superconductors.

The interaction among quasiparticles can be generally expressed as,

$$H' = \frac{1}{2} \sum_{\mathbf{k_1 k_2 k_3 k_4}} V_{\mathbf{k_1 k_2 k_3 k_4}} \delta_{\mathbf{k_1 - k_2 + k_3 - k_4}, G} C^{\dagger}_{\mathbf{k_1}} \sigma_3 C_{\mathbf{k_2}} C^{\dagger}_{\mathbf{k_3}} \sigma_3 C_{\mathbf{k_4}}, \tag{4.33}$$

where $C_{\mathbf{k}}$ is the Nambu spinor fermion operator

$$C_{\mathbf{k}} = \begin{pmatrix} c_{\mathbf{k}\uparrow} \\ c^{\dagger}_{-\mathbf{k}\downarrow} \end{pmatrix}, \tag{4.34}$$

and G is the reciprocal lattice vector. $V_{\mathbf{k_1 k_2 k_3 k_4}} = V_{\mathbf{k_3 k_4 k_1 k_2}}$ is the interaction vertex function, which is assumed to be real for convenience.

The correction from the quasiparticle scattering to the Green's function is determined by the Dyson equation, Eq. (4.6). By neglecting the correction to the chemical potential under the second order perturbation, the self-energy function is given by

$$\Sigma\left(\mathbf{k},\omega_n\right) = \frac{2}{\beta^2} \sum_{\omega_m\omega_t\omega_s} \sum_{\mathbf{k}_2\mathbf{k}_3\mathbf{k}_4} V_{\mathbf{k}\mathbf{k}_2\mathbf{k}_3\mathbf{k}_4}\delta_{\mathbf{k}-\mathbf{k}_2+\mathbf{k}_3-\mathbf{k}_4,G}\delta\left(\omega_m+\omega_s-\omega_t-\omega_n\right)$$

$$\sigma_3 G^{(0)}\left(\mathbf{k}_2,\omega_m\right)\sigma_3\left[V_{\mathbf{k}_2\mathbf{k}_3\mathbf{k}_4\mathbf{k}}G^{(0)}\left(\mathbf{k}_3,\omega_t\right)\sigma_3 G^{(0)}\left(\mathbf{k}_4,\omega_s\right)\sigma_3\right.$$

$$\left.-V_{\mathbf{k}_4\mathbf{k}_3\mathbf{k}_2\mathbf{k}}\mathrm{Tr}G^{(0)}\left(\mathbf{k}_3,\omega_t\right)\sigma_3 G^{(0)}\left(\mathbf{k}_4,\omega_s\right)\sigma_3\right]. \tag{4.35}$$

It describes the process that a Bogoliubov quasiparticle with momentum $\mathbf{k}$ decays to momentum $\mathbf{k}_2$ by breaking a Cooper pair into two Bogoliubov quasiparticles with momenta $\mathbf{k}_3$ and $\mathbf{k}_4$.

Substituting the expression of $G^{(0)}$, (4.4), into (4.35), we obtain

$$\Sigma\left(\mathbf{k},\omega_n\right)$$

$$= \frac{2}{\beta^2} \sum_{\omega_t\omega_s} \sum_{\mathbf{k}_2\mathbf{k}_3\mathbf{k}_4} \frac{V_{\mathbf{k}\mathbf{k}_2\mathbf{k}_3\mathbf{k}_4}\left(i\omega_n+i\omega_t-i\omega_s+\xi_{\mathbf{k}_2}\sigma_3-\Delta_{\mathbf{k}_2}\sigma_1\right)}{\left[\left(i\omega_n+i\omega_t-i\omega_s\right)^2-E_{\mathbf{k}_2}^2\right]\left[\left(i\omega_t\right)^2-E_{\mathbf{k}_3}^2\right]\left[\left(i\omega_s\right)^2-E_{\mathbf{k}_4}^2\right]}$$

$$\left[V_{\mathbf{k}_2\mathbf{k}_3\mathbf{k}_4\mathbf{k}}\left(i\omega_t+\xi_{\mathbf{k}_3}\sigma_3+\Delta_{\mathbf{k}_3}\sigma_1\right)\left(i\omega_s+\xi_{\mathbf{k}_4}\sigma_3-\Delta_{\mathbf{k}_4}\sigma_1\right)\right.$$

$$\left.-2V_{\mathbf{k}_4\mathbf{k}_3\mathbf{k}_2\mathbf{k}}\left(-\omega_t\omega_s+\xi_{\mathbf{k}_3}\xi_{\mathbf{k}_4}-\Delta_{\mathbf{k}_3}\Delta_{\mathbf{k}_4}\right)\right]\delta_{\mathbf{k}-\mathbf{k}_2+\mathbf{k}_3-\mathbf{k}_4,G}. \tag{4.36}$$

The summation over ω_s and ω_t can be obtained using the standard method. The calculation is generally lengthy and tedious. At zero temperature, the result is rather simple and the imaginary part, which is proportional to the electron scattering rate τ^{-1}, reads

$$\mathrm{Im}\Sigma\left(\mathbf{k},\omega>0\right)$$

$$= \sum_{\mathbf{k}_2\mathbf{k}_3\mathbf{k}_4} \frac{\pi V_{\mathbf{k}\mathbf{k}_2\mathbf{k}_3\mathbf{k}_4}}{4E_{\mathbf{k}_2}E_{\mathbf{k}_3}E_{\mathbf{k}_4}}\delta_{\mathbf{k}-\mathbf{k}_2+\mathbf{k}_3-\mathbf{k}_4,G}\delta\left(\omega-E_{\mathbf{k}_2}-E_{\mathbf{k}_3}-E_{\mathbf{k}_4}\right)$$

$$\left(E_{\mathbf{k}_2}+\xi_{\mathbf{k}_2}\sigma_3-\Delta_{\mathbf{k}_2}\sigma_1\right)\left[2V_{\mathbf{k}_4\mathbf{k}_3\mathbf{k}_2\mathbf{k}}\left(-E_{\mathbf{k}_3}E_{\mathbf{k}_4}+\xi_{\mathbf{k}_3}\xi_{\mathbf{k}_4}-\Delta_{\mathbf{k}_3}\Delta_{\mathbf{k}_4}\right)\right.$$

$$\left.-V_{\mathbf{k}_2\mathbf{k}_3\mathbf{k}_4\mathbf{k}}\left(-E_{\mathbf{k}_3}+\xi_{\mathbf{k}_3}\sigma_3+\Delta_{\mathbf{k}_3}\sigma_1\right)\left(E_{\mathbf{k}_4}+\xi_{\mathbf{k}_4}\sigma_3-\Delta_{\mathbf{k}_4}\sigma_1\right)\right]. \tag{4.37}$$

In an isotropic s-wave superconductor, there is a finite gap in the quasi-particle excitation spectrum, $E_{\mathbf{k}} > \Delta$. Equation (4.37) shows that the quasiparticle scattering has finite contribution to $\mathrm{Im}\Sigma$ only when $\omega > 3\Delta$. Hence, there is a threshold in the excitation energy of quasiparticles being scattered, $\omega = 3\Delta$, due to the opening of the pairing gap. This is because at zero temperature, the quasiparticle scattering needs to break a Cooper pair, and the quasiparticles after scattering can survive only when their energies are above the gap.

In a d-wave superconductor, the energy dependence of the scattering rate τ^{-1} behaves differently. In the low-energy limit, the momentum summation in Eq. (4.37) needs to be done just around the nodal region, in which the momentum dependence of the quasiparticle energy $E_{\mathbf{k}}$ is linear. In the absence of the Umklapp scattering, i.e. $G = 0$, one can do a linear transformation to separate ω from the summation.

If the interaction matrix elements $V_{k_1 k_2 k_3 k_4}$ are not strongly momentum dependent, one can show from dimensional analysis that the scattering rate scales cubically with ω [124]

$$\tau^{-1} = D_\tau \omega^3. \tag{4.38}$$

The ω-independent coefficient D_τ can be obtained by integrating over the momenta in Eq. (4.37). Compared with the energy dependence of τ^{-1} in the normal metal, $\tau^{-1} \sim \omega^2$, the scattering rate in the d-wave superconductors has one more power in ω. This extra ω comes from the linear density of states of quasiparticles.

At finite temperatures, the quasiparticle scattering rate can be obtained from Eq. (4.36). In the limit $\omega \to 0$, T (or more precisely $k_B T$) is the only parameter that has the dimension of energy. In this case, it is simple to show by dimensional analysis that, similar to Eq. (4.38), the quasiparticle lifetime scales cubically with temperature in the low temperature limit [125]

$$\tau^{-1} \sim T^3. \tag{4.39}$$

Thus the quasiparticle scattering rate τ^{-1} scales as ω^3 at zero temperature and as T^3 in the zero frequency limit. This is consistent with the temperature and frequency dependencies of the scattering rate obtained from ARPES and thermal conductivity measurements [126–128]. It shows that the low-energy scattering between quasiparticles is an important channel of scattering and should be considered seriously in the analysis of low-energy behaviors of quasiparticles.

However, the quasiparticle lifetime probed by the microwave conductivity does not equal that determined by the usual electron–electron scattering. The formulae $\tau^{-1} \sim T^3$ cannot be used to explain the microwave conductance measurements. Instead, the Umklapp scattering with $G \neq 0$ should be considered to account for the contribution of electron–electron scattering to the electric conductance.

In a d-wave superconductor, quasiparticles near the gap nodes generally do not satisfy the condition of Umklapp scattering. The Umklapp scattering emerges only when the total energy of two quasiparticles becomes larger than a threshold Δ_U. Hence the Umklapp scattering is thermally activated, and the quasiparticle scattering rate should scale exponentially with temperature [125]

$$\tau_t^{-1} \sim e^{-\Delta_U / k_B T}. \tag{4.40}$$

This exponential behavior of τ_t^{-1} agrees qualitatively with the measurement data of microwave conductivity for high-T_c superconductors [125, 129].

5

Tunneling Effect

5.1 Interface Scattering

Tunneling through a barrier is an important quantum phenomenon. It plays an important role in the study of microscopic mechanism of superconductivity. A superconducting junction is formed by a superconductor separated by a thin insulating layer or vacuum from a normal metal or another superconductor. Electric current flows across the junction under a small bias voltage. By measuring the electric current response to the applied bias, or the differential conductance, one can extract information on the density of states of superconducting quasiparticles as well as their interactions.

The contact between a superconductor and a metal or an insulating layer or film in a tunneling junction can be either in a face-to-face or in a point-to-face configuration. For example, the contact between the tip of a scanning tunneling microscope (STM) and the surface of a superconductor is of the point-to-face type. For a face-to-face contact, the tunneling current is distributed throughout the whole interface, and the differential conductance measures the average density of states of superconducting quasiparticles at the interface. For a point-to-face contact, the tunneling current concentrates around the probe. It measures the local density of states of superconducting quasiparticles around the contact. By scanning the tip across the entire sample, one can obtain the spatial distribution of the local density of states.

The tunneling effect of normal electrons relies strongly on the property of the tunneling interface. For an ideal contact between a metal and a superconductor, the interface can be approximately treated as an elastic scattering barrier. In this case, the tunneling effect can be understood by studying an elastic scattering problem of electrons at the interface. However, the scattering at the surface of a superconductor is more complicated than that at the surface of a normal metal. Besides the conventional reflection and transmission of normal electrons, there are also reflection of normal holes and transmission of hole quasiparticles generated by an off-diagonal scattering potential in the superconducting state. The reflection of holes is a peculiar feature of reflection on the surface of superconductors that was first pointed out by Andreev. This kind of reflection is now called the Andreev reflection [130]. It is important when the surface tunneling barrier is low. When the tunneling barrier is high, Andreev reflection is suppressed and the normal reflection of electrons becomes more important.

In the study of Andreev reflection and transmission of holes at the surface of superconductor, the semi-classical WKB approximation is usually adopted [131–133]. This approximation simplifies the steps in solving the self-consistent gap equation around the surface. However, it is not necessary to take this approximation if the self-consistency of the gap function is not strictly required.

5.1.1 Andreev Reflection

Let us investigate the electron scattering on the surface of a superconductor using the Hamiltonian defined by Eq. (3.71). We assume that the interface is perpendicular to the x-direction, and the scattering potential $U(x)$ depends only on x and is finite just in the vicinity of the interface. In addition, we assume that the system is translation invariant in the direction parallel to the interface and that the momentum parallel to the interface is conserved.

The Andreev reflection and transmission of holes on the surface of a supercon-ductor results from the superconducting condensation, which breaks the number conservation of normal electrons in the superconducting state. In the BCS theory, up-spin electrons are hybridized with down-spin holes. Thus both electrons and holes with opposite spins can be scattered on the superconductor surface. Figure 5.1 shows schematically the reflection and transmission wave vectors for both electrons and holes.

The Andreev reflection differs from the normal reflection of electrons, and leads to different physical effects. If the incident electron has an energy higher than the superconducting gap, it transmits almost completely through the surface barrier and becomes a quasiparticle propagating in the superconductor. In this case, the Andreev reflection is very weak and weakly affects the transmission current. In contrast, if the energy of the incident electron is smaller than the superconducting gap, the electron cannot transmit through the barrier and propagate as a quasiparticle. Nevertheless, if the incident electron captures another electron with opposite momentum to form a Cooper pair at the interface, they can cotunnel into the superconductor and become a pair of condensed electrons, doubling the transmission current. Thus the trans-mission wave contains a pair of electrons with zero total momentum. Considering the charge and momentum conservation, this implies that there must be a reflection wave of holes with a momentum nearly opposite to the incident wave vector on

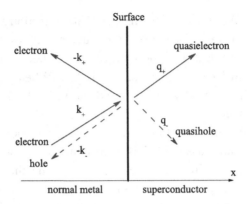

Figure 5.1 The electron and hole reflection and transmission at the superconductor surface. The incident electron is from the normal metal side, and the reflection and transmission waves can be either electron- or hole-like. The directions of the probability currents for the electron and hole are marked with arrows. The electric current directions are opposite to those of the probability currents of the hole-type reflection and transmission waves.

the normal metal side. Hence the Andreev reflection arises from the superconducting pairing of the transmitting electron with another electron at the interface. The ordinary electron reflection suppresses the transmission probability, while the Andreev reflection enhances the transmission probability because it corresponds to the transmission of a Cooper pair.

In the ordinary reflection, the momenta of incident and reflection electrons are conserved along the direction parallel to the interface, and take opposite values along the direction normal to the interface. In the Andreev reflection, however, the momentum of the reflected hole parallel to the interface equals the corresponding momentum of the incident electron but in the opposite direction. The momenta of the incident electron and the reflected hole normal to the interface have opposite sign, but their magnitudes are not equal to each other. Thus in the Andreev reflection, the momentum of the refection hole is different from the incident electron. But their energies with respect to the Fermi surface are the same, which is a consequence of elastic scattering.

Incident and Reflection Wavefunction

Due to the translation invariance along the yz-plane, the momentum along this plane is a good quantum number, denoted as $\mathbf{k}_\parallel$. However, the momentum along the x-direction is not conversed. On the normal metal side, the wavefunction is a superposition of the plane waves for the incident electron, the reflected electron, and the Andeev reflection hole. If an up-spin electron is emitted from $x = -\infty$ to the interface, the wavefunction on the normal metal side is then given by

$$\psi\left(\mathbf{r}\right) = e^{i\mathbf{k}_{\parallel}\cdot\mathbf{r}_{\parallel}} \begin{pmatrix} e^{ik_{+}x} + be^{-ik_{+}x} \\ ae^{ik_{-}x} \end{pmatrix}, \tag{5.1}$$

where k_{+} and k_{-} are the wave vectors of the incident electron and the Andreev reflection hole along the x-axis, respectively. The coefficient of the incident electron wavefunction is normalized to 1. a and b are the coefficients of the Andreev hole and normal electron reflections, respectively.

This wavefunction can be also written in a second quantized form as

$$|\psi\rangle = \left(c^{\dagger}_{k_{+},\mathbf{k}_{\parallel},\uparrow} + bc^{\dagger}_{-k_{+},\mathbf{k}_{\parallel},\uparrow} + ac_{-k_{-},-\mathbf{k}_{\parallel},\downarrow} \right) |0\rangle, \tag{5.2}$$

where $|0\rangle$ is the Fermi sea. In the coordinate representation, $|\psi\rangle$ becomes

$$\psi\left(\mathbf{r}\right) = e^{i\mathbf{k}_{\parallel}\cdot\mathbf{r}_{\parallel}} \left[\left(e^{ik_{+}x} + be^{-ik_{+}x} \right) c^{\dagger}_{\mathbf{r}\uparrow} + ae^{ik_{-}x} c_{\mathbf{r}\downarrow} \right] |0\rangle. \tag{5.3}$$

Creating a spin-up electron with momentum $\mathbf{k}$ in the Fermi sea is equivalent to annihilating a spin-down electron with momentum $-\mathbf{k}$.

The incident and reflection momenta of the electron are $(k_{+}, \mathbf{k}_{\parallel})$ and $(-k_{+}, \mathbf{k}_{\parallel})$, respectively. The momentum of the Andreev reflection hole, on the other hand, is $(-k_{-}, -\mathbf{k}_{\parallel})$. Given the values of energy $E > 0$ and the momentum parallel to the interface $\mathbf{k}_{\parallel}$ for the incident electron, $k_{\pm}$ are determined by the equation

$$E_{e} = E_{h} = E, \tag{5.4}$$

where

$$E_{e} = \frac{\hbar^{2}}{2m}\left(k_{+}^{2} + k_{\parallel}^{2}\right) - \mu, \tag{5.5}$$

$$E_{h} = \mu - \frac{\hbar^{2}}{2m}\left(k_{-}^{2} + k_{\parallel}^{2}\right), \tag{5.6}$$

are the excitation energies of electrons and holes in the normal metal. $k_{\parallel}$ is the amplitude of $\mathbf{k}_{\parallel}$. Substituting these expressions into Eq. (5.4), we find that

$$k_{\pm} = \sqrt{\frac{2m(\mu \pm E)}{\hbar^{2}} - k_{\parallel}^{2}}. \tag{5.7}$$

Transmission Wavefunction

On the superconducting side, the wavefunction of electrons is a plane-wave in the Nambu spinor representation,

$$\psi(\mathbf{r}) = e^{i\mathbf{q}\cdot\mathbf{r}} \begin{pmatrix} u_{\mathbf{q}} \\ v_{\mathbf{q}} \end{pmatrix}, \tag{5.8}$$

where $\mathbf{q}$ is the wave vector of superconducting quasiparticles. Since the momentum parallel to the interface is conserved, we have $\mathbf{q}_\parallel = \mathbf{k}_\parallel$. The coefficients $u_\mathbf{q}$ and $v_\mathbf{q}$ are determined by the BCS mean-field equations

$$\begin{pmatrix} \xi_\mathbf{q} & \Delta_\mathbf{q} \\ \Delta_\mathbf{q} & -\xi_\mathbf{q} \end{pmatrix} \begin{pmatrix} u_\mathbf{q} \\ v_\mathbf{q} \end{pmatrix} = E \begin{pmatrix} u_\mathbf{q} \\ v_\mathbf{q} \end{pmatrix}, \tag{5.9}$$

where

$$\xi_\mathbf{q} = \frac{\hbar^2}{2m} \mathbf{q}^2 - \mu, \tag{5.10}$$

and $\Delta_\mathbf{q}$ is the gap function.

Given E, $\mathbf{q}$ is determined by the eigenenergy of Eq. (5.9)

$$\xi_\mathbf{q}^2 + \Delta_\mathbf{q}^2 = E^2, \tag{5.11}$$

and $(u_\mathbf{q}, v_\mathbf{q})$ is given by Eq. (1.30) and Eq. (1.31). The associated creation operator of Bogoliubov quasiparticle is defined by

$$\gamma_{1,\mathbf{q}}^\dagger = u_\mathbf{q} c_{\mathbf{q}\uparrow}^\dagger + v_\mathbf{q} c_{-\mathbf{q}\downarrow}. \tag{5.12}$$

Applying $\gamma_{1,\mathbf{q}}$ to the BCS ground state creates a quasiparticle excitation of energy E. The matrix on the left-hand side of Eq. (5.9) has another eigenvalue $-E$, and the corresponding annihilation operator of superconducting quasiparticles is defined by

$$\gamma_{2,\mathbf{q}} = -v_\mathbf{q} c_{\mathbf{q}\uparrow}^\dagger + u_\mathbf{q} c_{-\mathbf{q}\downarrow}. \tag{5.13}$$

Applying $\gamma_{2,\mathbf{q}}^\dagger$ to the BCS ground state also creates a quasiparticle excitation of energy E. But the spin of $\gamma_{2,\mathbf{q}}^\dagger$ is orthogonal to the wavefunction of the incident electron, and does not need to be considered.

For the s- or d-wave superconductor, $|\Delta(q_x, \mathbf{k}_\parallel)| = |\Delta(-q_x, \mathbf{k}_\parallel)|$. In these cases, if $q_x > 0$ is a solution to the equation, so is $-q_x$. Similarly to the normal metal side, for each energy E, $|q_x|$ can take two different values, which are larger and smaller than the Fermi wave vector, respectively. These two values of $|q_x|$, if denoted by q_+ and q_-, satisfy the inequalities

$$q_+ > \sqrt{\frac{2m\mu}{\hbar^2} - k_\parallel^2}, \qquad q_- < \sqrt{\frac{2m\mu}{\hbar^2} - k_\parallel^2}. \tag{5.14}$$

Since the solutions corresponding to $\pm q_x$ are degenerate, there are four solutions for a given energy E. The solutions of $q_x = \pm q_+$ correspond to the electron-type excitations with the probability flux along $\pm x$-direction, and those of $q_x = \pm q_-$ correspond to the hole-type excitations with the probability flux along $\mp x$ direction. It should be emphasized that the direction of the probability flux of hole-type excitations is antiparallel to its electric current. For the scattering problem discussed here, it is sufficient just to consider the solution with the probability flux along the $+x$ direction. At the mean-field level, the quasiparticle probability is conserved.

Thus inside the superconductor, the transmission wave is a superposition of two solutions of $\mathbf{q} = (q_+, \mathbf{k}_\parallel)$ and $\mathbf{q} = (-q_-, \mathbf{k}_\parallel)$ and can be cast into the form

$$\psi(\mathbf{r}) = e^{i\mathbf{k}_\parallel \cdot \mathbf{r}_\parallel} \left[c e^{iq_+ x} \begin{pmatrix} u_+ \\ v_+ \end{pmatrix} + d e^{-iq_- x} \begin{pmatrix} u_- \\ v_- \end{pmatrix} \right], \qquad (5.15)$$

where c and d are the coefficients of two transmission waves.

If Δ_q does not depend on the value of $q_\pm$, such as in an isotropic s-wave superconductor, we have

$$\begin{pmatrix} u_- \\ v_- \end{pmatrix} = - \begin{pmatrix} v_+ \\ u_+ \end{pmatrix}. \qquad (5.16)$$

On the other hand, if $\Delta(-q_-, \mathbf{k}_\parallel) = -\Delta(q_+, \mathbf{k}_\parallel)$, we have

$$\begin{pmatrix} u_- \\ v_- \end{pmatrix} = \begin{pmatrix} -v_+ \\ u_+ \end{pmatrix}. \qquad (5.17)$$

In order to determine the reflection and transmission coefficients, we need to know the detailed shape of the scattering potential. For an arbitrary scattering potential, it is difficult to obtain an analytic solution. However, if the scattering potential $U(x)$ is simply a δ-function, we can obtain a relatively simple analytic solution.

5.2 Tunneling Conductance

In order to calculate the tunneling current of electrons at a metal–superconductor interface, let us evaluate the current contributed by the incident and reflected electrons and the Andreev reflection holes on the normal metal side. The tunneling current of normal electrons is determined by the difference between the currents with and without applying an external bias voltage to the tunneling junction.

In the normal metal, the electric current operator is defined by

$$\hat{J}_x = -e \sum_{\mathbf{k}\sigma} v_x(\mathbf{k}) c_{\mathbf{k}\sigma}^\dagger c_{\mathbf{k}\sigma}, \qquad (5.18)$$

where $v_x(\mathbf{k}) = \hbar k_x / m$ is the velocity of electrons along the x-axis. The wavefunction of an electron which is incident from $x \to -\infty$ and reflected on the superconductor surface is described by Eq. (5.2). Its contribution to the electric current is given by

$$J_x(k_+, \mathbf{k}_\parallel) = \langle \psi | \hat{J}_x | \psi \rangle = -\frac{e\hbar}{m} \left[k_+ \left(1 - |b|^2 \right) + k_- |a|^2 \right]. \qquad (5.19)$$

Here the vacuum is the electron Fermi sea and $|\psi\rangle$ must be normalized according to the incident electron wavefunction according to the boundary condition of the tunneling problem.

In the absence of an external voltage, an electron or hole can transmit from the normal metal side to the superconductor side, and vice versa. Thus the net current vanishes. However, after applying a voltage V, the motion of electrons in the two directions is no longer balanced, and the tunneling current becomes finite.

Upon applying an external voltage V, the chemical potential of the normal metal is increased by eV in comparison with that of the superconductor. If the quasiparticle eigenstates in the superconductor side are not affected by the applied voltage, the contribution from the electrons or holes transmitted from the superconducting state to the current in the normal metal side is not changed before and after applying the voltage. The net current density is therefore determined by the difference between the electric current in the normal metal with and without the external voltage, described by the formula

$$I_x = \frac{2e\hbar}{m} \int' \frac{d\mathbf{k}}{(2\pi)^3} [f(E - eV) - f(E)] \left[k_+ \left(1 - |b|^2\right) + k_-|a|^2\right], \quad (5.20)$$

where the factor 2 results from the spin degeneracy, and $\int' d\mathbf{k}$ represents the summation over permitted incident and reflection momenta that are determined by Eq. (5.7). The above analysis neglects the correction of the applied voltage to the microscopic electronic structure, which is a nonlinear effect. Hence Eq. (5.20) is just a result of the approximation of the linear response.

Equation (5.20) is consistent with the results obtained by Blonder–Tinkham–Klapwijk (BTK) [134]. But the derivation here is simpler. It allows us to see more clearly how good and reliable this formula can be used in the analysis of tunneling effect. In BTK's original derivation, the assumption of $k_+ = k_- = k_F$ is adopted. Equation (5.20) holds more generally than the formula derived by BTK.

Equation (5.20) is a general formula for the tunneling current. It is valid independent on the shape of the interface scattering potential, provided the scattering is elastic, which preserves the energy as well as the momenta parallel to the interface. However, if the surface barrier is very high or the scattering potential is very strong, it may not always be convenient and necessary to solve the Schrödinger equation to obtain the reflection coefficients a and b. In this case, as will be discussed in §5.5, it is more convenient to treat approximately the scattering potential by an energy-independent tunneling matrix.

The differential conductance of the tunneling junction is given by the derivative of the tunneling current with respect to the applied bias

$$G(V) = \frac{dI_x}{dV} = -\frac{2e^2\hbar}{m} \int' \frac{d\mathbf{k}}{(2\pi)^3} \frac{df(E - eV)}{dE} \left[k_+ \left(1 - |b|^2\right) + k_- |a|^2\right]$$

$$= \frac{2e^2\hbar}{m} \int' \frac{d\mathbf{k}}{(2\pi)^3} \delta(E - eV - \mu) \left[k_+ \left(1 - |b|^2\right) + k_- |a|^2\right]. \tag{5.21}$$

At zero temperature, it becomes

$$G(V) = \frac{2e^2}{(2\pi)^3\hbar} \int' d\mathbf{k}_\parallel \left[1 - |b(eV)|^2 + \frac{k_-(eV)}{k_+(eV)}|a(eV)|^2\right], \tag{5.22}$$

where the scattering coefficients and $k_\pm$ take values at $E = eV$. The $\int'$ represents integration only over permitted incoming and reflecting angles.

In order to compare with the tunneling conductance in a metal–metal tunneling junction, we define the normalized differential conductance

$$\sigma(V) = \frac{G(V)}{G(V \to \infty)}. \tag{5.23}$$

In the high bias limit $V \to \infty$, the effect of superconducting pairing potential can be neglected, and $G(\infty)$ contributes purely from normal electrons. The deviation of $\sigma(V)$ from constant 1 results from the effect of superconductivity.

5.3 Scattering from the δ-Function Interface Potential

If the scattering potential $U(x)$ is a δ-function,

$$U(x) = U\delta(x), \tag{5.24}$$

the boundary condition is relatively easy to handle and an analytic solution for the scattering coefficients can be obtained. From the solution, a great deal of useful information about the tunneling effect can be extracted, allowing us to understand more transparently the scattering effect. The conclusion thus obtained can also be applied qualitatively to other tunneling systems with more general scattering potentials.

Around the interface, the superconducting gap parameter gradually decays to zero from the superconductor to the normal metal side, within a characteristic length scale of the order of superconducting coherence length ξ. In order to rigorously solve the scattering problem of electrons at the superconducting surface, we need to solve the BdG equation self-consistently. This is difficult, and generally can only be done numerically. Nevertheless, if the coherence length ξ is very short, a natural approximation is to take the limit $\xi \to 0$. In this limit, we can neglect the small variation of the superconducting gap function at the interface, and assume the gap function to be entirely zero in the normal metal side and finite and uniform in the superconductor side. This is a common approximation used in the study of tunneling effect of superconducting quasiparticles, which can greatly simplify calculations.

The electron wavefunctions in the normal metal and superconductor sides are given by Eq. (5.1) and (5.15), respectively. At the interface, the continuous conditions for the wavefunction and its derivative are determined by the equations

$$\psi\left(x = 0^+\right) - \psi\left(x = 0^-\right) = 0, \tag{5.25}$$

$$\frac{\hbar^2}{2m}\psi'_x(x = 0^-) - \frac{\hbar^2}{2m}\psi'_x(x = 0^+) + U\psi(x = 0) = 0. \tag{5.26}$$

Based on these equations, we can further obtain the equations that determine the scattering coefficients

$$\begin{pmatrix} 0 & -1 & u_+ & u_- \\ -1 & 0 & v_+ & v_- \\ 0 & iw + k_+ & q_+u_+ & -q_-u_- \\ iw - k_- & 0 & q_+v_+ & -q_-v_- \end{pmatrix} \begin{pmatrix} a \\ b \\ c \\ d \end{pmatrix} = \begin{pmatrix} 1 \\ 0 \\ k_+ - i\omega \\ 0 \end{pmatrix}. \tag{5.27}$$

The solution for this set of linear equations is

$$\begin{pmatrix} a \\ b \\ c \\ d \end{pmatrix} = \frac{1}{p} \begin{pmatrix} -2k_+ (q_+ + q_-) v_+v_- \\ 2k_+(w_-v_-u_+ - w_+v_+u_-) - p \\ 2k_+w_-v_- \\ -2k_+w_+v_+ \end{pmatrix}, \tag{5.28}$$

where

$$w = \frac{2mU}{\hbar^2}, \tag{5.29}$$

$$w_+ = iw - k_- + q_+, \tag{5.30}$$

$$w_- = iw - k_- - q_-, \tag{5.31}$$

$$p = w_-(iw + k_+ + q_+)u_+v_- - w_+(iw + k_+ - q_-)v_+u_-. \tag{5.32}$$

From Eq. (5.20), it is clear that the tunneling current depends strongly on the incident direction. The contribution to the tunneling current from incident electrons with momenta almost parallel to the interface is very small, while that from incident electrons perpendicular to the interface is large. Thus for most problems related to the electron tunneling, we only need to consider the situation in which the incident electron kinetic energy $\hbar^2k_+^2/2m \gg E$. In this case, the deviations of the x-components of the momenta of the reflection and transmission electrons from the Fermi momentum are small, and can be approximately expressed as

$$k_\pm \approx k_{F_x}(1 \pm \delta_k), \tag{5.33}$$

$$q_\pm \approx k_{F_x}(1 \pm \delta_q), \tag{5.34}$$

where

$$\delta_k = \frac{mE}{\hbar^2 k_{F_x}^2},$$ (5.35)

$$\delta_q = \frac{m\sqrt{E^2 - \Delta_{\mathbf{k}_F}^2}}{\hbar^2 k_{F_x}^2},$$ (5.36)

and $\Delta_{\mathbf{k}_F}$ is the gap value at $\mathbf{k} = \mathbf{k}_F = (k_{F_x}, \mathbf{k}_{\parallel})$. Using these expressions, the scattering coefficients can be approximately written as

$$\begin{pmatrix} a \\ b \\ c \\ d \end{pmatrix} \approx \frac{1}{\gamma} \begin{pmatrix} 4(1 + \delta_k)v_+v_- \\ \delta_k u_+ v_- - (2\delta_k - \eta + 2)\eta(u_+v_- - u_-v_+) \\ 2(2 - \eta)(1 + \delta_k)v_- \\ 2\eta(1 + \delta_k)v_+ \end{pmatrix},$$ (5.37)

where

$$Z = \frac{2mU}{\hbar^2 k_F},$$ (5.38)

$$\gamma = 4u_+v_- - \eta^2(u_+v_- - u_-v_+),$$ (5.39)

$$\eta = iZ + \delta_q + \delta_k.$$ (5.40)

If the incident direction is nearly normal to the superconductor surface, k_{F_x} is almost equal to the Fermi wave vector k_F. In this case, δ_k and δ_q can be neglected, and the above result can be further simplified as

$$\begin{pmatrix} a \\ b \\ c \\ d \end{pmatrix} \approx \frac{1}{\gamma} \begin{pmatrix} 4v_+v_- \\ -(Z^2 + 2iZ)(u_+v_- - u_-v_+) \\ 2(2 - iZ)v_- \\ 2iZv_+ \end{pmatrix},$$ (5.41)

where $\gamma \approx 4u_+v_- + Z^2(u_+v_- - u_-v_+)$. This is just the result obtained by utilizing the semi-classical WKB approximation [132, 133] for the Andreev reflection. It is simply to verify that this solution satisfies the equation

$$|b|^2 + |c|^2 - |a|^2 - |d|^2 = 1,$$ (5.42)

which is just the current conservation equation along the x-direction. This is because the current is proportional to the product of the wave vector and the magnitude square of the scattering coefficients, and under the approximation $k_{F_x} \approx k_F$, the x-direction wave vectors for the reflection and transmission waves are all equal to k_{F_x}.

In the limit that the δ-function potential vanishes, i.e. $U = 0$, the scattering coefficients determined by Eq. (5.41) possess the following two features: (1) $b = d = 0$, namely there are only Andreev hole reflection and electron transmission, but no electron reflection or hole transmission; (2) The Andreev hole reflection coefficient has a simple form, $a = v_+/u_+$. Furthermore, when $E < |\Delta_{q_+}|$, then $\xi_{q_+} = \sqrt{E^2 - \Delta_{q_+}^2}$ is purely imaginary. As u_+ equals the complex conjugate of v_+, we have $|a| = 1$. In this case, the Andreev reflection doubles the electric current. Of course this is only an ideal case. When $U \neq 0$, the enhancement of the electric current due to the Andreev reflection can be either smaller or larger than 1.

The differential conductance of the metal–superconductor tunneling junction can be evaluated using Eq. (5.21) and the scattering coefficients derived above. To simplify the calculation, we only consider the limit where the incident direction is normal to the superconductor surface and the Fermi wave vector $k_F \to \infty$. In this case, the scattering coefficients are given by Eq. (5.37) and the zero temperature differential conductance can be expressed as

$$G(V) = \frac{2e^2}{(2\pi)^3 \hbar} \int' d\mathbf{k}_\| g(eV), \tag{5.43}$$

where

$$g(E) = 1 - |b(E)|^2 + |a(E)|^2. \tag{5.44}$$

The b-term contributes from the reflection wave. The a-term results from the Andreev scattering which enhances the current by the reflection of holes.

In the limit $E \to \infty$, the effect of superconducting pairing is very small and it can be shown that

$$g(\infty) = \frac{4}{4 + Z^2}, \tag{5.45}$$

independent of the momentum and energy of scattering electrons, and of the pairing symmetry.

For an isotropic s-wave superconductor, $\Delta_q = \Delta$ and $v_\pm = -u_\mp$. Then $u_\pm$ and $v_\pm$ depend only on E, but not on the component of the incident momentum parallel to the surface $\mathbf{k}_\|$. In this case, it can be shown that $g(E)$ is also independent of momentum $\mathbf{k}_\|$, and the tunneling conductance is proportional to $g(eV)$, i.e. $G(V) \propto g(eV)$.

The energy dependence of $g(E)$ is given by

$$g(E) = \begin{cases} \dfrac{4E}{2E + (2 + Z^2)\sqrt{E^2 - \Delta^2}} & \text{if } E > \Delta \\[4mm] \dfrac{8\Delta^2}{4E^2 + (2 + Z^2)^2(\Delta^2 - E^2)} & \text{if } E < \Delta \end{cases}. \tag{5.46}$$

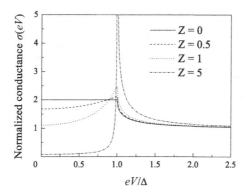

Figure 5.2 Normalized differential conductance $\sigma(V)$ for an isotropic s-wave superconductor tunneling junction.

From this expression, the bias dependence of the differential conductance can be obtained. Figure 5.2 shows the normalized differential conductance as a function of applied bias voltage at several different scattering potentials for the isotropic s-wave superconductor.

In the case of $Z = 0$, the conductance is independent of the bias voltage at $eV < \Delta$ and the tunneling current is twice the corresponding value in a metal–metal junction. This enhancement of electric currents in the small voltage is a characteristic feature of the Andreev reflection, resulting from the superconducting pairing of transmission electrons. At high bias $eV > \Delta$, $\xi_{q+} > 0$, then $|a|$ decreases with the increase of V and vanishes in the limit $V \to \infty$.

With the increase of scattering potential, the transmission probability decreases, and the normalized zero-bias differential conductance is given by

$$\sigma(0) = \frac{8 + 2Z^2}{(2 + Z^2)^2}. \tag{5.47}$$

When the bias voltage equals the gap value, σ increases with increasing Z,

$$\sigma(V = \Delta/e) = \frac{4 + Z^2}{2}. \tag{5.48}$$

In the limit of $Z \to \infty$, $\sigma(V)$ scales as

$$\sigma(V) = \begin{cases} \dfrac{eV}{\sqrt{e^2V^2 - \Delta^2}} & \text{if } eV > \Delta, \\[2mm] \dfrac{Z^2}{2} & \text{if } eV = \Delta, \\[2mm] \dfrac{2\Delta^2}{Z^2(\Delta^2 - e^2V^2)} & \text{if } eV \ll \Delta. \end{cases} \tag{5.49}$$

For a d-wave superconductor, the tunneling current depends on the orientation of the metal–superconductor interface. In the following, we only discuss the tunneling conductance when the x-axis is parallel to the nodal direction of the superconducting gap function. When the x-axis is parallel to the antinodal direction, i.e. the direction of the maximal gap, the differential conductance is qualitatively the same as for an isotropic s-wave superconductor.

When the x-axis is parallel to the nodal direction of the d-wave superconductor, the gap function is given by

$$\Delta_k = \Delta \tilde{k}_x \tilde{k}_y, \tag{5.50}$$

where $\tilde{k}_{x,y} = k_{x,y}/k_F$ is the normalized momentum. When the incident direction is nearly parallel to the x-axis, $|\tilde{k}_x| \approx 1$, the gap parameter is $\Delta_{q_+} \approx \Delta \tilde{k}_y$ for the quasielectron along the transmission direction and $\Delta_{q_-} \approx -\Delta \tilde{k}_y$ for the quasi-hole. Based on this observation, we have $v_{\pm} = \mp u_{\pm}$, and,

$$g(E) = \begin{cases} \dfrac{8E^2 + 4E(2 + Z^2)\sqrt{E^2 - \Delta^2 \tilde{k}_y^2}}{\left[(2 + Z^2)E + 2\sqrt{E^2 - \Delta^2 \tilde{k}_y^2}\right]^2}, & \text{if } E > |\Delta \tilde{k}_y| \\[6mm] \dfrac{8\Delta^2 \tilde{k}_y^2}{(2 + Z^2)^2 E^2 + 4(\Delta^2 \tilde{k}_y^2 - E^2)}, & \text{if } E < |\Delta \tilde{k}_y| \end{cases} \tag{5.51}$$

$G(V)$ is obtained by integrating $g(E)$ over the direction of $\tilde{k}_y$. By dividing $G(V \to \infty)$, we obtain the normalized differential conductance $\sigma(V)$. Figure 5.3 shows $\sigma(V)$ as a function of the bias voltage V for the d-wave superconductor at

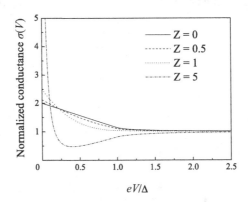

Figure 5.3 Normalized differential conductance $\sigma(V)$ for a d-wave superconductor tunneling junction with the nodal direction parallel to the x-axis.

several different scattering potentials. Unlike in an *s*-wave superconductor, $\sigma(V)$ increases as Z increases in the low bias limit. Furthermore, at $E = 0$, $g(0) = 2$ independent of $\tilde{k}_y$, thus the differential conductance at zero bias is given by

$$\sigma(V = 0) = \frac{4 + Z^2}{2}, \tag{5.52}$$

which diverges in the limit $Z \to \infty$.

This divergence is a manifestation of the surface resonance states of the *d*-wave superconductor. It provides a criterion to judge by experimental measurements whether there are gap nodes along the direction normal to the superconductor surface. The zero-energy resonance states appear when the denominator of the scattering coefficient given in (5.51) becomes zero, namely

$$(2 + Z^2)E + 2\sqrt{E^2 - \Delta^2 \tilde{k}_y^2} = 0. \tag{5.53}$$

This equation does not have real solutions, but it has a complex solution, given by

$$E = -\frac{2i\,\Delta\tilde{k}_y}{\sqrt{Z^4 + 4Z^2}}. \tag{5.54}$$

The real part of this complex solution is zero, thus the resonance states appear exactly at zero energy. The inverse of the imaginary part is the lifetime of the surface resonance state. When Z is finite, each zero energy surface state couples to the metallic continuum and becomes a resonance state with a finite lifetime. In the limit $Z \to \infty$, the zero energy surface state is asymptotically decoupled from the metallic states and becomes a sharp resonance state with an infinite lifetime.

When $eV = \Delta$, σ does not diverge. With increasing Z, the spectral weight transfers to the zero energy. Consequently, $\sigma(\Delta/e)$ decreases and approaches $\pi/4$ as $Z \to \infty$, unlike in an *s*-wave superconductor.

The tunneling measurement results of high-T_c superconductors agree qualitatively with the theoretical prediction. Figure 5.4 shows the differential conductance along the direction parallel to the CuO$_2$ plane for Bi$_2$Sr$_2$CaCu$_2$O$_8$ [135]. It verifies the existence of the zero energy resonance peak in high-T_c superconductors. Furthermore, it was also found that this zero energy resonance peak exists only in the superconducting phase [136], and disappears in the normal phase. The zero-bias conductance peaks in YBa$_2$Cu$_3$O$_{6+x}$ and other high-T_c cuprates have also been found by experimental measurements [137–140]. These tunneling measurements provide strong support to the *d*-wave pairing symmetry of high-T_c superconductivity.

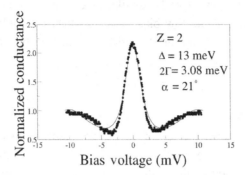

Figure 5.4 Tunneling differential conductance of $Bi_2Sr_2CaCu_2O_8$ along the direction of CuO_2 planes. The solid line is a fitting curve obtained based on the extended BTK formula. Γ is the inverse of the effective quasiparticle lifetime, and α is the angle between the tunneling current and the antinodal direction. (From Ref. [135])

5.4 Surface Bound State

The above discussion shows that in a tunneling junction of d-wave superconductor, if the interface is perpendicular to the nodal direction and the tunneling probability is very small ($Z \to \infty$), there exists a zero-energy resonance at the interface. This resonance state leads to a sharp zero-energy peak in the tunneling differential conductance, which can be measured by tunneling experiments. Naturally, one may ask how strongly the existence of this zero-energy resonance state depends on the detailed structure of the metal–superconductor interface, and whether they can be detected by other experimental techniques, such as photoelectron spectroscopy or Raman scattering. This is of particular interesting from the experimental point of view. In order to address this question, we study a d-wave superconductor which contacts directly with the vacuum or with an insulator. We will show that the zero-energy resonance state exists in all these cases and is an intrinsic property of d-wave superconductors.

To explore this problem, let us consider a d-wave superconductor with the nodal direction of the gap function perpendicular to the interface. Without loss of generality, we set the interface parallel to the y-direction. Under this setup, the gap function symmetry is d_{xy}. Owing to the translation symmetry along the y-direction, we solve the Bogoliubov–de Gennes Hamiltonian with a fixed wave vector k_y,

$$H = \sigma_z \left[-\frac{\hbar^2}{2m} \frac{d^2}{dx^2} + \frac{\hbar^2 k_y^2}{2m} - \mu(x) \right] - i\sigma_x \frac{\Delta \tilde{k}_y}{k_F} \frac{d}{dx}. \qquad (5.55)$$

The left-hand side of the junction ($x < 0$) is the vacuum with $\mu(x) = -\infty$, and the right-hand side ($x > 0$) is the d-wave superconductor with $\mu(x) = \mu = \frac{\hbar^2}{2m} k_F^2 > 0$.

We intend to solve the above Hamiltonian to find a zero energy bound state

$$H \begin{pmatrix} u(x) \\ v(x) \end{pmatrix} = 0. \tag{5.56}$$

Since $\sigma_y H = -H\sigma_y$, the zero energy state should be an eigenstate of σ_y with $v = \pm iu$. We have

$$\left[-\frac{\hbar^2}{2m}\frac{d^2}{dx^2} \pm \frac{\Delta \tilde{k}_y}{k_F}\frac{d}{dx} + \frac{\hbar^2 k_y^2}{2m} - \mu(x) \right] u(x) = 0. \tag{5.57}$$

The $\pm$ signs are determined by the boundary condition such that $u(x) \to 0$ at $x \to \infty$.

To find the bound state, we set

$$u(x) = e^{\beta x}, \qquad \text{Re}\beta < 0. \tag{5.58}$$

Inserting this expression into Eq. (5.57), we obtain the equation

$$\frac{\beta^2}{k_F^2} \mp \frac{\Delta \tilde{k}_y}{E_F}\frac{\beta}{k_F} + 1 - \tilde{k}_y^2 = 0. \tag{5.59}$$

In case $|\tilde{k}_y| \ll 1$, the above equation has a pair of complex roots

$$\beta = -\frac{|k_y|\Delta}{2E_F} \pm ik_F \sqrt{1 - \tilde{k}_y^2 - \left(\frac{\tilde{k}_y\Delta}{2E_F}\right)^2}. \tag{5.60}$$

The corresponding zero energy wavefunction that satisfies the boundary condition is

$$u(x) = e^{-|k_y|\Delta x/2E_F} \sin k_F \sqrt{1 - \tilde{k}_y^2 + \left(\frac{\Delta \tilde{k}_y}{2E_F}\right)^2}\, x. \tag{5.61}$$

It is localized near the surface within the length scale of $\xi/\tilde{k}_y$ with

$$\xi = \frac{\hbar v_f}{\Delta} \tag{5.62}$$

and oscillates with a wave vector close to k_F, and ξ is just the coherence length.

5.5 Tunneling Hamiltonian

When the metal–superconductor junction is separated by an insulating layer not in ideal contact, the scattering of electrons at the interface may not be elastic. In this case, the electron tunneling cannot be treated as an elastic process, and the theoretical study becomes complicated. A frequently used approach is to treat the tunneling matrix element as a phenomenological parameter under proper approximations. This phenomenological approach does not require the conservation of electron energy and momentum during the tunneling process. In comparison with the previous approach which assumes the tunneling process is elastic, the application range of this phenomenological approach is broader. This approach is often used to study the tunneling problem in combination with the Green's function method. However, it is difficult to calculate the tunneling matrix elements from microscopic models and to know how good the approximation is.

If the interface potential is not a δ-function, the scattering coefficients can usually only be obtained numerically. If the insulating layer is thick and the transmission coefficient is small, the tunneling junction can be approximately described by an effective tunneling Hamiltonian as

$$H = H_L + H_R + H_T. \tag{5.63}$$

H_L and H_R are the Hamiltonians for the left and right-hand side of the junction, respectively. The tunneling Hamiltonian H_T is defined as

$$H_T = \sum_{m,n} (T_{m,n} c_{R,n}^{\dagger} c_{L,m} + h.c.). \tag{5.64}$$

This describes the tunneling process of an electron from state m on the left-hand side of the junction to state n on the right-hand side of the junction. $T_{m,n}$ is the corresponding tunneling matrix element. The quantum states on the two sides of the junction can be very different.

If both sides of the tunneling junction are uniform conductors, the wave vector $\mathbf{k}$ and spin σ are good quantum numbers. In this case, $m = (\mathbf{k}, \sigma)$ and $T_{m,n} = T_{\mathbf{k}\sigma, \mathbf{k}'\sigma'}$ is the tunneling matrix element of electrons from the state $(\mathbf{k}, \sigma)$ to $(\mathbf{k}', \sigma')$. If the scattering potential is spin-independent and the characteristic energy scales, including temperature and frequency, are much smaller than the Fermi energies of the normal metal and the superconductor, we can approximate the tunneling matrix elements as a state-independent constant

$$T_{m,n} \approx T_{\mathbf{k}_F\sigma, \mathbf{k}'_F\sigma'} = T. \tag{5.65}$$

This approximation is broadly and successfully used in the study of tunneling problems, especially in the analysis of experimental results. Of course, this approximation is valid only when the tunneling matrix element is not sensitive to the momentum of electrons. Otherwise, it needs to be modified.

The tunneling Hamiltonian Eq. (5.63) treats the two sides of the junction as two relatively independent systems connected by the tunneling matrix, similar to the tight-binding approximation in the band theory. This model neglects the interplay between H_T and $H_{L,R}$, and is applicable only to the system with small tunneling probability.

The tunneling Hamiltonian was first established in the study of elastic scattering at the interface. It was extended to inelastic scattering systems just from phenomenological considerations. For an elastic scattering system, the tunneling matrix elements $T_{m,n}$ can be represented in terms of the interface barrier potential. However, for an inelastic scattering system, the tunneling matrix elements $T_{m,n}$ have to be taken as phenomenological fitting parameters.

Bardeen carried out the first microscopic investigation on the tunneling matrix elements [141]. Later, Harrison derived an explicit expression of $T_{m,n}$ under the WKB approximation [142]. Their works established a microscopic framework for the tunneling Hamiltonian, and provided a theoretical picture for understanding the physical meaning of the tunneling Hamiltonian (5.63).

Let us consider an elastic scattering system with the interface lying in $x_a < x < x_b$. We assume that the scattering potential $U(x)$ depends only on x, and the momentum parallel to the interface is conserved. In this case, we only need to consider the motion of electrons along the x-direction. The Hamiltonian is defined as

$$
H = \begin{cases} H_L(x) & x < x_a \\ -\dfrac{\hbar^2}{2m}\partial_x^2 + U(x) & x_a < x < x_b \\ H_R(x) & x > x_b \end{cases}.
\tag{5.66}
$$

It is not necessary to know the detailed expressions of $H_{L,R}(x)$. We assume that the interactions among electrons or superconducting quasiparticles are very weak and the tunneling can be treated as a single-particle problem.

To discuss the tunneling problem, we need first to understand the physical meaning of the electron operators $c_{L,m}$ and $c_{R,m}$ in the tunneling Hamiltonian. For this purpose, we assume that $\psi_{L,m}(x)$ is the wavefunction created by $c_{L,m}^{\dagger}$, which is defined in the entire space. Within the metal as well as the insulating layer side, $\psi_{L,m}$ is the eigenwavefunction of the Hamiltonian

$$
H\psi_{L,m} = E_m^L \psi_{L,m}, \qquad x < x_b.
\tag{5.67}
$$

Physically, we are interested in the scattering problem with energy $E_m^L \ll U(x)$. In this case, $\psi_{L,m}$ would decay exponentially in the insulating region $x_a < x < x_b$. When $x > x_b$, then $\psi_{L,m}$ is not the eigensolution of H_R. We do not need to know the concrete form of $\psi_{L,m}$. Nevertheless, the wavefunction needs to be continuous at $x = x_b$. Furthermore, $\psi_{L,m}$ decays rapidly in the region $x > x_b$, and vanishes in the limit $x \to +\infty$.

The state created by $c_{R,m}^{\dagger}$ is similarly defined. When $x > x_a$, $\psi_{R,m}$ is the solution to the eigenequation

$$H\psi_{R,m} = E_m^R \psi_{R,m}. \tag{5.68}$$

It decays exponentially with decreasing x in the range $x_a < x < x_b$. In the region $x < x_a$, $\psi_{R,m}$ is not the solution to H_L, but it satisfies the continuity condition at $x = x_a$. Again, $\psi_{R,m}$ decays rapidly in the region $x < x_a$, and becomes zero as $x \to -\infty$.

$\psi_{L,m}$ and $\psi_{R,m}$ together form a set of nonorthogonal but complete bases. Any eigenstate $\psi(x)$ of H can be represented as linear superpositions of $\psi_{L,m}$ and $\psi_{R,m}$ as

$$\psi = \sum_{\alpha,m} a_{\alpha,m} \psi_{\alpha,m}. \tag{5.69}$$

From the eigenequation of the Hamiltonian,

$$H\psi = E\psi, \tag{5.70}$$

we find that the eigenenergy E is determined by the equation

$$\det A_{\beta,n;\alpha,m} = 0, \tag{5.71}$$

where

$$A_{\beta,n;\alpha,m} = \langle \psi_{\beta,n} | H - E | \psi_{\alpha,m} \rangle. \tag{5.72}$$

In Eq. (5.72), the off-diagonal terms of $\langle \psi_{\beta,n} | H | \psi_{\alpha,m} \rangle$ and $\langle \psi_{\beta,n} | \psi_{\alpha,m} \rangle$ are small. The eigenvalues are predominately determined by the diagonal terms such that E is approximately equal to E_m^α. This implies that the tunneling takes place just between states with $E_m^L \approx E_n^R \approx E$. The corresponding tunneling matrix element is then given by

$$T_{m,n} = A_{L,m;R,n} \approx \langle \psi_{L,m} | H - E_n^R | \psi_{R,n} \rangle = \int_{-\infty}^{x_1} dx\, \psi_{L,m}^* (H - E_n^R) \psi_{R,n}, \tag{5.73}$$

where x_1 can take any value between x_a and x_b. Using the equation

$$\int_{-\infty}^{x_1} dx\, \psi_{R,n} (H - E_m^L) \psi_{L,m}^* = 0, \tag{5.74}$$

we find that the tunneling matrix element can be further expressed as

$$T_{m,n} \approx \int_{-\infty}^{x_1} dx\, \left(\psi_{L,m}^* H \psi_{R,n} - \psi_{R,n} H \psi_{L,m}^* \right), \tag{5.75}$$

where the approximation $E_m^L \approx E_n^R$ is used. If we assume

$$H_L(x) = -\frac{\hbar^2}{2m}\partial_x^2, \tag{5.76}$$

Eq. (5.75) can be further simplified as

$$
\begin{aligned}
T_{m,n} &\approx -\frac{\hbar^2}{2m}\int_{-\infty}^{x_1} dx \left(\psi_{L,m}^*\partial_x^2\psi_{R,n} - \psi_{R,n}\partial_x^2\psi_{L,m}^*\right) \\
&= -\frac{\hbar^2}{2m}\left[\psi_{L,m}^*(x_1)\partial_x\psi_{R,n}(x_1) - \psi_{R,n}(x_1)\partial_x\psi_{L,m}^*(x_1)\right].
\end{aligned} \tag{5.77}
$$

This is the formula that was first obtained by Bardeen [141], and Eq. (5.77) depends on $\psi_{R,n}(x)$ and $\psi_{L,m}(x)$, but not explicitly on $H_R(x)$.

Now let us derive the tunneling matrix elements between two normal metals using Eq. (5.77). For simplicity, we assume H_R to have the form

$$H_R(x) = -\frac{\hbar^2}{2m}\partial_x^2 + V_R, \tag{5.78}$$

where V_R is a constant.

$H_{L,R}(x)$ describe free electrons. Their eigenstates can be represented by wave vectors. In the left conductor, since the tunneling probability is small, the reflection probability is close to 1. Therefore, the electron wavefunction is approximately a superposition of incident and reflection waves with an equal weight, which can be represented as

$$\psi_{L,k_x} = c_L \sin(k_x x + \gamma_L), \qquad x < x_a, \tag{5.79}$$

where c_L is the normalization constant independent on k_x, γ_L is the phase shift due to the scattering potential and

$$k_x = \sqrt{\frac{2m E_x}{\hbar^2}}. \tag{5.80}$$

In the scattering region, it is difficult to obtain a rigorous solution. However, if $U(x)$ is a slowly varying potential, the WKB approximation is applicable, which gives

$$\psi_{L,k_x} = c_L\sqrt{\frac{k_x}{2p(x)}}\exp\left[-\int_{x_a}^x p(x)dx\right], \qquad x_a < x < x_b, \tag{5.81}$$

where

$$p(x) = \sqrt{\frac{2m[V(x) - E_x]}{\hbar^2}}. \tag{5.82}$$

Similarly, the electron wavefunction on the right conductor can be obtained using the WKB approximation,

$$
\psi_{R,q_x} = \begin{cases} c_R \sin(q_x x + \gamma_R) & x > x_b \\ c_R \sqrt{\dfrac{q_x}{2p(x)}} \exp\left[-\displaystyle\int_x^{x_b} p(x)dx\right] & x_a < x < x_b \end{cases},
\tag{5.83}
$$

where c_R is the normalization constant, γ_R is the scattering phase shift and

$$
q_x = \sqrt{\frac{2m(E_x - V_R)}{\hbar^2}}.
\tag{5.84}
$$

Substituting these results into Eq. (5.77), we find that the tunneling matrix elements can be expressed as [142]

$$
\begin{aligned}
T_{k_x,q_x} &= \frac{\hbar^2 c_L c_R \sqrt{k_x q_x}}{2m} \exp\left[-\int_{x_a}^{x_b} p(x)dx\right] \\
&= \frac{\hbar c_L c_R \sqrt{v_{L,x} v_{R,x}}}{2} \exp\left[-\int_{x_a}^{x_b} p(x)dx\right],
\end{aligned}
\tag{5.85}
$$

where $v_{L,x} = \hbar k_x/m$ and $v_{R,x} = \hbar q_x/m$ are the velocities of electrons in the left and right conductors, respectively.

Equation (5.85) shows that the tunneling probability is not only closely related to the scattering potential, but also depends on the velocities of the incident and transmission electrons. This result is obtained under the assumption that both sides of the tunneling junction are normal metals. When the tunneling probability is low, the hole reflection and transmission rates induced by the Cooper pairing are also small and negligible. This suggests that Eq. (5.85) can be also applied to a tunneling system with one or both sides of the junction being superconductors.

5.6 Tunneling Current

In an external bias, the tunneling circuit is actually a nonequilibrium system. To evaluate the tunneling current, one has to use the method of closed-time-path Green's functions which is generally difficult to implement. However, if the bias is small, the nonlinear effect is negligible and the tunneling current can be evaluated using the conventional perturbation theory based on equilibrium states [6].

We assume that the eigenstates in both the metallic and superconducting sides of the junction can be labeled by the momenta of electrons, and the tunneling Hamiltonian H_T, defined by Eq. (5.86), can be expressed as

$$
H_T = \sum_{\mathbf{kq}} \left(T_{\mathbf{k},\mathbf{q}} c_{L\mathbf{k}\sigma}^\dagger c_{R\mathbf{q}\sigma} + h.c. \right).
\tag{5.86}
$$

The corresponding tunneling current operator is defined by

$$\hat{I} = -\frac{2e}{\hbar} \text{Im} \sum_{\mathbf{kq}} T_{\mathbf{k,q}} c^{\dagger}_{L\mathbf{k}\sigma} c_{R\mathbf{q}\sigma}. \tag{5.87}$$

In many textbooks, the tunneling current operator is defined through the time derivative of the particle numbers in the left- or the right-hand side of the junction [6]. These two kinds of definitions are equivalent if H_L commutes with the particle number in the left conductor,

$$N_L = \sum_{\mathbf{k}\sigma} c^{\dagger}_{L\mathbf{k}\sigma} c_{L\mathbf{k},\sigma}. \tag{5.88}$$

However, if the left junction is also a superconductor, the BCS mean-field Hamiltonian breaks charge conservation. We cannot use it directly to calculate the time-derivative of the particle number operator. Rather the full interacting Hamiltonian of the left-hand side without taking the mean-field approximation, should be employed. This Hamiltonian maintains the charge conservation

$$[H_L, N_L] = 0, \tag{5.89}$$

and the electric current operator, defined through the time-derivative of N_L, is

$$-e\frac{dN_L}{dt} = \frac{-ie}{\hbar}[H, N_L] = -\frac{2e}{\hbar} \text{Im} \sum_{\mathbf{kq}} T_{\mathbf{k,q}} c^{\dagger}_{L\mathbf{k}\sigma} c_{R\mathbf{q}\sigma}, \tag{5.90}$$

consistent with Eq. (5.87).

We use H_T as a perturbation and treat the sum of H_L and H_R, $H_0 = H_L + H_R$, as the unperturbed Hamiltonian. In the interaction picture, the time-evolution of operators is defined by

$$B(t) = e^{iH_0 t} B e^{-iH_0 t}. \tag{5.91}$$

The left- and right-hand sides of the junction are in different equilibrium states. The thermodynamic average of this time-dependent operator is determined by the formula

$$\langle B \rangle_0 = \frac{Tre^{-\beta K_0} B}{Tre^{-\beta K_0}}, \tag{5.92}$$

where

$$K_0 = H_0 - \mu_L N_L - \mu_R N_R. \tag{5.93}$$

The chemical potential difference between the left and right conductors equals the external voltage

$$eV = \mu_L - \mu_R. \tag{5.94}$$

Thus in order to study the time-evolution of operators in this nonequilibrium system using the approach based on the equilibrium states, one should replace H_0 by K_0. This is equivalent to replacing B by a new operator $\tilde{B}$, defined by

$$B(t) = e^{iK_0t}\tilde{B}e^{-iK_0t} = e^{iH_0t}Be^{-iH_0t}, \tag{5.95}$$

where

$$\tilde{B} = e^{-iK_0t}e^{iH_0t}Be^{-iH_0t}e^{iK_0t}. \tag{5.96}$$

In case both H_L and H_R commute with the particle number operators N_L and N_R, $\tilde{B}$ becomes

$$\tilde{B} = e^{i(\mu_L N_L + \mu_R N_R)}Be^{-i(\mu_L N_L + \mu_R N_R)}. \tag{5.97}$$

Using Eq. (5.97), the tunneling Hamiltonian in the interaction picture can be expressed as

$$
\begin{aligned}
H_T(t) &= e^{iK_0t}e^{i(\mu_L N_L - \mu_R N_R)t}H_T e^{-i(\mu_L N_L - \mu_R N_R)t}e^{-iK_0t} \\
&= e^{iK_0t}\sum_{\mathbf{kq}\sigma}\left(T_{\mathbf{k,q}}e^{ieVt}c^{\dagger}_{Lk\sigma}c_{Rq\sigma} + T^*_{\mathbf{k,q}}e^{-ieVt}c^{\dagger}_{Rq\sigma}c_{Lk\sigma}\right)e^{-iK_0t} \\
&= e^{ieVt}A(t) + e^{-ieVt}A^{\dagger}(t), \tag{5.98}
\end{aligned}
$$

where

$$A(t) = \sum_{\mathbf{kq}\sigma}T_{\mathbf{k,q}}c^{\dagger}_{Lk\sigma}(t)c_{Rq\sigma}(t) \tag{5.99}$$

$$c_{\alpha k\sigma}(t) = e^{-iK_0t}c_{\alpha k\sigma}e^{iK_0t}. \tag{5.100}$$

Similarly, the tunneling current operator in the interaction picture is defined by

$$\hat{I}(t) = \frac{ie}{\hbar}\left[e^{ieVt}A(t) - e^{-ieVt}A^{\dagger}(t)\right]. \tag{5.101}$$

Up to the first order in H_T, the expectation value of the tunneling current can be expressed as

$$I(t) = \left\langle\hat{I}(t)\right\rangle_0 - \frac{i}{\hbar}\int_{-\infty}^{t}dt'\left\langle\left[\hat{I}(t), H_T(t')\right]\right\rangle_0. \tag{5.102}$$

There is no tunneling current in the absence of perturbation,

$$\left\langle\hat{I}(t)\right\rangle_0 = 0. \tag{5.103}$$

Thus $I(t)$ is completely determined by the second term in Eq. (5.102).

Substituting the expressions of $\hat{I}$ and $H_T(t)$ into Eq. (5.102), we find that $I(t)$ contains both the normal tunneling current due to quasiparticle tunneling and the

Josephson current due to the tunneling of Cooper pairs. If we use $I_Q(t)$ and $I_J(t)$ to represent respectively these two kinds of currents, then

$$I(t) = I_Q(t) + I_J(t),\tag{5.104}$$

$$I_Q(t) = -\frac{2e}{\hbar^2}\mathrm{Im}\int_{-\infty}^{\infty} dt'\, e^{ieV(t-t')} X_{ret}(t-t'),\tag{5.105}$$

$$I_J(t) = \frac{2e}{\hbar^2}\mathrm{Im}\int_{-\infty}^{\infty} dt'\, e^{-ieV(t+t')} Y_{ret}(t-t'),\tag{5.106}$$

where

$$X_{ret}(t-t') = -i\theta(t-t')\left\langle\left[A(t), A^\dagger(t')\right]\right\rangle_0.\tag{5.107}$$

$$Y_{ret}(t-t') = -i\theta(t-t')\left\langle\left[A^\dagger(t), A^\dagger(t')\right]\right\rangle_0.\tag{5.108}$$

$$A(t) = \sum_{\mathbf{kq}\sigma} T_{\mathbf{k,q}} c^\dagger_{L\mathbf{k}\sigma}(t) c_{R\mathbf{q}\sigma}(t).\tag{5.109}$$

If either or both sides are superconducting, the retarded Green's functions, X_{ret} and Y_{ret}, at the zeroth order should be defined with respect to the corresponding BCS mean-field Hamiltonians. X_{ret} at the zeroth order only involves the normal Green's function. In contrast, Y_{ret} involves the anomalous Green's function. Consequently, I_J exists only in the superconducting junction. There is no Josephson current if either side of the junction is a normal metal, i.e. $Y_{ret} = 0$.

Below we discuss the property of tunneling current of normal electrons, and leave the discussion on the Josephson tunneling current to Chapter 6.

5.7 Tunneling Effect of Quasiparticles

The tunneling current of quasiparticles, Eq. (5.105), can be also expressed as

$$I_Q = -\frac{2e}{\hbar^2}\mathrm{Im} X_{\mathrm{ret}}(eV),\tag{5.110}$$

where $X_{\mathrm{ret}}(eV)$ is the Fourier transform of $X_{\mathrm{ret}}(t)$ in the frequency space

$$X_{\mathrm{ret}}(eV) = \int_{-\infty}^{\infty} dt\, e^{ieVt} X_{\mathrm{ret}}(t).\tag{5.111}$$

$X_{\mathrm{ret}}(eV)$ can be calculated using the finite-temperature Green's function theory. The Matsubara Green's function corresponding to $X_{\mathrm{ret}}(t)$ is defined by

$$\begin{aligned}
X(\tau) &= -\left\langle T_\tau A(\tau) A^\dagger(0)\right\rangle_0 \\
&= \sum_{\mathbf{kq}} \left|T_{\mathbf{k,q}}\right|^2 \left[G_{L,11}(\mathbf{k}, -\tau)G_{R,11}(\mathbf{q}, \tau) + G_{L,22}(\mathbf{k}, \tau)G_{R,22}(\mathbf{q}, -\tau)\right].
\end{aligned}$$

$$\tag{5.112}$$

In the imaginary frequency space, it becomes

$$X(i\omega_n) = \int_0^\beta d\tau e^{i\omega_n \tau} X(\tau) = X_1(i\omega_n) + X_2(-i\omega_n), \qquad (5.113)$$

where

$$X_\alpha(i\omega_n) = \frac{1}{\beta} \sum_{\mathbf{k}\mathbf{q}p_n} |T_{\mathbf{k},\mathbf{q}}|^2 G_{L,\alpha\alpha}(\mathbf{k}, p_n) G_{R,\alpha\alpha}(\mathbf{q}, p_n + \omega_n). \qquad (5.114)$$

From the Rehman representation of the Matsubara Green's function Eq. (F.24), Eq. (5.114) can be represented using the retarded Green's function as

$$X_\alpha(i\omega_n) = \sum_{\mathbf{k}\mathbf{q}} \frac{|T_{\mathbf{k}\mathbf{q}}|^2}{\pi^2} \int d\omega_1 d\omega_2 \frac{f(\omega_1) - f(\omega_2)}{i\omega_n + \omega_1 - \omega_2}$$
$$\mathrm{Im}G^R_{L,\alpha\alpha}(\mathbf{k}, \omega_1) \mathrm{Im}G^R_{R,\alpha\alpha}(\mathbf{q}, \omega_2). \qquad (5.115)$$

Substituting Eq. (5.115) into Eq. (5.113) and taking the analytical continuation, i.e. $i\omega_n \to \omega + i0^+$, we obtain

$$I_Q(eV) = \frac{2e}{\pi\hbar^2} \sum_{\mathbf{k}\mathbf{q}} |T_{\mathbf{k}\mathbf{q}}|^2 [j_1(\mathbf{k}, \mathbf{q}, eV) - j_2(\mathbf{k}, \mathbf{q}, -eV)], \qquad (5.116)$$

where

$$j_\alpha(\mathbf{k}, \mathbf{q}, \omega) = \int d\omega_1 \mathrm{Im}G^R_{L,\alpha\alpha}(\mathbf{k}, \omega_1) \mathrm{Im}G^R_{R,\alpha\alpha}(\mathbf{q}, \omega_1 + \omega)$$
$$[f(\omega_1) - f(\omega_1 + \omega)]. \qquad (5.117)$$

5.7.1 Systems with Constant Tunneling Matrix Elements

If the tunneling matrix element $T_{\mathbf{k}\mathbf{q}}$ does not depend on $\mathbf{k}$ and $\mathbf{q}$, i.e. $T_{\mathbf{k}\mathbf{q}} = T_0$, then from the definition of quasiparticle density of states,

$$\rho(\omega) = -\frac{1}{\pi} \sum_{\mathbf{k}} \mathrm{Im}G^R_{11}(\mathbf{k}, \omega) = -\frac{1}{\pi} \sum_{\mathbf{k}} \mathrm{Im}G^R_{22}(\mathbf{k}, -\omega), \qquad (5.118)$$

we can reexpress $I_Q(eV)$ as

$$I_Q(eV) = \frac{4\pi e |T_0|^2}{\hbar^2} \int d\omega \rho_L(\omega) \rho_R(\omega + eV) [f(\omega) - f(\omega + eV)]. \qquad (5.119)$$

Hence the normal tunneling current is determined by the convolution of the density of states on the two sides of the junction. Equation (5.119) is a formula commonly used in the analysis of tunneling measurement results. At zero temperature, it reduces to

$$I_Q(eV) = \frac{4\pi e |T_0|^2}{\hbar^2} \int_0^{eV} d\omega \rho_L(\omega + eV)\rho_R(\omega).$$ (5.120)

Superconductor–Superconductor Junction

If both sides of the junction are superconductors, the above integral with respect to ω is typically an elliptic integral and there is no analytic solution. Nevertheless, it can be shown that the derivative of I_Q with respect to V, i.e. the differential conductance, reaches the maximum at $eV = \Delta_0^L + \Delta_0^R$, where Δ_0^L and Δ_0^R are the maximal gap values. If both sides are d-wave superconductors, the low-energy density of states is linear, then at low voltage,

$$\begin{aligned} I_Q(eV) &\approx \frac{4\pi e |T_0|^2 N_{L,F} N_{R,F}}{\hbar^2 \Delta_0^L \Delta_0^R} \int_0^{eV} d\omega (\omega + eV)\omega \\ &= \frac{10\pi e^4 |T_0|^2 N_{L,F} N_{R,F}}{3\hbar^2 \Delta_0^L \Delta_0^R} V^3, \end{aligned}$$ (5.121)

which varies cubically with V. Thus the low-voltage differential conductance is proportional to V^2. The quadratic power here is a consequence of the convolution of two linear density of states. If one side of the junction is an s-wave superconductor, and the other side is a d-wave superconductor, then the tunneling current is zero if eV is smaller than the s-wave superconducting gap Δ_s.

Normal Metal–Superconductor Junction

If one end of the tunneling junction is a superconductor and the other, say, the left-hand end, is a normal metal, then the low-energy electron density of states in the left-hand end can be approximated by the value at the Fermi energy $N_{L,F}$. In this case, the tunneling current at low bias becomes

$$I_Q(eV) = \frac{4\pi e |T_0|^2 N_{L,F}}{\hbar^2} \int_0^{eV} d\omega \rho_R(\omega),$$ (5.122)

and the corresponding differential conductance is given by

$$\frac{dI_Q}{dV} = \frac{4\pi e^2 |T_0|^2 N_{L,F}}{\hbar^2} \rho_R(eV),$$ (5.123)

which is proportional to the density of states of quasiparticles in the superconductor. Hence the density of states of quasiparticles in superconductors can be probed through the measurement of the tunneling current in a superconductor–insulator–metal junction.

5.7.2 Interlayer Tunneling in Cuprates

The above analysis is valid when the tunneling matrix elements are momentum independent. However, in high-T_c cuprates, T_{kq} cannot be treated as a constant since the electron velocity along the c-axis depends strongly on the in-plane momentum of electrons in the CuO$_2$ plane if the tunneling current is along the c-axis.

Normal Metal–Superconductor Junction

If the left-hand side of the junction is a high-T_c superconductor, and the right-hand side is a normal metal conductor, then

$$v_c(\mathbf{k}) \propto \cos^2(2\varphi_L), \tag{5.124}$$

$$\left|T_{\mathbf{kq}}\right|^2 = |T_0|^2 \cos^2(2\varphi_L), \tag{5.125}$$

where φ_L is the azimuthal angle of the in-plane component of the momentum $\mathbf{k}$, and T_0 is approximately a constant. In this case, the zero temperature tunneling current is determined by the formula

$$I_Q = -\frac{2e\pi N_F^R |T_0|^2}{\hbar^2} \sum_{\mathbf{k}} \cos^2 2\varphi_L$$

$$\int_0^{eV} d\omega \mathrm{Im}\left[G_{L,11}(\mathbf{k}, \omega - eV) + G_{L,22}(\mathbf{k}, \omega)\right]. \tag{5.126}$$

For an ideal d-wave superconductor, Eq. (5.126) can be simplified as

$$I_Q = \frac{2e\pi^2 N_F^R |T_0|^2}{\hbar^2} \sum_{\mathbf{k}} \cos^2 2\varphi_L \int_0^{eV} d\omega \delta\left(\omega - E_{\mathbf{k}}^L\right), \tag{5.127}$$

where $E_{\mathbf{k}}^L = \sqrt{\xi_{L,\mathbf{k}}^2 + \Delta_{L,\mathbf{k}}^2}$ is the quasiparticle spectrum. By changing the momentum summation into an integral over $\xi_{\mathbf{k}}$ and setting $x = \cos(2\varphi_L)$, the above expression can be simplified as

$$I_Q = \frac{4e\pi N_F^R N_F^L |T_0|^2}{\hbar^2} \int_0^1 \frac{dx x^2}{\sqrt{1-x^2}} \sqrt{(eV)^2 - \left(\Delta_{L,0} x\right)^2}. \tag{5.128}$$

The right-hand side is an elliptic integral. When $eV \ll \Delta_{L,0}$, Eq. (5.128) can be approximately integrated out, which gives

$$I_Q \approx \frac{\pi^2 N_F^R N_F^L |T_0|^2 e^5 V^4}{4\hbar^2 \Delta_{L,0}^3}. \tag{5.129}$$

The corresponding differential conductance [143] is

$$\frac{dI_Q}{dV} \approx \frac{\pi^2 N_F^R N_F^L |T_0|^2 e^5 V^3}{\hbar^2 \Delta_{L,0}^3}. \tag{5.130}$$

This V^3-dependence of the differential conductance is clearly different from the result obtained in the system where the tunneling matrix elements T_{kq} are momentum independent. In the latter case, the differential conductance is linearly proportional to the density of states and varies linearly with V. The difference shows that the tunneling matrix elements have a strong effect on the tunneling current and should be seriously considered in the analysis of tunneling experimental results in high-T_c superconductors. In fact, the cubic voltage dependence of the differential conductance can be obtained simply through dimensional analysis. The low-energy density of states in the d-wave superconductors is linear, hence proportional to the energy. The function $\cos(2\varphi_L)$ in the tunneling matrix element has the same momentum dependence as the gap function, and thus has the same dimension as the energy. This implies that the effective dimension of $\cos^2(2\varphi_L)$ is 2. This, in combination with the linear density of states, leads to the cubic power of V in the differential conductance.

Superconductor–Superconductor Junction

On the other hand, if both sides of the tunneling junction are high-T_c superconductors, then the tunneling matrix elements along the c-axis can be expressed as

$$|T_{kq}|^2 = |T_0|^2 \cos^2(2\varphi_L)\cos^2(2\varphi_R). \tag{5.131}$$

Since the contribution to the quasiparticle tunneling current from the two sides of the junction is independent, it is simple to show that the differential conductance at low voltage is proportional to V^6, based on the dimensional analysis. This is confirmed by more sophisticated calculations. The zero temperature tunneling current is determined by the formula

$$I_Q = \frac{\pi e|T_0|^2}{\hbar^2}\sum_{kq}\cos^2(2\varphi_L)\cos^2(2\varphi_R)\int_{-eV}^0 \delta(\omega + E_k^L)\delta(\omega + eV - E_q^R). \tag{5.132}$$

After integrating over energy, it becomes

$$I_Q = \frac{e|T_0|^2 N_F^L N_F^R}{4\pi\hbar^2}\int_0^{eV} d\omega\omega(eV - \omega)\int_0^{2\pi}d\varphi_L d\varphi_R \mathrm{Re}$$
$$\frac{\cos^2(2\varphi_L)}{\sqrt{(eV - \omega)^2 - \Delta_{L,0}^2\cos^2(2\varphi_L)}}\frac{\cos^2(2\varphi_R)}{\sqrt{\omega^2 - \Delta_{R,0}^2\cos^2(2\varphi_R)}}. \tag{5.133}$$

At low bias, only low-energy quasiparticles contribute to the tunneling current. The integrals over φ_L and φ_R can be calculated approximately, which yields

$$I_Q \approx \frac{\pi|T_0|^2 N_F^L N_F^R e^8 V^7}{560\hbar^2\Delta_{L,0}^3\Delta_{R,0}^3}. \tag{5.134}$$

The corresponding differential conductance is

$$\frac{\mathrm{d}I_Q}{\mathrm{d}V} = \frac{\pi |T_0|^2 N_F^L N_F^R e^8 V^6}{80 \hbar^2 \Delta_{L,0}^3 \Delta_{R,0}^3}.$$

(5.135)

Hence, as predicted, $\mathrm{d}I_Q/\mathrm{d}V$ is proportional to V^6 at small bias, which different from the result $\mathrm{d}I_Q/\mathrm{d}V \propto V^2$ in the system where T_{kq} is a constant.

It should be emphasized that Eqs. (5.130) and (5.135) are valid only for ideal superconductors with infinite quasiparticle lifetime. If the superconductor is affected by disorder potentials or other effects, the quasiparticle lifetime becomes finite. If the energy scale of the inverse lifetime is larger than or equal to eV, the effect of the angular factor $\cos^2(2\varphi)$ will be smeared out. The tunneling current will become the same as that obtained based on the assumption that T_{kq} is a constant. Hence when applying these formulae to analyze the tunneling experiments of high-T_c superconductors, we should take full account for the scattering effect induced by impurities or other elementary excitations.

6

Josephson Effect

6.1 Josephson Tunneling Current

The tunneling current between two superconductors separated by a thin insulating barrier may arise from the tunneling of Cooper pairs. This is called the Josephson effect. This is different from the normal quasiparticle tunneling which occurs only when the applied bias energy is above the gap function. Josephson pair tunneling is driven by the phase difference between the two superconductors, and therefore provides another phase-sensitive tool to probe the gap sign change in a d-wave superconductor, in addition to the zero-energy modes residing on the surface perpendicular to the gap nodal direction as explained in §5.4.

The Josephson current, defined by Eq. (5.106), can also be written as

$$I_J(t) = \frac{2e}{\hbar^2} \text{Im} \left[e^{-2ieVt} Y_{\text{ret}}(eV) \right], \tag{6.1}$$

where

$$Y_{\text{ret}}(\omega) = \int_{-\infty}^{\infty} dt' e^{i\omega t} Y_{\text{ret}}(t) \tag{6.2}$$

is the Fourier transform of $Y_{\text{ret}}(t)$. Unlike I_Q, I_J is time-dependent. It indicates that a constant voltage can generate a time-dependent tunneling current. This peculiar property of the Josephson effect is a typical manifestation of quantum phase coherence.

In the imaginary time representation, the Matsubara function corresponding to $Y_{\text{ret}}(t)$ is defined by

$$
\begin{aligned}
Y(\tau) &= -\left\langle T_\tau A^\dagger(\tau) A^\dagger(0) \right\rangle_0 \\
&= -\frac{1}{V^2} \sum_{\mathbf{kq}} T^*_{\mathbf{k,q}} T^*_{-\mathbf{k},-\mathbf{q}} \left[G_{L,12}(\mathbf{k},\tau) G_{R,21}(\mathbf{q},-\tau) \right.
\end{aligned}
$$

$$\left. + G_{L,12}(\mathbf{k},-\tau) G_{R,21}(\mathbf{q},\tau) \right], \tag{6.3}$$

where V is the volume of the system. The corresponding Fourier transform is

$$Y(i\omega) = \int_0^\beta d\tau e^{i\omega\tau} Y(\tau) = \tilde{Y}(i\omega) + \tilde{Y}(-i\omega), \tag{6.4}$$

where

$$\tilde{Y}(i\omega) = -\frac{1}{V^2\beta} \sum_{\mathbf{k}\mathbf{q}p_n} T^*_{\mathbf{k},\mathbf{q}} T^*_{-\mathbf{k},-\mathbf{q}} G^L_{12}(\mathbf{k}, ip_n + i\omega) G^R_{21}(\mathbf{q}, ip_n). \tag{6.5}$$

If one or both ends of the junction are $d_{x^2-y^2}$-superconductors and the tunneling matrix element $T^*_{\mathbf{k},\mathbf{q}} T^*_{-\mathbf{k},-\mathbf{q}}$ is invariant under $\pi/2$ rotation in momentum space, then

$$\sum_{\mathbf{k}} T^*_{\mathbf{k},\mathbf{q}} T^*_{-\mathbf{k},-\mathbf{q}} G_{L,12}(\mathbf{k}, p_n) = 0. \tag{6.6}$$

Consequently, the Josephson current vanishes

$$I_J(t) = 0. \tag{6.7}$$

This is a consequence of the d-wave pairing gap whose average over the Fermi surface is zero. It is also a common feature of all non-s-wave superconductors.

The absence of the Josephson tunneling current in a junction with one of the ends being a d-wave superconductors is a consequence of the first order perturbation. The contribution from higher order perturbations to the tunneling current is generally finite and should be evaluated in order to find the Josephson current of d-wave superconductors. Needless to say, the Josephson tunneling current between an s- and a d-wave superconductor is much smaller than that between two s-wave superconductors with the same tunneling matrix elements. This is a difference between these two kind of junctions.

However, if a Josephson junction is formed by coupling two d-wave superconductors through a weak link along a direction parallel to the ab-plane, the interface breaks the tetragonal symmetry so that $T^*_{\mathbf{k},\mathbf{q}} T^*_{-\mathbf{k},-\mathbf{q}}$ is not invariant under a $\pi/2$ rotation in space. In this case, the Josephson current is finite even in the first order perturbation. When both sides of the junctions are d-wave superconductors, a physically interesting case is shown in Fig. 6.1, where the crystalline axes on one side are different from those in the other side. For this kind of Josephson junction, a simple calculation based on Eqs. (5.85) and (6.5) yields

$$\tilde{Y}(i\omega) \propto -\frac{1}{\beta} \sum_{p_n} \chi(\theta_L, p_n + \omega)\chi(\theta_R, p_n)\Delta_{L,0}\Delta^*_{R,0}, \tag{6.8}$$

where

$$\chi(\theta_i, p_n) = \int_{\theta_i-\pi/2}^{\theta_i+\pi/2} d\phi \frac{\cos(2\phi)}{(ip_n)^2 - \varepsilon_k^2 - |\Delta_{i,0}|^2\cos^2(2\phi)}, \quad (i = L, R). \tag{6.9}$$

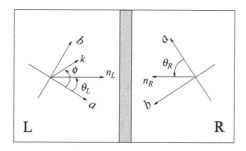

Figure 6.1 A Josephson junction of two superconductors connected along the *ab*-plane but with different crystalline orientations. θ_l and θ_R are the azimuthal angles between the *a*-axes and the normal direction on the left- and right-hand sides of the junction, respectively.

It can be shown that under the $\pi/2$ rotation of θ_i, $\chi(\theta_i, p_n)$ changes sign

$$\chi\left(\theta_i + \frac{\pi}{2}, p_n\right) = -\chi\left(\theta_i, p_n\right) \tag{6.10}$$

as a consequence of *d*-wave pairing. This implies that a $\pi/2$ rotation for a *d*-wave superconductor is equivalent to a phase change of π. If $\theta_R = \theta_L + \pi/2$, then $\chi(\theta_L, p_n)\chi(\theta_R, p_n) < 0$ at $i\omega = 0$. This generates a π-phase shift in the Josephson current.

In general, if φ_L and φ_R are the superconducting phases on the left- and right-hand sides of the junction, $\Delta_{L,0} = |\Delta_{L,0}| \exp(i\varphi_L)$ and $\Delta_{R,0} = |\Delta_{R,0}| \exp(i\varphi_R)$, $\tilde{Y}(i\omega)$ can then be expressed as

$$\tilde{Y}(i\omega) \propto C(\theta_L, \theta_R, i\omega)e^{-i(\varphi_L - \varphi_R)}. \tag{6.11}$$

In the absence of an applied bias or when the bias voltage is very low, if we neglect the frequency dependence in $C(\theta_L, \theta_R, i\omega)$, the above expression then becomes

$$\tilde{Y}(i\omega) \propto C(\theta_L, \theta_R)e^{i(\varphi_L - \varphi_R)}. \tag{6.12}$$

Substituting it into Eq. (6.4) and performing an analytic continuation, we obtain immediately the formula for the Josephson current between two *d*-wave superconductors

$$I_J = I_0 C(\theta_L, \theta_R) \sin\left(2eVt + \varphi_L - \varphi_R\right). \tag{6.13}$$

By applying an external magnetic field, instead of an electric field, to the junction, the above formula becomes

$$I_J = I_0 C(\theta_L, \theta_R) \sin \tilde{\varphi}_{LR}, \tag{6.14}$$

where

$$\tilde{\varphi}_{LR} = \varphi_L - \varphi_R - \int_L^R \frac{2e}{\hbar} \mathbf{A} \cdot \mathbf{dl} \tag{6.15}$$

is the gauge invariant phase difference with $\mathbf{A}$ the vector potential. Both Eqs. (6.13) and (6.14) are gauge invariant.

Compared with the corresponding formula for an s-wave superconductor, an extra geometrical factor $C(\theta_L, \theta_R)$, which is related to the orientation of the crystalline axes of the d-wave superconductor, emerges in Eq. (6.14). When the sign of $C(\theta_L, \theta_R)$ is changed, the phase of the Josephson current changes by π. This effect is absent in the Josephson junction of s-wave superconductors. In a circuit containing one or more Josephson junctions of d-wave superconductors, it is possible to spontaneously generate a half-quantum magnetic flux. Detection of this spontaneously generated half-quantum flux allows us to determine whether or not the superconductor in the circuit has the d-wave pairing symmetry. This will be discussed in the next section.

Similar to $\chi(\theta_i, p_n)$, $C(\theta_L, \theta_R)$ changes sign when θ_L or θ_R is rotated by $\pi/2$

$$C\left(\theta_L + \frac{\pi}{2}, \theta_R\right) = -C(\theta_L, \theta_R), \tag{6.16}$$

$$C\left(\theta_L, \theta_R + \frac{\pi}{2}\right) = -C(\theta_L, \theta_R). \tag{6.17}$$

Since the system is invariant under the reflection with respect to the line normal to the interface, we have

$$C(-\theta_L, -\theta_R) = C(\theta_L, \theta_R). \tag{6.18}$$

Using these symmetry properties, $C(\theta_L, \theta_R)$ can be expanded as [144]

$$C(\theta_L, \theta_R) = \sum_{nn'} \left\{ C_{n,n'} \cos[(4n + 2)\theta_L] \cos[(4n' + 2)\theta_R] \right.$$
$$\left. + D_{n,n'} \sin[(4n + 2)\theta_L] \sin[(4n' + 2)\theta_R] \right\}. \tag{6.19}$$

The cosine and sine terms do not mix due to the reflection symmetry. Both $C_{n,n'}$ and $D_{n,n'}$ depend on the band structures. However, in most cases, only the zeroth order terms ($n = n' = 0$) are important. Thus we have

$$C(\theta_L, \theta_R) \approx C_{0,0} \cos(2\theta_L) \cos(2\theta_R) + D_{0,0} \sin(2\theta_L) \sin(2\theta_R), \tag{6.20}$$

where the first term is that originally obtained by Sigrist and Rice [145]. If we further assume $D_{0,0} = -C_{0,0}$, the above expression reduces to the formula first obtained by Tsuei [146] for an extremely disordered Josephson junction

$$C(\theta_L, \theta_R) \approx C_{0,0} \cos[2(\theta_L + \theta_R)]. \tag{6.21}$$

6.2 Spontaneous Magnetic Flux Quantization

In a superconducting ring composed of a single or multiple Josephson junctions, for example, a superconducting quantum interference device (SQUID), if all the ends of the junctions are s-wave superconductors, then the flux trapped inside the ring is quantized, and the minimal flux quantum is given by

$$\Phi_0 = \frac{h}{2e} = 2 \times 10^{-15} \text{Wb} = 2.07 \times 10^{-7} \text{Gs} \cdot \text{cm}^2. \tag{6.22}$$

This flux quantization is a consequence of superconducting phase coherence, which is a macroscopic quantum phenomenon that can be generated by applying an external magnetic field or a supercurrent. However, if one or more of the superconductors in the ring are d-wave superconductors, then the condition for the flux quantization inside the ring is modified and a half-quantum flux may emerge spontaneously. This can be understood from the Ginzburg–Landau supercurrent formula (1.82)

$$\mathbf{J}_s = \frac{ie\hbar}{2m} \left(\psi^* \nabla \psi - \psi \nabla \psi^* \right) + \frac{2e^2}{m} \mathbf{A} \psi^* \psi. \tag{6.23}$$

In the superconducting state, the fluctuation of the pairing amplitude is weak, and only the spatial variation of the superconducting phase needs to be considered. Expressing $\psi(x) = |\psi(x)| \exp(i\varphi)$, the supercurrent becomes

$$\mathbf{J}_s = \frac{e\hbar}{m} |\psi|^2 \nabla \varphi - \frac{2e^2 |\psi|^2}{m} \mathbf{A}. \tag{6.24}$$

It can also be expressed as

$$\mathbf{A} + \frac{m}{2e^2 |\psi|^2} \mathbf{J}_s = \frac{\Phi_0}{2\pi} \nabla \varphi. \tag{6.25}$$

Taking a loop integration of the above equation around the ring, the left-hand side just equals the sum of the magnetic flux generated by the external magnetic field inside the ring, $\Phi_a = HS$ (S is the area enclosed by the loop), and the flux generated by the supercurrent around the ring, $\Phi_s = LI_s$ (L is inductance of the ring)

$$\oint_C d\mathbf{l} \cdot \left(\mathbf{A} + \frac{m}{2e^2 |\psi|^2} \mathbf{J} \right) = \Phi_a + \Phi_s. \tag{6.26}$$

On the right-hand side, the loop integral of $\nabla \varphi$, up to a multiple of 2π, is equal to the sum of the phase difference across each junction, hence

$$\Phi_a + \Phi_s = \frac{\Phi_0}{2\pi} \sum_{<ij>} \phi_{ij} + n\Phi_0. \tag{6.27}$$

The $\langle ij \rangle$ represents the Josephson junction formed by two neighboring superconductors i and j, and ϕ_{ij} is the phase difference across the junction, and n is an integer.

In the absence of the external voltage, the supercurrent inside the ring can be expressed in terms of the junction parameters according to Eq. (6.13) as

$$I_s = |I_{ij}^0| \sin(\phi_{ij} + \delta_{ij}), \tag{6.28}$$

where δ_{ij} equals either 0 or π, depending on the sign of the coefficient $C(\theta_L, \theta_R)$ in Eq. (6.13). If $\delta_{ij} = \pi$, this junction is called a π-junction.

As far as the spontaneous flux quantization is concerned, the supercurrent I_s around the ring is generally very small and the magnetic flux it generated is just of the order of Φ_0. However, the critical current I_{ij}^0 at each junction is generally much larger than I_s, i.e. $|I_{ij}^0| \gg I_s$. This implies that $L|I_{ij}^0| \gg \Phi_0$ and $|\sin(\phi_{ij} + \delta_{ij})| \ll 1$. Thus the sine function in Eq. (6.28) can be expanded up to the leading order as

$$\frac{I_s}{|I_{ij}^0|} \approx \phi_{ij} + \delta_{ij} + 2\pi m_{ij}, \tag{6.29}$$

where m_{ij} is an integer satisfying the inequality

$$|\phi_{ij} + \delta_{ij} + 2\pi m_{ij}| \ll \pi. \tag{6.30}$$

Substituting Eq. (6.29) into Eq. (6.27), we have

$$2\pi \frac{\Phi_a + \Phi_s}{\Phi_0} + \sum_{<ij>} \frac{I_s}{|I_{ij}^0|} \approx \sum_{<ij>} \delta_{ij} + 2\pi m, \tag{6.31}$$

where m is an integer. After neglecting the second term on the left-hand side, this equation becomes

$$\Phi_a + \Phi_s \approx \left(\sum_{<ij>} \delta_{ij} + 2\pi m \right) \frac{\Phi_0}{2\pi}. \tag{6.32}$$

If the number of π-junctions inside the ring is odd, the sum over δ_{ij} is equal to an odd number times of π. The flux quantization condition then becomes

$$\Phi_a + \Phi_s = \left(m + \frac{1}{2} \right) \Phi_0, \tag{6.33}$$

which is fundamentally different from the system without π-junctions.

It should be emphasized that the half-quantum flux relies on the phase change of the pairing order parameter. Only when the pairing phase is momentum dependent, for example, in a d-wave superconductor, can the half-quantum flux emerge. It does not appear in a ring with only isotropic s-wave superconductors. Therefore, the detection of the half-quantum flux can be used to judge *decisively* whether the superconducting pairing is of *d*-wave symmetry. Furthermore, in the absence of an external field, $\Phi_a = 0$, Eq. (6.33) shows that the ring encloses a finite flux whose minimal value equals half of the flux quantum Φ_0. This is a spontaneous generation of the half-quantum flux in a ring consisting of an odd number of π-junctions. It reveals a fundamental difference between *s*- and *d*-wave superconductors.

On the other hand, if the number of π-junctions is even, the sum over the phase difference δ_{ij} equals an integer multiple of 2π. In this case, the magnetic flux quantization condition is the same as for a ring formed purely by *s*-wave superconductors

$$\Phi_a + \Phi_s = m\Phi_0. \tag{6.34}$$

In the absence of an external magnetic field, the state with $m = 0$ generally has a minimal energy, and the system has no spontaneously generated magnetic flux.

6.3 Phase-Sensitive Experiments

Pairing symmetry of superconducting electrons can be detected by the measurement of quantum interference effects of Josephson junctions. Unlike thermodynamic or electromagnetic transport measurements, this class of experiments depends on the phases of superconducting gap functions, but not on their magnitudes. In other words, they are phase-sensitive, and the measurement results depend entirely on the macroscopic interference between supercurrents from different Josephson junctions, regardless of the microscopic details of specific materials.

The phase-sensitive experiments of high-T_c superconductors focus mainly on the measurement of quantum interference effects of single or double Josephson junctions and the detection of spontaneous magnetic fluxes in a circuit of Josephson junctions. A detailed discussion of the experimental results can be found from Refs. [147, 148].

6.3.1 Quantum Interference of Josephson Junctions

One of the most important applications of the Josephson effect is to fabricate SQUIDs. The simplest SQUID is a loop circuit composed of two parallel connected Josephson junctions. The supercurrent in the circuit is very sensitive to the magnetic flux enclosed by the loop, providing an ideal tool to probe weak magnetic fields. Furthermore, the quantum interference effect with two Josephson junctions can be used to detect pairing symmetry. It plays an important role in the study of high-T_c superconductivity.

A typical SQUID used in high-T_c quantum interference experiments, as depicted in Fig. 6.2(a), is constructed by connecting a high-T_c superconductor, say YBCO,

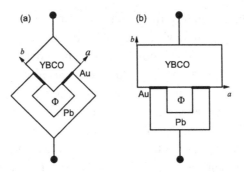

Figure 6.2 A SQUID consisting of an s-wave superconductor, Pb, and a high T_c superconductor, YBCO. The two superconductors are linked by the gold film. The two tunneling junctions are located respectively on the ac- and bc-surfaces of YBCO for the corner SQUID (a), and on the same ac-surface of YBCO for the edge SQUID (b).

with an s-wave superconductor through a weak link. It is called a corner SQUID, since the two junctions are located on the two edges of a corner, touching the ac and bc surfaces of a YBCO crystal, respectively. We denote these two junctions as a and b, and the corresponding phase differences across the junctions as ϕ_a and ϕ_b, respectively. The supercurrent in the system is the sum of the Josephson tunneling currents through these two junctions

$$I_s = I_{s,a} + I_{s,b}, \qquad (6.35)$$

$$I_{s,a} = |I_a| \sin(\phi_a + \delta_a),$$

$$I_{s,b} = |I_b| \sin(\phi_b + \delta_b),$$

where $\delta_{a,b}$ equals zero for a zero junction, or π for a π-junction.

Taking the integral over Eq. (6.24) along the loop of this SQUID, it can be shown that the phase difference between ϕ_a and ϕ_b satisfies the equation,

$$\phi_a - \phi_b = 2\pi \frac{\Phi_a}{\Phi_0} + 2\pi \frac{I_{s,a}L_a - I_{s,b}L_b}{\Phi_0} + 2\pi n, \qquad (6.36)$$

where L_a and L_b are the self-inductances of the a and b junctions, respectively. For a symmetric SQUID, $I_{s,a}L_a = I_{s,b}L_b$, the second term on the right-hand side of Eq. (6.36) vanishes and the above equation becomes

$$\phi_a - \phi_b = 2\pi \frac{\Phi_a}{\Phi_0} + 2\pi n. \qquad (6.37)$$

If the maximal tunneling currents of these two junctions are equal to each other, i.e. $|I_a| = |I_b|$, then the total supercurrent is

$$I_s = 2|I_a| \sin \gamma_0 \cos \frac{1}{2}(\phi_a - \phi_b + \delta_a - \delta_b), \qquad (6.38)$$

where γ_0 is an adjustable phase factor

$$\gamma_0 = \frac{1}{2}(\phi_a + \delta_a + \phi_b + \delta_b). \tag{6.39}$$

The maximal value of I_s takes place at $\gamma_0 = \pi/2$ and is given by

$$I_{max} = 2|I_a| \cos \left(\frac{\pi \Phi_a}{\Phi_0} + \frac{\delta_a - \delta_b}{2} \right). \tag{6.40}$$

For both a- and b-junctions, if one end of the junction is an s-wave superconductor and the other is a $d_{x^2-y^2}$-wave superconductor, and the tunneling directions are along the two principal axes of the $d_{x^2-y^2}$-wave superconductor, then $|\delta_a - \delta_b| = \pi$ as shown in Fig. 6.2(a). In this case, $|I_{max}|$ reaches the maxima and minima when Φ_a equals the half-integer and integer flux quanta, respectively. On the other hand, if the two junctions are formed by connecting the s-wave and the $d_{x^2-y^2}$ superconductors on the same surface through weak links, as shown in Fig. 6.2(b), then $|\delta_a - \delta_b| = 0$. Now $|I_{max}|$ reaches maxima when Φ_a equals zero or an integer flux quantum, and reaches minima when Φ_a equals an half-integer flux quantum.

The above discussion indicates that, in a weak external magnetic field, the quantum interference current in a SQUID composed of an s- and a d-wave superconductor depends on the geometric configuration of the junctions. When the two junctions lie on the same edge of the d-wave superconductor, the maximal supercurrent appears when the external magnetic flux is zero. In contrast, when the two junctions lie on the two adjacent edges, the maximal supercurrent appears when the external flux equals $\pm\Phi_0/2$. Clearly, this property can be used to detect the pairing symmetry.

Based on the above discussions, Wollman et al. carried out the first phase-sensitive measurement for high-T_c superconductors [149]. They measured the maximal bias current as a function of the external flux in two different SQUIDs shown in Fig. 6.2. The results are shown in Fig. 6.3. In their experiments, what they measured was the periodic modulation of the resistance with an applied magnetic field in the fluctuation regime of the critical current. The variation of the minimal resistance with the external flux exhibits a phase shift at each given bias current. For the two SQUIDs shown in Fig. 6.2, it can be shown that the phase shifts equal $\Phi_0/2$ and 0 in the limit of zero bias current, respectively. In order to determine this phase shift, they extrapolated the data to the limit of zero bias current and found that the resulting shift is around $0.3 \sim 0.6 \, \Phi_0$ for the corner-SQUID shown in Fig. 6.2(a). On the other hand, for the edge-SQUID shown in Fig. 6.2(b), they found that the phase shift in the zero bias current limit is around zero, which is qualitatively different from the previous case. Their results do not ensure that these two phase shifts are precisely located at 0 and $\Phi_0/2$, but the trend it revealed provides strong evidence in support of the d-wave pairing symmetry of high-T_c superconductivity.

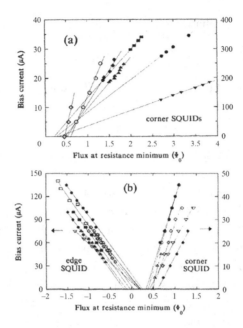

Figure 6.3 Extrapolations of the magnetic flux at the minimal resistance as a function of the bias current for the two SQUIDs shown in Fig. 6.2. Different symbols stand for different samples. For the corner SQUID shown in figure 6.2(a), the flux in the zero current limit should take the half-quantum value $\Phi_0/2$ for the d-wave superconductor. In contrast, for the edge SQUID shown in Fig. 6.2(b), the flux in the zero current limit should occur at zero flux regardless of the pairing symmetry. [149]

In Fig. 6.3, the extrapolated results exhibit a considerable variation for different samples. The variance may arise from two effects. First, the SQUID may not be as symmetric as expected due to the twin crystal structure or the orthogonal distortion of the YBCO crystals, which leads to the difference between the tunneling matrix elements and structures of the two junctions. Second, there may exist residual fluxes inside the loop. Both of them can cause the phase shifts to deviate from their ideal values.

In order to eliminate the uncertainty resulting from the asymmetry as well as the residual flux in SQUIDs, Wollman et al., proposed to use a single Josephson junction to detect the pairing symmetry [149, 150]. Similar to the SQUIDs shown in Fig. 6.2, they probe the pairing symmetry by measuring the interference effect caused by the phase difference between the tunneling currents from the two adjacent edges. The structure of this Josephson tunneling junction is illustrated in Fig. 6.4. It is a single Josephson tunneling junction because the same junction touches both ac and bc surfaces of YBCO, unlike the SQUID shown in Fig. 6.2(a).

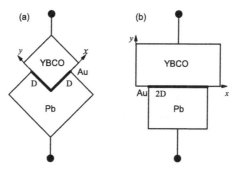

Figure 6.4 Josephson junction consisting of an *s*-wave superconductor and a *d*-wave superconductor. (a) a corner sharing junction; (b) an edge-sharing junction.

The Josephson junction shown in Fig. 6.4(b) is not much different from the usual one. For this kind of junction, the phase difference between the two superconductors can be adjusted by an external magnetic field. The tunneling currents vary in a pattern very similar to the single slit Fraunhofer diffraction in optics. For a perpendicular magnetic field, if we assume the length of the junction is smaller than the London penetration depth, then within this length scale of the junction, the magnetic field distribution is nearly uniform.

The total Josephson tunneling current can in fact be calculated in a gauge invariant way. Let us consider a Josephson junction along the *y*-axis, The gauge invariant phase change is

$$\tilde{\phi}_{ab}(x) = \phi(\mathbf{r}_{a1}) - \phi(\mathbf{r}_{b1}) - \frac{2\pi}{\Phi_0} \int_{-\infty}^{\infty} dy\, A_y(x, y), \qquad (6.41)$$

where $\mathbf{r}_{a1} = (x, +\infty)$ and $\mathbf{r}_{b1} = (x, -\infty)$ are two coordinates at the two ends of the junction. The change of this gauge invariant phase is

$$\begin{aligned}
&\tilde{\phi}_{ab}(x + \Delta x) - \tilde{\phi}_{ab}(x) \\
&= \phi(\mathbf{r}_{a2}) - \phi(\mathbf{r}_{b2}) - \frac{2\pi}{\Phi_0} \int_{-\infty}^{\infty} dy\, A_y(x + \Delta x, y) \\
&\quad - \phi(\mathbf{r}_{a1}) + \phi(\mathbf{r}_{b1}) + \frac{2\pi}{\Phi_0} \int_{-\infty}^{\infty} dy\, A_y(x, y),
\end{aligned} \qquad (6.42)$$

where $\mathbf{r}_{a2} = (x + \Delta x, +\infty)$, $\mathbf{r}_{b2} = (x + \Delta x, -\infty)$. Since there is no tunneling current along the *x*-direction, we have

$$\phi(\mathbf{r}_{a2}) - \phi(\mathbf{r}_{a1}) = -\frac{2\pi}{\Phi_0} \Delta x A_x(x, -\infty),$$

$$\phi(\mathbf{r}_{b2}) - \phi_b(\mathbf{r}_{b1}) = -\frac{2\pi}{\Phi_0} \Delta x A_x(x, +\infty). \qquad (6.43)$$

Hence,

$$\tilde{\phi}_{ab}(x + \Delta x) - \tilde{\phi}_{ab}(x) = \frac{2\pi}{\Phi_0} \oint d\mathbf{l} \cdot \mathbf{A} = \frac{B_0 L \Delta x}{\Phi_0}, \tag{6.44}$$

which equals the flux penetrating the loops surrounding $\mathbf{r}_{a1}$, $\mathbf{r}_{b1}$, $\mathbf{r}_{b2}$, and $\mathbf{r}_{a2}$. $L = d + l_a + l_b$ equals the summation of the thickness d of the junction and the penetration lengths on both sides. According to Eq. (6.14), the total Josephson tunneling current is

$$\begin{aligned} I_s &= j_0 \int_{-D}^{D} dx \sin\left(\phi_a - \phi_b - \frac{B_0 L x}{\Phi_0}\right) \\ &= 2 D j_0 \sin(\phi_a - \phi_b) \frac{\sin(\pi \Phi/\Phi_0)}{\pi \Phi/\Phi_0}, \end{aligned} \tag{6.45}$$

where $\phi_{a,b}$ are the superconducting phases at $x = 0$ deeply inside the a and b sides, respectively. The dependence of the critical tunneling current on the magnetic flux is

$$I_{\max} = I_0 \left| \frac{\sin(\pi \Phi/\Phi_0)}{\pi \Phi/\Phi_0} \right|, \tag{6.46}$$

where $\Phi = 2 B_0 L D$, and $I_{\max}$ reaches the maximum when $\Phi_a = 0$.

For the Josephson junction shown in Fig. 6.4(a), an s-wave superconductor is connected to the two adjacent surfaces, i.e. the ac and bc surfaces, of a high-T_c superconductor. For convenience, we take different gauges for the vector potentials on the two sides of the junction, and calculate the tunneling currents on the ac- and bc-surfaces separately. The total current is simply given by the sum of these two tunneling currents. It should be noted that the directions of currents from the two surfaces are orthogonal to each other, and only the components along the vertical direction contribute to the total current. If the superconducting electrons are s-wave paired, then the total tunneling current is given by the formula

$$\begin{aligned} I_s &= \frac{j_0}{\sqrt{2}} \left\{ \int_0^D dx \sin\left(\phi_a - \phi_b - \frac{B_0 L x}{\Phi_0}\right) + \int_0^D dy \sin\left(\phi_a - \phi_b + \frac{B_0 L y}{\Phi_0}\right) \right\} \\ &= \sqrt{2} j_0 D \sin(\phi_a - \phi_b) \frac{\sin(\pi \Phi/\Phi_0)}{\pi \Phi/\Phi_0}. \end{aligned} \tag{6.47}$$

Except for an extra factor $\sqrt{2}$, this result is the same as for the Josephson junction shown in Fig 6.4(b).

On the other hand, if the superconductor possesses the $d_{x^2-y^2}$-wave pairing symmetry, the tunneling currents from the ac- and bc-surfaces exhibit a π-phase difference and the total tunneling current becomes

$$I_s = \frac{1}{\sqrt{2}} j_0 \left\{ \int_0^D dx \sin\left(\phi_a - \phi_b - \frac{B_0 L x}{\Phi_0}\right) - \int_0^D dy \sin\left(\phi_a - \phi_b + \frac{B_0 L y}{\Phi_0}\right) \right\}$$

$$= -\sqrt{2} j_0 D \cos(\phi_a - \phi_b) \frac{\sin^2(\pi\Phi/2\Phi_0)}{\pi\Phi/\Phi_0}. \tag{6.48}$$

It reaches the maximum when $\cos(\phi_a - \phi_b) = 1$,

$$I_{max} = I_0 \frac{\sin^2(\pi\Phi/2\Phi_0)}{\pi\Phi/\Phi_0}. \tag{6.49}$$

Unlike the Josephson junction formed by two s-wave superconductors, $I_{max} = 0$, rather than taking the maximal value, at $\Phi = 0$. From the derivative, it can be shown that the maximum of I_{max} is determined by the equation

$$\pi\Phi = \tan(\pi\Phi/2\Phi_0). \tag{6.50}$$

Solving this equation numerically, we find that the maximum of I_{max} occurs at $\Phi \approx 0.742\Phi_0$.

Wollman et al. measured the field dependence of the critical tunneling currents for these two Josephson junctions. Figure 6.5 shows their results [150]. For the edge Josephson junction illustrated in Fig. 6.4(b), the maximal critical current occurs at zero field. In contrast, for the corner Josephson junction in Fig. 6.4(a), the critical

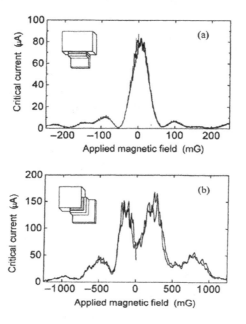

Figure 6.5 Magnetic field dependence of the tunneling current for the two Josephson junctions shown in Fig. 6.4. (From Ref. [150])

current at zero field is a local minimum. It does not reach zero, probably because the junction is not completely symmetric: YBCO does not possess the tetragonal symmetry due to the existence of the CuO chains. This suggests that the pairing symmetry in YBCO cannot be purely d-wave; it should also contain a small s-wave component, which contributes a finite critical current at zero field. This set of experimental results agrees with the theoretical prediction for d-wave superconductors. It lends strong support to the theory that the high-T_c electrons are d-wave paired.

6.3.2 *Spontaneous Quantized Flux*

In a superconducting quantum interference ring consisting of an odd number of π-junctions, there exists a spontaneously generated half-quantum flux, which can be used to judge whether high-T_c electrons have the d-wave pairing symmetry or not. The appearance of a spontaneous quantized flux relies only on the sign change of the superconducting gap function along different directions, rather than on the gap amplitude. This is again a phase-sensitive probe for the pairing gap, similar to the measurement of the interference effect in a Josephson junction or SQUID. It provides a powerful tool to determine qualitatively the pairing symmetry. Furthermore, the presence of the spontaneous half-quantum flux is not affected by the symmetry in the tunneling parameters of Josephson junctions or SQUIDs. If the orientation of the crystalline axes can be delicately designed to generate a spontaneous half-quantum magnetic flux, then not only the sign change of the gap function under the spatial rotation, but also the direction of the nodal lines can be accurately determined.

Tsuei and Kirtley of IBM carried out the first experimental measurement for the spontaneous half-quantum flux. They grew epitaxially a high quality high-T_c film of YBCO on a tricrystal substrate of SiTrO$_3$ with three delicately designed orientations. As shown in Fig. 6.6, the [100]-directions of the upper two crystals are rotated $30°$ and $60°$ with respect to the horizontal axis counterclockwise, respectively. Since a $90°$ rotation is a symmetry operation of the tetragonal system, this is a $\pi/2$-disclination topological defect of the crystalline configuration. Four Josephson rings were etched on the film as shown in Fig. 6.6. Because of the difference in the orientations of the crystalline axes, the phase differences at different crystal interfaces are different. The central ring crosses three interfaces. Since the crystalline axes rotate $\pi/2$ around the center, the superconducting phase changes by π in total. Hence it is a π-junction. For the other three rings, two of them cross the same interface twice, and the third one does not cross any interface. The phase changes around these rings are zero, and thus there are no spontaneously generated magnetic fluxes.

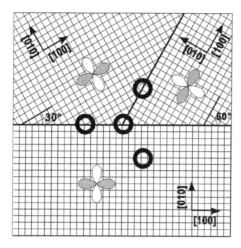

Figure 6.6 YBCO film epitaxially grown on the (100) surface of SiTiO$_3$: orientations of the crystalline axes and the phase patterns of the $d_{x^2-y^2}$ order parameters (From Ref. [151]). The tricrystal substrate SiTiO$_3$ is composed of three domains with different crystalline orientations. The crystalline axes of the YBCO film are aligned by the crystalline axes of SiTiO$_3$. Tunneling junctions naturally formed at the interfaces exhibit different phase differences. The thickness of the film is 1 200 Å. The four etched rings each has an inner radius 48 μm and a width 10 μm. If YBCO is a d-wave superconductor, the central ring should be a π-ring and contains a spontaneously generated half-quantum flux. For the other three rings, the accumulated phase differences are zero and there are no spontaneously generated magnetic fluxes.

Tsuei and Kirtley measured the fluxes in the four rings using the scanning SQUID microscope. Their results are shown in Fig. 6.7 [146, 148]. In the absence of external magnetic field, they found that the flux enclosed by the central tri-junction ring equals $\Phi_0/2$ within the experimental error, while the fluxes through the other three junctions are zero. Their results are fully consistent with the theoretical prediction of d-wave superconductors. They measured systematically how the magnetic fluxes change using different tri-crystal films by varying the crystalline orientations of the substrate. All the results they obtained support YBCO to have the d-wave pairing symmetry.

They also systematically investigated properties of spontaneously generated half-quantum fluxes in other high-T$_c$ superconductors, including both hole- and electron-doped ones [148, 151–156]. Their results are all consistent with the d-wave pairing theory, and suggest that the d-wave pairing is a universal feature of high-T$_c$ superconductivity.

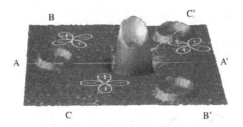

Figure 6.7 The scanning SQUID image for the tricrystal Josephson rings shown in Fig. 6.6. Within the experimental error, the flux through the central three-junction ring equals $\Phi_0/2$, while for other three rings the fluxes are zero. [148]

6.4 Paramagnetic Meissner Effect

The preceding discussion shows that there are spontaneously generated orbital currents or half-quantum fluxes in a Josephson junction ring composed of d-wave superconductors. This spontaneously generated flux can also exist in granular superconductors. It may induce the so-called paramagnetic Meissner effect, or the anti-Meissner effect, leading to a positive magnetic susceptibility in a weak magnetic field [145, 157].

The paramagnetic Meissner effect arises from the spontaneous fluxes generated in the granular superconductor. It depends on the granular structure. Two adjacent grains can be regarded as a weakly coupled Josephson junction. A large number of weakly coupled grains form a Josephson junction network which contains numerous loops. In a d-wave superconductor, some of the loops contain zero or an even number of π-junctions and there are no spontaneously generated magnetic fluxes. The other loops contain an odd number of π-junctions and have finite magnetic fluxes. In the absence of an applied magnetic field, these spontaneous magnetic fluxes (orbital moments) are randomly oriented and the net flux is zero. However, an applied magnetic field can polarize these fluxes. If the polarized magnetic moments from the π-loops surpass the contribution from the superconducting diamagnetic current, the granular superconductor becomes paramagnetic.

However, the magnetic moments of these spontaneously generated fluxes are generally very small. A weak magnetic field can completely polarize them. Hence the paramagnetic susceptibility is very weak. It becomes visible only in the vicinity of the superconducting transition temperature and in the weak field limit. Moreover, the paramagnetic susceptibility decreases with increasing field. It drops to zero when the paramagnetic moments contributed from π-loops become fully polarized. On the other hand, the superconducting diamagnetic current increases with the applied magnetic field. When the magnetic field is above a threshold, the diamagnetic moments dominate and the granular superconductor becomes diamagnetic.

This transition field from the paramagnetic to diamagnetic phases is very small, typically less than 1 gauss. For comparison, the earth magnetic field on average is about 0.5 gauss. Thus the magnetic field used to measure the paramagnetic Meissner effect should be as weak as possible, provided that the background contribution to the fluctuating magnetic moment can be well screened.

To measure the paramagnetic Meissner effect, one needs to cool down the granular system from the normal phase to the superconducting phase. There are two situations that should be distinguished. One is the so-called field cooling. This is to apply a magnetic field before cooling down. The other is the zero-field cooling. This is to apply a magnetic field after cooling.

In the case of field cooling, the spontaneous fluxes generated by the Josephson loops below the superconducting transition temperature are aligned with the applied field. In a weak applied field, the superconducting diamagnetic current is small, and the paramagnetic response is stronger than the diamagnetic one. The system is paramagnetic. With the increase of field, the diamagnetic effect becomes stronger and stronger, and the system eventually becomes diamagnetic. Figure 6.8 shows the susceptibility of the granular $Bi_2Sr_2CaCu_2O_8$ superconductor under field cooling as a function of temperature at several different external magnetic fields. As expected, the magnetic susceptibility is positive, or paramagnetic, in weak fields. It decreases with increasing field and becomes diamagnetic after the magnetic field exceeds a threshold at about 1 gauss. As shown in Fig. 6.8, in the field cooling case, the paramagnetic susceptibility grows with lowering temperature and becomes saturated at low temperatures as long as the field is week.

In the case of zero field cooling, the magnetic field is applied after the temperature has fallen far below the superconducting transition temperature and the susceptibility is measured with increasing temperature. As the measurement time is

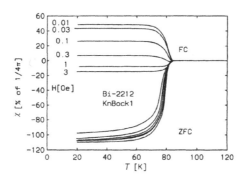

Figure 6.8 Temperature dependence of the magnetic susceptibility at various magnetic fields under field cooling (FC) and under zero field cooling (ZFC) for $Bi_2Sr_2CaCu_2O_8$ granular superconductors. (From Ref. [158])

usually very short, there is not enough time for the orbital moments spontaneously generated in the π-loops to relax toward the direction of the applied magnetic field. The orbital current in each π-loop is randomly oriented as in the zero field and the orbital moments cancel each other, yielding an extremely small net paramagnetic moment. As a result, the diamagnetic effect dominates and the susceptibility is always negative. Nevertheless, the difference between the susceptibility under high and low magnetic fields,

$$\delta\chi(T, H < H^*) = \frac{M(T, H)}{H} - \frac{M(T, H^*)}{H^*}, \qquad (6.51)$$

still contains information about the paramagnetism in the π-Josephson loops. $M(T, H)$ is the total magnetic moment of the system. In the experiment, H^* is typically around a few gauss, much higher than the field to fully polarize the paramagnetic orbital moments so that the nonlinear effect of paramagnetic moments can be ignored. Thus $M(T, H^*)/H^*$ is contributed almost entirely by the diamagnetic current. The relative susceptibility $\delta\chi(T, H < H^*)$ defined above subtracts the contribution from the nearly field-independent diamagnetic susceptibility. It represents the paramagnetic response from the π-rings in the weak external magnetic field.

Figure 6.9 shows how the relative susceptibility of $Bi_2Sr_2CaCu_2O_8$ granular superconductors $\delta\chi(T, H < H^*)$ varies with temperature and applied magnetic field. A peak in $\delta\chi(T, H < H^*)$ appears below T_c and its height grows with decreasing H. The presence of this peak is not difficult to understand physically. At low temperatures, the spontaneously generated Josephson currents are frozen and difficult to be flipped by a weak magnetic field. Thus the paramagnetic response is vanishingly small and $\delta\chi$ should decrease with decreasing temperature. Close to T_c, the spontaneously generated magnetic moments decrease and become zero exactly at T_c. Thus $\delta\chi$ should decrease with increasing temperature just below T_c.

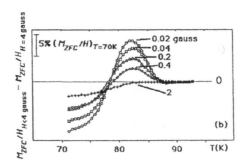

Figure 6.9 Temperature dependence of the relative susceptibility $\delta\chi(T, H < H^*)$ for $Bi_2Sr_2CaCu_2O_8$ granular superconductors under zero field cooling. $H^* = 4$ Gauss. (From Ref. [159])

This indicates that there must be a peak in $\delta\chi(T, H < H^*)$ between low temperature and T_c. The decrease of the peak height is a natural consequence of the decrease in the spontaneous magnetization of Josephson orbital moments in high fields.

It should be pointed out that the paramagnetic Meissner effect occurs not just in granular superconductors with d- or other non-s-wave pairing symmetry. It was also observed in some bulk s-wave superconductors. For example, weak paramagnetic Meissner effects were observed in Nb [160] and Al [161] superconductors. In these materials, the paramagnetic Meissner effect is likely to result from the sample inhomogeneity, especially on their surfaces. Compared with the paramagnetic Meissner effect in the d-wave granular superconductor [162], the s-wave counterpart is much weaker, and the observed paramagnetic susceptibility is also much smaller than that of the BSCCO granular superconductor. In addition, the paramagnetic Meissner effect in the d-wave granular superconductor is stable. It is an equilibrium property. However, in s-wave superconductors, the paramagnetic Meissner effect is a property of the metastable state. It decays with time. These two major differences can be used to determine whether the paramagnetic Meissner effect is due to the spontaneous magnetization of Josephson π-loops.

7

Single Impurity Scattering

7.1 Nonmagnetic Impurity Scattering

Impurity scattering is a ubiquitous and important physical effect in real materials. In most circumstances, for example, in the preparation of high quality samples, impurity is a factor that we do not want but cannot avoid. Vacancies, defects, and dislocations in materials are typical sources of impurity scattering. Those impurities with strong disorder characteristics are not produced on purpose. In fact, they emerge randomly and it is difficult to precisely control them. On the other hand, it is a common practice to intentionally change the material property by systematic doping with impurities. The intrinsic physical properties can be understood by measuring and analyzing the effects induced by impurities. Historically, impurity scattering has played a crucial role in the study of semiconductors. It has also been an important and active subject of high-T_c study.

In the study of high-T_c superconductivity, it is common to use zinc, nickel, or other dopants to partially replace some of the atoms in high-T_c materials. For different chemical and crystal structure and different dopants, the substituted atoms and their locations are different. Their effects on superconductivity are also different. Clearly, the effect is stronger if it is the Cooper atoms in the CuO_2 planes that are substituted. Otherwise, the effect is weaker. Which atom is substituted by a dopant depends on how close the ionic radii and chemical valences of two kinds of atom are. Doping of zinc or nickel atoms is mainly to replace copper atoms in high-T_c cuprates since they are all divalent cations and have similar ionic radii. Their effects on the superconducting properties are significant and in a certain sense universal. They can be detected by experimental measurements. Besides there are also many investigations with lithium, magnesium, and aluminum doping. These atoms substitute the copper atoms in the CuO_2 planes as well.

The Zn^{2+} cation exhibits a fully occupied valence electron configuration $3d^{10}$, and it is a nonmagnetic ion. Because of strong antiferromagnetic fluctuations in high-T_c superconductors, a non-magnetic impurity is often viewed as a magnetic one. However, this viewpoint lacks convincing theoretical justification. It is based on an implicit assumption that the zinc atom only replaces a spin but has no effect on surrounding antiferromagnetic correlations. It might be correct in the low hole doping limit, but it fails in general. Moreover, it cannot explain why zinc has qualitatively the same effect on physical properties in both underdoped and overdoped high-T_c superconductors, while antiferromagnetic fluctuations in the overdoped regime are much weaker than those in the underdoped regime. The valence electron configuration of Ni^{2+} is $3d^8$, which has eight electrons in the $3d$ orbitals. Due to the Hund's rule coupling, these $3d$ electrons are bounded into a spin 1 cation. As a result, Ni^{2+} is magnetic. In order to study the effect of the zinc and nickel impurities, one should start with a low-energy effective Hamiltonian derived in §2.8, and consider antiferromagnetic fluctuations and the impurity scattering effect in a unified way.

In an s-wave superconductor, the nonmagnetic impurity scattering may eliminate the gap anisotropy and slightly reduce the superconducting transition temperature T_c. However, nonmagnetic impurity scattering itself does not change T_c significantly, and the same for other physical properties. This is just a statement of the Anderson theorem [163] on nonmagnetic impurity scattering in an s-wave superconductor. A proof of this theorem with the condition for its validity is given in Appendix D. In contrast, a magnetic impurity serves as a pair-breaker in the superconducting state. It flips spins and breaks Cooper pairs, affecting strongly on the superconducting energy gap.

For a d-wave superconductor, the Anderson theorem is no longer valid. Similarly to a magnetic impurity in an s-wave superconductor, a nonmagnetic impurity is also a pair-breaker for d-wave paired electrons, and a thorough investigation of its effects is helpful for understanding physical properties of d-wave superconductors. In fact, both magnetic and non-magnetic impurity scattering effects have played an important role in the study of high-T_c superconductors, especially in the determination of pairing symmetry. In some cases, the single-impurity problem can be exactly solved, which is also useful for the understanding of many-impurity scattering problems discussed in the next chapter.

7.1.1 Impurity Scattering in a Metal

As an introduction to the T-matrix method, we first consider the single-impurity scattering problem in normal metals. The Hamiltonian reads

$$H = H_0 + H_{\text{imp}}, \tag{7.1}$$

where

$$H_0 = -\frac{\hbar^2 \nabla^2}{2m}, \qquad H_{\text{imp}} = U(\mathbf{r}). \tag{7.2}$$

Owing to the screening of conduction electrons, the range of the impurity potential is usually confined within a few lattice constants. Ideally, we assume the impurity potential

$$U(\mathbf{r}) = V_0 \delta(\mathbf{r}), \tag{7.3}$$

which is localized at the origin $\mathbf{r} = 0$.

The single-particle Green's function G is defined by the formula

$$(\omega - H)G(\omega) = I, \tag{7.4}$$

where I is the identity operator and ω is generally a complex frequency. The retarded Green's function corresponds to the limit where ω is a real frequency plus an infinitesimal small imaginary part. Equivalently, the Green's function can be expressed as

$$G(\omega) = \frac{1}{\omega - H}. \tag{7.5}$$

Expanding using the impurity potential term U, we obtain the Dyson equation

$$G(\omega) = G^{(0)}(\omega) + G^{(0)}(\omega)UG(\omega) = G^{(0)}(\omega)\frac{1}{1 - UG^{(0)}(\omega)}, \tag{7.6}$$

where $G^{(0)}$ is the Green's function of the free particle

$$G^{(0)}(\omega) = \frac{1}{\omega - H_0}. \tag{7.7}$$

The Dyson equation can be also written using the T-matrix as

$$G(\omega) = G^{(0)}(\omega) + G^{(0)}T(\omega)G^{(0)}(\omega), \tag{7.8}$$

where

$$T(\omega) = \frac{1}{1 - UG^{(0)}}U. \tag{7.9}$$

The imaginary part of the retarded Green's function is the density of states

$$\rho(\omega) = -\frac{1}{\pi}\text{Im}\text{Tr}G(\omega + i0^+) \tag{7.10}$$

at a real frequency ω. The impurity density of states is defined by the difference in the density of states with or without the impurity contribution, i.e.

$$\Delta\rho(\omega) = -\frac{1}{\pi}\text{Im}\text{Tr}\left[G(\omega + i0^+) - G^{(0)}(\omega + i0^+)\right]. \tag{7.11}$$

Using the identity

$$\mathrm{Tr}G(\omega) = -\frac{\partial}{\partial\omega}\mathrm{Tr}\ln G(\omega) = -\frac{\partial}{\partial\omega}\ln\det G(\omega), \qquad (7.12)$$

we can further represent the impurity density of states as

$$\Delta\rho(\omega) = -\frac{1}{\pi}\mathrm{Im}\frac{\partial}{\partial\omega}\ln\det G(\omega)\left[G^{(0)}(\omega)\right]^{-1}. \qquad (7.13)$$

Furthermore, using the Dyson equation and the fact that U is real, we can simplify the above expression as

$$\Delta\rho(\omega) = \frac{1}{\pi}\frac{\partial}{\partial\omega}\mathrm{Im}\ln\det T(\omega). \qquad (7.14)$$

Compared to the relation Eq. (2.61) for $\Delta\rho$ and the phase shift $\eta(\omega)$, the expression of the phase shift in terms of the T-matrix is

$$\eta(\omega) = \arg\left[\det T(\omega)\right]. \qquad (7.15)$$

For the δ-potential U defined in Eq. (7.3), there exists only the s-wave scattering, and the scattering problem can be exactly solved. In real space, the Dyson equation becomes

$$G(\mathbf{r},\mathbf{r}',\omega) = G^{(0)}(\mathbf{r}-\mathbf{r}',\omega) + G^{(0)}(\mathbf{r},\omega)V_0 G(0,\mathbf{r}',\omega). \qquad (7.16)$$

A graphical representation of this equation is shown in Fig. 7.1. Expressing using the T-matrix, it becomes

$$G(\mathbf{r},\mathbf{r}',\omega) = G^{(0)}(\mathbf{r}-\mathbf{r}',\omega) + G^{(0)}(\mathbf{r},\omega)T(\omega)G^{(0)}(-\mathbf{r}',\omega). \qquad (7.17)$$

For the δ-function potential, T-matrix is a function of ω only, and is determined by the equation

$$T(\omega) = V_0 + V_0 G^{(0)}(0,\omega)T(\omega). \qquad (7.18)$$

The solution is

$$T(\omega) = \frac{1}{V_0^{-1} - G^{(0)}(0,\omega)}. \qquad (7.19)$$

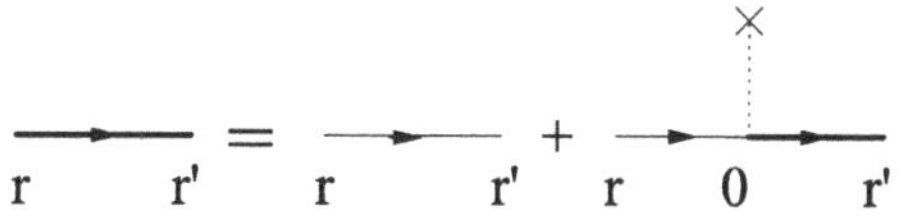

Figure 7.1 Feynman diagrams for the single-impurity scattering problem.

Now let us introduce a dimensionless Green's function at $\mathbf{r} = 0$

$$G_0(\omega) = \frac{1}{\pi N_F} G^0(0, \omega + i0^+). \tag{7.20}$$

This can be further expressed as

$$\begin{aligned}
G_0(\omega) &= \frac{1}{\pi N_F V} \sum_k \frac{1}{\omega - \xi_k + i0^+} \\
&= \frac{1}{\pi N_F V} \sum_k \frac{P}{\omega - \xi_k} - \frac{i}{N_F V} \sum_k \delta(\omega - \xi_k) \\
&= \Lambda_1(\omega) - i\Lambda_2(\omega),
\end{aligned} \tag{7.21}$$

where

$$\Lambda_1(\omega) \approx \frac{1}{\pi} \ln \left| \frac{\xi_2 - \omega}{\xi_1 - \omega} \right|, \qquad \Lambda_2(\omega) \approx 1, \tag{7.22}$$

and $\xi_{1,2}$ are the band top and bottom energies, respectively. The phase shift $\eta(\omega)$ is then found to be

$$\cot \eta(\omega) = \frac{1 - V_0 \pi N_F \Lambda_1(\omega)}{\pi N_F V_0 \Lambda_2(\omega)}. \tag{7.23}$$

The s-wave phase shift is determined by the scattering on the Fermi surface, i.e. $\eta(\omega = 0) = \eta_0$

$$\cot \eta_0 = \frac{1 - N_F V_0 \ln |\xi_2/\xi_1|}{\pi V_0 N_F}. \tag{7.24}$$

If the system is particle–hole symmetric, then the phase shift is simply determined by V_0 and the density of states of normal electrons at the Fermi level

$$\cot \eta_0 = c, \tag{7.25}$$

with

$$c = \frac{1}{\pi N_F V_0}. \tag{7.26}$$

7.1.2 Scattering in a Superconductor

In a superconductor, the electronic structure of an impurity is different from the surrounding atoms. The interacting potential between the impurity and the surrounding conduction electrons takes the Coulomb form:

$$H_{\text{imp}} = \int d\mathbf{r} \psi^\dagger(\mathbf{r}) U(\mathbf{r}) \psi(\mathbf{r}), \tag{7.27}$$

where $\psi^T = (c_\uparrow, c_\downarrow^\dagger)$ is the Nambu spinor.

Again for the δ-potential, the nonmagnetic single-impurity problem in a super-conductor can be solved exactly. On the other hand, if the range of the impurity scattering potential is finite [164], the scattering channels with nonzero angular momenta contribute as well. This may cause difficulty in theoretical calculations. In the discussion below, we only consider the impurity scattering problem of δ-function potential

$$U(\mathbf{r}) = V_0 \sigma_3 \delta(\mathbf{r}). \tag{7.28}$$

In the absence of the impurity, the Green's function of an electron is given by Eq. (4.4). In the complex frequency space, it becomes

$$G^{(0)}(\mathbf{k}, \omega) = \frac{\omega + \xi_{\mathbf{k}} \sigma_3 + \Delta_{\mathbf{k}} \sigma_1}{\omega^2 - \xi_{\mathbf{k}}^2 - \Delta_{\mathbf{k}}^2}. \tag{7.29}$$

The single-impurity scattering is still governed by the Feynman diagrams shown in Fig 7.1. Again, there are no crossing diagrams and the self-consistent Dyson equation, expressed in the coordinate representation, is given by

$$G(\mathbf{r}, \mathbf{r}', \omega) = G^{(0)}(\mathbf{r} - \mathbf{r}', \omega) + G^{(0)}(\mathbf{r}, \omega) V_0 \sigma_3 G(0, \mathbf{r}', \omega). \tag{7.30}$$

Setting $\mathbf{r} = 0$, the above equation becomes

$$G(0, \mathbf{r}', \omega) = G^{(0)}(-\mathbf{r}', \omega) + G^{(0)}(0, \omega) V_0 \sigma_3 G(0, \mathbf{r}', \omega). \tag{7.31}$$

From this equation, we find that

$$G(0, \mathbf{r}', \omega) = \frac{c}{c - G_0(\omega) \sigma_3} G^{(0)}(-\mathbf{r}', \omega), \tag{7.32}$$

where $G_0(\omega)$ is the dimensionless Green's function

$$G_0(\omega) = \frac{1}{\pi N_F} G^{(0)}(\mathbf{r} = 0, \omega) = \frac{1}{\pi N_F V} \sum_{\mathbf{k}} G^{(0)}(\mathbf{k}, \omega), \tag{7.33}$$

and c is the parameter defined in Eq. (7.26).

Substituting Eq. (7.32) into Eq. (7.30), $G(\mathbf{r}, \mathbf{r}', \omega)$ is found to be

$$G(\mathbf{r}, \mathbf{r}', \omega) = G^{(0)}(\mathbf{r} - \mathbf{r}', \omega) + G^{(0)}(\mathbf{r}, \omega) T(\omega) G^{(0)}(-\mathbf{r}', \omega), \tag{7.34}$$

where

$$T(\omega) = V_0 \sigma_3 \frac{c}{c - G_0(\omega) \sigma_3}, \tag{7.35}$$

is the T-matrix describing the impurity scattering.

For the d-wave superconductor, we assume that $\Delta_{\mathbf{k}} = \Delta_\varphi = \Delta_0 \cos 2\varphi$ depends only on the azimuthal angle φ of the wave vector, and $\xi_{\mathbf{k}}$ depends only on the amplitude of $\mathbf{k}$ but not on its angle φ. Now the radial integral of $G_0(\omega)$ over $\mathbf{k}$ defined in Eq. (7.33) can be evaluated, and the result is

$$G_0(\omega) = \frac{1}{2\pi^2} \int_0^{2\pi} d\varphi \int_{-\infty}^{\infty} d\xi \frac{\omega + \xi\sigma_3 + \Delta_\varphi\sigma_1}{\omega^2 - \xi^2 - \Delta_\varphi^2}. \tag{7.36}$$

The σ_3-term is odd with respect to ξ, and the σ_1-term exhibits d-wave symmetry with respect to φ, hence, both disappear after integration. Taking the integral over ξ, we obtain

$$G_0(\omega) = -\frac{1}{2\pi} \int_0^{2\pi} d\varphi \frac{\omega\,\theta\left(\mathrm{Re}\sqrt{\Delta_\varphi^2 - \omega^2}\right)}{\sqrt{\Delta_\varphi^2 - \omega^2}}, \tag{7.37}$$

where $\theta(x)$ is the standard Heaviside step function. It should be emphasized that in Eq. (7.36), the real part of $\sqrt{\Delta_\varphi^2 - \omega^2}$ must be nonnegative in order to avoid the ambiguity in the evaluation of the square root of this complex number. In deriving Eq. (7.37), we have used the property that the average of Δ_φ on the Fermi surface is zero. The particle–hole symmetry and the wide bandwidth assumption are also used in the integration of ξ. By adding an infinitesimal imaginary part to ω, Eq. (7.37) becomes

$$G_0(\omega + i0^+) = -\frac{1}{2\pi} \int_0^{2\pi} d\varphi \left[\mathrm{Re}\frac{\omega}{\sqrt{\Delta_\varphi^2 - \omega^2}} + i\mathrm{Re}\frac{|\omega|}{\sqrt{\omega^2 - \Delta_\varphi^2}} \right]. \tag{7.38}$$

The right-hand side of Eq. (7.37) is an elliptic integral. In case $0 < |\omega| \ll \Delta_0$,

$$G_0(\omega) \approx -\frac{2\omega}{\pi\Delta_0} \ln\frac{2\Delta_0}{-i\omega}. \tag{7.39}$$

In the limit of the imaginary part of ω approaching zero, the corresponding retarded Green's function is

$$G_0(\omega + i0^+) \approx -\frac{2\omega}{\pi\Delta_0} \ln\frac{2\Delta_0}{|\omega|} - i\frac{|\omega|}{\Delta_0}. \tag{7.40}$$

In a d-wave superconductor, $G_0(\omega)$ depends only on ω because $\Delta_\mathbf{k}$ averages to zero over the Fermi surface. In this case, T is a diagonal matrix, given by

$$T(\omega) = T_0 + T_3\sigma_3, \tag{7.41}$$

where

$$T_0 = \frac{1}{\pi N_F} \frac{G_0(\omega)}{c^2 - G_0^2(\omega)}, \tag{7.42}$$

$$T_3 = \frac{1}{\pi N_F} \frac{c}{c^2 - G_0^2(\omega)}. \tag{7.43}$$

In the discussion of nonmagnetic impurity scattering, two limits should be considered. One is the strong scattering limit, which is also called the resonance scattering limit, or the unitary scattering limit. The other is the weak scattering limit, which is also called the Born scattering limit. The d-wave superconductors under these two limits exhibit different behaviors, both qualitatively and quantitatively. The strong scattering limit corresponds to the limit $c \to 0$ with the phase shift $\eta_0 \to \pi/2$. The weak scattering Born limit corresponds to $c \to \infty$ or $\eta_0 \to 0$.

7.2 Resonance State

The pole of the T-matrix corresponds to the frequency of a resonance state induced by the impurity scattering. It is determined by

$$G_0(\Omega) = +c. \tag{7.44}$$

The pole Ω has no real solution. It has only complex solutions when $c \ll 1$. Equation (7.44) is difficult to solve exactly in general, but it can be solved approximately when $\mathrm{Re}\,\Omega \ll \Delta_0$. Substituting the expression of $G_0(\omega)$ given by Eq. (7.40) into Eq. (7.44), we have

$$\bar{\Omega}\left(\ln\frac{4}{\pi c} \pm i\frac{\pi}{2} - \ln\bar{\Omega}\right) = 1, \tag{7.45}$$

where $\bar{\Omega} = \pm 2\Omega/\pi\Delta_0 c$. After taking the logarithm of both sides, it becomes

$$\ln\bar{\Omega} + \ln\left(\ln\frac{4}{\pi c} \pm i\frac{\pi}{2} - \ln\bar{\Omega}\right) = 0. \tag{7.46}$$

Supposing $\ln\bar{\Omega}$ is very small, Ω can be solved by expanding the left-hand side with respect to $\ln\bar{\Omega}$. Up to the first order terms in $\ln\bar{\Omega}$, the solution is

$$\Omega \approx \pm\frac{\pi c\Delta_0}{2}\exp\left(-\frac{a}{a-1}\ln a\right), \tag{7.47}$$

$$a = \ln\frac{4}{\pi c} \pm i\frac{\pi}{2}.$$

In the limit $|c| \to 0$, $|a| \gg 1$, Eq. (7.47) is approximately given by

$$\Omega \approx \pm\Omega_0 - i\Gamma_0, \tag{7.48}$$

where

$$\Omega_0 = -\frac{2\pi|c|\Delta_0\ln(\pi|c|/4)}{4\ln^2(\pi|c|/4) + \pi^2}, \tag{7.49}$$

$$\Gamma_0 = \frac{\pi^2|c|\Delta_0}{4\ln^2(\pi|c|/4) + \pi^2}, \tag{7.50}$$

where $\Gamma_0/\Omega_0 \sim -1/\ln(\pi|c|/4)$. In the unitary limit, i.e. $c \to 0$, we have $\Gamma_0 \sim \Omega_0/|\ln \Omega_0|$. The retarded Green's function $G(\mathbf{r},\mathbf{r}',\omega)$ is analytic in the upper half complex plane of ω. The pole of the T-matrix only exists in the lower half-plane.

The complex solution of Ω indicates that the pole of the T-matrix corresponds to a resonance state, not a bound state. This is due to the hybridization between the impurity state with the low-energy bulk excitations in d-wave superconductors. The resonance frequency decreases with increasing c. In the unitary scattering limit $c \to 0$, both the resonance frequency and Γ_0 becomes 0. The fact that $\Gamma_0/\Omega_0 \to 1/|\ln \Omega_0|$ shows that the resonance state is marginally long-lived, similar to a bound state. It should be noted that the resonance state only exists in the strong scattering regime. In the limit of weak scattering ($c \to \infty$), there is no resonance state. In fact, when $c \sim 1$, $\Omega_0 \sim \Delta_0$ and $\Gamma_0 \sim \Omega_0$, it is no longer meaningful to interpret the pole of the T-matrix as a resonance state.

In a superconductor, there are two characteristic length scales. One is the Fermi wavelength l_F related to the Fermi wave vector $l_F = 2\pi/k_F$. The other is the Cooper pair coherence length determined by the Fermi velocity v_F and the energy gap Δ_0 through $\xi_0 = \hbar v_F/\Delta_0$. In the presence of a resonance state, another characteristic length appears related to the impurity resonance state $l_0 = \hbar v_F/\Omega_0$. When the condition $l_F \ll \xi_0 \ll l_0$ is satisfied, the influence from the resonance state becomes important.

The appearance of resonance divides the energy space into three characteristic regimes. The first is the weak scattering regime at $\omega \ll \Omega_0$, so that $|G_0(\omega)| \ll |c|$, and T_3 dominates over T_0. The second is the high frequency regime, where $|\omega| \gg \Omega_0$, $G_0(\omega)| \gg |c|$, and the T_0-term dominates over the T_3 one. The third is the resonance regime $|\omega| \sim \Omega_0$, where $T_\pm$ changes rapidly with ω and the resonance state appears at $|\omega| = \Omega_0$. In the resonance regime, the impurity potential scattering plays an important role. Thus in the discussion of a specific problem, in order to capture the key points with a simple and correct method, it is crucial to know which regime the system is in.

In high-T_c cuprates, a zinc impurity is a strong scattering center, as already explained in §2.8. A zinc impurity is expected to generate a sharp peak near the Fermi surface. This resonance peak was observed experimentally [165]. Figure 7.2 shows the scanning tunneling microscope spectroscopy above the zinc impurity in $Bi_2Si_2CaCu_2O_8$. The resonance state energy generated by the zinc impurity is 1.5 meV, much smaller than the superconducting gap $\Delta_0 \sim 48$ meV. According to Eq. (7.49), the scattering parameter is found to be $c = \cot \eta_0 = 0.07$. This corresponds to a scattering phase shift $\eta_0 = 0.48\pi$. This phase shift is very close to $\pi/2$, indicating that the zinc impurity scattering potential is indeed in the unitary limit. According to Eq. (7.26), the zinc scattering potential can be expressed in a dimensionless way using c as

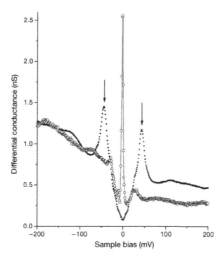

Figure 7.2 Scanning tunneling differential conductance, measured at (open circles) and far away (solid dots) from the Zn impurity site in the $Bi_2Si_2CaCu_2O_{8+\delta}$ high T_c superconductor. [165]

$$N_F V_0 = \frac{1}{\pi c} \sim 5, \qquad (7.51)$$

where the density of states N_F is inversely proportional to the bandwidth. This value is much larger than 1, hence, it is in the strong scattering regime. For the $Bi_2Si_2CaCu_2O_8$ superconductor, N_F is estimated to be 1.5 $(eV)^{-1}$ from the specific heat measurement [166]. Correspondingly, the zinc scattering potential strength is estimated as $V_0 \sim 3.3$ eV. This value, as discussed in §2.8, is just the energy of the Zhang–Rice singlet, much larger than the bandwidth.

7.3 Correction to the Density of States

The density of states of superconducting electrons is determined by the imaginary part of the retarded Green's function

$$\rho(\mathbf{r}, \omega) = -\frac{1}{\pi} \text{Im} G_{11}(\mathbf{r}, \mathbf{r}, \omega + i0^+) = \rho_0(\omega) + \delta\rho(\mathbf{r}, \omega), \qquad (7.52)$$

where

$$\rho_0(\omega) = -\frac{1}{\pi} \text{Im} G_{11}^{(0)}(\mathbf{r} = 0, \omega + i0^+), \qquad (7.53)$$

$$\delta\rho(\mathbf{r}, \omega) = -\frac{1}{\pi} \text{Im} G_{1\alpha}^{(0)}(\mathbf{r}, \omega + i0^+) T_{\alpha\alpha}(\omega + i0^+) G_{\alpha 1}^{(0)}(-\mathbf{r}, \omega + i0^+). \qquad (7.54)$$

ρ_0 is the electron density of states in the absence of impurity, independent of the spatial coordinate. $\delta\rho(\mathbf{r},\omega)$ is the correction to the density of states induced by the impurity. The potential scattering does not change the total number of states on each site, and the summation of $\delta\rho(\mathbf{r},\omega)$ over frequency is 0, satisfying the sum rule ([3])

$$\int d\omega\,\delta\rho(\mathbf{r},\omega) = 0. \tag{7.55}$$

If the impurity scattering is in the Born limit, $c \to \infty$, the correction to the density of states is inversely proportional to c and can be neglected. In the presence of resonance states, the correction to the density of states remains small in the low and high energy regimes, but it becomes significant in the resonance energy regime.

In the following, we focus on the properties of the impurity density of states in the limit of $c \to 0$, where the analytic behavior of the density of states can be relatively simple to derive. The results obtained in this limit, nevertheless, hold qualitatively even when c is not so small. It should be emphasized that in the analysis of the density of states in the zero energy limit, we should first take the limit $c \to 0$ and then set $\omega \to 0$. The sequence of these two limits should not be swapped. Otherwise, $\mathrm{Im}T(\omega \to 0) = 0$ at any finite $c \neq 0$, and the contribution of the resonance state is entirely ignored.

At the impurity site, the correction to the density of states is given by

$$\delta\rho(0,\omega) = -N_F\mathrm{Im}\frac{G_0^3(\omega)}{c^2 - G_0^2(\omega)}. \tag{7.56}$$

At the resonance frequency Ω, the unperturbed Green's function is

$$G_0(\Omega_0) \approx -|c| - \frac{i\pi|c|}{2\ln(4/\pi|c|)}. \tag{7.57}$$

Clearly, $|\mathrm{Im}G_0| \ll |\mathrm{Re}G_0|$, so we have

$$\delta\rho(0,\Omega_0) = -N_F\frac{(\mathrm{Re}G(\Omega_0))^2}{\mathrm{Im}G(\Omega_0)} = \frac{N_F|c|\ln(4/\pi|c|)}{\pi}. \tag{7.58}$$

This shows that at $\omega = \pm\Omega_0$, $\delta\rho \sim |c|\ln^{-1}|c|$ has a weak peak at the resonance frequency. The weight of this peak (namely the area enclosed by the peak) is approximately given by

$$w(\mathbf{r} = 0) \sim \delta\rho(0,\Omega_0)\Gamma_0 \approx \frac{\pi N_F \Delta_0 c^2}{4}\ln^{-3}\frac{4}{\pi|c|}. \tag{7.59}$$

Figure 7.3 shows the correction to the density of states at the impurity site. Two resonance peaks induced by the impurity emerge in the density of states. As shown by the figure, the smaller is c, the closer is the resonance energy to zero. A larger c gives rise to a higher resonance peak in the density of states. Note that

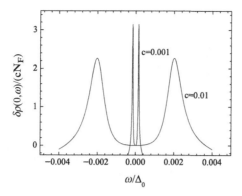

Figure 7.3 Impurity correction to the quasiparticle density of states $\delta\rho(0,\omega)/(cN_F)$ around the resonance energy in the unitary limit.

$\delta\rho(0,\omega)/(cN_F)$ is normalized by c in this figure. Without this normalization, the resonance peak of $c = 0.01$ would be almost 10 times higher than that of $c = 0.001$.

In the limit $c \to 0$ and $\omega \to 0$, the T-matrix exhibits a singularity. However, $G^{(0)}(\mathbf{r},\omega \to 0) \neq 0$ at $\mathbf{r} \neq 0$, and $G^{(0)}(\mathbf{r},\omega)$ is regular as a function of ω. It becomes real in the limit $\omega \to 0$,

$$\lim_{\omega\to 0} G^{(0)}(\pm\mathbf{r},\omega + i0^+) = -\frac{1}{V}\sum_{\mathbf{k}} \frac{\xi_{\mathbf{k}}\sigma_3 + \Delta_{\mathbf{k}}\sigma_1}{\xi_{\mathbf{k}}^2 + \Delta_{\mathbf{k}}^2} \cos(\mathbf{k}\cdot\mathbf{r}). \tag{7.60}$$

In this case,

$$\delta\rho(\mathbf{r},\omega) \approx -\frac{1}{\pi}G^{(0)}_{1\alpha}(\mathbf{r},0)\mathrm{Im}T_{\alpha\alpha}(\omega)G^{(0)}_{\alpha 1}(-\mathbf{r},0). \tag{7.61}$$

At the resonance frequency Ω_0,

$$\mathrm{Im}T(\Omega_0) \approx -\frac{1}{\pi N_F c}\begin{pmatrix} \dfrac{\pi}{8\ln(4/\pi c)} & 0 \\ 0 & \dfrac{2\ln(4/\pi c)}{\pi} \end{pmatrix}. \tag{7.62}$$

Thus $\delta\rho(\mathbf{r},\Omega_0)$ diverges as $|c|^{-1}$, and the spectral weight of the resonance peak on the neighboring sites is c^{-2} times stronger than that on the impurity site. This means that the resonance effect is significantly stronger on the four neighboring sites of the impurity. To detect the resonance state experimentally, one should therefore concentrate on measuring the density of states not just on the impurity site, but more on its neighboring sites.

In the limit $c \to 0$, $T_0 \approx 1/\pi N_F G_0(\omega)$, and T_3 is small compared to T_0. Therefore, when $\Delta_0 \gg \omega \gg \Omega_0$, $\delta\rho(\mathbf{r},\omega)$ is approximately given by

$$\delta\rho(\mathbf{r},\omega) = \frac{1}{\pi^2 N_F}\left[G^{(0)}_{11}(\mathbf{r},0)\right]^2 \mathrm{Im}G^{-1}_0(\omega). \tag{7.63}$$

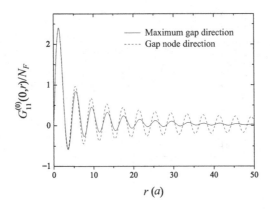

Figure 7.4 $G_{11}^{(0)}(r,0)/N_F$ as a function of r along the gap nodal (dashed line) and antinodal (solid line) directions, respectively. $k_F = \pi/2a$, $\xi_0 = 5\pi a$, and a is the lattice constant.

Hence the ω dependence of $\delta\rho(\mathbf{r}, \omega)$ is determined by the imaginary part of $G_0^{-1}(\omega)$, and its $\mathbf{r}$ dependence is determined by the square of $G_{11}^{(0)}(\mathbf{r}, 0)$. The imaginary part of G_0^{-1}, according to Eq. (7.40), is

$$\mathrm{Im}\, G_0^{-1}(\omega + i0^+) \simeq \frac{\pi^2 \Delta_0}{|\omega| \left(4 \ln^2 \frac{2\Delta_0}{|\omega|} + \pi^2\right)}. \tag{7.64}$$

The dependence of $G_{11}^{(0)}(\mathbf{r}, 0)$ on $\mathbf{r}$ is more complicated. But when $\varepsilon \gg \Delta_0$ and $r > \xi_0$, $G_0(\mathbf{r}, 0)$ in Eq. (7.60) contributes mainly from the momentum summation in the region of $\mathbf{k} \perp \mathbf{r}$, due to the rapid oscillation of $\cos(\mathbf{k} \cdot \mathbf{r})$.

If $\mathbf{r}$ is along the antinodal direction,

$$G_{11}^{(0)}(\mathbf{r}, 0) = 2N_F \int_0^1 dx \frac{1}{\sqrt{1-x^2}} \exp\left[-\frac{\Delta_0 \bar{r}}{\varepsilon_F} |2x^2 - 1| x\right] \sin(\bar{r}x), \tag{7.65}$$

where $\bar{r} = k_F r$. On the other hand, if $\mathbf{r}$ is along the nodal direction,

$$G_{11}^{(0)}(\mathbf{r}, 0) = 2N_F \int_0^1 dx \frac{1}{\sqrt{1-x^2}} \exp\left(-\frac{2\Delta_0 \bar{r}}{\varepsilon_F} \sqrt{1-x^2}x^2\right) \sin(\bar{r}x). \tag{7.66}$$

Figure 7.4 shows the behavior of $G_{11}^{(0)}(\mathbf{r}, 0)$ as a function of r along these two directions under the condition $\varepsilon_F \gg \Delta_0$. $G_{11}^{(0)}(\mathbf{r}, 0)$ takes a maximal value at $\bar{r} \sim 2$, and then decreases with increasing r. The decay along the antinodal direction is faster than that along the nodal direction.

7.4 Tunneling Spectrum of Zinc Impurity

The impurity effect on the superconducting electron density of states can be measured by the scanning tunneling microscope (STM). From the result shown in §7.3, we know that the impurity correction to the density of states at the resonance energy is much weaker on the impurity site than on its four nearest neighbors. This result is not consistent with the experimental observations around the zinc impurity in high-T_c superconductors. What the experiment found [165], as shown in Fig. 7.5, is that the impurity induced STM spectral weight is the strongest at the zinc impurity site and very weak at the four nearest neighboring sites. This dislocation of the spectral weight seems to contradict the theory which takes Zn as a strong nonmagnetic impurity. However, if the effect of the anisotropy in the c-axis tunneling matrix element is taken into account, it can be shown that the spatial dislocation of the STM spectroscopy is in fact not a negative, but an affirmative evidence for the nonmagnetic resonance impurity theory of Zn.

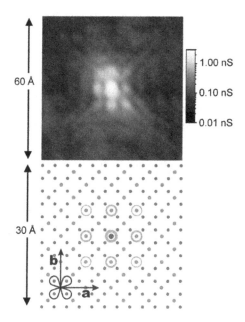

Figure 7.5 Spatial distribution of the differential tunneling conductance of the Zn resonance state in $Bi_2Si_2CaCu_2O_{8+}$. The differential conductance is largest at the zinc site (i.e. the brightest spot), and then exhibits alternating dark and bright spots on the sites away from the impurity. The tunneling differential conductance shows local minima at four nearest neighboring sites (dark spots), and local maxima at four next-nearest neighboring Cu sites (bright spots). The a- and b-axes are rotated from the crystal axes by $\pi/4$. (From Ref. [165])

In general, the differential conductance measured by STM is proportional to the electron density of states. This is a basic assumption commonly used in the analysis of STM experimental data. However, this assumption is not valid in all cases. In fact, the tunneling conductance depends not just on the electron density of states, but also on the tunneling matrix elements. If the tunneling matrix elements do not exhibit an apparent anisotropy (or momentum dependence) so that it can be approximately treated as a constant, the differential tunneling conductance is indeed determined purely by the density of states. On the contrary, if the tunneling matrix elements are strongly momentum dependent, the tunneling conductance is strongly modified and no longer proportional to the electron density of states. This is just the situation encountered in the STM measurement of the zinc impurity in high-T_c superconductors.

High-T_c cuprates are quasi-two-dimensional materials with weak interlayer couplings. It is easier to cleave a high-T_c sample along a surface parallel to the CuO_2 planes. Most of the STM experiments place the probe tips perpendicular to the CuO_2 planes, and measure the tunneling currents along the c-axis direction. However, the c-axis hopping of electrons is highly anisotropic because the hopping of the oxygen holes in the $2p$ orbitals along the c-axis needs the assistance of the copper $4s$ orbital in cuprates. As the overlap between the oxygen antibonding p orbitals and the copper $4s$-orbital possesses the $d_{x^2-y^2}$ symmetry, the electron hopping depends strongly on the in-plane momentum of electrons on the CuO_2 planes.

Following the discussion given in §2.7, one can show by the symmetry argument that the tunneling Hamiltonian along c-axis is given by [103, 167, 168],

$$H_t = \sum_i t_\perp c_i^\dagger D_i + H.c., \tag{7.67}$$

where $t_\perp$ is the tunneling constant, and c_i is the electron operator in the metallic probe tip. D_i is an effective electron operator defined in the superconductor

$$D_i = \sum_j F_{i,j} \left(\delta_{j,i_0} + \frac{1}{\sqrt{2}} \delta_{j \neq i_0} \right) d_j, \tag{7.68}$$

where d_j is the annihilation operator of electron at site j and i_0 is the coordinate of the impurity site, and

$$F_{i,j} = \frac{1}{N} \sum_k \frac{\cos k_x - \cos k_y}{\sqrt{\cos^2(k_x/2) + \cos^2(k_y/2)}} e^{i\mathbf{k}\cdot(i-j)}. \tag{7.69}$$

The form factor $\cos k_x - \cos k_y$ arises from the $d_{x^2-y^2}$ symmetry of the hybridization matrix elements between the Cu $4s$ and O $2p$ antibonding orbitals as explained in

§2.7. From the second order perturbation of H_t, it can be shown that the tunneling conductance g_i at site i is proportional to the density of states of the effective electron operator D_i,

$$g_i(V) \propto \left\langle D_i \delta(eV - H_{imp}) D_i^\dagger \right\rangle, \qquad (7.70)$$

where V is the bias voltage and H_{imp} is the system Hamiltonian defined by Eq. (2.68) for a zinc impurity.

In Eq. (7.70), if $D_i = d_i$, $g_i(V)$ is proportional to the local density of states of electrons $\left\langle d_i \delta(eV - H_{imp}) d_i^\dagger \right\rangle$. However, in high-$T_c$ superconductors, $D_i \neq d_i$. From the expression of $F_{i,j}$, it is simple to show that: (1) $F_{i,j} = 0$ if $i = j$ due to the d-wave symmetry of the c-axis hopping form factor; (2) $|F_{i,j}|$ reaches the maximum if i and j are nearest neighbors. These properties of $F_{i,j}$ lead to the following results: (1) The quasiparticle excitation at site i has no contribution to the tunneling conductance at that site; (2) the tunneling conductance at site i is contributed to mainly by the quasiparticle excitations on its four nearest neighboring sites.

Based on these properties, it is not difficult to understand why the spatial dislocation occurs in the STM spectroscopy of the impurity resonance state. The tunneling current on the impurity site comes mainly from the density of states on its four nearest neighbor sites, giving rise to the strongest STM spectra. In contrast, on the four nearest neighboring sites of the zinc impurity, the STM spectroscopy comes mainly from their nearest neighbors on which the weights of the resonance states are very small. Thus the spatial dislocation of the zinc impurity resonance state is caused by the anisotropic tunneling matrix elements along the c-axis ([168]), consistent with the theory that Zn is a strong nonmagnetic scattering center.

7.5 Comparison with Anisotropic s-Wave Pairing State

The previous discussion indicates that the unitary nonmagnetic impurity scattering has a strong impact on the d-wave pairing states. However, it is not clear whether this effect is due to the gap symmetry or simply due to the gap anisotropy. Can a strongly anisotropic s-wave superconductor exhibit the same effect? This is an important question that needs to be addressed in order to understand the effect of impurity scattering in high-T_c superconductors. To answer this question, let us compare the impurity scattering in a d-wave superconductor with that in a strongly anisotropic s-wave superconductor whose gap function is defined by

$$\Delta(\mathbf{k}) = \Delta_0 |\cos(2\phi)|. \qquad (7.71)$$

It has the same amplitude as the d-wave superconducting gap, but without changing sign on the Fermi surface.

The quasiparticle Green's function and the Dyson equation for the anisotropic s-wave superconductor can be similarly derived as for the d-wave superconductor. The difference is that the summation of $G^{(0)}$ over momentum $\mathbf{k}$ on the right-hand side of Eq. (7.33) now has an extra off-diagonal term in additional to the $G_0(\omega)$ term previously obtained for the d-wave superconductor

$$G_0^s(\omega) = \frac{1}{\pi N_F V} \sum_{\mathbf{k}} G^{(0)}(\mathbf{k}, \omega) = G_0(\omega) + G_1(\omega)\sigma_1, \qquad (7.72)$$

where

$$G_1(\omega) = \frac{1}{\pi N_F V} \sum_{\mathbf{k}} \frac{\Delta_0 |\cos(2\varphi)|}{\omega^2 - \xi_{\mathbf{k}}^2 - \Delta_{\mathbf{k}}^2}. \qquad (7.73)$$

Now the T-matrix becomes

$$T(\omega) = V_0\sigma_3 \frac{c}{c - G_0^s(\omega)\sigma_3} = \frac{1}{\pi N_F} \frac{G_0(\omega) - c\sigma_3 - G_1(\omega)\sigma_1}{c^2 + G_1^2(\omega) - G_0^2(\omega)}, \qquad (7.74)$$

and the poles of the T-matrix are determined by the equation

$$G_0^2(\omega) = c^2 + G_1^2(\omega). \qquad (7.75)$$

In comparison with Eq. (7.44), the resonance states previously discussed for the d-wave superconductor can be straightforwardly generalized to the s-wave case. We only need to replace c^2 in the d-wave case by $c^2 + G_1^2(\omega)$. Such a replacement indicates that the impurity scattering effect is weakened by the G_1 term in the anisotropic s-wave superconductor. In particular, the smaller is c, the stronger is the correction from the G_1 term. This correction becomes dominant in the unitary scattering limit.

In the limit $\omega \ll \Delta_0$, G_1 is

$$G_1(\omega) \approx -\frac{2}{\pi} \left(\sqrt{1 - \frac{\omega^2}{\Delta_0^2}} + i\frac{|\omega|}{\Delta_0} \right). \qquad (7.76)$$

Since $G_1(\omega \to 0) \to -2/\pi$ is finite, it suggests that in the low energy limit, the effective scattering parameter c in the anisotropic s-wave superconductor is equivalent to $\sqrt{c^2 + 4/\pi^2}$ in the d-wave superconductor. Thus no matter how strong the scattering potential is, the impurity scattering in the s-wave superconductor is not in the resonance scattering limit and there is no low energy resonance state. This is just a consequence of the Anderson theorem ([163]).

The above analysis indicates that the low energy resonance state is caused by the gap symmetry, rather than the gap anisotropy. For the s-wave pairing state, the average value of the gap function on the full Fermi surface is finite. A nonmagnetic impurity cannot significantly suppress the gap function. In contrast, in the d-wave case, since the gap function averages to zero, even a nonmagnetic impurity is a strong pair-breaker. In other words, if the non-magnetic impurity induced resonance state is observed in a superconductor, the superconducting electrons are not s-wave paired no matter whether the gap function is isotropic or anisotropic. Therefore, the experimental observation of the zinc impurity induced resonance state is a direct evidence in support of non-s-wave pairing state in high-T_c superconductors.

7.6 Classical Spin Scattering

In a conventional metal, a classical spin acts like a potential scatterer. However, in a singlet pairing superconductor, the scattering effect of a classical spin is different. This problem was first studied by Lu Yu. He investigated the effect of a paramagnetic spin in an s-wave superconductor, and predicted that the magnetic impurity induces a bound state inside the superconducting gap [169]. Later on, this problem was further explored by Shiba [170], who provided a unified description for both the weak and strong scattering limits.

The interaction between a classical spin and an electron is defined by the Hamiltonian

$$H_{\text{int}} = J \int d\mathbf{r} \delta(\mathbf{r}) c_{\mathbf{r}}^{\dagger} \frac{\sigma}{2} c_{\mathbf{r}} \cdot \mathbf{S}, \tag{7.77}$$

where $\sigma = (\sigma_1, \sigma_2, \sigma_3)$ are the Pauli matrices. In the treatment of the spin-flip terms, it is more convenient to double the dimension of the Nambu spinor from two to four by defining the following spinor operators,

$$c_{\mathbf{k}} = \begin{pmatrix} c_{\mathbf{k}\uparrow} \\ c_{\mathbf{k}\downarrow} \end{pmatrix}, \qquad d_{\mathbf{k}} = \begin{pmatrix} c_{\mathbf{k}} \\ c_{-\mathbf{k}}^{\dagger} \end{pmatrix}. \tag{7.78}$$

$d_{\mathbf{k}}$ is a four-dimensional spinor. In this new representation, the BCS Hamiltonian (1.25) becomes

$$H_{BCS} = \frac{1}{2} \sum_{\mathbf{k}} d_{\mathbf{k}}^{\dagger} \begin{pmatrix} \xi_{\mathbf{k}} & i\Delta_{\mathbf{k}}\sigma_2 \\ -i\Delta_{\mathbf{k}}\sigma_2 & -\xi_{\mathbf{k}} \end{pmatrix} d_k. \tag{7.79}$$

The corresponding free Green's function is

$$G^{(0)}(\mathbf{k}, \omega) = \left[\omega - \begin{pmatrix} \xi_{\mathbf{k}} & i\Delta_{\mathbf{k}}\sigma_2 \\ -i\Delta_{\mathbf{k}}\sigma_2 & -\xi_{\mathbf{k}} \end{pmatrix} \right]^{-1}, \tag{7.80}$$

where

$$\sigma_2 = \begin{pmatrix} 0 & -i \\ i & 0 \end{pmatrix} \tag{7.81}$$

is the charge conjugation matrix acting in the spin space. Using the identity

$$c_{\mathbf{r}}^{\dagger} \sigma_{\alpha} c_{\mathbf{r}} = c_{\mathbf{r}} \sigma_2 \sigma_{\alpha} \sigma_2 c_{\mathbf{r}}^{\dagger}, \tag{7.82}$$

the interaction Hamiltonian Eq. (7.77) can be expressed as

$$H_{\text{int}} = \frac{1}{4} J \int d\mathbf{r} \delta(\mathbf{r}) d_{\mathbf{r}}^{\dagger} \mathcal{A} d_{\mathbf{r}}, \tag{7.83}$$

where

$$\mathcal{A} = \begin{pmatrix} \sigma \cdot \mathbf{S} & 0 \\ 0 & \sigma_2 \sigma \cdot \mathbf{S} \sigma_2 \end{pmatrix}. \tag{7.84}$$

This classical spin scattering problem can be similarly solved as for a nonmagnetic impurity. The Dyson equation of the single-particle Green's function is now given by

$$G(\mathbf{r}, \mathbf{r}', \omega) = G^{(0)}(\mathbf{r} - \mathbf{r}', \omega) + G^{(0)}(\mathbf{r}, \omega) \frac{1}{4} J \mathcal{A} G(0, \mathbf{r}', \omega). \tag{7.85}$$

From this equation, we find that

$$G(0, \mathbf{r}', \omega) = \left[1 - \frac{1}{4} J F(\omega) \mathcal{A} \right]^{-1} G^{(0)}(-\mathbf{r}', \omega), \tag{7.86}$$

$$F(\omega) = G^{(0)}(\mathbf{r} = 0, \omega). \tag{7.87}$$

Substituting Eq. (7.86) into Eq. (7.85), we obtain

$$G(\mathbf{r}, \mathbf{r}', \omega) = G^{(0)}(\mathbf{r} - \mathbf{r}', \omega) + G^{(0)}(\mathbf{r}, \omega) T(\omega) G^{(0)}(-\mathbf{r}', \omega). \tag{7.88}$$

It has the same form as Eq. (7.34), but the T-matrix now becomes

$$T(\omega) = \frac{1}{4} J \mathcal{A} \left[1 - \frac{1}{4} J F(\omega) \mathcal{A} \right]^{-1}. \tag{7.89}$$

H_{BCS} is invariant under spin rotations. This implies that F and $\mathcal{A}$ satisfy the commutation relation

$$\mathcal{A} F(\omega) = F(\omega) \mathcal{A}. \tag{7.90}$$

Using this result and the identity $\mathcal{A}^2 = S^2 I$, the T-matrix can be simplified as,

$$T(\omega) = \left[\frac{1}{4} J \mathcal{A} + \left(\frac{1}{4} J S \right)^2 F(\omega) \right] \frac{1}{1 - \left(\frac{1}{4} J S \right)^2 F^2(\omega)}. \tag{7.91}$$

As the average of $\mathcal{A}$ over the direction of the impurity spin is zero, i.e. $\langle \mathcal{A} \rangle = 0$, the spin averaged T-matrix is found to be

$$\langle T(\omega) \rangle = \left(\frac{1}{4}JS\right)^2 \frac{F(\omega)}{1 - \left(\frac{1}{4}JS\right)^2 F^2(\omega)}, \tag{7.92}$$

independent on the pairing symmetry.

We now apply the above result to the magnetic impurity states and make a comparison between the isotropic s-wave and the d-wave superconductors. In an isotropic s-wave superconductor, $\Delta_\mathbf{k} = \Delta$,

$$F(\omega) = -\frac{\pi N_F}{\sqrt{\Delta^2 - \omega^2}}\left[\omega + i\Delta\sigma_2 \begin{pmatrix} 0 & 1 \\ -1 & 0 \end{pmatrix}\right]. \tag{7.93}$$

From the eigenvalues of $F(\omega)$, which are given by $-\pi N_F/\sqrt{\Delta^2 - \omega^2}$, we find the poles of the T-matrix

$$\omega = \pm\frac{1-\alpha}{1+\alpha}\Delta, \qquad \alpha = \left(\frac{\pi J S N_F}{4}\right)^2. \tag{7.94}$$

Clearly, $|\omega| < \Delta$. This pair of poles falls on the real axis, and their absolute values are smaller than the pairing gap. Therefore, there is a pair of bound states generated by a classical paramagnetic impurity inside the s-wave superconducting gap. The binding energy depends on the density of states in the normal state and the coupling constant J. When J is very small, the bound state approaches the gap edge. With the increase of J, the bound state moves close to the Fermi level.

However, in a d-wave superconductor,

$$F(\omega) = \pi N_F G_0(\omega) \tag{7.95}$$

is a constant matrix, and the poles of the T-matrix are determined by the equation

$$G_0^2(\omega) = c^2, \qquad c^{-1} = \frac{\pi J S N_F}{4}. \tag{7.96}$$

It is simply the equation that determines the poles in a nonmagnetic impurity system, i.e. Eq. (7.44), except that parameter c is redefined. The results previously obtained on the resonance state induced by a nonmagnetic impurity can be directly used here. The correction to the density of states by a magnetic impurity is not the same as that for a nonmagnetic one. The T-matrix after average over the impurity spin is simple. It contains only the T_0 term, not the T_3 term.

7.7 Kondo Effect

The interaction of a quantum magnetic impurity with electrons in a metal may destabilize the Fermi surface and change qualitatively the behaviors of resistivity, magnetic susceptibility, and other physical quantities. This is just the so-called Kondo effect [171]. The Kondo interaction is described by the Hamiltonian, defined by Eq. (7.77), but **S** is now a quantum spin operator rather than a classical one. The Kondo effect exists in a d-wave superconductor. It can also strongly affect the physical properties of superconducting quasiparticles.

In a normal metal, the Kondo effect arises from the screening of the magnetic impurity by the conduction electrons below a characteristic temperature scale called the Kondo temperature T_K. It generates a Kondo resonance state at the Fermi level. This effect was first discovered by Kondo through a third order perturbation calculation. The enhanced magnetic scattering on the Fermi surface has a large impact on transport properties, causing a logarithmic increase of the resistivity as shown in experiments around T_K. Below T_K, the Kondo problem lies in the strong coupling regime and becomes nonperturbative. A small Kondo coupling may lead to a qualitative change in the ground state. This problem was well studied in the 1960s and 1970s through the poor-man scaling [172], numerical renormalization group [173], and other methods. In a normal metal, the Kondo system has two fixed points. One is the unstable fixed point at $J = 0$, and the other is the stable fixed point at $J \to \infty$. They correspond to the weak and strong coupling limits, or equivalently high or low temperature limits, respectively. In the high temperature region, the impurity spin is decoupled from the conduction electrons and behaves like a free magnetic moment without screening. It gives rise to a Curie–Weiss-like impurity magnetic susceptibility. In the zero temperature limit, on the other hand, the coupling between the impurity and conduction electrons is strong. The impurity spin is completely screened and the system behaves like a normal Fermi liquid. The magnetic susceptibility is Pauli paramagnetic, just like in a normal metal.

In a superconductor, the Kondo effect is greatly weakened by the superconducting energy gap. Nevertheless, the Kondo effect exists when the exchange coupling between the impurity spin and superconducting electrons is sufficiently strong. It may also screen the impurity spin and change the interaction between superconducting Cooper pairs. In particular, the Kondo effect can suppress the superconducting gap near the impurity. However, a self-consistent calculation for this screening effect is rather difficult. In most of calculations, the correction of the Kondo effect to the superconducting gap function is either ignored or just considered for the average gap in the whole space. These analyses are not based on the self-consistent solution. It is not clear to what extent they can be applied to real materials.

In a conventional isotropic s-wave superconductor, since the quasiparticle excitations on the Fermi surface are fully gapped, the Kondo effect is completely suppressed and the impurity magnetic moment is not screened, provided that the Kondo
coupling is not significantly larger than the energy gap [174].

In a d-wave superconductor, the zero energy quasiparticle density of states is zero
and the impurity moment is also unscreened in the weak coupling regime. However,
as the quasiparticle density of states varies linearly with energy, the screening exists
if the Kondo coupling is strong enough. When the characteristic Kondo temperature
T_K is much smaller than the superconducting transition temperature T_c, the superconducting correlation is stronger than the Kondo screening effect and the system is
in the weak coupling limit. On the other hand, if $T_K \gg T_c$, Kondo screening takes
place before the superconducting transition, and the screening survives even in the
superconducting phase. This implies that there is a critical Kondo coupling J_c in
a d-wave superconductor: the impurity moment is unscreened when J is below J_c
and screened with a Kondo resonance state on the Fermi level when J is above J_c.

The existence of the Kondo effect above a critical coupling in a d-wave superconductor was confirmed by mean-field calculations based on the large-N expansion
[175–177]. Similar to the Kondo problem in a metal, the impurity magnetic moment
behaves differently in the unscreened and screened phases. The transition from the
unscreened phase to the screened one is a quantum phase transition. The critical
coupling constant J_c depends on the superconducting gap Δ_0. It increases with
increasing Δ_0. When the impurity concentration of the system becomes finite, the
impurity induced quasiparticle density of states on the Fermi surface also becomes
finite. This enhances the Kondo effect and drives J_c to zero [175].

In high-T_c superconductors, the magnetic impurity may exist either in or out
of the CuO_2 planes. The Kondo coupling directly in the CuO_2 plane is generally
stronger. It has a higher possibility of inducing the Kondo screening. The interaction
between a magnetic impurity out of the CuO_2 planes and conduction electrons is
relatively weak, and the impurity moment has less chance of being screened. When
the Kondo screening happens, a Kondo resonance state emerges at the Fermi energy,
which may account for the observed resonance state in the nickel or other magnetic
impurity doped high-T_c superconductors. Refs. [177, 178] offer a more detailed
discussion of this topic.

7.8 Quasiparticle Interference

STM has received a great deal of attention in revealing electronic structures of high-
T_c cuprates [179, 180]. In previous sections, we have discussed the resonance or

bound states induced by strong impurity scattering centers. If the scattering effect is relatively weak, no resonance states would appear. Instead, it induces an energy-dependent modulation of local density of states in space, which is referred to as quasiparticle interference (QPI) [181–185].

Unlike ARPES, STM is a local probe. The tunneling differential conductance measured by STM is approximately proportional to the local density of states. For a system without impurities, the crystalline symmetry would be perfectly preserved and reflected in the spatial pattern of the STM spectra. However, this symmetry is fragile against impurities or defects. In the presence of impurities or defects, the local density of states would inevitably become spatially inhomogeneous. An elastic scattering caused by an impurity potential excites a quasiparticle from wave vector $\mathbf{k}_1$ to $\mathbf{k}_2$ on the equal-energy surface. This induces a hybridization between these two wave vectors and yields an interference pattern at the wave vector $\mathbf{q} = \mathbf{k}_1 - \mathbf{k}_2$, which could be observed from the momentum-space representation of the density of states obtained from the Fourier transform of the STM spectra [185, 186]. A typical example is the $2k_F$ Friedel oscillation around an impurity in a metal or semiconductor [187]. This oscillation was observed by STM [188]. It results from the interplay between the impurity and the Fermi surface.

For high-T_c materials, the QPI pattern, or the Fourier transform of the local density of states, at various tunneling biases exhibits a number of characteristic scattering peaks in the superconducting state. For example, for $Bi_2Sr_2CaCu_2O_{8+\delta}$ [181], two sets of characteristic peaks, along the diagonal and CuO bond direction, respectively, were observed in the QPI spectra. In particular, the magnitude of the modulation wave vector $\mathbf{q}$ along the diagonal direction increases with the energy (or biased voltage), while that along the CuO bond direction decreases with the energy. The rate of change in $\mathbf{q}$ as a function of energy is faster along the diagonal direction. Similar features were observed in other high-T_c compounds [182, 183].

The evolution of the characteristic QPI peaks results from the quantum interference between quasiparticles induced by impurity scattering [185]. Roughly speaking, the spectral weight of QPI at a wave vector $\mathbf{q}$ is proportional to the integral of the joint density of states at momenta $\mathbf{k}$ and $\mathbf{k} + \mathbf{q}$ over the full equal-energy surface. A peak emerges if the momentum difference between two local maxima of density of states (more precisely, the local spectral weight) happens to equal $\mathbf{q}$.

Figure 7.6 (a) shows schematically the equal-energy contour of the d-wave quasiparticle excitations. At zero energy, there are only four gap nodes. With increasing energy, each node expands into a banana-shaped curve. The dispersion of d-wave quasiparticles is highly anisotropic. Its rate of change is determined by the Fermi velocity v_F along the diagonal direction and by the gap sloop v_Δ along the direction perpendicular. The local maxima of the spectral weight are located at banana tips because $v_F \gg v_\Delta$. Hence the local peaks appear at wave vectors connecting the banana tips. In Fig. 7.6 (a), two representative scattering wave vectors are marked

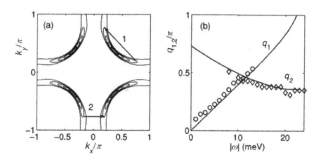

Figure 7.6 (a) Sketches of the equal-energy contours of the d-wave quasiparticle excitations in high-T_c superconductors. The scattering wave vectors of processes 1 and 2 are along the diagonal directions and the CuO bond directions, respectively. (b) Energy dependence of the modulation wave vectors, q_1 and q_2, associated with the scattering processes 1 and 2. (From Wang et al. [185])

as 1 and 2, respectively. These two wave vectors, represented by $\mathbf{q}_1$ and $\mathbf{q}_2$ in Fig. 7.6 (b), are along the diagonal and the CuO bond directions, respectively, consistent with the peak positions observed by experiments [181–183]. Furthermore, $\mathbf{q}_1$ increases and $\mathbf{q}_2$ decreases as the tunneling energy increases. It is also clear that the variation of $\mathbf{q}_1$ with energy is faster than $\mathbf{q}_2$.

Corresponding to the eight tips of the four banana-shaped equal-energy contours, there are in total seven different nonequivalent scattering wave vectors connecting these tips. In addition to $\mathbf{q}_1$ and $\mathbf{q}_2$, the QPI peaks corresponding to the other five scattering wave vectors are weaker in intensity. Nevertheless, they are also observed in experiments [182].

The above intuitive picture is justified by the calculation based on the T-matrix approximation. We include both the scalar and classic magnetic impurity potentials in the calculation. For simplicity, only the diagonal part of the magnetic scattering is considered. The QPI spectrum is given by the Fourier transform of the impurity-induced local density of states

$$\delta\rho(\mathbf{q},\omega) = \int d^2\mathbf{r}\, e^{-i\mathbf{q}\mathbf{r}} \delta\rho(\mathbf{r},\omega) = \frac{1}{2\pi}[A_{11}(\mathbf{q},\omega) + A_{22}(-\mathbf{q}, -\omega)], \qquad (7.97)$$

where

$$A(\mathbf{q},\omega) = \sum_{\mathbf{k}} \text{Im}\left[G^{(0)}(\mathbf{k}+\mathbf{q},\omega)T(\omega)G^{(0)}(\mathbf{k},\omega)\right]. \qquad (7.98)$$

$G^{(0)}(\mathbf{k},\omega)$ is the retarded single-particle Green's function (4.4). The T-matrix is given by

$$T^{-1}(\omega) = \frac{1}{V_0\sigma_3 + V_m} - \int \frac{d^2\mathbf{k}}{(2\pi)^2} G^{(0)}(\mathbf{k},\omega), \qquad (7.99)$$

where V_0 is the scalar scattering potential, and V_m is the magnetic scattering potential. Under the "on-shell" approximation, the integral in (7.98) is predominately determined by the poles of $G^{(0)}(\mathbf{k}+\mathbf{q},\omega)$ and $G^{(0)}(\mathbf{k},\omega)$. In particular, a peak emerges when both poles appear simultaneously at the same energy. This is just the condition for the appearance of a large joint density of states connected by the scattering wave vector $\mathbf{q}$.

QPI can also be used to extract the phase information of the superconducting gap functions via the coherence factor effect. The scattering probability from $\mathbf{k}$ to $\mathbf{k}+\mathbf{q}$ by the scalar and magnetic potential is proportional to

$$|u_k u_{k+q} \mp v_k v_{k+q}|^2, \tag{7.100}$$

respectively. The expressions of u and v are given in Eqs. (1.30) and (1.31). For the scalar potential, the scattering probability is enhanced if the gap function changes sign at $\mathbf{k}$ and $\mathbf{k}+\mathbf{q}$, i.e. $\Delta_k \Delta_{k+q} < 0$, and suppressed otherwise. On the contrary, the magnetic scattering is enhanced if the gap function takes the same sign at $\mathbf{k}$ and $\mathbf{k}+\mathbf{q}$, i.e. $\Delta_k \Delta_{k+q} > 0$. For high-$T_c$ cuprates, the scattering along the diagonal direction connects the gap function with opposite signs, while that along the CuO bond direction connects the gap function with the same signs. However, as the scattering is a relatively weak effect, it is still difficult to see clearly the difference resulting from the sign change of the gap function simply from the QPI patterns.

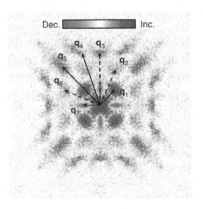

Figure 7.7 Changes in the QPI spectral weight induced by an applied magnetic field, $Z(\mathbf{q}, E, B = 11T) - Z(\mathbf{q}, E, B = 0)$, at $E = 4.4$ meV. The wave vectors of the enhanced peaks at $\mathbf{q}_1$, $\mathbf{q}_4$, and $\mathbf{q}_5$ result from the scattering between two banana tips with the same gap signs. The wave vectors of the suppressed peaks at $\mathbf{q}_2$, $\mathbf{q}_3$, $\mathbf{q}_6$, and $\mathbf{q}_7$ result from the scattering between two banana tips with opposite gap signs. (Please note that the nomenclature of $\mathbf{q}_1$ and $\mathbf{q}_2$ is opposite to the one shown in Figure 7.6.) (From Hanaguri et al. [184])

One way to enhance the phase effect of the superconducting energy gap in the QPI spectrum is to use superconducting vortices as additional scattering centers. This kind of experiment was first performed by Hanaguri et al. [184]. To sharpen the QPI signals from the background, they analyzed the QPI pattern for the ratio between the tunneling conductances at $+E$ and $-E$

$$Z(\mathbf{r}, E) = \frac{g(\mathbf{r}, E)}{g(\mathbf{r}, -E)}. \tag{7.101}$$

Figure 7.7 shows the difference between $Z(\mathbf{q}, E)$ in the vortex phase at $B = 11$ T and that at zero field. It is extremely difficult to give a quantitative prediction for QPI in the presence of vortices due to their nonlocal structures of phase winding. Nevertheless, since the time-reversal symmetry is broken, qualitatively the coherence factors should behave similarly to the case of magnetic potential scattering. It is expected that the scattering peaks will be enhanced if their characteristic wave vectors are connected by the gap functions with the same signs, and suppressed otherwise. This is precisely what was observed in the experiment.

8

Many-Impurity Scattering

8.1 Scattering Potential and Disorder Average

Compared to single-impurity scattering, many-impurity scattering is considerably more difficult to handle. It is impossible to exactly solve the scattering problem for a many-impurity system with random impurity configurations. Usually two kinds of approximations are adopted in the analytic calculations: (1) the impurity concentration is very low so that the interaction among impurities is negligible; (2) the screening from electrons to the impurity potential is so strong that it is short-ranged and isotropic. Furthermore, it is assumed that s-wave scattering plays the leading role, and the contribution from higher angular momentum channels can be neglected. The self-consistent T-matrix theory of impurity scattering in a d-wave superconductor is just established based on these approximations.

Physical properties of the many-impurity system are governed by two parameters: the scattering phase shift, and the impurity concentration. Both are closely related to the two approximations mentioned above [164]. The s-wave scattering approximation neglects the anisotropy of the impurity state in a d-wave superconductor, especially the anisotropic behavior of the impurity resonance state along the gap nodal and antinodal directions. This anisotropy may affect the interaction between impurities. A self-consistent treatment of this problem is rather difficult. It should be confessed that our current understanding on the many-impurity scattering problem is incomplete or even incorrect in some limits.

We assume that the impurity scattering is described by a δ-function potential,

$$U(\mathbf{r}) = V_0 \delta(\mathbf{r}). \tag{8.1}$$

This is equivalent to taking the s-wave scattering approximation for an arbitrary impurity scattering potential – if we start from an arbitrary scattering potential but take the s-wave approximation, we should obtain the same result with a properly adjusted V_0. However, the derivation from the δ-function potential is simple.

It avoids tedious and sometimes obscure discussions, and allows us to grasp more clearly the physics governing the many-impurity scattering problem. In fact, s-wave scattering is an approximation that we have to take in order to treat analytically the impurity scattering problem. To go beyond this approximation, the vertex or other corrections to the Green's function need to be considered. These corrections are difficult to calculate. It is also difficult to gain intuitive physical insights from this kind of calculation.

Assuming the impurities are located at $\{R_l\} = \{R_1, ... R_l, ...\}$, the interaction between impurities and electrons can be expressed as

$$H_i = \sum_{il} U(\mathbf{r}_i - \mathbf{R}_l)c_i^\dagger c_i = \sum_{\mathbf{q}} V_0 \rho_i(\mathbf{q}) \psi_{\mathbf{k}+\mathbf{q}}^\dagger \sigma_3 \psi_{\mathbf{k}}, \tag{8.2}$$

where $\psi_{\mathbf{k}}^\dagger = (c_{\mathbf{k},\uparrow}^\dagger, c_{-\mathbf{k},\downarrow})$ and

$$\rho_i(\mathbf{q}) = \frac{1}{V} \sum_l e^{i\mathbf{q}\cdot\mathbf{R}_l} \tag{8.3}$$

is a function of the impurity configuration.

The effect of impurity scattering on a d-wave superconductor can be solved through the perturbation expansion. A key step is to take the random average for the impurity scattering potentials. At the nth order of perturbation, the random average of the scattering potential is given by

$$f_n(\mathbf{q}_1, \mathbf{q}_2 \cdots, \mathbf{q}_n) = \langle \rho_i(\mathbf{q}_1)\rho_i(\mathbf{q}_2) \cdots \rho_i(\mathbf{q}_n) \rangle_{\text{imp}} . \tag{8.4}$$

It is difficult to evaluate this average rigorously. However, in the limit the impurity concentration is very low and the total impurity number N_i is very large, the interaction among impurities is very small and negligible. A common approximation adopted is to ignore the interference effect between different impurities and take

$$f_n(\mathbf{q}_1 \cdots \mathbf{q}_n) \approx n_i \delta(\Sigma' \mathbf{q}) + n_i^2 \delta(\Sigma' \mathbf{q})\delta(\Sigma' \mathbf{q}) + n_i^3 \delta(\Sigma' \mathbf{q})\delta(\Sigma' \mathbf{q})\delta(\Sigma' \mathbf{q}) + \cdots , \tag{8.5}$$

where $n_i = N_i/V$ is the impurity concentration. $\Sigma' \mathbf{q}$ is the sum over all or part of $\mathbf{q}_i$. More precisely, in the first term, $\Sigma' \mathbf{q}$ is the sum over all $\mathbf{q}_i$. In the second term, $\Sigma' \mathbf{q}$ appears twice. In this case, all $\mathbf{q}_i$ are divided into two groups. The first $\Sigma' \mathbf{q}$ is the sum over all the q_i in the first group, and the second $\Sigma' \mathbf{q}$ is the sum over all the q_i in the second group. $(\Sigma' \mathbf{q})$'s in higher order terms should be similarly understood. This approximation neglects the interference effect between different impurities. It assumes that the scattering to electrons by different impurities is independent such that the momentum of an incident electron, after scattered once or multiple times by one impurity, is not changed. This is equivalent to just keeping the reducible Feynman diagrams of impurity scattering, as shown in Fig. 8.1(a), and neglecting all the irreducible diagrams, as shown in Fig. 8.1(b).

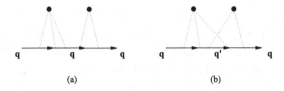

Figure 8.1 (a) A reducible impurity scattering Feynman diagram in which the scattering of electrons by different impurities is independent. (b) An irreducible impurity scattering Feynman diagram, which contains the interaction between impurities. In the self-consistent T-matrix or the single-impurity approximation, this kind of diagram is neglected.

Under this approximation, the many-impurity scattering problem is reduced effectively to a single-impurity one. The multi-impurity effect is reflected in the factor n_i associated with each single-impurity scattering term. However, unlike the single-impurity problem discussed in the preceding chapter, the system becomes translation invariant after taking the disorder average. In addition, after the disorder average, the divergence of the density of states induced by the single-impurity resonance state in the unitary limit is smeared out. The divergence is replaced by a broadened and lowered peak. This result is obtained based on the T-matrix approximation. It implies that the impurity resonance state is difficult to detect by measuring an impurity-averaged quantity.

8.2 Self-Energy Function

A major effect of disorder on superconductivity is to change the self-energy of quasiparticles. For a d-wave superconductor, the self-energy needs to be determined self-consistently from the Green's function of electrons. Under the approximations previously introduced, the Dyson equation of the single-particle Green's function, as illustrated in Fig. 8.2, is given by

$$G(\mathbf{q}, \omega) = G^{(0)}(\mathbf{q}, \omega) + G^{(0)}(\mathbf{q}, \omega) \Sigma(\omega) G(\mathbf{q}, \omega), \tag{8.6}$$

where $\Sigma(\omega)$ is the self-energy function. The self-energy is momentum-independent. This is due to the use of the δ-function potential. The solution to this equation is

$$G(\mathbf{q}, \omega) = \frac{1}{\omega - \xi_{\mathbf{q}}\sigma_3 - \Delta_{\mathbf{q}}\sigma_1 - \Sigma(\omega)}. \tag{8.7}$$

In the single-impurity approximation, $\Sigma(\omega)$ is determined by the Feynman diagrams shown in Fig. 8.3. The first order correction to the self-energy is given by

$$\Sigma^{(1)}(\omega) = n_i V_0 \sigma_3. \tag{8.8}$$

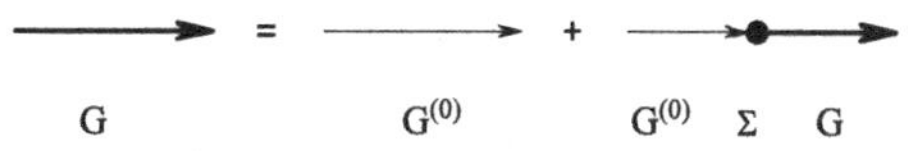

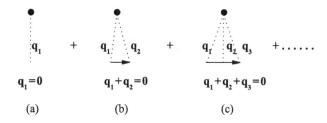

Figure 8.2 Diagrammatic representation of the Dyson equation.

$q_1 = 0$ $q_1 + q_2 = 0$ $q_1 + q_2 + q_3 = 0$

(a)   (b)   (c)

Figure 8.3 The self-energy correction by the impurity scattering. Feynman diagrams for the first (a), second (b), and third (c) order correction to the self-energy under the single-impurity approximation.

The second order correction is

$$\Sigma^{(2)}(\omega) = n_i V_0 \sigma_3 c^{-1} \underline{G}(\omega) \sigma_3, \tag{8.9}$$

where $c = 1/\pi N_F V_0$, and

$$\underline{G}(\omega) = \frac{1}{\pi N_F V} \sum_q G(\mathbf{q}, \omega) \tag{8.10}$$

is the momentum averaged Green's function. $\underline{G}$ is similar to G_0 defined in Eq. (7.33), but $\underline{G}$ is determined by the full Green's function, not just $G^{(0)}$. c is related to the scattering phase shift η_0 by the equation

$$c = \cot \eta_0, \tag{8.11}$$

which is the same as in the single-impurity system.

From the Feynman diagrams shown in Fig. 8.3, it is not difficult to show that the nth order correction to the self-energy has the form

$$\Sigma^n(\omega) = n_i V_0 \sigma_3 \left(c^{-1} \underline{G}(\omega) \sigma_3 \right)^{n-1}. \tag{8.12}$$

Therefore, the total self-energy is

$$\Sigma(\omega) = \Sigma^{(1)}(\omega) + \Sigma^{(2)}(\omega) + \Sigma^{(3)}(\omega) + \cdots = n_i V_0 c \sigma_3 \frac{1}{c - \underline{G}(\omega)\sigma_3}. \tag{8.13}$$

Equations (8.7), (8.10) and (8.13) form a set of self-consistent equations for determining the single-particle Green's function. Expressing the electron Green's function $\underline{G}$ and the self-energy in the basis space spanned by $(I, \sigma_1, \sigma_2, \sigma_3)$,

$$\underline{G}(\omega) = \underline{G}_0(\omega) + \underline{G}_1(\omega)\sigma_1 + \underline{G}_2(\omega)\sigma_2 + \underline{G}_3(\omega)\sigma_3, \tag{8.14}$$

$$\Sigma(\omega) = \Sigma_0(\omega) + \Sigma_1(\omega)\sigma_1 + \Sigma_2(\omega)\sigma_2 + \Sigma_3(\omega)\sigma_3, \tag{8.15}$$

and using the identity

$$(A + \mathbf{B} \cdot \boldsymbol{\sigma})^{-1} = \frac{A - \mathbf{B} \cdot \boldsymbol{\sigma}}{A^2 - |B|^2},$$ (8.16)

we find that Σ's are determined by the equations

$$\begin{pmatrix} \Sigma_0(\omega) & \Sigma_1(\omega) \\ \Sigma_2(\omega) & \Sigma_3(\omega) \end{pmatrix} = -n_i c V_0 \begin{pmatrix} \underline{G}_0(\omega) & \underline{G}_1(\omega) \\ \underline{G}_2(\omega) & c - \underline{G}_3(\omega) \end{pmatrix} A(\omega),$$ (8.17)

where

$$A(\omega) = \frac{1}{\underline{G}_0^2(\omega) - \underline{G}_1^2(\omega) - \underline{G}_2^2(\omega) - \left[\underline{G}_3(\omega) - c\right]^2}.$$ (8.18)

Similarly, from Eq. (8.10), $\underline{G}$ is found to be

$$\underline{G}_0(\omega) = [\omega - \Sigma_0(\omega)] \sum_{\mathbf{q}} B(\mathbf{q}, \omega),$$ (8.19)

$$\underline{G}_1(\omega) = \sum_{\mathbf{q}} B(\mathbf{q}, \omega) \left[\Delta_{\mathbf{q}} + \Sigma_1(\omega)\right],$$ (8.20)

$$\underline{G}_2(\omega) = \Sigma_2(\omega) \sum_{\mathbf{q}} B(\mathbf{q}, \omega),$$ (8.21)

$$\underline{G}_0(\omega) = \sum_{\mathbf{q}} \left[\xi_{\mathbf{q}} + \Sigma_4(\omega)\right] \sum_{\mathbf{q}} B(\mathbf{q}, \omega),$$ (8.22)

where

$$B(\mathbf{q}, \omega) = \frac{1}{\pi N_F V} \frac{1}{(\omega - \Sigma_0)^2 - \left(\xi_{\mathbf{q}} + \Sigma_3\right)^2 - \left(\Delta_{\mathbf{q}} + \Sigma_1\right)^2 - \Sigma_2^2}.$$ (8.23)

For a d-wave superconductor, the momentum summation of the energy gap vanishes, i.e. $\sum_{\mathbf{k}} \Delta_{\mathbf{k}} = 0$. It is simple to show that

$$\Sigma_1 = \underline{G}_1 = \Sigma_2 = \underline{G}_2 = 0$$ (8.24)

is a self-consistent solution to these equations [189]. In this case, $\underline{G}_3$ reduces to

$$\underline{G}_3(\omega) = \frac{1}{\pi N_F} \int \frac{d\varphi}{2\pi} \int_{-\infty}^{\infty} d\xi \rho_N(\xi)$$
$$\frac{\xi + \Sigma_3(\omega)}{[\omega - \Sigma_0(\omega)]^2 - [\xi + \Sigma_3(\omega)]^2 - \Delta_{\varphi}^2}.$$ (8.25)

In obtaining this expression, we assumed that $\xi(\mathbf{k})$ depends only on $|\mathbf{k}|$ and that $\Delta_{\mathbf{k}} = \Delta_{\varphi}$ depends only on φ. If the system is particle–hole symmetric and $\rho_N(\xi)$ does not strongly depend on ξ, the above integral can be simplified as

$$\underline{G}_3(\omega) = \frac{1}{\pi} \int \frac{d\varphi}{2\pi} \int_{-\infty-\Sigma_3}^{\infty-\Sigma_3} d\xi \frac{\xi}{[\omega - \Sigma_0(\omega)]^2 - \xi^2 - \Delta_\varphi^2}. \tag{8.26}$$

Σ_3 is a complex number, and the integral on the right-hand side does not need to be zero. Its value depends on the poles of the integrand

$$\xi_\pm = \pm\sqrt{[\omega - \Sigma_0(\omega)]^2 - \Delta_\varphi^2}. \tag{8.27}$$

From the self-consistent equations, it can be shown that the absolute value of the imaginary part of $\xi_\pm$ is larger than the absolute value of the imaginary part of Σ_3. Therefore, there are no poles in the complex ξ-plane enclosed by the real axis and the line of $\mathrm{Im}\xi = \mathrm{Im}\Sigma_3$. Since the integrand is an odd function of ξ along the real axis, the integral along the real axis is zero. Consequently, $\underline{G}_3$ should also be zero [189].

The above result shows that $\underline{G} = \underline{G}_0$ is a constant matrix. Substituting this result into the self-consistent equations, we find that

$$\Sigma_0(\omega) = \frac{n_i V_0 c \underline{G}_0(\omega)}{c^2 - \underline{G}_0^2(\omega)}, \qquad \Sigma_3(\omega) = \frac{n_i V_0 c^2}{c^2 - \underline{G}_0^2(\omega)}, \tag{8.28}$$

and

$$\underline{G}_0(\omega) = \frac{\omega - \Sigma_0(\omega)}{\pi N_F V} \sum_q \frac{1}{[\omega - \Sigma_0(\omega)]^2 - [\xi_q + \Sigma_3(\omega)]^2 - \Delta_q^2}. \tag{8.29}$$

At $\omega = 0$, $\Sigma_0(0) \equiv -i\Gamma_0$ is purely an imaginary number, and $\Sigma_3(0)$ is real. Now we have

$$G_0(0) = -i\Gamma_0 \int \frac{d\varphi}{2\pi} \frac{1}{\sqrt{\Gamma_0^2 + \Delta_\varphi^2}}, \tag{8.30}$$

where Γ_0 is proportional to the quasiparticle scattering rate. If τ is the quasiparticle lifetime at the zero frequency, then

$$\tau^{-1} = 2\Gamma_0. \tag{8.31}$$

In the normal state, $\Delta_k = 0$, the integral in Eq. (8.29) can be integrated out rigorously. It can be shown that $\underline{G}_0(\omega)$ is frequency-independent

$$\underline{G}_0(\omega) = \underline{G}(\omega) = -i. \tag{8.32}$$

Therefore the self-energy in the normal state is

$$\Sigma_N(\omega) = \frac{n_i V_0 c}{1 + c^2}(-i + c\sigma_3). \tag{8.33}$$

The electron scattering rate in the normal state is twice that of the imaginary part of the self-energy

$$\Gamma_N = -2\text{Im}\Sigma_N(\omega) = \frac{2n_i V_0 c}{1 + c^2}.$$ (8.34)

In the discussion of physical properties of disordered d-wave superconductors, two scattering limits need to be paid more attention. One is the unitary limit, which is also called the resonance limit. The other is the Born scattering limit. They correspond to the strong and weak scattering limits, respectively. In the unitary limit, $c \to 0$, and the corresponding scattering phase shift $\eta_0 = \pi/2$. This is the largest phase shift that can be taken in the s-wave scattering channel. In the Born scattering limit, $c \to \infty$, the corresponding phase shift η_0 is very small, close to zero. Many physical properties of d-wave superconductors are different, not just quantitatively, but also qualitatively, in these two limits. The scattering rate, for example, behaves as

$$\Gamma_N \approx 2\pi n_i V_0^2 N_F,$$ (8.35)

or $c \to \infty$ in the Born limit, and

$$\Gamma_N \approx \frac{2n_i}{\pi N_F},$$ (8.36)

or $c \to 0$ in the unitary limit, respectively.

The problem of impurity scattering in a d-wave superconductor is difficult to solve analytically. A general solution to the Green's function in the whole frequency or temperature range can be obtained only through numerical calculations. However, in the unitary or Born scattering limit, the analysis is greatly simplified.

The zinc impurity has a large effect on high-T_c superconducting properties. The scattering induced by the zinc impurity is generally believed to be in the resonance scattering limit. The disorder effect induced by the structure inhomogeneity or defects on the interface or CuO$_2$ planes is more complicated. It usually lies between the Born and unitary scattering limits. The effect of nonmagnetic impurity scattering can in principle be understood by interpolating between the unitary and Born scattering limits. Thus a thorough understanding of these two scattering limits is helpful for understanding more comprehensively the impurity effect in d-wave superconductors.

8.3 Born Scattering Limit

In the Born scattering limit, $c \to \infty$, the electron self-energy is given by

$$\Sigma_0(\omega) = \frac{1}{2}\Gamma_N \underline{G}_0(\omega), \qquad \Sigma_3(\omega) = n_i V_0.$$ (8.37)

Σ_3 is ω-independent. It can be absorbed into the chemical potential by redefining $\xi_{\mathbf{k}}$. Thus only Σ_0 needs to be determined self-consistently. In this case,

$$\underline{G}_0(\omega) = \frac{\omega - \Sigma_0(\omega)}{\pi} \int \frac{d\varphi}{2\pi} \int_{-\infty}^{\infty} d\xi \frac{1}{[\omega - \Sigma_0(\omega)]^2 - \xi^2 - \Delta_\varphi^2}. \qquad (8.38)$$

After integrating over ξ using the formula

$$\int_{-\infty}^{+\infty} \frac{dx}{a - x^2 + i0^+} = \frac{-i\pi}{\sqrt{a}} \theta\left(\mathrm{Im}\sqrt{a}\right), \qquad (8.39)$$

we obtain

$$\underline{G}_0(\omega) = -i\,[\omega - \Sigma_0(\omega)] \int \frac{d\varphi}{2\pi} \frac{\theta\left(\mathrm{Im}\sqrt{[\omega - \Sigma_0(\omega)]^2 - \Delta_\varphi^2}\right)}{\sqrt{[\omega - \Sigma_0(\omega)]^2 - \Delta_\varphi^2}}. \qquad (8.40)$$

For the $d_{x^2-y^2}$ superconductor, $\Delta_\varphi = \Delta_0 \cos 2\varphi$, the above integral can be reexpressed as

$$\underline{G}_0(\omega) = -\frac{i2\bar{\omega}}{\pi} \int_0^1 \frac{dx}{\sqrt{1 - x^2}} \frac{\theta\left(\mathrm{Im}\sqrt{\bar{\omega}^2 - x^2}\right)}{\sqrt{\bar{\omega}^2 - x^2}}, \qquad (8.41)$$

by setting $x = \cos 2\varphi$ and $\bar{\omega} = [\omega - \Sigma_0(\omega)]/\Delta_0$. This is an elliptic integral.

The above self-consistent equations (8.37) and (8.41) can be solved numerically. Figures 8.4 and 8.5 show the real and imaginary parts of the self-energy as a function of frequency for three different values of Γ_N, respectively.

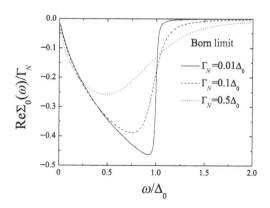

Figure 8.4 Frequency dependence of the real part of the self-energy $\Sigma_0(\omega)$ in the Born scattering limit ($c \to \infty$).

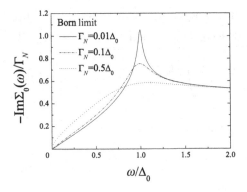

Figure 8.5 Frequency dependence of the imaginary part of the self-energy $\Sigma_0(\omega)$ in the Born scattering limit ($c \to \infty$).

When $|\bar{\omega}| \ll 1$, the integral contributes mainly from the domain of $x \ll 1$ in which $1/\sqrt{1-x^2}$ can be approximately replaced by 1, we then have

$$\underline{G}_0(\omega) \approx -\frac{2\bar{\omega}}{\pi} \ln \frac{1+\sqrt{1-\bar{\omega}^2}}{\sqrt{-\bar{\omega}^2}} \approx -\frac{2\bar{\omega}}{\pi} \ln \frac{2}{\sqrt{-\bar{\omega}^2}} \qquad (|\bar{\omega}| \ll 1). \qquad (8.42)$$

The self-consistent equation of Σ_0 now becomes

$$\sqrt{-\bar{\omega}^2} = 2\exp\left(\frac{\pi\Sigma_0}{\Gamma_N\bar{\omega}}\right). \qquad (8.43)$$

At zero frequency, $\omega = 0$, the solution is

$$\Sigma_0(0) = -i2\Delta_0 \exp\left(-\frac{\pi\Delta_0}{\Gamma_N}\right). \qquad (8.44)$$

The low frequency behavior of $\Sigma_0(\omega)$ can be solved by the series expansion. Up to the second order terms of ω,

$$\Sigma_0(\omega) \approx \Sigma_0(0) + \left(1 - \frac{\pi\Delta_0}{\Gamma_N}\right)\omega - \frac{\pi^2\Delta_0^2}{2\Gamma_N^2\Sigma_0(0)}\omega^2 + o(\omega^2). \qquad (8.45)$$

In high-T_c superconductors, Γ_N is generally much smaller than Δ_0, and $\Sigma_0(0)$ is also very small. Therefore, the disorder scattering only slightly modifies the superconducting properties in this limit. Nevertheless, Σ_0 is finite. It can affect the conducting behavior of d-wave superconductors at low temperatures.

In the above discussion, if the self-consistent condition is not implemented in the determination of the Green's function, then $\bar{\omega} = \omega/\Delta_0$ and in the low frequency limit $|\omega| \ll \Delta_0$,

$$\Sigma_0(\omega) \approx -\frac{\Gamma_N\omega}{\pi\Delta_0}\left(\ln\frac{2\Delta_0}{|\omega|} + i\frac{\pi}{2}\right). \qquad (8.46)$$

Now $\Sigma_0(0) = 0$, which is different from the self-consistent solution. This difference shows the importance of self-consistent treatment to the Green's function. However, in the intermediate frequency regime, $\Gamma_N \ll |\omega| \ll \Delta_0$, the higher order corrections from the impurity scattering become less important. In this case, there is not much difference between the self-consistent and non-self-consistent results, and Eq. (8.46) is valid.

8.4 Resonant Scattering Limit

In the resonant scattering limit, $c \to 0$. The self-consistent equation of the self-energy becomes

$$\Sigma_0(\omega) = -\frac{n_i V_0 c}{\underline{G}_0(\omega)} = -\frac{\Gamma_N}{2\underline{G}_0(\omega)}, \qquad \Sigma_3(\omega) = -\frac{n_i V_0 c^2}{\underline{G}_0^2(\omega)}. \qquad (8.47)$$

Unlike in the Born scattering limit, $\underline{G}_0(\omega)$ now appears in the denominator. Compared to Σ_0, Σ_3 is a higher order small quantity and can be neglected. $\underline{G}_0(\omega)$ is still determined by Eq. (8.41).

Figures 8.6 and 8.7 show the numerical solutions of the self-consistent equations. For small Γ_N, the real and imaginary parts of Σ_0 are very small, close to zero, at $\omega = \Delta_0$, different from the results in the Born scattering limit.

In the limit $|\bar{\omega}| \ll 1$, the self-consistent equation is approximately given by

$$4\bar{\omega}\Sigma_0(\omega)\left(\ln 2 - \ln\sqrt{-\bar{\omega}^2}\right) \approx \pi\Gamma_N. \qquad (8.48)$$

It is still difficult to solve this equation analytically. If we neglect the slowly varying logarithmic term, this equation is simplified to

$$[\omega - \Sigma_0(\omega)]\Sigma_0(\omega) \approx \frac{\pi\Gamma_N\Delta_0}{4\ln 2}. \qquad (8.49)$$

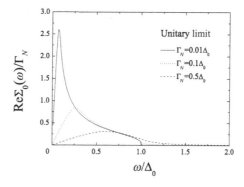

Figure 8.6 Frequency dependence of the real part of the self-energy $\Sigma_0(\omega)$ in the unitary scattering limit ($c \to 0$).

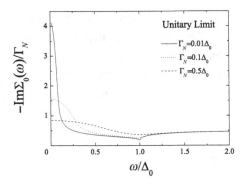

Figure 8.7 Frequency dependence of the imaginary part of the self-energy $\Sigma_0(\omega)$ in the unitary scattering limit ($c \to 0$).

The solution is

$$\Sigma_0(\omega) \approx \frac{\omega \pm \sqrt{\omega^2 - \pi\Gamma_N\Delta_0/\ln 2}}{2}. \tag{8.50}$$

For these two solutions, the one with the plus sign has a positive imaginary part, corresponding to the advanced Green's function and does not need to be considered.

At low frequency, $\Sigma_0(\omega)$ can be expanded in terms of ω. The first three terms are

$$\Sigma_0(\omega) = \Sigma_0(0) + \frac{\omega}{2} + \frac{\omega^2}{8\Sigma_0(0)} + o(\omega^2), \tag{8.51}$$

where $\Sigma_0(0)$ is the self-energy at $\omega = 0$,

$$\Sigma_0(0) \approx -\frac{i\sqrt{\pi\Gamma_N\Delta_0}}{2\sqrt{\ln 2}}. \tag{8.52}$$

By comparison, we find that $|\Sigma_0(0)|$ drops with increasing Γ_N in the Born scattering limit, but rises with Γ_N in the resonance scattering limit. This indicates that the scattering effect is stronger in the resonant scattering limit than in the Born scattering limit in d-wave superconductors.

In the intermediate frequency regime, $\Delta_0 \gg \omega \gg |\Sigma_0(0)|$, the self-consistency becomes less important, similarly to the Born scattering limit. In this case, $\bar{\omega} = \omega/\Delta_0$ and

$$\Sigma_0(\omega) \approx \frac{\pi\Delta_0\Gamma_N}{4\omega\left(\ln\frac{2\Delta_0}{|\omega|} + i\frac{\pi}{2}\right)}. \tag{8.53}$$

$\Sigma_0(\omega)$ exhibits different asymptotic behaviors at $\omega \ll |\Sigma_0(0)|$ and $\Delta_0 \gg \omega \gg |\Sigma_0(0)|$. This difference implies that the disorder scattering behaves quite differently for the energy smaller or larger than $|\Sigma_0(0)|$. If the energy is lower than $|\Sigma_0(0)|$, the impurity scattering strongly affects the physical properties of d-wave superconductors, similarly to a gapless s-wave superconductor induced by magnetic impurities. On the other hand, if the energy is higher than $|\Sigma_0(0)|$, the impurity scattering is small, and the impurity correction to the superconducting properties is negligible.

In the literature, an energy interval is regarded as being in a gapless region if $\omega \ll |\Sigma_0(0)|$, or an intrinsic region if $\omega \gg |\Sigma_0(0)|$. The effect of impurity scattering is strong in the gapless region, but weak in the intrinsic region.

The single-particle Green's function can be represented using a unified formula for both the Born and unitary scattering limits. In the gapless region, $\omega \ll \Gamma_0$, if the self-energy is expanded up to the first order in ω, the retarded Green's function in these two limits is found to be

$$G^R(\mathbf{k}, \omega) = \frac{1}{a\omega + i\Gamma_0 - \xi_k \sigma_3 - \Delta_k \sigma_1}, \tag{8.54}$$

where $a = 1 - \Sigma_0'(0)$. In the Born scattering limit,

$$a = \frac{\pi \Delta_0}{\Gamma_N}, \tag{8.55}$$

while in the unitary limit

$$a = \frac{1}{2}. \tag{8.56}$$

Here $1/a$ is not the quasiparticle weight since Eq. (8.54) is valid only in the limit $\omega \ll \Gamma_0$. The fermion spectral function remains normalized when integrating over the frequency.

In the intrinsic region, $\Gamma_0 \ll \omega \ll \Delta_0$, after neglecting the impurity correction to ω, the Green's function is approximately given by

$$G^R(\mathbf{k}, \omega) = \frac{1}{\omega + i\Gamma(\omega) - \xi_k \sigma_3 - \Delta_k \sigma_1}, \tag{8.57}$$

where $\Gamma(\omega)$ is a frequency-dependent quasiparticle scattering rate. In the Born scattering limit

$$\Gamma(\omega) \approx \frac{\Gamma_N \omega}{2\Delta_0}, \tag{8.58}$$

while in the unitary limit,

$$\Gamma(\omega) \approx \frac{\pi^2 \Delta_0 \Gamma_N}{8\omega \ln^2(2\Delta_0/|\omega|)}. \tag{8.59}$$

8.5 Correction to the Superconducting Critical Temperature

Impurity scattering does not affect the superconducting transition temperature T_c for an s-wave superconductor. This is a consequence of the Anderson theorem [163] (see Appendix D), which holds for all conventional metal-based superconductors. However, for d-wave superconductors, the Anderson theorem does not hold anymore, and there is a finite correction to the critical transition temperature from the impurity scattering.

In order to study the impurity correction to the transition temperature T_c, it is more convenient to use the Matsubara Green's functions. Under the T-matrix approximation, the Green's function of electrons is given by

$$G(\mathbf{k}, i\omega_n) = \frac{1}{i\tilde{\omega}_n - \tilde{\xi}_{\mathbf{k}}\sigma_3 - \Delta_{\mathbf{k}}\sigma_1}, \tag{8.60}$$

where

$$i\tilde{\omega}_n = i\omega_n - \Sigma_0(i\omega_n), \tag{8.61}$$

$$\tilde{\xi}_{\mathbf{k}} = \xi_{\mathbf{k}} + \Sigma_3(i\omega_n). \tag{8.62}$$

In terms of Green's function, the gap equation

$$\Delta_0 = -\frac{g}{2} \sum_{\mathbf{k}} \phi_{\mathbf{k}} \langle c_\uparrow^\dagger(\mathbf{k}) c_\downarrow^\dagger(-\mathbf{k}) \rangle \tag{8.63}$$

is represented as

$$\Delta_0 = -\frac{g k_B T}{2} \sum_{\mathbf{k}, \omega_n} \phi_{\mathbf{k}} \mathrm{Tr}\sigma_1 G(\mathbf{k}, i\omega_n). \tag{8.64}$$

Substituting Eq. (8.60) into the gap equation, we obtain

$$1 = -g k_B T \sum_{\mathbf{k}, \omega_n} \frac{\phi_{\mathbf{k}}^2}{(i\tilde{\omega}_n)^2 - \Delta_{\mathbf{k}}^2 - \tilde{\xi}_{\mathbf{k}}^2}. \tag{8.65}$$

At the critical point, $\Delta_{\mathbf{k}} = 0$, from the expressions of $\Sigma_0(i\omega_N)$ and $\underline{G}_0(i\omega_n)$ given before, it can be shown that in the normal state,

$$\Sigma_{0,N}(i\omega_n) = -i\Gamma_N \mathrm{sgn}(\omega_n). \tag{8.66}$$

It is different from the corresponding expression in the real frequency representation, $\Sigma_{0,N}(\omega) = -i\Gamma_N$, which is ω independent. Inserting it into (8.65) and taking the average of $\phi_{\mathbf{k}}^2$ over the Fermi surface, we obtain the equation that determines T_c

$$K(\Gamma_N) = 1, \tag{8.67}$$

where

$$K(\Gamma) = \frac{g k_B T_c N_F}{2} \sum_{\omega_n} \int d\xi \frac{1}{(|\omega_n| + \Gamma/2)^2 + \xi^2}. \tag{8.68}$$

The summation over ω_n or the integration over ξ must have finite lower and upper bounds. Otherwise, the right-hand side of the above equation diverges. This requirement is physically correct because electrons can form superconducting pairs only within a finite energy interval. The divergence can be removed by imposing finite upper and lower bounds either to the summation over ω_n or to the integral over ξ.

In the limit $\Gamma_N = 0$, it is more convenient to impose the restriction on the lower and upper bounds for the integral over ξ, but not on the Matsubara frequency summation. In this case, the summation over ω_n is simply given by

$$\frac{1}{\beta} \sum_n \frac{1}{\omega_n^2 + \xi^2} = \frac{1}{\xi} \tanh \beta \xi. \tag{8.69}$$

This leads to

$$K(0) = \frac{g N_F}{2} \int_0^{\omega_0} d\xi \frac{1}{\xi} \tanh \frac{\xi}{k_B T_c}, \tag{8.70}$$

where ω_0 is the characteristic frequency of electron pairing. The integral over ξ can be completed using the method introduced in §3.2, which yields

$$K(0) = \frac{g N_F}{2} \int_0^{\omega_0} d\xi \frac{1}{\xi} \tanh \frac{\xi}{2 k_B T_c} \approx \frac{g N_F}{2} \ln \frac{1.134 \omega_0}{k_B T_c}. \tag{8.71}$$

In a pure system without impurities, the superconducting T_{c0} is determined by the equation

$$\frac{2}{g N_F} = \ln \frac{1.134 \omega_0}{k_B T_{c0}}. \tag{8.72}$$

Using this formula, Eq. (8.67) can be rewritten as

$$\ln \frac{T_{c0}}{T_c} + \frac{2}{g N_F} [K(\Gamma_N) - K(0)] = 0. \tag{8.73}$$

The difference between $K(\Gamma_N)$ and $K(0)$, i.e. $K(\Gamma_N) - K(0)$, is a regular function of ω_n and ξ, because the divergent terms in both $K(0)$ and $K(\Gamma_N)$ are subtracted. Thus if we directly calculate this difference, both the upper and lower bounds in the integral over ξ, as well as in the summation over ω_n, can be taken to infinity. After taking the integral over ξ, the above equation becomes

$$\ln \frac{T_{c0}}{T_c} + \sum_{n \geqslant 0} \left(\frac{1}{n + \frac{1}{2} + \frac{\Gamma_N}{4 \pi k_B T_c}} - \frac{1}{n + \frac{1}{2}} \right) = 0. \tag{8.74}$$

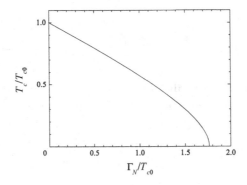

Figure 8.8 Correction to T_c by the impurity scattering: A universal plot of T_c/T_{c0} as a function of Γ_N/T_{c0}.

The summation over n can be expressed using the digamma function $\psi(x)$ as

$$\sum_{n\geqslant 0}\left(\frac{1}{n+\frac{1}{2}+\frac{\Gamma_N}{4\pi k_B T_c}}-\frac{1}{n+\frac{1}{2}}\right)=\psi\left(\frac{1}{2}\right)-\psi\left(\frac{1}{2}+\frac{\Gamma_N}{4\pi k_B T_c}\right), \qquad (8.75)$$

where $\psi(x)$ is defined as

$$\psi(x)=-0.577-\frac{1}{x}+\sum_{n=1}^{\infty}\left(\frac{1}{n}-\frac{1}{n+x}\right). \qquad (8.76)$$

Thus the equation for determining the transition temperature T_c is given by

$$\ln\frac{T_{c0}}{T_c}=\psi\left(\frac{1}{2}+\frac{\Gamma_N}{4\pi k_B T_c}\right)-\psi\left(\frac{1}{2}\right). \qquad (8.77)$$

Both the coupling constant g and the normal state density of states N_F do not appear explicitly in this equation. T_c is determined purely by T_{c0} and Γ_N. This implies that the impurity correction to the transition temperature is universal, determined just by Γ_N. Figure 8.8 shows the numerical solution for Eq. (8.77). At $\Gamma_N \sim 1.764T_{c0}$, the impurity scattering suppresses T_c completely.

Figure 8.9 shows the experimental results of the transition temperature T_c for $YBa_2(Cu_{1-z}M_z)_3O_7$ and other high-T_c superconductors as a function of the nickel or zinc impurity concentration z. For all the cases shown in the figure, T_c decreases linearly with z within the measurement errors. However, the slopes are different for different superconductors. The experimental results agree with the theoretical prediction for the impurity correction to T_c for d-wave superconductors.

At low doping, Γ_N is proportional to z. For $YBa_2Cu_3O_7$, 1% zinc concentration suppresses T_c by 1/7 of its maximal value. From the result shown in Fig. 8.8, we estimate $\Gamma_N \sim 2T_{c0}/7$ for 1% zinc. Substituting this value of Γ_N and $c = 0.07$

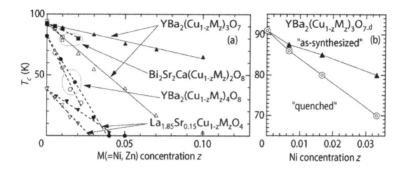

Figure 8.9 T_c versus the impurity concentration z for $YBa_2(Cu_{1-z}M_z)_3O_7$ and other high-T_c superconductors, M = Ni or Zn. The solid and open symbols represent T_c of Ni- and Zn-substituted superconductors, respectively. The solid and dashed lines in (a) are linear fits to the experimental data using the formula $T_c = T_{c0} - \alpha_M z$, where α_M is the fitting parameter. For Y123, Bi2212, Y124, and La214 superconductors, α_{Ni}/α_{Zn} equal 0.26, 0.46, 0.80, and 0.62, respectively. (b) T_c versus z for both as grown and quenched $YBa_2(Cu_{1-z}Ni_z)_3O_7$ samples, respectively. (Taken from Ref. [190])

extracted from the STM experiment [165] into Eq. (8.34), the zinc impurity potential strength V_0 is estimated to be $V_0 \sim 1.8$ eV. It is smaller than, but of the same order as, that estimated from the density of states. It shows that the zinc impurity is really a strong scattering center. The suppression of T_c by Ni is only $\frac{1}{4}$ that of Zn. The effective value of c for nickel is also larger. Therefore, the scattering from nickel is much weaker than from zinc.

In $YBa_2Cu_4O_8$ and La_2CuO_4, the difference in the suppression of T_c induced by nickel and zinc impurities is smaller than in $YBa_2Cu_3O_7$. This is probably due to the fact that the doped zinc or nickel atoms may not all lie on the CuO_2 planes in the former two compounds.

8.6 Density of States

The density of states $\rho(\omega)$ is determined by the imaginary part of the Green's function,

$$\rho(\omega) = -\frac{1}{\pi V} \text{Im} \sum_{\mathbf{k}} G_{11}(\mathbf{k}, \omega)$$

$$= -\frac{1}{\pi V} \text{Im} \sum_{\mathbf{k}} \frac{\omega - \Sigma_0(\omega)}{[\omega - \Sigma_0(\omega)]^2 - [\xi_{\mathbf{k}} + \Sigma_3(\omega)]^2 - \Delta_{\mathbf{k}}^2}$$

$$= -N_F \text{Im} \underline{G}_0(\omega). \tag{8.78}$$

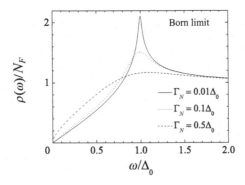

Figure 8.10 Quasiparticle density of states in the Born scattering limit.

Again the summation over momentum cannot be solved exactly. Nevertheless, in the Born or unitary scattering limit, $\Sigma_3(\omega)$ can be absorbed into the chemical potential and omitted. Under this approximation, the above expression can be simplified as

$$\rho(\omega) = -\frac{N_F}{\pi V}\text{Im}\int\frac{d\varphi}{2\pi}\int d\xi\frac{\omega - \Sigma_0(\omega)}{[\omega - \Sigma_0(\omega)]^2 - \xi^2 - \Delta_\varphi^2}$$

$$= \frac{2N_F}{\pi}\text{Im}\bar{\omega}\int_0^1\frac{dx}{\sqrt{1-x^2}}\frac{\theta\left(\text{Re}\sqrt{x^2-\bar{\omega}^2}\right)}{\sqrt{x^2-\bar{\omega}^2}}. \tag{8.79}$$

In the limit $|\bar{\omega}| \ll 1$, the integral is contributed mainly from the region $x \ll 1$, thus $\sqrt{1-x^2}$ is approximately equal to 1 and

$$\rho(\omega) \approx \frac{2N_F}{\pi}\text{Re}\bar{\omega}\int_0^1 dx\frac{-i}{\sqrt{x^2-\bar{\omega}^2}} = \frac{2N_F}{\pi}\text{Im}\left(\bar{\omega}\ln\frac{1+\sqrt{1-\bar{\omega}^2}}{\sqrt{-\bar{\omega}^2}}\right). \tag{8.80}$$

From this expression, we find that the impurity induced density of states is finite at zero energy

$$\rho(0) \approx \frac{2N_F\Gamma_0}{\pi\Delta_0}\ln\frac{\Delta_0 + \sqrt{\Delta_0^2 + \Gamma_0^2}}{\Gamma_0}, \tag{8.81}$$

where $\Gamma_0 = i\Sigma_0(0)$. This is different than in the ideal d-wave superconductor where the zero-energy density of states vanishes. In the Born scattering limit,

$$\Gamma_0 = 2\Delta_0\exp\left(-\frac{\pi\Delta_0}{\Gamma_N}\right), \tag{8.82}$$

while in the unitary limit

$$\Gamma_0 = \sqrt{\frac{\pi\Gamma_N\Delta_0}{4\ln 2}}. \tag{8.83}$$

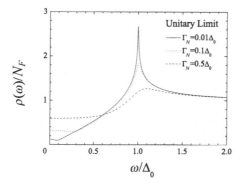

Figure 8.11 Quasiparticle density of states in the unitary scattering limit.

In the gapless region, $\rho(\omega) \ll N_F$, $\rho(\omega)$ can be solved by the Taylor expansion with respect to ω. It is simple to show that the first order term in the Taylor series of ω is zero in both the Born and unitary scattering limits. Thus $\rho(\omega)$ varies as ω^2 in the low energy limit.

In the unitary scattering limit, $\rho(\omega)$ is approximately given by

$$\rho(\omega) \approx \rho(0) \left(1 - \frac{\omega^2}{4\Gamma_0^2} \right), \qquad (|\omega| \ll \Gamma_0). \tag{8.84}$$

However, in the Born scattering limit, it changes to

$$\rho(\omega) \approx \rho(0) \left(1 + \frac{\pi^2 \Delta_0^2 \omega^2}{2\Gamma_0^2 \Gamma_N^2} \right), \qquad (|\omega| \ll \Gamma_0). \tag{8.85}$$

Clearly, $\rho(\omega)$ behaves differently in these two limits. In the Born scattering limit, $\rho(\omega)$ increases monotonically with ω. However, in the unitary scattering limit, $\rho(\omega)$ decreases with ω in the low energy limit, giving rise to a local maximum at $\omega = 0$. This local maximum results from the quasiparticle resonance states induced by the impurity scattering on the Fermi surface, but the divergence in the resonant density of states is smeared out by the disorder average.

In the intrinsic region, $\Delta \gg \omega \gg \Gamma_0$, $\rho(\omega)$ behaves similarly as in a disorder-free system. The disorder scattering only induces a small correction to $\rho(\omega)$,

$$\rho(\omega) \approx \frac{N_F \omega}{\Delta_0} \left[1 + \frac{2\Gamma_0}{\pi \omega} \left(\ln \frac{2\Delta_0}{|\omega|} - 1 \right) \right]. \tag{8.86}$$

The nonzero average of $\rho(\omega)$ on the Fermi surface is an important feature of disordered d-wave superconductors. It results from the sign change of the d-wave superconducting gap function on the Fermi surface. If the gap function only has nodes but does not change sign on the Fermi surface, for example, in an anisotropic s-wave superconductor with the gap function $\Delta_{\mathbf{k}} \propto |\cos k_x - \cos k_y|$ or $\Delta_{\mathbf{k}} \propto (\cos k_x - \cos k_y)^2$, the disorder scattering tends to turn the gap function more isotropic. Therefore, the correction to the gap function by the self-energy is finite, and the disorder scattering cannot yield a finite density of states at the Fermi energy. In contrast, it will eliminate the gap nodes and make the quasiparticle excitations fully gapped on the Fermi surface. Therefore, the change of impurity scattering to the superconducting properties in a d-wave superconductor differs qualitatively from that in an anisotropic s-wave superconductor. This can be used to determine whether the superconducting gap of high-T_c superconductors has the d-wave or anisotropic s-wave pairing symmetry.

8.7 Entropy and Specific Heat

Momentum is not a good quantum number in a disordered system. Nevertheless, the superconducting quasiparticles, as the solutions of the BdG equation, are still the energy eigenstates. If $(E_n, |\phi_n\rangle)$ is an eigenpair of the Hamiltonian for a disordered superconductor

$$H|\phi_n\rangle = E_n|\phi_n\rangle, \tag{8.87}$$

then the entropy of this system is defined by

$$S = -k_B \sum_n \{f(E_n)\ln f(E_n) + [1 - f(E_n)]\ln[1 - f(E_n)]\}$$

$$= -k_B \int_{-\infty}^{\infty} d\omega \rho(\omega)\{f(\omega)\ln f(\omega) + [1 - f(\omega)]\ln[1 - f(\omega)]\}. \tag{8.88}$$

Taking the derivative with respect to temperature, the specific heat is found to be

$$C = T\frac{\partial S}{\partial T} = -k_B \int_{-\infty}^{\infty} d\omega \beta \omega^2 \rho(\omega) \frac{\partial f}{\partial \omega}. \tag{8.89}$$

In obtaining this expression, the temperature derivative of $\rho(\omega)$ is omitted.

In the unitary scattering limit, from the result of $\rho(\omega)$ previously obtained, we find the specific heat in the gapless region ($k_B T \ll \Gamma_0$) to be

$$C \approx \frac{\pi^2 \rho(0)k_B^2 T}{3}\left(1 - \frac{7\pi^2 k_B^2 T^2}{20\Gamma_0^2}\right). \tag{8.90}$$

The corresponding result in the Born scattering limit is

$$C \approx \frac{\pi^2 \rho(0)k_B^2 T}{3}\left(1 + \frac{7\pi^4 \Delta_0^2 k_B^2 T^2}{10\Gamma_0^2 \Gamma_N^2}\right). \tag{8.91}$$

In either limit, the specific heat of a disordered d-wave superconductor is a linear function of temperature at low temperatures. This is a direct consequence of finite density of states at the Fermi level. The temperature dependence of C/T is similar to the energy dependence of the density of states. In the resonant scattering limit, C/T shows a peak at zero temperature. This peak, however, does not exist in the Born scattering limit. Therefore, the measurement of low temperature specific heat allows us to determine which limit the impurity scattering is in. The next order correction to the specific heat by the impurity scattering is proportional to T^3, similar to the temperature dependence of the specific heat contributed by phonons.

In the intrinsic temperature region, $T \gg \Gamma_0$, the disorder scattering does not significantly change the temperature dependence of the specific heat in the d-wave superconductor. C behaves almost the same as in a pure d-wave superconductor.

9

Superfluid Response

9.1 Linear Response Theory of Superfluids

When an external magnetic field is applied to a superconductor, the magnetic field will be expelled from the interior of the superconductor by the screening effect of the supercurrent. This is the diamagnetic response of the superconductor to an external magnetic field, a fundamental property of superconducting states. A correct and microscopic description of this electromagnetic response is crucial toward a comprehensive understanding of experimental results. In high-T_c superconductors, a variety of experimental evidence for the existence of gap nodes in the pairing function are obtained through the measurement of electromagnetic response functions.

The superfluid density is an important quantity to describe the supercurrent response, and to connect microscopic mechanism of superconductivity with macroscopic electromagnetism. As revealed by the London equation, Eq. (1.1), the superfluid density is inversely proportional to the square of the penetration depth. These two quantities are intimately connected by the Meissner effect. In a superconductor, the larger is the superfluid density, the stronger is the screening effect of the supercurrent, hence the shorter is the magnetic penetration depth. The variance of the superfluid density with temperature is determined purely by the thermally excited superconducting quasiparticles. It is different if the pairing symmetry is different. At low temperatures, the superfluid density varies exponentially in an isotropic s-wave superconductor, but linearly with temperature in a d-wave superconductor. Thus we can acquire a great deal of valuable information of quasiparticle excitations through the measurement of the superfluid density or the magnetic penetration depth.

In the linear response theory, the response of the system to an external field is determined by the Kubo formula [6]

$$J_\mu(\mathbf{q}, \omega) = - \sum_\nu K_{\mu\nu}(\mathbf{q}, \omega) A_\nu(\mathbf{q}, \omega), \tag{9.1}$$

where $K_{\mu\nu}$ is the response function of the electric current J_μ to the vector field applied A_ν. A derivation of this formula is presented in Appendix G.

$K_{\mu\nu}$ is determined by the electron effective mass and the current–current correlation function

$$K_{\mu\nu}(\mathbf{q}, \omega) = \frac{e^2}{V\hbar^2} \sum_{\mathbf{k}} \left\langle \frac{\partial^2 \varepsilon_{\mathbf{k}}}{\partial k_\mu \partial k_\nu} \right\rangle + \Pi_{\mu\nu}(\mathbf{q}, \omega), \qquad (9.2)$$

where $\Pi_{\mu\nu}$ is the current–current correlation function defined by

$$\Pi_{\mu\nu}(\mathbf{q}, \omega) = -\frac{i}{V\hbar^2} \int_0^\infty dt e^{i\omega t} \left\langle \left[J_\mu\left(\mathbf{q}, t\right), J_\nu\left(-\mathbf{q}, 0\right) \right] \right\rangle. \qquad (9.3)$$

J_μ is the electron current operator

$$J_\mu(\mathbf{q}) = e \sum_{\mathbf{k}} \frac{\partial \varepsilon_{\mathbf{k}+\mathbf{q}/2}}{\partial k_\mu} c_{\mathbf{k}+\mathbf{q}}^\dagger c_{\mathbf{q}}. \qquad (9.4)$$

In an isotropic and homogeneous system

$$\varepsilon_{\mathbf{k}} = \frac{\hbar^2 \mathbf{k}^2}{2m}, \qquad (9.5)$$

the first term of $K_{\mu\nu}$ is simply given by

$$\frac{e^2}{V\hbar^2} \sum_{\mathbf{k}} \left\langle \frac{\partial^2 \varepsilon_{\mathbf{k}}}{\partial k_\mu \partial k_\nu} \right\rangle = \frac{e^2 n}{m} \delta_{\mu,\nu}. \qquad (9.6)$$

It is proportional to the ratio between the electron density and the effective mass, independent of temperature. However, in real materials, this term depends on the band structure and can vary with temperature.

When an external electromagnetic field with a wave vector $\mathbf{q}$ and frequency ω is applied to the system, a current is generated. The response function connecting the current and the electromagnetic field is the conductivity tensor $\sigma(\mathbf{q}, \omega)$ defined by

$$J_\mu(\mathbf{r}, t) = \sum_\nu \sigma_{\mu\nu}(\mathbf{q}, \omega) E_\nu(\mathbf{r}, t). \qquad (9.7)$$

The electric field is related to the vector potential by the equation

$$A_\mu(\mathbf{r}, t) = -\frac{i}{\omega} E_\mu(\mathbf{r}, t). \qquad (9.8)$$

From Eq. (9.1), it is simple to show that $\sigma_{\mu\nu}$ is proportional to the response function $K_{\mu\nu}$

$$\sigma_{\mu\nu}(\mathbf{q}, \omega) = \frac{i}{\omega} K_{\mu\nu}(\mathbf{q}, \omega). \qquad (9.9)$$

The direct-current (DC) conductivity is the value of $\sigma_{\mu\nu}(\mathbf{q}, \omega)$ in a uniform electric field, $\mathbf{q} = 0$, and in the zero frequency limit, i.e. $\sigma_{\mu\nu}(0, \omega \to 0)$.

The current–current correlation function $\Pi_{\mu\nu}(\mathbf{q}, \omega)$ can be obtained by first calculating the corresponding Matsubara Green's function in the imaginary frequency space, $\tilde{\Pi}_{\mu\nu}(\mathbf{q}, i\omega_n)$, before taking the analytic continuation. $\tilde{\Pi}_{\mu\nu}(\mathbf{q}, i\omega_n)$ is defined by

$$\tilde{\Pi}_{\mu\nu}(\mathbf{q}, i\omega_n) = -\frac{1}{V\hbar^2} \int_0^\beta d\tau e^{i\omega_n\tau} \langle J_\mu(\mathbf{q}, t) J_\nu(-\mathbf{q}, 0) \rangle. \tag{9.10}$$

Substituting the expression of $J_\mu(\mathbf{q}, t)$ in Eq. (9.4) into Eq. (9.10) and after a simple derivation, one can express $\tilde{\Pi}_{\mu\nu}$ using the single-particle Green's function as

$$\tilde{\Pi}_{\mu\nu}(\mathbf{q}, i\omega_n) = \frac{e^2}{\beta V\hbar^2} \sum_{\mathbf{k}, \omega_m} \frac{\partial \varepsilon_{\mathbf{k}+\mathbf{q}/2}}{\partial k_\mu} \frac{\partial \varepsilon_{\mathbf{k}+\mathbf{q}/2}}{\partial k_\nu} \mathrm{Tr} G(\mathbf{k}, i\omega_m) G(\mathbf{k} + \mathbf{q}, i\omega_n + i\omega_m). \tag{9.11}$$

Given $\tilde{\Pi}_{\mu\nu}$, the real frequency current–current correlation function $\Pi_{\mu\nu}$ can be obtained through analytic continuation

$$\Pi_{\mu\nu}(\mathbf{q}, \omega) = \tilde{\Pi}_{\mu\nu}(\mathbf{q}, i\omega_n \to \omega + i0^+). \tag{9.12}$$

To perform the analytic continuation, it is usually more convenient to represent the Matsubara Green's function as an integral of the retarded Green's function $G^R(\mathbf{k}, \omega)$ in the spectral representation

$$G(\mathbf{k}, i\omega_n) = -\int_{-\infty}^\infty \frac{d\omega}{\pi} \frac{\mathrm{Im} G^R(\mathbf{k}, \omega)}{i\omega_n - \omega}. \tag{9.13}$$

By substituting this expression into Eq. (9.11), and summing over the frequency, we can further express $\tilde{\Pi}_{\mu\nu}$ as

$$\tilde{\Pi}_{\mu\nu}(\mathbf{q}, i\omega_n) = \frac{e^2}{\pi^2 V\hbar^2} \sum_{\mathbf{k}} \int_{-\infty}^\infty d\omega_1 d\omega_2 \frac{\partial \varepsilon_{\mathbf{k}+\mathbf{q}/2}}{\partial k_\mu} \frac{\partial \varepsilon_{\mathbf{k}+\mathbf{q}/2}}{\partial k_\nu} [f(\omega_1) - f(\omega_2)]$$
$$\frac{\mathrm{TrIm} G^R(\mathbf{k}, \omega_1) \mathrm{Im} G^R(\mathbf{k} + \mathbf{q}, \omega_2)}{i\omega_n + \omega_1 - \omega_2}. \tag{9.14}$$

The retarded current–current correlation function is then found to be

$$\Pi_{\mu\nu}(\mathbf{q}, \omega) = \frac{e^2}{\pi^2 V\hbar^2} \sum_{\mathbf{k}} \int_{-\infty}^\infty d\omega_1 d\omega_2 \frac{\partial \varepsilon_{\mathbf{k}+\mathbf{q}/2}}{\partial k_\mu} \frac{\partial \varepsilon_{\mathbf{k}+\mathbf{q}/2}}{\partial k_\nu} [f(\omega_1) - f(\omega_2)]$$
$$\times \frac{\mathrm{TrIm} G^R(\mathbf{k}, \omega_1) \mathrm{Im} G^R(\mathbf{k} + \mathbf{q}, \omega_2)}{\omega + \omega_1 - \omega_2 + i0^+}. \tag{9.15}$$

The current–current correlation function $\Pi_{\mu\nu}$ is a complex function. Its imaginary part is proportional to the alternating-current conductivity, namely the optical conductivity. In an ideal superconductor without energy dissipation, both the resistivity and the optical conductivity vanish. In the presence of elastic or inelastic disorder scatterings, the resistivity is still zero in the superconducting state, but the optical conductivity becomes finite.

$\Pi_{\mu\nu}$ contributes mainly from the electrons around the Fermi surface. In an isotropic two-dimensional system, if we approximate the electron velocity, $v \propto [(\partial\varepsilon_\mathbf{k}/\partial k_x)^2 + (\partial\varepsilon_\mathbf{k}/\partial k_y)^2]^{1/2}$, by the Fermi velocity v_F, then the in-plane correlation function $\Pi_{ab} = \delta_{ab}\Pi_\|$ and

$$\Pi_\|(\mathbf{q},\omega) = \frac{e^2 v_F^2}{2\pi^2 V} \sum_\mathbf{k} \int_{-\infty}^{\infty} d\omega_1 d\omega_2 \, [f(\omega_1) - f(\omega_2)]$$
$$\times \frac{\mathrm{Tr}\,\mathrm{Im}G^R(\mathbf{k},\omega_1)\mathrm{Im}G^R(\mathbf{k}+\mathbf{q},\omega_2)}{\omega + \omega_1 - \omega_2 + i0^+}. \tag{9.16}$$

Based on the discussion in §12, we find that $\Pi_\|(\mathbf{q},\omega)$ differs from the spin–spin correlation function $\chi_{zz}(\mathbf{q},\omega)$ defined in Eq. (12.8) just by a constant factor

$$\Pi_\|(\mathbf{q},\omega) = -\frac{2e^2 v_F^2}{\gamma_e^2 \hbar^2} \chi_{zz}(\mathbf{q},\omega). \tag{9.17}$$

It shows that the electromagnetic response function $\Pi_\|(\mathbf{q},\omega)$ and the spin response function $\chi_{zz}(\mathbf{q},\omega)$ are closely connected.

In real experimental measurements, the ranges of momentum and frequency measured are different for different physical quantities. There are two particularly interesting limits: the long wavelength limit, $q \to 0$, and the low-frequency limit, $\omega \to 0$. Depending on which limit is taken, the response function corresponds to different measurement quantities:

(1) The imaginary part of the response function in the long wavelength limit, i.e. $\mathrm{Re}\Pi_{\mu\mu}(\mathbf{q} \to 0, \omega)$, is proportional to the optical conductivity.
(2) The real part of the diagonal response function at the zero frequency and long wavelength limit, i.e. $\mathrm{Re}\Pi_{\mu\mu}(\mathbf{q} \to 0, \omega = 0)$, is the paramagnetic contribution to the superfluid density.
(3) The measurement of the spin–spin correlation function depends on experimental methods. The Knight shift of the nuclear magnetic resonance (NMR) is proportional to the real part of the spin–spin correlation function at zero momentum and zero frequency, $\mathrm{Re}\chi_{zz}(\mathbf{q} \to 0, \omega = 0)$. It is also proportional to the paramagnetic response function of the superfluid density. Since the diamagnetic part of the superfluid density is roughly temperature independent, the Knight shift and the superfluid density should exhibit similar temperature dependence.

The spin-relaxation rate $1/T_1$ measures the dynamic response of spins in the static limit and is determined by the spin–spin correlation functions over the entire momentum space. In comparison, the energy and momentum ranges measured by the neutron scattering spectroscopy are much larger than NMR, but the resolution is lower. These experimental methods are complementary to each other and play an important role in the study of physical properties of d-wave superconductors.

The above electromagnetic response function is derived under the linear approximation. In principle, it is valid only in the zero field limit $H \to 0$. However, for the d-wave superconductor, the nonlinear response of the superconducting state to the magnetic field is important at low temperatures due to the presence of pairing gap nodes. On the other hand, the nonlocal effect could also become important at low temperatures, since the effective coherence length along the nodal direction is infinite. Both the nonlinear and nonlocal effects can strongly affect the low temperature electromagnetic response functions. This is different than in an s-wave superconducting state, where the superconducting energy gap is finite and the linear response holds even in a weak but finite magnetic field.

9.2 Superfluid Density

The superfluid density is a fundamental quantity characterizing superconductivity. It is proportional to the energy scale that Cooper pairs form phase coherence. The superfluid density can be taken as an order parameter of superconductivity, because it is zero in the nonsuperconducting phase and finite in the superconducting phase. In the standard theory of superconductivity, the pairing energy gap is considered as the superconducting order parameter. This choice of order parameter is strictly speaking not that rigorous. It is only valid when the phase fluctuation is small. In the presence of strong phase fluctuations, electrons are often paired (namely to have a finite pairing gap) but do not exhibit long-range phase coherence. However, the long-range phase coherence is a prerequisite of superconductivity. The larger the superfluid density, the higher is the energy cost of phase fluctuations. Thus the superfluid density is a measure of the robustness of phase coherence, which is also called the phase stiffness.

The superfluid density n_s^μ is inversely proportional to the square of the magnetic penetration depth λ. It can be obtained from the real part of $K_{\mu\mu}$. According to the London equation, Eq. (1.1), and the definition of conductivity, Eq. (9.7), it can be shown that n_s^μ is proportional to the real part of $K_{\mu\mu}$ in the long wavelength limit

$$\mathrm{Re} K_{\mu\mu}(\mathbf{q}_\perp \to 0, q_\parallel = 0, \omega = 0) = \frac{e^2 n_s^\mu}{m_\mu} = \frac{1}{\mu_0 \lambda^2}, \qquad (9.18)$$

where $\mathbf{q}_\perp$ and $q_\parallel$ are the components of $\mathbf{q}$ perpendicular and parallel to the external magnetic field, respectively. This expression shows that in the calculation of the superfluid density, one should first take the limit of $\omega \to 0$, and then the limit $\mathbf{q} \to 0$. This sequence of limits is opposite to that taken in the calculation of DC conductivity. The reason for this is not difficult to understand. The DC conductivity is the coefficient measuring the current response to an applied electric field, while the superfluid density is the coefficient measuring the supercurrent response to an applied magnetic field.

For an ideal d-wave superconductor, the single-particle Green's function is given by Eq. (4.4). The frequency summation in Eq. (9.11) can be evaluated rigorously and $\Pi_{\mu\nu}$ in the limit of $(\omega = 0, q \to 0)$ is

$$\Pi_{\mu\nu}(\mathbf{q} \to 0, 0) = \lim_{q \to 0} \frac{2e^2}{V\hbar^2} \sum_{\mathbf{k}} \frac{\partial \varepsilon_{\mathbf{k}}}{\partial k_\mu} \frac{\partial \varepsilon_{\mathbf{k}}}{\partial k_\nu} \frac{f(E_{\mathbf{k}}) - f(E_{\mathbf{k}+\mathbf{q}})}{E_{\mathbf{k}} - E_{\mathbf{k}+\mathbf{q}}}. \tag{9.19}$$

In the limit $q \to 0$

$$\lim_{q \to 0} \frac{f(E_{\mathbf{k}}) - f(E_{\mathbf{k}+\mathbf{q}})}{E_{\mathbf{k}} - E_{\mathbf{k}+\mathbf{q}}} = \frac{\partial f(E_{\mathbf{k}})}{\partial E_{\mathbf{k}}}, \tag{9.20}$$

so that

$$\Pi_{\mu\nu}(\mathbf{q} \to 0, 0) = \frac{2e^2}{V\hbar^2} \sum_{\mathbf{k}} \frac{\partial \varepsilon_{\mathbf{k}}}{\partial k_\mu} \frac{\partial \varepsilon_{\mathbf{k}}}{\partial k_\nu} \frac{\partial f(E_{\mathbf{k}})}{\partial E_{\mathbf{k}}}. \tag{9.21}$$

Hence, the superfluid density along the μ-direction is

$$n_s^\mu = \frac{m_\mu}{V\hbar^2} \sum_{\mathbf{k}} \left[\left\langle \frac{\partial^2 \varepsilon_{\mathbf{k}}}{\partial k_\mu^2} \right\rangle + 2 \left(\frac{\partial \varepsilon_{\mathbf{k}}}{\partial k_\mu} \right)^2 \frac{\partial f(E_{\mathbf{k}})}{\partial E_{\mathbf{k}}} \right]. \tag{9.22}$$

Experimentally, the superfluid density, or the magnetic penetration depth λ, can be measured by the microwave attenuation on the surface of superconductor, μSR, infrared spectroscopy, and AC magnetic susceptibility. The microwave attenuation can accurately measure the relative values of λ at different temperatures, but not their absolute values. μSR and the AC magnetic susceptibility, on the other hand, can measure directly the absolute values of λ. However, these experiments need large samples, and the measurement errors are relatively large.

The penetration depths of high-T_c superconductors are much larger than the coherence lengths of Cooper pairs. Hence even the surface microwave experiments probe the bulk property of high-T_c superconductors, not just their surface properties. Furthermore, the analysis of the penetration depth measurement data is relatively simple, since it is affected just by the superconducting electrons, not by the normal quasiparticles or other thermal excitations, like phonons.

At low temperatures, there are corrections to the above formula from the non-linear and nonlocal effects in d-wave superconductors. These effects change the temperature as well as the magnetic field dependence of the penetration depth. In particular, they alter the temperature dependence of the penetration depth from T to T^2 at very low temperature, which is actually important for the stability of d-wave superconductors. Otherwise, if the penetration depth remains a linear function of temperature down to zero temperature, the entropy would become finite at zero temperature, violating the third law of thermodynamics, i.e. the Nernst law [191, 192]. This is of course nonphysical. We should be cautious of extending the linear or the local approximation to the zero temperature limit.

In the d-wave superconductor, the nonlinear and nonlocal effects have a common feature. They are strongly anisotropic, depending on the relative angle between the nodal direction and the superconductor surface. This dependence can be used to probe the nodal direction on the Fermi surface, which in turn can be used to determine the pairing symmetry. In real materials, the nonlinear and nonlocal effects are intertwined. Within certain parameter regimes, or under certain boundary conditions, the anisotropy induced by one effect could be weakened by another. A comprehensive analysis of these two effects is needed in order to analyze correctly the experimental results.

9.3 Superfluid Response in Cuprate Superconductors

9.3.1 In-Plane Superfluid Density

In high-T_c superconductors, the electron energy–momentum dispersion relation is approximately given by [99, 193, 194]

$$\varepsilon_{\mathbf{k}} \simeq \frac{\hbar^2(k_x^2 + k_y^2)}{2m} - \frac{\hbar^2}{m_c} \cos k_z \cos^2(2\varphi), \tag{9.23}$$

where m and m_c are the effective masses of electrons parallel and perpendicular to the CuO_2 plane, respectively. φ is the azimuthal angle of the in-plane momentum (k_x, k_y). In this expression, the anisotropy of the band structure in the CuO_2 plane is neglected, and the dependence of the c-axis dispersion on the in-plane momentum (k_x, k_y) is also simplified. These simplifications do not change the physical results qualitatively, but do change the coefficients of the temperature dependence of the superfluid density.

In the CuO_2 plane, if we neglect the effect of the m_c-term on the electron dispersion in the ab-plane, then

$$\left(\frac{\partial}{\partial k_x}, \frac{\partial}{\partial k_y} \right) \varepsilon_{\mathbf{k}} = \frac{\hbar^2}{m} \left(k_x, k_y \right), \qquad \frac{\partial^2 \varepsilon_{\mathbf{k}}}{\partial k_x^2} = \frac{\partial^2 \varepsilon_{\mathbf{k}}}{\partial k_x^2} = \frac{\hbar^2}{m}. \tag{9.24}$$

Substituting them into Eq. (9.22), the superfluid density along the ab plane is reduced to

$$n_s^{ab} = n + \frac{\hbar^2}{mV} \sum_{\mathbf{k}} (k_x^2 + k_y^2) \frac{\partial f(E_{\mathbf{k}})}{\partial E_{\mathbf{k}}}, \tag{9.25}$$

where n is the electron concentration. The second term is the paramagnetic contribution to the superfluid density, resulting from the thermal excitation of quasiparticles. At zero temperature, the superfluid density equals the electron concentration. This is a result which is valid for a parabolic band. In a lattice system, the Galilean invariance is broken and the superfluid density is not equal to the electron concentration even at zero temperature. As the momentum summation contributes mainly from the electrons around the Fermi surface, we can take the approximation $k_x^2 + k_y^2 \approx k_F^2$ and simplify the above expression as

$$n_s^{ab} \simeq n - \frac{\hbar^2 k_F^2}{m} Y(T), \tag{9.26}$$

where

$$Y(T) = -\frac{1}{V} \sum_{\mathbf{k}} \frac{\partial f(E_{\mathbf{k}})}{\partial E_{\mathbf{k}}} = -\int_{-\infty}^{\infty} dE \rho(E) \frac{\partial f(E)}{\partial E} \tag{9.27}$$

is the Yoshida function. Due to the anisotropy of the d-wave energy gap, the integral on the right-hand side of Eq. (9.27) cannot be solved analytically. At $T \ll \Delta$, the contribution arises mainly from the terms with $E \sim k_B T$. In this case, $\rho(E)$ is approximately given by Eq. (3.49), and

$$n_s^{ab} \simeq n + \frac{2\hbar^2 k_F^2 N_F}{m \Delta_0} \int_0^{\infty} dE E \frac{\partial f(E)}{\partial E} = n \left[1 - \frac{(2 \ln 2) k_B T}{\Delta_0} \right]. \tag{9.28}$$

The low temperature dependence of the magnetic penetration depth in the CuO_2 plane is then found to be

$$\lambda_{ab}(T) \simeq \lambda_{ab}(0) \left[1 + \frac{(\ln 2) k_B T}{\Delta_0} \right]. \tag{9.29}$$

Thus the penetration depth varies linearly with temperature at low temperatures. This is an important feature of the d-wave or other superconductors with linear density of states in low energies. Generally it can be shown that the exponent in the leading temperature-dependence of the penetration depth is equal to the exponent in the leading energy-dependence of the density of states. Hence the latter can be obtained by measuring the penetration depth.

In an isotropic s-wave superconductor, there are very few quasiparticles that can be excited above the energy gap at low temperatures. The penetration depth is thermally activated and exhibits an exponential temperature dependence [2]

$$\lambda_{ab}(T) - \lambda_{ab}(0) \sim e^{-\Delta/k_B T}. \tag{9.30}$$

The difference in the low temperature dependence between the s- and d-wave superconductors is an important criterion in determining the pairing symmetry of high-T_c superconductors from the penetration depth.

In the early stages of the high-T_c study, due to the relatively poor sample quality, the experimental errors were large. Even though high-T_c superconductors were predicted to possess the d-wave pairing symmetry, it was not supported by the penetration depth measurements.

In 1993, Hardy and coworkers found for the first time the linear temperature dependence of λ in the high quality YBCO twin crystals [195]. Their results are consistent with the d-wave pairing symmetry. Their experimental results, demonstrated the importance of the penetration depth measurement in the study of high-T_c superconducting mechanism. The improvement of sample quality is crucial because the impurity scattering has a strong effect on the low temperature superfluid density. After their work, many experimental groups made great efforts to measure accurately the penetration depth. The linear temperature dependence of λ_{ab}

$$\lambda_{ab}(T) \simeq \lambda_{ab}(0) + \alpha_{ab} T, \tag{9.31}$$

has now been verified in the YBCO single crystal [196, 197], YBCO twin crystal [198, 199], YBCO twin crystal film [200], BSCCO single crystal [201–203], HgBa-CaCuO and TlBaCuO powders [204], and many other high-T_c compounds. It shows that even at low temperatures, there are still quasiparticle excitations, indicating the existence of gap nodes. This is a strong support to the $d_{x^2-y^2}$-wave pairing symmetry of high-T_c superconductivity.

The value of the zero temperature penetration depth $\lambda_{ab}(0)$ depends on the sample quality. The better the sample quality, the stronger is the supercurrent screening and the smaller is $\lambda_{ab}(0)$. The experimental results are consistent with this expectation. The sample quality can be judged based on the measurement of normal state resistance and properties in the critical region of the superconducting phase transition. For high quality samples, the structure is more homogeneous and the critical transition range is narrower. The corresponding residual microwave resistance and the normal state resistance are also small.

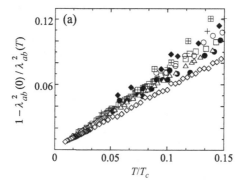

Figure 9.1 (a) Temperature dependence of the normalized superfluid density in the CuO_2 plane for $HgBa_2Ca_2Cu_3O_{8+\delta}$ (hollow diamond) [205], $HgBa_2CuO_{4+\delta}$ (triangle) [205], $YBa_2Cu_3O_7$ (solid circle) [206, 207], $YBa_2Cu_3O_{6.7}$ (hollow circle) [207], $YBa_2Cu_3O_{6.57}$(square) [207], $La_{2-x}Sr_xCuO_4$ [208] with $x = 0.2$ (cross), 0.22 (solid diamond), and 0.24 (square with cross). (From Ref. [209])

Figure 9.1 shows the temperature dependence of the superfluid density in the CuO_2 plane for a class of representative high-T_c superconductors [209]. As mentioned before, the low temperature superfluid density varies linearly with temperature within measurement errors. In addition, the slopes of the normalized superfluid density $1 - \lambda_{ab}^2(0K)/\lambda_{ab}(T)$ are nearly the same for all the samples shown in the figure, independent of the doping level (e.g. underdoping, optimal doping, and overdoping). From the slope and the low temperature behavior of the normalized superfluid density

$$1 - \frac{\lambda_{ab}^2(0)}{\lambda_{ab}^2(T)} \approx \frac{(2 \ln 2)k_B T}{\Delta_0}, \tag{9.32}$$

the amplitude of Δ_0 can be extracted. Δ_0 is determined by fitting the experimental data with the above equation, and scales almost linearly with T_c for several cuprate superconductors (Figure 9.2) [209], consistent with the d-wave BCS mean-field result, $\Delta \simeq 2.14T_c$. It should be emphasized that Δ_0 in Eq. (9.32) is only the gap amplitude around the nodal points, or more precisely, the slope of the gap function at the nodal points. It is not necessary to be the maximal energy gap. In the underdoped superconductors, due to the presence of the pseudogap, the momentum dependence of the gap function is not as simple as $\Delta_k = \Delta_0 \cos(2\varphi)$, and the maximal energy gap is larger than the value predicted by the BCS theory.

The approximate linear scaling behavior between Δ_0 and T_c, as shown in Fig. 9.2, differs significantly from the behavior between the maximal energy gap and T_c in the underdoped regime. The existence of this scaling behavior is likely an intrinsic property of high-T_c superconductors. It suggests that in the vicinity of gap nodes,

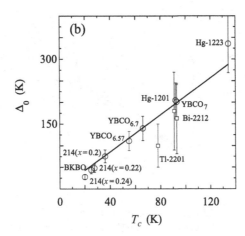

Figure 9.2 The scaling relation between the gap magnitude in the nodal region and the superconducting transition temperature T_c. (From Ref. [209]) In addition to the superconductors shown in Fig. 9.1, the results for $Ba_2Sr_2CaCu_2O_{8+\delta}$ [201, 202], $Tl_2Ba_2CuO_{6+\delta}$ [210], and the s-wave superconductors $Bi_{0.6}K_{0.4}BiO_3$ [204] are also included. The solid line is the theoretical result for the weak coupling d-wave superconductor, $\Delta_0 = 2.14T_c$.

the quasiparticle excitations can still be described by the BCS mean-field theory even in the underdoped regime. This is related to the Fermi surface structure of high-T_c cuprates. In underdoped materials, the ARPES measurements found that there exist arc-like Fermi surfaces, which are centered around ($\pm\pi/2$, $\pm\pi/2$) and extend toward ($\pm\pi,0$) and ($0, \pm\pi$) [120, 121]. The pseudogap, on the other hand, begins to spread from the antinodal regions around ($\pm\pi,0$) and ($0, \pm\pi$) toward the nodal points only at some temperature not much higher than T_c. Thus the pseudogap will not completely suppress the Fermi surface at T_c and the pairing on the remnant Fermi arcs will dominate low energy excitations in the superconducting state, leading to a linear behavior between the gap slope at the nodal point, Δ_0, and T_c.

However, it should be pointed out that theoretical explanations of the experimental results of linear penetration depth are not unified. Either the pairing phase fluctuation [32, 211] or the proximity effect could also be invoked to explain the linear temperature dependence of λ_{ab}. It is difficult, however, to use these effects to understand why the impurity scattering can change the temperature dependence of λ_{ab} from T to T^2. It should also be noted that the in-plane superfluid density depends only on the quasiparticle density of states and is not sensitive to the phase of the gap function and the locations of the gap nodes. It is inadequate to fully determine the pairing symmetry only based on the measurement of the in-plane penetration depth.

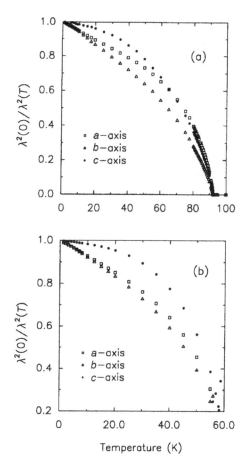

Figure 9.3 Temperature dependence of the normalized superfluid density for (a) YBa$_2$Cu$_3$O$_{6.95}$ ($T_c = 93.5$ K) and (b) YBa$_2$Cu$_3$O$_{6.6}$ ($T_c = 59$ K) along three principal axes. (From Ref. [197])

In YBa$_2$Cu$_3$O$_{7-x}$, there are CuO chains and the magnetic penetration depth along the chain direction (the b-axis) differs from that along the direction perpendicular (the a-axis) [197, 212]. For the optimally doped YBCO superconductor, $\lambda_a(0) = 1\,600$ Å is about 1.6 times larger than $\lambda_b(0) = 1\,030$ Å [197, 212]. For the underdoped YBa$_2$Cu$_3$O$_{7-x}$ superconductors, the difference between $\lambda_a(0)$ and $\lambda_b(0)$ is smaller [212]. For YBa$_2$Cu$_4$O$_8$, the difference between $\lambda_a(0)$ and $\lambda_b(0)$ is larger, $\lambda_a(0) = 2\,000$ Å and $\lambda_b(0) = 800$ Å, hence $\lambda_a(0)$ is about 2.5 times larger than $\lambda_b(0)$ [212]. Nevertheless, as shown in Fig. 9.3, no matter how large the difference between $\lambda_a(0)$ and $\lambda_b(0)$, the temperature derivatives of $\lambda_\mu(T)$ ($\mu = a, b$), after being normalized by λ ($T=0$), are nearly the same along these two directions.

The difference between the lattice constants along the a- and b-axes of YBCO is less than 2.5%, thus the difference between the superfluid densities comes mainly from the contribution of CuO chains. A CuO chain can be viewed as a quasi-one-dimensional system. It contributes to λ_b, but barely to λ_a. If we attribute the origin of the difference between λ_a and λ_b completely to the CuO chains, then in the optimally doped YBa$_2$Cu$_3$O$_{7-x}$ sample, about 60% of the b-axis superfluid density comes from the contribution of CuO chains. For YBa$_2$Cu$_4$O$_8$, the chain contribution even increases to 80%. For the underdoped YBa$_2$Cu$_3$O$_{7-x}$, electrons on the CuO chains are easy to localize by impurity scattering. Their contribution to the superfluid density is weakened, hence the difference between λ_a and λ_b is suppressed. In the normal state, the CuO chains also have significant contributions to the electric [213, 214] and thermal conductivities [215]. For example, in the single crystal of YBa$_2$Cu$_3$O$_{6.95}$, the electric and thermal conductivities along the b-axis are about twice larger than those along the a-axis. The anisotropic ratios are nearly the same as for the superfluid density.

9.3.2 Interlayer Superfluid Density

Along the c-axis, the energy dispersion of electrons and its derivatives depend on the in-plane momentum,

$$\frac{\partial \varepsilon_\mathbf{k}}{\partial k_z} = \frac{\hbar^2}{m_c} \sin k_z \cos^2(2\varphi), \qquad \frac{\partial^2 \varepsilon_\mathbf{k}}{\partial k_z^2} = \frac{\hbar^2}{m_c} \cos k_z \cos^2(2\varphi). \qquad (9.33)$$

The temperature dependence of the superfluid density now becomes complicated and is determined by the equation

$$n_s^c = \frac{1}{V} \sum_\mathbf{k} \left[-\cos k_z \cos^2(2\varphi) \frac{\xi_\mathbf{k}}{E_\mathbf{k}} \tanh \frac{\beta E_\mathbf{k}}{2} \right.$$
$$\left. + \frac{2\hbar^2}{m_c} \sin^2 k_z \cos^4(2\varphi) \frac{\partial f(E_\mathbf{k})}{\partial E_\mathbf{k}} \right]. \qquad (9.34)$$

The first term is the diamagnetic contribution. It is nearly temperature-independent if the bandwidth is much larger than Δ. In this case, it equals the superfluid density at zero temperature

$$n_s^c(0) = \frac{3\hbar^2 N_F}{16 m_c}. \qquad (9.35)$$

The second term in Eq. (9.34) is paramagnetic and temperature-dependent. It is difficult to calculate this term analytically. Nevertheless, in the limit $k_B T \ll \Delta_0$, the momentum summation in Eq. (9.34) can be approximately evaluated, from which the low temperature superfluid density is obtained as [99, 100]

$$n_s^c(T) \approx n_s^c(0) \left[1 - 450\zeta(5) \left(\frac{k_B T}{\Delta_0} \right)^5 \right], \tag{9.36}$$

where $\zeta(x)$ is the zeta function

$$\zeta(x) = \sum_n \frac{1}{n^x}. \tag{9.37}$$

$\zeta(5) = \pi^6/945 \approx 1.04$. The terms higher than T^5 are neglected.

The T^5-dependence of n_s^c is caused by the $\cos 2\varphi$ term in the c-axis dispersion relation. It shows that the anisotropy of the electron structure in the CuO_2 plane has a significant effect on the c-axis superfluid density. This is a peculiar property of high-T_c cuprates. In fact, the power of the temperature dependence of n_s^c can be obtained by simple dimensional analysis. The quasiparticle density of states in the d-wave superconductor is linear, which contributes one power of T. The additional power of T^4 is generated by $\cos^4(2\varphi)$ because $\Delta \cos(2\varphi)$ is of the dimension of energy.

The T^5-dependence is a consequence of the coincidence of the anisotropy of the c-axis hopping matrix element and the anisotropy of the d-wave pairing gap function. In particular, the zeros of the hopping constant coincide with the gap nodes. Thus through the measurement of the c-axis superfluid density, not only can we determine if there are nodes in the gap function, but also determine the locations of these nodes on the Fermi surface.

Due to the quasi-two-dimensional nature of high-T_c superconductors, λ_c is about one or two orders larger than λ_{ab} [197, 201]. As the change of λ_c with temperature is very weak at low temperatures, it is quite difficult to determine the power of the temperature dependence of λ_c. Experimentally, the T^5-temperature dependence of λ_c was first confirmed in $HgBa_2CuO_{4+\delta}$. This intrinsic temperature dependence of the penetration depth along the c-axis could be measured because the anisotropy between the c-axis and the ab-plane penetration depth is small and the coherent interlayer hopping is relatively large in this superconductor. Figure 9.4 shows the experimental data obtained based on the AC magnetic susceptibility measurement for $HgBa_2CuO_{4+\delta}$. Later on, this T^5-behavior was also observed in the Bi2212 samples with larger anisotropy [216]. The agreement between the theoretical and experimental results indicates that as long as the sample is clean, the c-axis hopping of electrons is predominantly coherent. In other words, the contribution from the

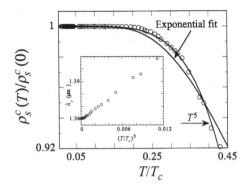

Figure 9.4 Temperature dependence of the normalized c-axis superfluid density, $\rho_s^c(T)/\rho_s^c(0K)$, for $HgBa_2CuO_{4+\delta}$ at low temperatures (From Ref. [100]). The hollow circles are the experimental data, and the solid lines are the theoretical fitting curves. The experiment data are consistent with the T^5 behavior predicted by theoretical calculation, which can be more clearly seen from the linear dependence of the penetration depth on $(T/T_c)^5$ in the inset.

incoherent hopping to the c-axis transport properties is small and electrons are not dynamically confined within each CuO_2 plane. This agreement also shows that there are gap nodes on the Fermi surface and the nodes are located along the diagonal directions, in support of the $d_{x^2-y^2}$-wave pairing symmetry.

The anisotropy between the c-axis and the ab-plane is relatively small in $YBa_2Cu_3O_{6+\delta}$. The coherent interlayer hopping of electrons should also be important for the low temperature electromagnetic response functions along the c-axis. However, due to the buckling of the CuO_2 planes and the existence of CuO chains, the a- and b-axes are not symmetric. The interlayer hopping matrix element is not simply proportional to $t_c \propto (\cos k_x - \cos k_y)^2$, and its zeros do not coincide with the d-wave pairing gap nodes. Thus the temperature dependence of λ_c is changed. It does not possess the T^5-temperature dependence as in a tetragonal compound.

9.4 Impurity Correction

Impurity scattering strongly affects the low temperature behavior of superfluid density. When the impurity scattering is sufficiently strong, it can change the temperature dependence of the in-plane superfluid density from T to T^2 [217]. This property is important in the analysis of experiment data, especially in the case where the sample quality is not that good. Experiments in the early days of the high-T_c study found that the temperature dependence of the in-plane penetration depth is very weak at low temperatures, which deviates from the intrinsic linear behavior of d-wave superconductors but resembles the exponential behavior of s-wave superconductors. This was once regarded as evidence for the s-wave symmetry of high-T_c pairing. However, after a more careful analysis by subtracting the contribution from the paramagnetic impurities, it was found that the low-temperature superfluid density varies quadratically instead of exponentially with temperature. This quadratic temperature dependence of the superfluid density is a common feature of disordered d-wave superconductors.

As discussed previously, $K_{\mu\mu}$ is determined by the sum of the following two terms,

$$K_{\mu\mu}(\mathbf{q} \to 0, 0) = \frac{e^2}{\hbar^2 V} \sum_{\mathbf{k}} \left\langle \frac{\partial^2 \varepsilon_{\mathbf{k}}}{\partial k_{\mu}^2} \right\rangle + \Pi_{\mu\mu}(\mathbf{q} \to 0, 0). \qquad (9.38)$$

The first term is diamagnetic, proportional to the average of the inverse effective mass of electrons in the Brillouin zone. This term is barely influenced by the impurity scattering and nearly temperature independent. It approximately equals the value in the clean system at zero temperature. The second term is paramagnetic. It is affected by the impurity scattering, leading to a finite correction to the superfluid density at zero temperature. Thus the zero temperature superfluid density is not completely the contribution from the first term.

The second term is determined by the current–current correlation function

$$\Pi_{\mu\mu}(\mathbf{q} \to 0, 0) = \frac{e^2}{\pi^2 \hbar^2 V} \sum_{\mathbf{k}} \left(\frac{\partial \varepsilon_{\mathbf{k}}}{\partial k_{\mu}} \right)^2 Z_{\mathbf{k}}, \qquad (9.39)$$

where

$$Z_{\mathbf{k}} = \int_{-\infty}^{\infty} d\omega_1 d\omega_2 \frac{f(\omega_1) - f(\omega_2)}{\omega_1 - \omega_2 + i0^+} \mathrm{Tr} \mathrm{Im} G^R(\mathbf{k}, \omega_1) \mathrm{Im} G^R(\mathbf{k}, \omega_2). \qquad (9.40)$$

At low temperatures, the temperature dependence of the superfluid density reflects the energy dependence of low-energy quasiparticle density of states. Disorder scattering changes the density of states around the nodal points, which in turn changes the temperature dependence of the superfluid density.

The retarded Green's function $G^R(\mathbf{k}, \omega)$ and the advanced Green's function $G^{R*}(\mathbf{k}, \omega)$ are analytic in the upper and lower half complex plane of ω, respectively. From the residue theorem, it is simple to show that the following equations are valid

$$\int_{-\infty}^{\infty} d\omega_1 \frac{G^R(\mathbf{k}, \omega_1)}{\omega - \omega_1 + i0^+} = -2\pi i G^R(\mathbf{k}, \omega), \qquad (9.41)$$

$$\int_{-\infty}^{\infty} d\omega_1 \frac{G^{R*}(\mathbf{k}, \omega_1)}{\omega_1 - \omega + i0^+} = -2\pi i G^{R*}(\mathbf{k}, \omega), \qquad (9.42)$$

$$\int_{-\infty}^{\infty} d\omega_1 \frac{G^R(\mathbf{k}, \omega_1)}{\omega_1 - \omega + i0^+} = \int_{-\infty}^{\infty} d\omega_1 \frac{G^{R*}(\mathbf{k}, \omega_1)}{\omega - \omega_1 + i0^+} = 0. \qquad (9.43)$$

Substituting them into Eq. (9.44) and after simplification, we find that

$$Z_\mathbf{k} = -\pi \int_{-\infty}^{\infty} d\omega f(\omega) \, \mathrm{Im Tr} G^R(\mathbf{k}, \omega) G^R(\mathbf{k}, \omega). \qquad (9.44)$$

In the gapless regime, $G^R(\mathbf{k}, \omega)$ is determined by Eq. (8.54). Substituting (8.54) into (9.44) gives

$$Z_\mathbf{k} = -2\pi \mathrm{Im} \int_{-\infty}^{\infty} d\omega f(\omega) \frac{(a\omega + i\Gamma_0)^2 + E_\mathbf{k}^2}{\left[(a\omega + i\Gamma_0)^2 - E_\mathbf{k}^2\right]^2}. \qquad (9.45)$$

The integral over ω can be performed through the Sommerfeld expansion introduced in Appendix E. The two leading terms in the expansion are

$$Z_\mathbf{k} = -2\pi \left[\frac{\Gamma_0}{a\left(\Gamma_0^2 + E_\mathbf{k}^2\right)} - \frac{a\Gamma_0\left(\Gamma_0^2 - 3E_\mathbf{k}^2\right)k_B^2 T^2}{3\left(\Gamma_0^2 + E_\mathbf{k}^2\right)^3} + o\left(T^4\right) \right]. \qquad (9.46)$$

The first term is temperature independent, which is the correction from the paramagnetic current–current correlation function to the zero temperature superfluid density. The second term is proportional to T^2. It leads to the T^2-dependence of the low temperature superfluid density, different than the linear temperature dependence in the intrinsic d-wave superconductor. The T^2-dependence results from the fact that the density of states of electrons on the Fermi surface in the disordered d-wave superconductor is finite and the Sommerfeld expansion can be used in the limit $T \ll \Gamma_0$. This is also a universal behavior of d-wave superconductors, independent of the detailed scattering potential.

In the CuO$_2$ plane, the momentum summation of $Z_\mathbf{k}$ can be done using the standard method. In the limit $\Gamma_0 \ll \Delta_0$, the result is approximately given by

$$\frac{1}{V} \sum_\mathbf{k} Z_\mathbf{k} \approx -4\pi N_F \left[\frac{\Gamma_0}{a\Delta_0} \ln \frac{2\Delta_0}{\Gamma_0} + \frac{ak_B^2 T^2}{6\Gamma_0\Delta_0} + o\left(T^4\right) \right]. \qquad (9.47)$$

The temperature dependence of the ab-plane superfluid density is then obtained as

$$n_s^{ab} \approx n \left[1 - \frac{2\Gamma_0}{\pi a\Delta_0} \ln \frac{2\Delta_0}{\Gamma_0} - \frac{ak_B^2 T^2}{3\pi\Gamma_0\Delta_0} + o\left(T^4\right) \right], \qquad (9.48)$$

where $n = N_F m v_F^2$ is the concentration of electrons. The corresponding magnetic penetration depth is [217]

$$\lambda_{ab} \approx \lambda_0 \left[1 + \frac{\Gamma_0}{\pi a\Delta_0} \ln \frac{2\Delta_0}{\Gamma_0} + \frac{ak_B^2 T^2}{6\pi\Gamma_0\Delta_0} + o\left(T^4\right) \right], \qquad (9.49)$$

where λ_0 is the penetration depth of the intrinsic d-wave superconductor at zero temperature. As already mentioned, λ_{ab} varies quadratically with temperature.

When the temperature becomes much higher than the disorder energy scale, i.e. $\Gamma_0 \ll T$, the system enters the intrinsic regime. In this regime, the disorder effect is weakened and λ varies linearly with T, just as in an intrinsic d-wave superconductor. The correction from the disorder scattering to λ is a higher-order quantity of Γ_0/T.

For high-T_c superconductors, besides the experimental confirmation of the intrinsic linear temperature dependence of the magnetic penetration depth, the T^2-dependence of the penetration depth in the disordered d-wave superconductors were also supported by a vast range of experimental measurements. In the Zn or Ni-doped YBCO single crystals [206, 218], or other superconducting films or single crystals without Zn or Ni doping but with relatively poor quality [200, 219, 220], it was indeed found that λ_{ab} varies as T^2 at low temperatures, consistent with the prediction for disordered d-wave superconductors.

Along the c-axis, substituting the expression of $\partial\varepsilon_{\mathbf{k}}/\partial k_z$ and Eq. (9.46) into Eq. (9.39), the momentum summation of $(\partial\varepsilon_{\mathbf{k}}/\partial k_z)^2 Z_{\mathbf{k}}$ in the gapless regime is approximately found to be

$$\frac{1}{V}\sum_{\mathbf{k}}\left(\frac{\partial\varepsilon_{\mathbf{k}}}{\partial k_z}\right)^2 Z_{\mathbf{k}} \approx -\frac{2\hbar^4 N_F}{m_c^2}\left[\frac{\pi\Gamma_0}{4\Delta_0 a} + \frac{\pi a\Gamma_0 k_B^2 T^2}{4\Delta_0^3} + o\left(T^4\right)\right]. \tag{9.50}$$

The corresponding superfluid density in the gapless regime is

$$n_s^c = \frac{3\hbar^2 N_F}{16 m_c}\left[1 - \frac{8\Gamma_0}{3\pi\Delta_0 a} - \frac{8a\Gamma_0 k_B^2 T^2}{3\pi\Delta_0^3} + o\left(T^4\right)\right]. \tag{9.51}$$

In obtaining this expression, the correction of the impurity scattering to the diamagnetic term is neglected [100]. n_s^c scales as T^2 at low temperatures, showing a stronger temperature dependence than the T^5-dependence in the intrinsic d-wave superconductor.

In comparison to λ_{ab}, λ_c is more strongly affected by the impurity scattering because the interlayer hopping matrix elements are much smaller than the intralayer ones. When the disorder effect dominates, the T^5-law of λ_c is no longer valid. Equation (9.36) should be replaced by Eq. (9.51), and λ_c scales as T^2 at low temperatures. This quadratic temperature dependence of λ_c has been found in most high-T_c superconductors experimentally.

The above results are derived under the assumption that the interlayer hopping is coherent, namely the momentum is conserved during the hopping. This assumption is violated in the presence of disordered interlayer potentials. In particular, if the impurity concentration becomes significantly high, the correction to the superfluid density from the incoherent interlayer scattering induced by the impurities needs to be considered.

9.5 Superfluid Response in a Weakly Coupled Two-Band Superconductor

We have discussed the superfluid response of a single-band system. However, in many superconductors, there exist two or even more Fermi surfaces, whose superconducting response functions are significantly different from the single-band case. A thorough understanding of this difference is essential to the understanding of experimental results.

Among the multiband systems, we often meet a class of superconductors in which the interband coupling is weak. A typical weakly coupled multiband superconductor is MgB_2 [221, 222], which has two electron-like Fermi surfaces and one hole-like Fermi surface. One of these three bands has stronger pairing interaction, which drives the system into the superconducting phase at 39 K. The pairing interactions in the other two bands are relatively weak. As the two main bands that are responsible for superconductivity carry different parity numbers, the coupling between them is very weak [223].

The superfluid density of a weakly coupled two-band superconductor may exhibit quite different temperature dependence than a single-band one. In a single band superconductor, it exhibits a common feature that the curvature of the superfluid density as a function of temperature, $d^2\rho_s(T)/dT^2$, is always negative. In a two-band system with strong interband coupling, if the superfluid responses from the two bands are locked, then the curvature of the superfluid density remains negative. However, in a weakly coupled two-band superconductor, the curvature is modified. It may become positive over a particular temperature interval. This is a characteristic feature of the two-band superconductor with weak interband coupling. A simple physical picture for understanding this is illustrated in Fig. 9.5. In the absence of interband coupling, let us assume that T_{c1} and T_{c2} ($> T_{c1}$)

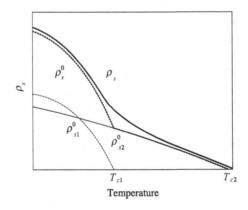

Figure 9.5 Temperature dependence of the superfluid density of a weakly coupled two-band superconductor. (From Ref. [224])

are the transition temperatures for the two bands, respectively. The corresponding superfluid densities are denoted as ρ_{s1}^0 and ρ_{s2}^0, respectively. The curvatures of ρ_{s1}^0 and ρ_{s2}^0 are all negative as in the single-band case. The total superfluid density is just $\rho_s^0 = \rho_{s1}^0 + \rho_{s2}^0$, which exhibits a cusp at $T = T_{c1}$. Once the interband coupling is turned on, only the superconducting transition around T_{c2} survives. The transition at T_{c1} is now rounded off, and the corresponding temperature dependence curve of the superfluid density is also smoothed. However, the tendency toward a positive curvature around T_{c1} cannot be completely wiped out. It is replaced by a smooth curve with a positive curvature. This picture holds for all weakly coupled two-band superconductors, irrespective of the band structures and the pairing interactions.

The weakly coupled two-band superconductor can be described by a reduced two-band BCS model as

$$
\begin{aligned}
H = & \sum_{ik\sigma} \xi_{ik} c_{ik\sigma}^\dagger c_{ik\sigma} + \sum_{ikk'} V_{ikk'} c_{ik'\uparrow}^\dagger c_{i-k'\downarrow}^\dagger c_{i-k\uparrow} c_{ik\downarrow} \\
& + \sum_{kk'} \left(V_{3kk'} c_{1k'\uparrow}^\dagger c_{1-k'\downarrow}^\dagger c_{2-k\uparrow} c_{2k\downarrow} + h.c. \right),
\end{aligned}
\tag{9.52}
$$

where $c_{ik\sigma}$ ($i = 1, 2$) is the annihilation operator of electrons in the ith band. $V_{1kk'}$ and $V_{2kk'}$ are the pairing potentials of band 1 and 2, respectively. $V_{3kk'}$ is the interband pairing potential. A commonly used approximation is to assume that $V_{ikk'}$ ($i = 1, 2, 3$) are factorizable, i.e.

$$
V_{1kk'} = g_1 \gamma_{1k} \gamma_{1k'},
\tag{9.53}
$$

$$
V_{2kk'} = g_2 \gamma_{2k} \gamma_{2k'},
\tag{9.54}
$$

$$
V_{3kk'} = g_3 \gamma_{1k} \gamma_{2k'},
\tag{9.55}
$$

where g_1, g_2, and g_3 are the coupling constants, and γ_{1k} and γ_{2k} are the pairing symmetry functions of band 1 and 2, respectively. In principle, γ_{1k} and γ_{2k} could be different.

In real superconductors, the collective modes that induce the superconducting pairing of electrons could couple to both bands, giving rise to superconducting instabilities in both bands. Nevertheless, the superconducting transition temperature is determined mainly by the band with stronger pairing interaction. Another possibility is that the pairing interaction only explicitly shows up in one of the bands. The other band is not superconducting itself, but can become superconducting via the interband coupling. In this case, one of the intra-band pairing interactions, $V_{1kk'}$ and $V_{2kk'}$, equals zero.

In Eq. (9.52), the interband coupling is achieved through pair hopping. In real superconductors, the two bands could also be coupled through the single particle hopping, or hybridization. Which coupling is more important is determined by the pairing mechanism. In some systems, the pair hopping is stronger, while in other systems the single particle coupling is stronger. The role of the interband single particle hopping is to renormalize the energy dispersion; it can be absorbed into the kinetic energy terms by redefining the band dispersions as well as the pairing potentials. Hence, although the interband hybridization term is not explicitly included in Eq. (9.52), the band structure $\xi_{ik\sigma}$ and the pairing interaction $V_{ikk'}$ could be understood as that they already contain the correction from this hybridization.

The pairing interactions in the model described by Eq. (9.52) can be decoupled following the BCS mean-field theory. Under this approximation, the quasiparticle excitation spectrum of band i is given by

$$E_{ik} = \sqrt{\xi_{ik}^2 + \Delta_i^2 \gamma_{\mathbf{k}}^2}. \tag{9.56}$$

In obtaining the above expression, $\gamma_{1k} = \gamma_{2k} = \gamma_{\mathbf{k}}$ is assumed. Δ_i is the gap amplitude, determined by the self-consistent equations

$$\Delta_1 = \sum_{\mathbf{k}} \gamma_{\mathbf{k}} \left(g_1 \langle c_{1-\mathbf{k}\downarrow} c_{1\mathbf{k}\uparrow} \rangle + g_3 \langle c_{2-\mathbf{k}\downarrow} c_{2\mathbf{k}\uparrow} \rangle \right), \tag{9.57}$$

$$\Delta_2 = \sum_{\mathbf{k}} \gamma_{\mathbf{k}} \left(g_2 \langle c_{2-\mathbf{k}\downarrow} c_{2\mathbf{k}\uparrow} \rangle + g_3 \langle c_{1-\mathbf{k}\downarrow} c_{1\mathbf{k}\uparrow} \rangle \right), \tag{9.58}$$

where $\langle \cdots \rangle$ represents the thermodynamic average.

Under the above mean-field approximation, the total superfluid density is just the sum of the superfluid densities in these two bands

$$\rho_s = \rho_{s1} + \rho_{s2}. \tag{9.59}$$

ρ_{s1} and ρ_{s2} are correlated through the above two gap equations. Once the temperature dependences of Δ_1 and Δ_2 are obtained, ρ_{s1} and ρ_{s2} can be determined using the formulae previously introduced for the single-band superconductor.

YBCO contains both the CuO_2 planes and the CuO chains, and is a two-band system. Besides providing charge carriers, the CuO chains also contribute to superconductivity. There are two kinds of interactions that can drive the CuO chains to superconduct. One is the intrinsic intrachain pairing interaction, and the other is that induced by the proximity effect from the pairing interaction in the CuO_2 plane [224]. The temperature dependence of the penetration depth along the chain λ_b is very different in these two cases. In the former case, λ_b varies slowly with temperature at low temperatures, behaving just like in a single-band system. In the latter case, the contribution of the CuO chains to the superfluid density becomes prominent at low temperatures due to the proximity effect. Consequently, λ_b decrease quickly with decreasing temperature, and the curvature of the superfluid density changes to negative [99].

For $YBa_2Cu_3O_{7-x}$, the negative curvature of $\lambda_b(T)$ has not been observed experimentally. Thus it is unlikely that the chain superconductivity is induced by the

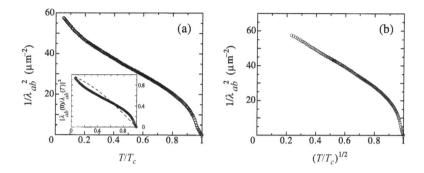

Figure 9.6 (a) The curve of $1/\lambda_{ab}^2$ vs. the reduced temperature T/T_c for YBa$_2$Cu$_4$O$_8$; (b) $1/\lambda_{ab}^2$ vs. $(T/T_c)^{1/2}$. (From Ref. [225])

proximity effect in this material. This shows that in YBa$_2$Cu$_3$O$_{7-x}$, the CuO chains couple strongly with the CuO$_2$ planes and their contributions to the superfluid density are not independent. Hence this material should not be treated as a weakly coupled two-band system.

However, the situation for YBa$_2$Cu$_4$O$_8$ is different. YBa$_2$Cu$_4$O$_8$ is intrinsically underdoped. Each unit cell along the c-axis contains two layers of CuO chains. These two chains are offset by half of the lattice constant along the chain direction (the b-axis), reducing significantly the chain–plane coupling. This leads to the strong anisotropy between the a- and b-axis penetration depths. It implies that the contributions from the CuO chains and the CuO$_2$ planes are nearly independent, and this material can be considered as a weakly coupled two-band system, and the superconductivity is mainly driven by the electrons in the CuO$_2$ planes. According to the discussion above, it is natural to predict that the curvature of the in-plane superfluid density as a function of temperature is positive at low temperatures. This prediction was confirmed experimentally [225]. Figure 9.6 shows that at temperature dependence of the superfluid density for YBa$_2$Cu$_4$O$_8$. Indeed the curvature of $1/\lambda_{ab}^2$ is positive at low temperatures. A further analysis shows that at low temperature $1/\lambda_{ab}^2$ scales as $T^{1/2}$. This is also consistent with the prediction made based on the proximity effect [224].

Besides YBa$_2$Cu$_4$O$_8$, the electron-doped high-T$_c$ superconductors are also two-band systems with weak interband coupling [226]. This will be discussed in §9.6. Unlike YBa$_2$Cu$_4$O$_8$, it is not so simple to justify that the electron-doped high-T$_c$ cuprates are weakly coupled two-band systems. Nevertheless, this scenario is supported by the penetration depth as well as many other experimental measurements. Many seemingly contradictory phenomena observed in the electron-doped high-T$_c$ cuprates can be naturally and consistently interpreted in the framework of the weakly coupled two-band model.

9.6 Electron-Doped Cuprate Superconductors

It is now commonly accepted that hole-doped high-T_c superconductors possess $d_{x^2-y^2}$ pairing symmetry. However, for electron-doped cuprates, it is still under debate whether the superconducting pairing has the $d_{x^2-y^2}$ or other symmetry. Although most of the experimental measurements, including ARPES [227, 228], Raman scattering [229], and phase-sensitive measurements [155, 230], support the theory of $d_{x^2-y^2}$-wave pairing symmetry, nevertheless, there are still discrepancies in the interpretation of other experimental results [231–235]. Kokales et al. [232] and Prozorov et al. [233, 234] measured the magnetic penetration depth and found a T^2-dependence in the low temperature superfluid density, consistent with the results of disordered d-wave superconductors. However, the temperature dependence of the superfluid density observed by Kim et al. [235] is more complicated. It does scale as T^2 in the underdoped regime, but in the overdoped regime it shows a temperature dependence that is more close to the exponential behavior of s-wave superconductors. Based on this observation, they speculated that there is a phase transition from the d-wave pairing state in the underdoped regime to the anisotropic s-wave pairing state in the overdoped regime. However, the analysis of Kim et al. [235] was made based on the single band assumption, which is not applicable to weakly coupled two-band systems [226].

A common feature of electron-doped high-T_c superconductors is that the superfluid density exhibits a positive curvature close to T_c [235–238]. The temperature range for the positive curvature is narrow in the underdoped regime, but very broad in the overdoped regime. Clearly, this is not due to the superconducting fluctuations. The appearance of the positive curvature in the superfluid density, as explained before, is a characteristic feature of weakly coupled two-band superconductors. In contrast, the curvature of the superfluid density of the single-band system is always negative below T_c. This indicates that, unlike hole-doped high-T_c superconductors, the electron-doped cuprates are weakly coupled two-band systems [226].

In addition to the superfluid density, many other experimental results also support the two-band picture for electron-doped high-T_c superconductors. The most direct one is ARPES. For the $Nd_{2-x}Ce_xCuO_4$ (NCCO) superconductor at low doping, a small Fermi surface of electrons appears around $(\pi, 0)$ in the Brillouin zone [239]. However, in the hole-doped case, the Fermi surface of holes first appears around $(\pi/2, \pi/2)$ [240]. The difference in the momenta of the Fermi surfaces in these two cases is due to the sign change of the next-nearest-neighbor hopping constant t' in the corresponding effective $t–J$ model [241, 242], which breaks the particle–hole symmetry. With the increase of the doping level, another small Fermi surface appears around $(\pi/2, \pi/2)$ in electron-doped cuprates. The superconducting phase emerges only when this Fermi surface rises above the Fermi level, similarly to the hole-doped materials [242]. It indicates that the emergence of superconductivity is closely related to the pairing correlation of electrons around $(\pi/2, \pi/2)$. In addition, a variety of transport measurements have shown that in order to understand comprehensively the temperature dependences of the Hall coefficient, magnetoresistance, and other physical quantities, one needs to assume that there are two kinds of charge carriers, i.e. electrons and holes, in electron-doped materials [243–245]. This would also imply that the electron-doped cuprates are two-band systems.

The two disconnected small Fermi surfaces in electron-doped high-T_c cuprates may hail from the upper and lower Hubbard bands, respectively [242, 246]. They may also arise from band folding due to the antiferromagnetic correlation [247, 248]. In either case, these two small Fermi surfaces can be described by a two-band model with a weak inter-band coupling [246, 248]. The interband coupling is weak because these two bands are not directly coupled by the main interaction, i.e. the antiferromagnetic interaction.

Let us denote the bands around $(\pi/2, \pi/2)$ and $(\pi, 0)$ as band 1 and 2, respectively. Similarly to hole-doped superconductors, we assume that the superconducting pairing of band 1 electrons possesses $d_{x^2-y^2}$-wave symmetry. At low temperatures, the contribution to the superfluid density from band 1 should be the same as in the single-band d-wave superconductor, exhibiting a linear temperature dependence,

$$\rho_{s,1}(T) \sim \rho_{s,1}(0) \left(1 - \frac{T}{T_c}\right). \tag{9.60}$$

If the Cooper pairing in band 2 is induced by the proximity effect from band 1, the pairing symmetry in band 2 should also be $d_{x^2-y^2}$. As the Fermi surfaces of band 2 and the gap nodal lines do not intersect, the quasiparticle excitations of band 2 are gapped even if it has the $d_{x^2-y^2}$-wave pairing symmetry. Thus the contribution to the low-temperature superfluid density from band 2, $\rho_{s,2}(T)$, is thermally activated as in the s-wave superconductor. It varies exponentially with temperature as,

$$\rho_{s,2}(T) \sim \rho_{s,2}(0) \left(1 - ae^{-\Delta_1'/k_B T}\right), \tag{9.61}$$

where Δ_1' is the minimal gap on the Fermi surfaces of band 2, and a is a doping dependent constant.

Under the mean-field approximation, the total superfluid density equals the sum of $\rho_{s,1}$ and $\rho_{s,2}$

$$\rho_s(T) = \rho_{s,1}(T) + \rho_{s,2}(T). \tag{9.62}$$

Thus, in the limit $T \ll T_c$, the normalized superfluid density is approximately given by

$$\frac{\rho_s(T)}{\rho_s(0)} \approx 1 - \frac{\rho_{s,1}(0)}{\rho_s(0)} \frac{T}{T_c} - \frac{\rho_{s,2}(0)}{\rho_s(0)} a e^{-\Delta_1'/k_B T}, \tag{9.63}$$

where $\rho_s(0) = \rho_{s,1}(0) + \rho_{s,2}(0)$. At low temperatures, $\rho_s(T)/\rho_s(0)$ is predominantly determined by the linear T term. However, this linear-T coefficient is now normalized by the factor $\rho_{s,1}(0)/\rho_s(0)$. At zero temperature, the superfluid density of $\rho_{s,i}(0)$ is proportional to the electron density and inversely proportional to the effective mass of band i. As doped electrons appear first at band 2, and then at band 1 at relatively high doping level, $\rho_{s,2}(0)$ is expected to be much larger than $\rho_{s,1}(0)$. This implies that $\rho_{s,1}(0)/\rho_s(0) \ll 1$ and the linear T-term in $\rho_s(T)/\rho_s(0)$ is significantly suppressed compared to the single-band system.

In real samples, the temperature dependence of $\rho_{s,1}(T)$ is further changed by the impurity scattering. At low temperatures, it is no longer a linear function of T. Instead, it varies as T^2

$$\rho_{s,1}(T) \sim \rho_{s,1}(0)\left(1 - \frac{k_B^2 T^2}{6\pi \Gamma_0 \Delta_2}\right), \tag{9.64}$$

where Γ_0 is the impurity scattering rate. In this case, $\rho_s(T)/\rho_s(0)$ is given by

$$\frac{\rho_s(T)}{\rho_s(0)} \approx 1 - \frac{\rho_{s,1}(0)}{\rho_s(0)}\frac{k_B^2 T^2}{6\pi \Gamma_0 \Delta_2} - \frac{\rho_{s,2}(0)}{\rho_s(0)}ae^{-\Delta_1'/k_B T}. \tag{9.65}$$

It shows that the temperature dependence of $\rho_s(T)/\rho_s(0)$ can be further suppressed by impurity scatterings. Thus the power-law temperature dependence of the superfluid density in the electron-doped high-T_c superconductors is weakened at low temperatures, and the overall temperature dependence is dominated by the thermally activated behavior. This set a barrier in identifying the pairing symmetry from the measurement data and led to the discrepancy among the interpretations from different experiment groups .

Figure 9.7 compares the experimental data with the fitting curves obtained using Eq. (9.65). The agreement is very good. It indicates that Eq. (9.65) catches correctly the intrinsic feature of low temperature superfluid density for electron-doped high-T_c cuprates.

Around T_c, it is difficult to obtain an analytic expression for the superfluid density, and numerical calculations are needed. With a reasonable assumption on the energy–momentum dispersions for these two bands [246], the temperature dependence of the superfluid density can be evaluated by first solving the gap equations (9.57) and (9.58). As shown in Fig. 9.8, the theoretical results [226] agree with experimental ones [235]. Here the impurity correction to the superfluid density is not considered in the theoretical calculations. The agreement between the theoretical and experimental results could be further improved if this correction were included.

The above analysis shows that even though the electron-doped and hole-doped cuprate superconductors behave quite differently in terms of the superfluid density,

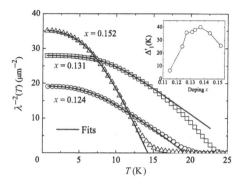

Figure 9.7 The low temperature superfluid density of the electron-doped high-T_c superconductor PCCO. The fitting is based on Eq. (9.65). (From Ref. [226]) The doping levels of $x = 0.124, 0.131$, and 0.152 are in the underdoping, optimal doping, and overdoping regimes, respectively. The circle, square, and triangle represent experimental results. The inset shows the doping dependence of Δ'_1.

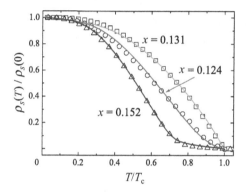

Figure 9.8 The experiment results of the normalized superfluid density of PCCO in comparison with theoretical results based on the weakly coupled two-band model. (The experiment data comes from Ref. [235])

both have the $d_{x^2-y^2}$-wave pairing symmetry. This implies that the pairing mechanisms are not fundamentally different in these two kinds of high-T_c superconductors.

9.7 Nonlinear Effect

In a small but finite external magnetic field, the nonlinear response of the d-wave superconductor becomes strong at low temperatures. The supercurrent generated by an external magnetic field induces a Doppler shift in the quasiparticle energy spectrum. This changes the energy dispersion around the gap nodes, which eliminates some nodal points by opening a small gap, and at the same time broadens other nodal points into a small gapless area. These features, particularly the latter one, change significantly the properties of low energy excitations, giving rise to a strong nonlinear correction to the low temperature electromagnetic response function.

The characteristic energy scale of the nonlinear effect is given by

$$E_{\text{nonlin}} = \hbar k_F v_s, \tag{9.66}$$

where v_s is the superfluid velocity. When the temperature is much smaller than E_{nonlin}, i.e. $k_B T \ll E_{\text{nonlin}}$, the nonlinear effect becomes important.

The Hamiltonian of superconducting quasiparticles in an external magnetic field is defined by

$$H(\mathbf{r}, \mathbf{r}') = \begin{pmatrix} \dfrac{1}{2m} \left(\mathbf{p} - \dfrac{e}{c} \mathbf{A} \right)^2 - \varepsilon_F & \Delta(\mathbf{r}, \mathbf{r}') e^{i\phi(\mathbf{R})} \\ \Delta(\mathbf{r}, \mathbf{r}') e^{-i\phi(\mathbf{R})} & -\dfrac{1}{2m} \left(\mathbf{p} + \dfrac{e}{c} \mathbf{A} \right)^2 + \varepsilon_F \end{pmatrix}, \tag{9.67}$$

where $\mathbf{R} = (\mathbf{r} + \mathbf{r}')/2$. $\phi(\mathbf{R})$ is the phase factor of the gap function. It can be gauged away by taking the following unitary transformation

$$U(\mathbf{r}, \mathbf{r}') = \begin{pmatrix} e^{i\phi(\mathbf{R})/2} & 0 \\ 0 & e^{-i\phi(\mathbf{R})/2} \end{pmatrix}. \tag{9.68}$$

This leads to an equivalent Hamiltonian given by

$$H(\mathbf{r}, \mathbf{r}') = i\hbar \mathbf{v}_s \cdot \nabla + \left(-\frac{\hbar^2 \nabla^2}{2m} + \frac{m}{2} v_s^2 - \varepsilon_F \right) \sigma_3 + \Delta(\mathbf{r}, \mathbf{r}') \sigma_1. \tag{9.69}$$

In obtaining this expression, $\nabla \cdot \mathbf{v}_s = 0$ is taken and

$$\mathbf{v}_s = \frac{\hbar}{2m} \nabla \phi - \frac{e\mathbf{A}}{mc} \tag{9.70}$$

is the superfluid velocity.

The transformation matrix $U(\mathbf{r}, \mathbf{r}')$ is single-valued only if $\phi(\mathbf{R})$ is single-valued. If the gap function is singular at certain points in real space, for example, in the presence of the magnetic vortex line, $\phi(\mathbf{R})$ does not return to itself after winding a vertex line. Instead it changes to $\phi(\mathbf{R}) \pm 2\pi$. In this case, the above transformation is singular, and special care is needed in order to handle properly the boundary condition. Here we assume that the system is in the Meissner phase without vortex lines so that the above transformation is single-valued. For simplicity we take the London gauge $\nabla \phi = 0$.

In Eq. (9.69), if we assume that the supercurrent velocity $\mathbf{v}_s$ is spatially independent, and the gap function Δ possesses the d-wave symmetry, then the Hamiltonian is diagonal in momentum space

$$H = \sum_{\mathbf{k}} \left[\hbar \mathbf{k} \cdot \mathbf{v}_s + \left(\xi_{\mathbf{k}} + \frac{m}{2} v_s^2 \right) \sigma_3 + \Delta_{\mathbf{k}} \sigma_1 \right], \tag{9.71}$$

where $\xi_{\mathbf{k}} = \hbar^2 k^2 / 2m - \varepsilon_F$. The v_s^2 term can be viewed as a correction to the chemical potential and absorbed into ε_F. The above equation is then simplified to

$$H = \sum_{\mathbf{k}} \left(\hbar \mathbf{k} \cdot \mathbf{v}_s + \xi_{\mathbf{k}} \sigma_3 + \Delta_{\mathbf{k}} \sigma_1 \right). \tag{9.72}$$

Its eigenspectra are given by

$$W_{\pm,\mathbf{k}} = \hbar \mathbf{k} \cdot \mathbf{v}_s \pm \sqrt{\xi_{\mathbf{k}}^2 + \Delta_{\mathbf{k}}^2}. \tag{9.73}$$

Compared to the zero field dispersion, the quasiparticle spectra are shifted by $\hbar \mathbf{k} \cdot \mathbf{v}_s$, which is dubbed as the Doppler shift.

Under the Doppler shift approximation, the quasiparticle contribution to the free energy is given by

$$F_q = -\frac{1}{\beta V} \sum_{\mathbf{k}} \ln \left(1 + e^{-\beta W_{+,\mathbf{k}}} \right) \left(1 + e^{-\beta W_{-,\mathbf{k}}} \right). \tag{9.74}$$

After taking derivatives, the current vector of quasiparticles is found to be

$$\mathbf{j}_q = -\frac{e}{m} \frac{\partial F_q}{\partial \mathbf{v}_s} = -\frac{e\hbar}{mV} \sum_{\mathbf{k}} \mathbf{k} \left[f \left(W_{+,\mathbf{k}} \right) + f \left(W_{-,\mathbf{k}} \right) \right]. \tag{9.75}$$

Using the identity that $f(x) + f(-x) = 1$, this expression becomes

$$\mathbf{j}_q = \frac{e\hbar}{mV} \sum_{\mathbf{k}} \mathbf{k} \left[f \left(-W_{-,\mathbf{k}} \right) - f \left(W_{+,\mathbf{k}} \right) \right]. \tag{9.76}$$

It is simple to verify that the quasiparticle contribution to the supercurrent is opposite to the direction of the supercurrent arising from the condensate. The total supercurrent equals the sum of these two terms

$$\mathbf{j}_s = -en\mathbf{v}_s + \frac{e\hbar}{mV} \sum_{\mathbf{k}} \mathbf{k} \left[f \left(E_k - \hbar \mathbf{k} \cdot \mathbf{v}_s \right) - f \left(E_k + \hbar \mathbf{k} \cdot \mathbf{v}_s \right) \right]. \tag{9.77}$$

Clearly, the dependence of the supercurrent on the external field or the superfluid velocity is nonlinear.

The above discussion shows that the Doppler shift induces corrections to the quasiparticle excitation spectra. It also induces a nonlinear correction to the electromagnetic response function. This correction is negligible in s-wave superconductors, but becomes significantly important in d-wave superconductors due to the presence of gap nodes.

In order to describe correctly the nonlinear effect, let us consider the superfluid response of the system to an external magnetic field. At zero temperature, the ground state energy F_q is simply a sum of the quasiparticle energies below the Fermi level,

$$F_q = \int_{-\infty}^{0} d\omega \, [\rho_+(\omega) + \rho_-(\omega)] \, \omega. \tag{9.78}$$

where

$$\rho_+(\omega) = \frac{1}{4} \sum_i \rho \left(\omega - \hbar k_F v_{s,i} \right) \theta(\omega - \hbar k_F v_{s,i}), \tag{9.79}$$

$$\rho_-(\omega) = \frac{1}{4} \sum_i \rho \left(\hbar k_F v_{s,i} - \omega \right) \theta(\hbar k_F v_{s,i} - \omega), \tag{9.80}$$

are the quasiparticle density of states corresponding to $W_\pm$.

Substituting these equations into Eq. (9.78) and using the linear behavior of $\rho(\omega)$ at low frequencies, we find F_q at the low field to be

$$F_q = -\frac{N_F}{12\Delta_0} \sum_i \left(\hbar k_F v_{s,i} \right)^3 \theta \left(v_{s,i} \right) - \int_0^\infty d\omega \rho\left(\omega \right) \omega. \tag{9.81}$$

The cubic dependence on $v_{s,i}$ in the first term results from the linear density of states. The Doppler shift affects mainly the quasiparticle excitations around the nodes with a radius proportional to $v_{s,i}$. Thus the total energy of the modes within these pockets scales with the cube of $v_{s,i}$, i.e. the area of the pockets contributes a square of $v_{s,i}$, and the energy itself contributes an additional linear power. The second term on the right-hand side of the equation is the quasiparticle contribution to the free energy in the absence of external magnetic field.

Taking the derivative of F_q with respect to v_s, we obtain the quasiparticle correction to the supercurrent

$$j_q = -\frac{e}{m} \frac{\partial F_q}{\partial v_s} = \begin{cases} \dfrac{e n v_s^2}{2 v_c}, & \mathbf{v}_s \parallel \text{node}, \\[4mm] \dfrac{e n v_s^2}{2\sqrt{2} v_c}, & \mathbf{v}_s \parallel \text{antinode}. \end{cases} \tag{9.82}$$

v_c is a characteristic velocity of Bogoliubov quasiparticles at the gap nodes along the direction perpendicular to the Fermi surface, which is proportional to the gap amplitude Δ_0. Combined with the contribution from the superfluid condensate, we find the supercurrent at zero temperature

$$j_s(v_s) = \begin{cases} -e n v_s \left(1 - \dfrac{v_s}{2 v_c} \right), & \mathbf{v}_s \parallel \text{node}, \\[4mm] -e n v_s \left(1 - \dfrac{v_s}{2\sqrt{2} v_c} \right), & \mathbf{v}_s \parallel \text{antinode}. \end{cases} \tag{9.83}$$

From the definition of the superfluid density, $\mathbf{j}_s = -en_s\mathbf{v}_s$, we obtain immediately the superfluid density in the ab-plane at zero temperature

$$n_s^{ab} = \begin{cases} n\left(1 - \dfrac{v_s}{2v_c}\right), & \mathbf{v}_s \parallel \text{node}, \\[3mm] n\left(1 - \dfrac{v_s}{2\sqrt{2}v_c}\right), & \mathbf{v}_s \parallel \text{antinode}. \end{cases} \tag{9.84}$$

This is the result that was first obtained by Yip and Sauls [249].

Since v_s is proportional to H, Eq. (9.84) shows that $n_{ab}^s(0)$ varies linearly with H at zero temperature. This linear field dependence results from the nodal structure of the d-wave pairing gap with the linear low-energy density of states. In the gapped s-wave superconductors, the field dependence of $n_{ab}^s(0)$ is zero at low fields. Hence through the measurement of the field dependence of $n_{ab}^s(0)$, one can determine whether there exist gap nodes or not. Moreover, the change of n_{ab}^s with the magnetic field varies along different directions. By measuring this direction dependence, one can also determine the positions of gap nodes on the Fermi surface.

At finite temperatures, the supercurrent vector can be decomposed into the zero temperature contribution and the finite temperature correction as

$$\mathbf{j}_s(T, v_s) = \mathbf{j}_s(0, v_s) + \int_0^T dT \frac{\partial \mathbf{j}(T, v_s)}{\partial T}, \tag{9.85}$$

where $\mathbf{j}_s = j_s\hat{\mathbf{v}}_s$ ($\hat{\mathbf{v}}_s$ is the unit vector along the direction of $\mathbf{v}_s$). The second term on the right-hand side is mainly the contribution of quasiparticles. Using the property of the Fermi distribution function,

$$\frac{\partial f(-x)}{\partial T} = -\frac{\partial f(x)}{\partial T}, \tag{9.86}$$

we find that the temperature derivative of j_s equals

$$\frac{\partial j_s(T, v_s)}{\partial T} = -\frac{2e\hbar}{mV} \sum_{\mathbf{k}} \mathbf{k} \cdot \hat{\mathbf{v}}_s \frac{\partial}{\partial T} f(\hbar\mathbf{k} \cdot \mathbf{v}_s + E_{\mathbf{k}}). \tag{9.87}$$

In the low temperature limit $k_B T \ll E_{\text{nonlin}}$, the quasiparticle correction to the supercurrent contributes mainly from the nodal area. Around each nodal point, $E_{\mathbf{k}}$ can be linearized and expressed as

$$E_{\mathbf{k}} = \sqrt{\left(\hbar v_F k_\parallel\right)^2 + (2\hbar v_c k_\perp)^2}, \tag{9.88}$$

where $k_\parallel$ and $k_\perp$ are the momentum components parallel and perpendicular to the Fermi surface at the nodal point, respectively. The momentum summation in

Eq. (9.87) can now be readily evaluated. By further integrating over T, the value of j_s, up to the leading order terms in temperature, is found to be

$$
j_s(T, v_s) \approx \begin{cases} j_s(0, v_s) + \dfrac{\pi e k_B^2 T^2}{12 \hbar^2 v_c}, & \mathbf{v}_s \parallel \text{node} \\[4mm] j_s(0, v_s) + \dfrac{\sqrt{2} \pi e k_B^2 T^2}{12 \hbar^2 v_c}, & \mathbf{v}_s \parallel \text{antinode} \end{cases}
\tag{9.89}
$$

Note that the sign convention of the electron charge here is $-e$, the T^2-correction is to reduce the supercurrent. The next order correction is proportional to $T^3 v_s$. This result shows that, at low temperatures, the dependences of the supercurrent on the magnetic field and temperature are nearly independent. Fixing temperature, j_s exhibits the same magnetic field dependence as at zero temperature. On the other hand, fixing magnetic field, j_s varies quadratically with temperature.

When the temperature becomes much larger than the nonlinear energy scale E_{nonlin} but much smaller than the maximal gap value Δ_0, $x = \beta \hbar \mathbf{k} \cdot \mathbf{v}_s$ is a small quantity. The supercurrent given in Eq. (9.77) can be expanded according to the power of this parameter as

$$
\mathbf{j}_s = -e n \mathbf{v}_s - \frac{2 e \hbar}{m V} \sum_{n, \mathbf{k}} \frac{\mathbf{k}}{n!} \frac{\partial^n f(E_\mathbf{k})}{\partial E_\mathbf{k}^n} (\hbar \mathbf{k} \cdot \mathbf{v}_s)^n .
\tag{9.90}
$$

It is easy to show that the even power terms of $\mathbf{v}_s$ vanish after the momentum summation, and only the odd power terms have contribution to $\mathbf{j}_s$. The leading order term, which is correct up to the linear order of $\mathbf{v}_s$, is just the result obtained with the linear approximation. The lowest order nonlinear correction starts from the cubic order of v_s. Up to this order of correction, the supercurrent is given by

$$
\mathbf{j}_s \approx -e n \mathbf{v}_s - \frac{2 e \hbar}{m V} \sum_{n, \mathbf{k}} \mathbf{k} \left[\frac{\partial f(E_\mathbf{k})}{\partial E_\mathbf{k}} \hbar \mathbf{k} \cdot \mathbf{v}_s + \frac{1}{3!} \frac{\partial^3 f(E_\mathbf{k})}{\partial E_\mathbf{k}^3} (\hbar \mathbf{k} \cdot \mathbf{v}_s)^3 \right] .
\tag{9.91}
$$

Hence, if $E_{\mathrm{nonlin}} \ll k_B T \ll \Delta_0$, the nonlinear correction to the supercurrent function $\mathbf{j}_s$ is proportional to H^3, and its correction to the superfluid density is proportional to H^2. In the s-wave superconductor, there is a finite energy gap in the quasiparticle excitations, the quasiparticle population is proportional to $\exp(-\Delta / k_B T)$ and the nonlinear effect is much weaker than in the d-wave case.

The Doppler shift defines an energy scale for the nonlinear effect

$$
E_{\mathrm{nonlin}} = \hbar k_F v_s \sim \frac{\hbar k_F e \lambda H_0}{mc},
\tag{9.92}
$$

and Eq. (9.84) is valid when $T \ll E_{\mathrm{nonlin}}$. Thus in order to observe the nonlinear field dependence of the superfluid density and its spatial anisotropy, temperature has to be much lower than E_{nonlin}. On the other hand, the measurement of the nonlinear effect can only be performed in the Meissner phase; that is the magnetic field must be below the lower critical field. Otherwise, magnetic vortex lines emerge and the above results are no longer valid. This means that E_{nonlin} is upper bounded by the lower critical field. For the YBCO superconductor, the lower critical field is at a few hundred gauss, and E_{nonlin} is estimated around 1 K (10^{-4} eV). Hence, the nonlinear effect can only be observed in this material at very low temperatures.

Experimentally, the anisotropy of the magnetic penetration depth induced by the nonlinear effect has not been observed yet in high-T_c superconductors [250, 251]. There are probably two reasons for this. First, the measurement temperature was still not low enough to observe the nonlinear effect. Second, the magnetic field was too high so that the system is in the mixed state rather than in the Meissner state. Owing to the existence of vortex lines, the supercurrent velocity distributions are broadened. In the Meissner state the supercurrent only flows parallel to the surface. Both could reduce the anisotropy of the penetration depth along different directions. In order to probe the anisotropy induced by the nonlinear effect, the measurement of the penetration depth needs to be performed in the Meissner phase at very low temperatures.

9.8 Non-linear Correction to the Penetration Depth

The above discussion was done under the assumption that v_s is spatially indepen-dent. However, in real systems, the magnetic field decays at the surface of the superconductor, and the characteristic length scale is just the penetration depth. Thus v_s cannot be a quantity that is spatially homogeneous. Nevertheless, if the spatial variation of v_s is not too sharp, the Hamiltonian (9.72) is still locally valid, and the quasiparticle spectra obtained with this Hamiltonian are just those at the corresponding spatial point. Below we discuss the spatial variation of the magnetic field under this approximation, and then determine how the magnetic penetration depth varies in a superconductor.

Under the linear approximation, the penetration depth is inversely proportional to the square root of the superfluid density and is spatially independent. The decay of the magnetic field along the direction perpendicular to a semi-infinitely large superconducting plate is determined by the London equation, Eq. (1.1). The solution is given by Eq. (1.11)

$$H^{(0)}(x) = H_0 e^{-x/\lambda}, \tag{9.93}$$

where H_0 is the magnetic field on the surface of the superconductor ($x = 0$). The superscript (0) here is to emphasize that $H^{(0)}(x)$ is the solution of the London equation without considering the nonlinear correction. The corresponding vector potential is

$$A^{(0)}(x) = H^{(0)}(x)\lambda, \tag{9.94}$$

and the spatial dependence of the supercurrent velocity is found to be

$$v_s^{(0)} = \frac{e\lambda H^{(0)}(x)}{mc} \tag{9.95}$$

within the linear approximation.

Without considering the correction from the nonlinear effect, v_s is proportional to H. Thus the magnetic field decay is equivalent to the decay of supercurrent velocity, and the London equation can also be expressed as the equation for the supercurrent velocity

$$\frac{\partial^2 v_s}{\partial x^2} = \frac{v_s}{\lambda^2}. \tag{9.96}$$

Considering the nonlinear effect, this equation clearly needs to be corrected. Under the local approximation, λ^{-2} should be replaced by Eq. (9.84). This leads to the equation of the supercurrent velocity corrected by considering the nonlinear effect

$$\frac{\partial^2 v_s}{\partial x^2} = \frac{v_s}{\lambda^2}(1 - \alpha v_s), \tag{9.97}$$

where α is a direction dependent quantity. $\alpha = 1/2v_c$ or $1/2\sqrt{2}v_c$ if v_s is along the nodal or antinodal direction.

The nonlinear equation Eq. (9.97) does not have analytic solutions. In order to solve the magnetic penetration depth and compare with experiments, usually numeric solutions need to be performed. However, on the superconducting surface, if we use v_s obtained under the linear approximation to replace v_s in the parentheses on the right-hand side of Eq. (9.97), the magnetic penetration depth on the superconductor surface is approximately given by

$$\lambda_{\text{eff}}(x = 0) \approx \lambda \left(1 - \frac{\alpha e\lambda H_0}{mc}\right)^{-1/2}. \tag{9.98}$$

9.9 Nonlocal Effect

The preceding discussion on the electromagnetic response function is made based on the local approximation. This approximation is valid when the coherence length $\xi_0 = v_F/\pi\Delta_0$ is much shorter than the penetration depth, $\xi_0 \ll \lambda_0$. However, in high-T_c superconductors, the electron coherence length is anisotropic. It depends on the direction of electron momentum. If the momentum is along the nodal direction, the effective coherence length of $\xi_k = v_F/\pi\Delta_k$ diverges. This implies that if $|\Delta_{\mathbf{k}}| \sim (\xi_0/\lambda_0)\Delta_0$ near the gap nodes, the condition for the local approximation $\xi_k \ll \lambda_0$ is not satisfied. The characteristic energy scale of nonlocal effect is given by

$$E_{\text{nonloc}} \sim |\Delta_{\mathbf{k}}| \sim \frac{\xi_0\Delta_0}{\lambda_0}, \tag{9.99}$$

below which the nonlocal effect becomes important and must be considered [252, 253].

E_{nonloc} defines an energy scale around the Fermi surface within which the quasiparticle excitations have contribution to the nonlocal effect. Corresponding to E_{nonloc}, one can also define a characteristic temperature to describe the nonlocal effect as

$$T_{\text{nonloc}} = \frac{E_{\text{nonloc}}}{k_B}. \tag{9.100}$$

In high-T_c superconductors, the region of the Fermi surface that contributes to the nonlocal effect is small compared to the entire Fermi surface as $\xi_0 \ll \lambda_0$. Thus the correction due to the nonlocal effect is typically small and negligible. However, when $T \ll T_{nonloc}$, the quasiparticle excitations are contributed mainly by the electrons around the nodal points, and the nonlocal effect is no longer negligible. In real materials, this effect could be dramatically suppressed by the impurity scattering.

The value of T_{nonloc} is determined by the three fundamental parameters of superconductors ($\lambda_0, \xi_0, \Delta_0$). For high-$T_c$ superconductors, the typical values of these parameters are

$$\lambda_0 \sim 1-3 \times 10^3 \text{ Å}, \quad \xi_0 \sim 15-30 \text{ Å}, \quad \Delta_0/k_B \sim 200-300 \text{ K}. \tag{9.101}$$

The corresponding value of T_{nonloc} is estimated to be $1 \sim 3$ K.

The response function Eq. (9.2) is defined in an infinitely large medium. In order to calculate the correction from the nonlocal effect to the magnetic penetration depth, the boundary condition of the magnetic field on the surface of superconductor must be handled carefully. In order to study a semi-infinite superconductor using the results of an infinite system, a commonly used approximation is to use a zero thickness current sheet to replace the superconducting surface. The magnetic fields thus generated point in opposite directions on the two sides of the boundary, and the corresponding vector potentials are mirror symmetric with respect to the boundary layer. In this case, electrons are completely reflected on the surface. The magnetic penetration depth is determined by the following integral [2]

$$\lambda_\alpha^{\text{spec}} = \frac{2}{\pi} \int_0^\infty \frac{dq}{\mu_0 K_{\alpha\alpha}(q\hat{n}, 0) + q^2}, \tag{9.102}$$

where $\hat{n}$ is the unit vector normal to the superconducting surface, and $K_{\alpha\alpha}(\mathbf{q}, \omega)$ is the electromagnetic response function defined in Eq. (9.2). If $K(\mathbf{q}, 0)$ is q-independent,

$$K(\mathbf{q}, 0) = \frac{1}{\mu_0 \lambda_L^2}, \tag{9.103}$$

then

$$\lambda_\alpha^{\text{spec}} = \lambda_L. \tag{9.104}$$

This is just the result that is obtained under the local approximation.

Another frequently used method to effectively handle the surface problem is to assume that the electron motion on the surface is diffusive. In this case, the formula for calculating the magnetic penetration depth is different from Eq. (9.102). These two methods yield qualitatively the same results although quantitatively they are different.

For a given wave vector $\mathbf{q}$, the current–current correlation function at zero frequency is determined by

$$\Pi_{xx}(\mathbf{q}, \omega = 0) = \frac{2e^2}{\beta\hbar^2 V} \sum_{\mathbf{k}\omega_m} \left(\frac{\partial \xi_\mathbf{k}}{\partial k_x}\right)^2 \frac{(i\omega_m)^2 + \xi_-\xi_+ + \Delta_-\Delta_+}{\left[(i\omega_m)^2 - E_+^2\right]\left[(i\omega_m)^2 - E_-^2\right]}, \tag{9.105}$$

where $\xi_\pm = \xi_{\mathbf{k}\pm\mathbf{q}/2}$, $\Delta_\pm = \Delta_{\mathbf{k}\pm\mathbf{q}/2}$, and $E_\pm = E_{\mathbf{k}\pm\mathbf{q}/2}$. When $|\mathbf{q}|$ is very small, these quantities can be expanded in terms of $\mathbf{q}$. Correct to the first order of $\mathbf{q}$, we have

$$\xi_\pm = \xi_\mathbf{k} \pm \nabla_\mathbf{k}\xi_\mathbf{k} \cdot \frac{\mathbf{q}}{2}, \tag{9.106}$$

$$\Delta_\pm = \Delta_\mathbf{k} \pm \nabla_\mathbf{k}\Delta_\mathbf{k} \cdot \frac{\mathbf{q}}{2}, \tag{9.107}$$

$$E_\pm = E_\mathbf{k} \pm \nabla_\mathbf{k}E_\mathbf{k} \cdot \frac{\mathbf{q}}{2}. \tag{9.108}$$

By employing the identities

$$E_\mathbf{k}\nabla_\mathbf{k}E_\mathbf{k} = \xi_\mathbf{k}\nabla_\mathbf{k}\xi_\mathbf{k} + \Delta_\mathbf{k}\nabla_\mathbf{k}\Delta_\mathbf{k}, \tag{9.109}$$

$$(\nabla_\mathbf{k}E_\mathbf{k})^2 = (\nabla_\mathbf{k}\xi_\mathbf{k})^2 + (\nabla_\mathbf{k}\Delta_\mathbf{k})^2, \tag{9.110}$$

Eq. (9.105) is simplified to

$$\Pi_{xx}(\mathbf{q}, 0) = \frac{2e^2}{\beta\hbar^2 V} \sum_{\mathbf{k}\omega_m} \left(\frac{\partial \xi_\mathbf{k}}{\partial k_x}\right)^2 \frac{(i\omega_m)^2 + W_+W_-}{\left[(i\omega_m)^2 - W_+^2\right]\left[(i\omega_m)^2 - W_-^2\right]}, \tag{9.111}$$

where

$$W_\pm = E_\mathbf{k} \pm \frac{1}{2}\alpha, \qquad \alpha = |\nabla_\mathbf{k}E_\mathbf{k} \cdot q\hat{n}|. \tag{9.112}$$

Performing the frequency summation, we arrive at

$$\begin{aligned}
\Pi_{xx}(\mathbf{q}, 0) &= \frac{2e^2}{\hbar^2 V} \sum_\mathbf{k} \left(\frac{\partial \xi_\mathbf{k}}{\partial k_x}\right)^2 \frac{f\left(E_\mathbf{k} + \frac{1}{2}\alpha\right) - f\left(E_\mathbf{k} - \frac{1}{2}\alpha\right)}{\alpha} \\
&= \frac{e^2 v_F^2}{\hbar^2} \int_0^\infty d\omega \rho(\omega) \frac{f\left(\omega + \frac{1}{2}\alpha\right) - f\left(\omega - \frac{1}{2}\alpha\right)}{\alpha}. \tag{9.113}
\end{aligned}$$

In the calculation below, we assume that the superconducting surface is perpendicular to the x-axis, and the magnetic field is along the z-axis. In this case, the low energy excitations concentrate in the nodal area, $\alpha = |q\partial_{k_x}E_{k_{\text{node}}}|$ is approximately the same around the four nodal points.

As $\rho(\omega)$ is linear in the low energy limit, we have

$$\Pi_{xx}(\mathbf{q},0) \approx \frac{e^2 v_F^2 N_F}{\alpha \hbar^2 \Delta_0} \int_0^\infty d\omega\omega \left[f\left(\omega + \frac{1}{2}\alpha\right) - f\left(\omega - \frac{1}{2}\alpha\right) \right]. \qquad (9.114)$$

After simplification, this can be further expressed as

$$\Pi_{xx}(\mathbf{q},0) \approx -\frac{e^2 v_F^2 N_F}{\alpha \hbar^2 \Delta_0} \left[\frac{2}{\beta^2} \int_0^{\beta\alpha/2} dx \frac{x}{e^x + 1} + \frac{\alpha^2}{8} + \frac{\alpha}{\beta} \ln\left(1 + e^{-\alpha\beta/2}\right) \right]. \qquad (9.115)$$

In the temperature range $k_B T \gg \alpha$, the nonlocal effect becomes negligible and

$$\Pi_{xx}(\mathbf{q},0) \approx \frac{e^2 v_F^2 N_F}{\hbar^2 \Delta_0} \int_0^\infty d\omega\omega \frac{\partial f}{\partial\omega} \qquad (9.116)$$

is momentum independent. This recovers the result obtained under the local approximation, as expected. When $k_B T \ll \alpha$, however, the nonlocal effect becomes strong. In this case, the upper limit of the integral on the right-hand side of Eq. (9.115) can be set to ∞, and then

$$\Pi_{xx}(\mathbf{q},0) \approx -\frac{e^2 v_F^2 N_F}{\alpha \hbar^2 \Delta_0} \left(\frac{\alpha^2}{8} + \frac{\pi^2 k_B^2 T^2}{6} \right). \qquad (9.117)$$

The first term on the right-hand side is temperature independent, which is a correction to the zero temperature superfluid density, or zero temperature magnetic penetration depth λ_0. Compared with the results under the local approximation, Π_{xx} approximately varies as T^2. Using Eq. (3.79) and Eq. (9.102), it can be also shown that λ_x^{spec} varies as T^2 at low temperatures. Thus similarly to the nonlinear effect, the nonlocal effect can also change the low temperature behavior of λ [253]. This change, as mentioned before, is very important in maintaining the stability of the d-wave superconducting state without violating the third law of thermodynamics.

10

Optical and Thermal Conductivities

10.1 Optical Conductivity

In an ideal superconductor, there is no energy dissipation, both the resistivity and the light absorption below the superconducting gap vanish at zero temperature. In real materials, however, there are always dissipations caused by impurity or other scattering effects. Due to the existence of superfluid in a superconductor, the direct-current resistivity remains zero, but the light absorption becomes finite. The optical and thermal conductivities are two fundamental quantities for characterizing the transport properties of electrons in a superconductor. They measure the responses of superconducting quasiparticles to an applied electromagnetic field and a temperature gradient, respectively. The experimental measurement and theoretical analysis of these quantities have played an important role in the study of high-T_c superconductivity. Not only can it be used to probe the pairing symmetry, but also to provide vital information on the interaction between superconducting quasiparticles and other low-lying excitations.

Both Cooper pairs and quasiparticles could be disturbed by a shining light. A Cooper pair could become two normal quasiparticles by absorbing a photon. This light absorption happens only when the frequency of the photon exceeds the pairing bound energy, namely twice of the single particle energy gap.

The light absorption rate, or the optical conductivity, is closely related to the mean free path l of normal electrons and the coherence length ξ of Cooper pairs. There are two limits that deserve special attention. One is the dirty limit with $l \ll \xi$. In this limit, the depaired electrons may encounter multiple impurity scattering and lose their initial momentum correlations within a characteristic time scale ξ/v_f. The other is the clean limit with $l \gg \xi$. In this limit, there is almost no impurity scattering within the time scale ξ/v_f and the electron momenta are essentially conserved.

In most of metal-based superconductors, the coherence length is much larger than the mean free path, $\xi \gg l$. Hence these superconductors are in the dirty limit.

Indeed, the measured infrared absorption spectra in these superconductors agree well with the theoretical calculation obtained in the dirty limit. In contrast, in high-T_c superconductors, the coherence length, $\xi \approx 20\sim30$ Å, is less than the mean free path and is generally considered to be in the clean limit. However, this judgement of clean limit may not always be correct due to the strong anisotropy of d-wave pairing function whose coherence length diverges along the nodal direction, much longer than the mean free path. Thus the quasiparticles around the gap nodes are always in the dirty limit. This suggests that in the low frequency or temperature regime where the contribution of nodal quasiparticle excitations dominate, the optical conductivity of high-T_c superconductors is in the dirty limit. By contrast, if the contribution from high energy quasiparticles dominates, the system is in the clean limit. In this limit, the light absorption depends on the scattering processes of electrons. In addition to the elastic scattering of impurities, the characteristic energy scale of inelastic scattering from spin excitations is large and comparable to $k_B T_c$, which can also affect the infrared absorption.

In the dirty limit, the momentum correlations between depaired electrons are completely destructed by impurity scattering. As the detailed scattering process is not important, the light absorption with the corresponding optical conductivity at finite frequencies can be accurately calculated. In the nonlocal electromagnetic response theory, the conductivity depends on the initial and final coordinates of electron, $\mathbf{r}$ and $\mathbf{r}'$. The correlation length between these two coordinates ξ_c is determined by the mean free path l and the coherence length in the absence of disorder ξ,

$$\frac{1}{\xi_c} = \frac{1}{l} + \frac{1}{\xi}. \tag{10.1}$$

In the limit $l \ll \xi$, $\xi_c \sim l$ is the characteristic length scale of the electromagnetic response function. It is also the coherence length of excited electrons. Within this length scale, the optical absorption is approximately dissipationless. As the characteristic length scale of dissipationless systems is ξ, which is significantly larger than l, the optical conductivity σ in the dirty limit is approximately proportional to the corresponding conductivity $\sigma^{(0)}$ in a dissipationless system in the limit $(\mathbf{r}' - \mathbf{r}) \to 0$ [254]

$$\sigma_\mu(\omega) \propto \sigma_\mu^{(0)}(\mathbf{r}' - \mathbf{r} \to 0, \omega) = \frac{1}{V} \sum_{\mathbf{q}} \sigma_\mu^{(0)}(\mathbf{q}, \omega). \tag{10.2}$$

Hence σ_μ is given by the average of $\sigma_\mu^{(0)}(\mathbf{q}, \omega)$ over all the momenta $\mathbf{q}$. It reflects the uncertainty of the momenta of excited electron pairs after scattering in the limit $l \ll \xi$.

In the clean limit, there is no universal theory to describe the optical conductivity of superconductors. In particular, different scattering centers or processes affect the optical absorption differently, and their effects need to be studied independently. Generally speaking, our understanding of the conducting behavior of superconducting quasiparticles is incomplete and there is not a commonly accepted microscopic theory. Both the optical absorption theory of Mattis and Bardeen in the dirty limit [254] and all semi-phenomenological theories of optical conductivity proposed in the clean limit have their own limitations. They should not be blindly used before a comprehensive understanding of physical properties of the system is achieved.

10.2 Optical Sum Rule

Compared with the normal state, the optical absorption spectral weight at finite frequencies is reduced in the superconducting state. The reduced weight is transferred to the zero frequency and becomes the superfluid density. Nevertheless, the total spectral weight, including the zero frequency part, is conserved. This is just a statement of the optical sum rule for the superconducting state, which is also called the FGT (Ferrell, Glover, Tinkham) optical sum rule [255, 256].

The FGT sum rule, or the general conductivity sum rule represented by Eq. (10.9) given below, results from the charge conservation. The proof of this sum rule is straightforward. To do this, let us start from Eq. (9.10). By inserting a complete set of eigenbases of the Hamiltonian, we can express the current–current correlation function as

$$\tilde{\Pi}_{\mu\nu}(\mathbf{q}, i\omega_m)$$

$$= -\frac{1}{V\hbar^2 Z} \sum_{nm} \int_0^\beta d\tau e^{i\omega_m \tau} e^{(E_n - E_m)\tau} e^{-\beta E_n} \langle n| J_\mu(\mathbf{q}) |m\rangle \langle m| J_\nu(-\mathbf{q}) |n\rangle$$

$$= -\frac{1}{V\hbar^2 Z} \sum_{nm} \frac{e^{-E_m \beta} - e^{-\beta E_n}}{i\omega_m + E_n - E_m} \langle n| J_\mu(\mathbf{q}) |m\rangle \langle m| J_\nu(-\mathbf{q}) |n\rangle. \qquad (10.3)$$

To convert the imaginary frequency to the real frequency by the analytic continuation, we then obtain the retarded current–current correlation function

$$\Pi_{\mu\nu}(\mathbf{q}, \omega) = -\frac{1}{V\hbar^2 Z} \sum_{nm} \frac{e^{-E_m \beta} - e^{-\beta E_n}}{\omega + E_n - E_m + i0^+}$$

$$\langle n| J_\mu(\mathbf{q}) |m\rangle \langle m| J_\nu(-\mathbf{q}) |n\rangle. \qquad (10.4)$$

Using the identity

$$\frac{1}{x + i0^+} = \frac{1}{x} - i\pi \delta(x), \qquad (10.5)$$

we obtain the following equation

$$\int_{-\infty}^\infty d\omega \frac{\mathrm{Im}\Pi_{\mu\nu}(\mathbf{q}, \omega)}{\omega} = \pi \mathrm{Re}\Pi_{\mu\nu}(\mathbf{q}, 0). \qquad (10.6)$$

From Eq. (9.9), the complex conductivity is found to be

$$\sigma_{\mu\nu}(\mathbf{q}, \omega) = \frac{i}{\omega + i0^+} \left[\frac{e^2}{V\hbar^2} \sum_{\mathbf{k}} \left\langle \frac{\partial^2 \varepsilon_{\mathbf{k}}}{\partial k_\mu \partial k_\nu} \right\rangle + \Pi_{\mu\nu}(\mathbf{q}, \omega) \right]. \tag{10.7}$$

Its real part is given by

$$\mathrm{Re}\sigma_{\mu\nu}(\mathbf{q}, \omega) = \left[\frac{\pi e^2}{V\hbar^2} \sum_{\mathbf{k}} \left\langle \frac{\partial^2 \varepsilon_{\mathbf{k}}}{\partial k_\mu \partial k_\nu} \right\rangle + \pi \mathrm{Re}\Pi_{\mu\nu}(\mathbf{q}, \omega) \right] \delta(\omega)$$
$$- \frac{\mathrm{Im}\Pi_{\mu\nu}(\mathbf{q}, \omega)}{\omega}. \tag{10.8}$$

Using Eq. (10.6) and performing the ω integration, we obtain

$$\int_{-\infty}^{\infty} d\omega \mathrm{Re}\sigma_{\mu\nu}(\mathbf{q}, \omega) = \frac{\pi e^2}{V\hbar^2} \sum_{\mathbf{k}} \left\langle \frac{\partial^2 \varepsilon_{\mathbf{k}}}{\partial k_\mu \partial k_\nu} \right\rangle. \tag{10.9}$$

This is just the generalized sum rule of electric conductivity, which is valid for any momentum $\mathbf{q}$.

The optical conductivity is the diagonal component of $\sigma_{\mu\nu}(\mathbf{q}, \omega)$ in the long wavelength limit ($q \to 0$). Because $\mathrm{Re}\sigma_{\mu\mu}(\omega) = \mathrm{Re}\sigma_{\mu\mu}(0, \omega)$ is an even function of ω, Eq. (10.9) is simplified as

$$\int_0^{\infty} d\omega \mathrm{Re}\sigma_{\mu\mu}(\omega) = \frac{\pi e^2}{2V\hbar^2} \sum_{\mathbf{k}} \left\langle \frac{\partial^2 \varepsilon_{\mathbf{k}}}{\partial k_\mu^2} \right\rangle. \tag{10.10}$$

In the superconducting state, the conductivity contains two terms, contributed by the superconducting electrons and the normal quasiparticles, respectively

$$\mathrm{Re}\sigma_{\mu\mu}(\omega) = \frac{\pi e^2 n_s^\mu}{m_\mu} \delta(\omega) + \mathrm{Re}\sigma_\mu(\omega). \tag{10.11}$$

The first term is the contribution of superfluid, and n_s^μ is the superfluid density. $\sigma_\mu(\omega)$ is the contribution of normal electrons and is regular at $\omega = 0$. From the above equations, we then obtain the FGT sum rule

$$\int_{0^+}^{\infty} d\omega \mathrm{Re}\sigma_\mu(\omega) + \frac{\pi e^2 n_s^\mu}{2m_\mu} = \frac{\pi e^2}{2V\hbar^2} \sum_{\mathbf{k}} \left\langle \frac{\partial^2 \varepsilon_{\mathbf{k}}}{\partial k_\mu^2} \right\rangle. \tag{10.12}$$

If the conductivity $\mathrm{Re}\sigma_\mu$ is measured at two different temperatures, below ($T < T_c$) and above ($T > T_c$) the superconducting transition temperature, respectively, then the difference of $\mathrm{Re}\sigma_\mu$ between these two temperatures, $\delta\sigma_\mu$, satisfies the equation

$$\int_{0^+}^{\infty} d\omega\, \delta\sigma_\mu(\omega) = \frac{\pi e^2 n_s^\mu}{2m_\mu} + \frac{\pi e^2}{2V\hbar^2} \sum_{\mathbf{k}} \left(\left\langle \frac{\partial^2 \varepsilon_{\mathbf{k}}}{\partial k_\mu^2} \right\rangle_n - \left\langle \frac{\partial^2 \varepsilon_{\mathbf{k}}}{\partial k_\mu^2} \right\rangle_s \right). \tag{10.13}$$

Dividing both sides by the first term on the right-hand side gives

$$R(\omega \to +\infty) = 1 + \frac{m_\mu}{V\hbar^2 n_s^\mu} \sum_{\mathbf{k}} \left(\left\langle \frac{\partial^2 \varepsilon_{\mathbf{k}}}{\partial k_\mu^2} \right\rangle_n - \left\langle \frac{\partial^2 \varepsilon_{\mathbf{k}}}{\partial k_\mu^2} \right\rangle_s \right), \tag{10.14}$$

where

$$R(\omega) = \frac{2m_\mu}{\pi e^2 n_s^\mu} \int_{0^+}^{\omega} d\omega \, \delta\sigma_\mu(\omega). \tag{10.15}$$

In an isotropic free electron system, the effective mass $m_\mu = m$ is momentum independent and

$$\varepsilon_k = \frac{\hbar^2 \mathbf{k}^2}{2m}. \tag{10.16}$$

Equation (10.12) becomes

$$\int_{0^+}^{\infty} d\omega \mathrm{Re}\sigma_\mu(\omega) + \frac{\pi e^2 n_s^\mu}{2m} = \frac{\pi e^2 n}{2m}. \tag{10.17}$$

This is the expression of the optical sum rule most frequently used in the literature [2]. But this equation is valid only when the energy dispersion of an electron is defined by Eq. (10.16). In this case, Eq. (10.14) simply becomes

$$R(\omega \to +\infty) = 1. \tag{10.18}$$

Equation (10.18) indicates that the integral of $\delta\sigma_\mu$ times the factor $2m_\mu/\pi e^2 n_s^\mu$ equals 1 in the limit $\omega \to \infty$ if the energy dispersion is given by Eq. (10.16). In real superconductors, only low energy electrons participate in the Cooper pairing and $\delta\sigma_\mu$ is very small when ω is much larger than the maximal gap value Δ_0. Generally $R(\omega)$ is close to 1 when $\omega \sim 6\Delta_0$ [257]. Thus the FGT sum rule can be tested through the measurement of low frequency conductivity.

For the optimally doped high-T_c superconductor, it was found from experimental measurements that $R(\omega)$ does indeed reach the saturated value at a relatively low frequency (~ 800 cm^{-1}), in good agreement with the theoretical prediction [257, 258]. However, for underdoped high-T_c superconductors, the infrared spectroscopy measurements show that the integral of the c-axis optical conductivity $R(\omega)$ remains much less than the theoretical prediction even when the integral upper limit is taken much larger than $6\Delta_0$. It does not even show any tendency toward saturation [258]. This indicates that high energy spectra also have a contribution to the superconducting condensation [259] and that high-T_c superconductivity is a phenomenon of multiple energy scales even though the superconducting condensation occurs at low temperatures. Conventional quantum field theory is established based on the idea of renormalization group. It investigates an effective low energy model (e.g. the t–J model) by integrating out all high energy excitation states. The involvement of high-energy states in low-energy physics of high-T_c copper oxides seems to suggest that the high-T_c superconductivity is not a renormalizable phenomenon.

The violation of the optical sum rule with the missing low frequency spectral weight in the underdoped high-T_c cuprates is closely related to the loss of low energy entropy or the pseudogap effect introduced in §3.4. Currently, there is not a satisfactory explanation for this phenomenon. For high-T_c cuprates, the energy–momentum dispersion relation deviates strongly from the free electron dispersion defined in Eq. (10.16), thus the correction from the second term of Eq. (10.14) to Eq. (10.18) is nonnegligible. This may be part of the reason for the violation of the optical sum rule in the CuO$_2$ plane [259]. However, along the c-axis, $R(\omega)$ of the underdoped superconductor can only reach 50% of the expectation even if ω reaches the infrared regime. It is definitely not sufficient to only consider the correction from the band structure to Eq. (10.18) [260].

10.3 Light Absorption in the Dirty Limit

In the dirty limit, the frequency dependence of the optical conductivity is determined by Eq. (10.2). In order to determine its value, we need to first calculate the conductivity $\sigma_\mu^{(0)}(\mathbf{q}, \omega)$ in an ideal BCS superconductor.

Substituting the free Green's function Eq. (4.4) into Eq. (9.11), we obtain the following current–current correlation function

$$\tilde{\Pi}_{\mu\nu}(\mathbf{q}, i\omega_n) = \frac{e^2}{V\hbar^2} \sum_{\mathbf{k}} \frac{\partial \varepsilon_{\mathbf{k}+\frac{\mathbf{q}}{2}}}{\partial k_\mu} \frac{\partial \varepsilon_{\mathbf{k}+\frac{\mathbf{q}}{2}}}{\partial k_\nu} A(\mathbf{k}, \mathbf{q}, i\omega_n), \qquad (10.19)$$

where

$$A(\mathbf{k}, \mathbf{q}, i\omega_n) = \frac{1}{\beta} \sum_{\omega_m} \mathrm{Tr} G^{(0)}(\mathbf{k}, i\omega_m) G^{(0)}(\mathbf{k}+\mathbf{q}, i\omega_m + i\omega_n)$$

$$= \frac{E_{\mathbf{k}+\mathbf{q}}(E_{\mathbf{k}+\mathbf{q}} - i\omega_n) + \xi_{\mathbf{k}}\xi_{\mathbf{k}+\mathbf{q}} + \Delta_{\mathbf{k}}\Delta_{\mathbf{k}+\mathbf{q}}}{E_{\mathbf{k}+\mathbf{q}}\left[(E_{\mathbf{k}+\mathbf{q}} - i\omega_n)^2 - E_{\mathbf{k}}^2\right]} f(E_{\mathbf{k}+\mathbf{q}})$$

$$- \frac{E_{\mathbf{k}+\mathbf{q}}(E_{\mathbf{k}+\mathbf{q}} + i\omega_n) + \xi_{\mathbf{k}}\xi_{\mathbf{k}+\mathbf{q}} + \Delta_{\mathbf{k}}\Delta_{\mathbf{k}+\mathbf{q}}}{E_{\mathbf{k}+\mathbf{q}}\left[(E_{\mathbf{k}+\mathbf{q}} + i\omega_n)^2 - E_{\mathbf{k}}^2\right]} [1 - f(E_{\mathbf{k}+\mathbf{q}})]$$

$$+ \frac{E_{\mathbf{k}}(E_{\mathbf{k}} + i\omega_n) + \xi_{\mathbf{k}}\xi_{\mathbf{k}+\mathbf{q}} + \Delta_{\mathbf{k}}\Delta_{\mathbf{k}+\mathbf{q}}}{E_{\mathbf{k}}\left[(E_{\mathbf{k}} + i\omega_n)^2 - E_{\mathbf{k}+\mathbf{q}}^2\right]} f(E_{\mathbf{k}})$$

$$- \frac{E_{\mathbf{k}}(E_{\mathbf{k}} - i\omega_n) + \xi_{\mathbf{k}}\xi_{\mathbf{k}+\mathbf{q}} + \Delta_{\mathbf{k}}\Delta_{\mathbf{k}+\mathbf{q}}}{E_{\mathbf{k}}\left[(E_{\mathbf{k}} - i\omega_n)^2 - E_{\mathbf{k}+\mathbf{q}}^2\right]} [1 - f(E_{\mathbf{k}})]. \qquad (10.20)$$

At zero temperature, $f(E_{\mathbf{k}}) = 0$, the above expression reduces to

$$A(\mathbf{k},\mathbf{q},i\omega_n) = -\frac{E_{\mathbf{k}}(E_{\mathbf{k}} - i\omega_n) + \xi_{\mathbf{k}}\xi_{\mathbf{k}+\mathbf{q}} + \Delta_{\mathbf{k}}\Delta_{\mathbf{k}+\mathbf{q}}}{E_{\mathbf{k}}\left[(E_{\mathbf{k}} - i\omega_n)^2 - E_{\mathbf{k}+\mathbf{q}}^2\right]}$$

$$-\frac{E_{\mathbf{k}+\mathbf{q}}(E_{\mathbf{k}+\mathbf{q}} + i\omega_n) + \xi_{\mathbf{k}}\xi_{\mathbf{k}+\mathbf{q}} + \Delta_{\mathbf{k}}\Delta_{\mathbf{k}+\mathbf{q}}}{E_{\mathbf{k}+\mathbf{q}}\left[(E_{\mathbf{k}+\mathbf{q}} + i\omega_n)^2 - E_{\mathbf{k}}^2\right]}. \quad (10.21)$$

Substituting it into the expression for $\tilde{\Pi}_{\mu\nu}$ and performing the analytic continuation yields

$$\Pi_{\mu\nu}(\mathbf{q},\omega) = -\frac{e^2}{V\hbar^2}\sum_{\mathbf{k}}\frac{\partial\varepsilon_{\mathbf{k}+\frac{\mathbf{q}}{2}}}{\partial k_\mu}\frac{\partial\varepsilon_{\mathbf{k}+\frac{\mathbf{q}}{2}}}{\partial k_\nu}\left(\frac{E_{\mathbf{k}}(E_{\mathbf{k}} - \omega) + \xi_{\mathbf{k}}\xi_{\mathbf{k}+\mathbf{q}} + \Delta_{\mathbf{k}}\Delta_{\mathbf{k}+\mathbf{q}}}{E_{\mathbf{k}}\left[(E_{\mathbf{k}} - \omega - i\delta)^2 - E_{\mathbf{k}+\mathbf{q}}^2\right]}\right.$$

$$\left.+\frac{E_{\mathbf{k}+\mathbf{q}}(E_{\mathbf{k}+\mathbf{q}} + \omega) + \xi_{\mathbf{k}}\xi_{\mathbf{k}+\mathbf{q}} + \Delta_{\mathbf{k}}\Delta_{\mathbf{k}+\mathbf{q}}}{E_{\mathbf{k}+\mathbf{q}}\left[(E_{\mathbf{k}+\mathbf{q}} + \omega + i\delta)^2 - E_{\mathbf{k}}^2\right]}\right). \quad (10.22)$$

Its imaginary part is

$$\mathrm{Im}\Pi_{\mu\nu}(\mathbf{q},\omega) = \frac{\pi e^2}{2V\hbar^2}\sum_{\mathbf{k}}\frac{\partial\varepsilon_{\mathbf{k}+\frac{\mathbf{q}}{2}}}{\partial k_\mu}\frac{\partial\varepsilon_{\mathbf{k}+\frac{\mathbf{q}}{2}}}{\partial k_\nu}\frac{E_{\mathbf{k}}E_{\mathbf{k}+\mathbf{q}} - \xi_{\mathbf{k}}\xi_{\mathbf{k}+\mathbf{q}} - \Delta_{\mathbf{k}}\Delta_{\mathbf{k}+\mathbf{q}}}{E_{\mathbf{k}}E_{\mathbf{k}+\mathbf{q}}}$$

$$\left[\delta\left(\omega + E_{\mathbf{k}} + E_{\mathbf{k}+\mathbf{q}}\right) + \delta\left(\omega - E_{\mathbf{k}} - E_{\mathbf{k}+\mathbf{q}}\right)\right]. \quad (10.23)$$

The real part of the electric conductivity $\sigma_\mu^{(0)}(\mathbf{q},\omega)$, which is proportional to $\mathrm{Im}\Pi_{\mu\mu}(\mathbf{q},\omega)$, is

$$\mathrm{Re}\sigma_\mu^{(0)}(\mathbf{q},\omega) = \frac{\pi e^2}{2\hbar^2\omega}\sum_{\mathbf{k}}\left(\frac{\partial\varepsilon_{\mathbf{k}+\frac{\mathbf{q}}{2}}}{\partial k_\mu}\right)^2\frac{E_{\mathbf{k}}E_{\mathbf{k}+\mathbf{q}} - \xi_{\mathbf{k}}\xi_{\mathbf{k}+\mathbf{q}} - \Delta_{\mathbf{k}}\Delta_{\mathbf{k}+\mathbf{q}}}{E_{\mathbf{k}}E_{\mathbf{k}+\mathbf{q}}}$$

$$\delta\left(\omega - E_{\mathbf{k}} - E_{\mathbf{k}+\mathbf{q}}\right). \quad (10.24)$$

Substituting this into Eq. (10.2), we obtain the zero temperature optical conductivity in the dirty limit

$$\mathrm{Re}\sigma_\mu(\omega) = \frac{\pi e^2 v_F^2}{2d\omega}\sum_{\mathbf{k},\mathbf{q}}\left(1 - \frac{\xi_{\mathbf{k}}\xi_{\mathbf{q}} + \Delta_{\mathbf{k}}\Delta_{\mathbf{q}}}{E_{\mathbf{k}}E_{\mathbf{q}}}\right)\delta\left(\omega - E_{\mathbf{k}} - E_{\mathbf{q}}\right), \quad (10.25)$$

where d is the spatial dimension. The term proportional to $\xi_{\mathbf{k}}\xi_{\mathbf{k}+\mathbf{q}}$ is an odd function of both $\xi_{\mathbf{k}}$ and $\xi_{\mathbf{q}}$. As it is zero after momentum summation, the above equation becomes

$$\mathrm{Re}\sigma_\mu(\omega) = \frac{\pi e^2 v_F^2}{2d\omega}\sum_{\mathbf{k},\mathbf{q}}\left(1 - \frac{\Delta_{\mathbf{k}}\Delta_{\mathbf{q}}}{E_{\mathbf{k}}E_{\mathbf{q}}}\right)\delta\left(\omega - E_{\mathbf{k}} - E_{\mathbf{q}}\right). \quad (10.26)$$

10.3.1 Isotropic s-Wave Superconductors

In an isotropic s-wave superconductor, $\Delta_{\mathbf{k}} = \Delta$, Eq. (10.26) becomes

$$\text{Re}\sigma\left(\omega\right) = \frac{\pi e^2 v_F^2 N_F^2}{2d\omega} \int_{-\infty}^{\infty} d\xi_2 d\xi_1 \frac{E_1 E_2 - \Delta^2}{E_1 E_2} \delta\left(\omega - E_1 - E_2\right), \qquad (10.27)$$

where $E_i = \sqrt{\xi_i^2 + \Delta^2}$ and N_F is the density of states of electrons at the Fermi level in the normal state. Using the identity

$$\int_{-\infty}^{\infty} d\xi_i = 2 \int_{\Delta}^{\infty} dE_i \frac{E_i}{\sqrt{E_i^2 - \Delta^2}}, \qquad (10.28)$$

it can be further expressed as

$$\text{Re}\sigma\left(\omega\right) = \frac{2\pi e^2 v_F^2 N_F^2}{d\omega} \theta\left(\omega - 2\Delta\right)$$
$$\int_{\Delta}^{\omega-\Delta} dE \frac{E\left(\omega - E\right) - \Delta^2}{\sqrt{E^2 - \Delta^2}\sqrt{(\omega - E)^2 - \Delta^2}}. \qquad (10.29)$$

Hence $\text{Re}\sigma(\omega)$ has a threshold or an absorption edge. $\text{Re}\sigma(\omega)$ is finite only when $\omega > 2\Delta$. This is an important property of s-wave superconductors. It is also an important criterion for testing and measuring the energy gap of an s-wave superconductor. Physically, this is because there are no quasiparticle excitations at zero temperature, and light can be absorbed by electrons only when the light frequency exceeds the binding energy of Cooper pairs.

To define

$$E = \frac{\omega + x\left(\omega - 2\Delta\right)}{2}, \qquad (10.30)$$

we can rewrite $\text{Re}\sigma(\omega)$ as

$$\text{Re}\sigma\left(\omega\right) = \frac{2\pi e^2 v_F^2 N_F^2\left(\omega - 2\Delta\right)}{\hbar^2 d\omega} \theta\left(\omega - 2\Delta\right) F\left(\alpha\right), \qquad (10.31)$$

where

$$F\left(\alpha\right) = \int_0^1 dx \frac{1 - \alpha x^2}{\sqrt{1 - x^2}\sqrt{1 - \alpha^2 x^2}}, \qquad \alpha = \frac{\omega - 2\Delta}{\omega + 2\Delta}. \qquad (10.32)$$

In the normal state, $\Delta = 0$ and $\alpha = 1$, hence $F(1) = 1$. Thus in the dirty limit, the normal state optical conductivity $\text{Re}\sigma(\omega)$ is ω-independent. The ratio between the optical conductivity in the superconducting state, $\text{Re}\sigma^s$, and that in the normal state, $\text{Re}\sigma^n$, is given by

$$\frac{\text{Re}\sigma^s\left(\omega\right)}{\text{Re}\sigma^n\left(\omega\right)} = \frac{\omega - 2\Delta}{\omega} F\left(\alpha\right) \theta\left(\omega - 2\Delta\right). \qquad (10.33)$$

This is actually the formula that is commonly used in the analysis of optical conductivity of isotropic s-wave superconductors. It is valid in the dirty limit. This formula was first derived by Mattis and Bardeen [254]. It shows that $\mathrm{Re}\sigma^s(\omega)/\mathrm{Re}\sigma^n(\omega)$ is a universal function of ω/Δ, independent of the details of scattering processes.

10.3.2 d-Wave Superconductors

In a d-wave superconductor, as the gap function $\Delta_{\mathbf{k}}$ changes sign when momentum $\mathbf{k}$ is rotated by $\pi/2$, the momentum summation of the $\Delta_{\mathbf{k}}\Delta_{\mathbf{q}}$ term in Eq. (10.26) also becomes zero. This simplifies the optical conductivity

$$\mathrm{Re}\sigma(\omega) = \frac{\pi e^2 v_F^2 N_F^2}{4\omega} \int_{-\infty}^{\infty} d\xi_1 d\xi_2 \int_0^{2\pi} \frac{d\varphi_1}{2\pi} \frac{d\varphi_2}{2\pi} \delta(\omega - E_1 - E_2), \quad (10.34)$$

where

$$E_i = \sqrt{\xi_i^2 + \Delta^2 \cos^2 2\varphi_i}, \quad (i = 1, 2). \quad (10.35)$$

Using the expression of the quasiparticle density of states in the d-wave superconductor

$$\rho(E) = N_F \int_{-\infty}^{\infty} d\xi \int_0^{2\pi} \frac{d\varphi_1}{2\pi} \delta\left(E - \sqrt{\xi^2 + \Delta^2 \cos^2 2\varphi}\right), \quad (10.36)$$

we can further express $\mathrm{Re}\sigma(\omega)$ as

$$\begin{aligned}
\mathrm{Re}\sigma(\omega) &= \frac{\pi e^2 v_F^2}{4\omega} \int_0^{\infty} dE_1 \int_0^{\infty} dE_2 \delta(\omega - E_1 - E_2) \rho(E_1) \rho(E_2) \\
&= \frac{\pi e^2 v_F^2}{4\omega} \int_0^{\omega} dE \rho(E) \rho(\omega - E).
\end{aligned} \quad (10.37)$$

Hence $\mathrm{Re}\sigma(\omega)$ is determined by the convolution of the density of states of d-wave quasiparticles. It equals the sum of the probabilities for exciting two quasiparticles to the energies at E and $\omega - E$, respectively. There is no absorption edge due to the presence of the nodal points in the d-wave gap function.

In the low frequency limit $E < \omega \ll \Delta$, $\rho(E)$ is approximately a linear function of E,

$$\rho(E) \approx a_0 E. \quad (10.38)$$

In this case, the optical conductivity varies quadratically with ω

$$\mathrm{Re}\sigma_\mu(\omega) \approx \frac{\pi e^2 v_F^2 a_0^2}{4\omega} \int_0^{\omega} dE \left(\omega E - E^2\right) = \frac{\pi e^2 v_F^2 a_0^2 \omega^2}{24}, \quad (10.39)$$

to the leading order of approximation.

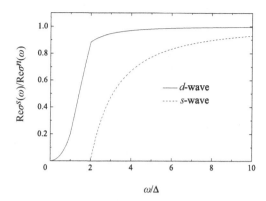

Figure 10.1 The zero temperature infrared conductivity as a function of frequency for the s- and d-wave superconductors in the dirty limit.

In the normal state, $\Delta = 0$, ρ equals approximately the density of states at the Fermi level, independent of ω. $\mathrm{Re}\sigma$ is also ω-independent. The ratio between the optical conductivity in the superconducting state and that in the normal state is given by

$$\frac{\mathrm{Re}\sigma^s(\omega)}{\mathrm{Re}\sigma^n(\omega)} = \frac{1}{N_F^2\omega}\int_0^\omega dE\,\rho(E)\,\rho(\omega - E). \qquad (10.40)$$

Figure 10.1 compares the frequency dependence of the optical conductivity for the s- and d-wave superconductors in the dirty limit. The difference lies mainly in the low frequency regime. There is an absorption edge in the s-wave superconductor but not in the d-wave one. In the d-wave superconductor, $\mathrm{Re}\sigma$ drops almost linearly with decreasing ω when $\Delta < \omega < 2\Delta$, and changes gradually to an ω^2-dependence when $\omega < \Delta$. When ω is slightly larger than 2Δ, the optical conductivity of the d-wave superconductor is already approaching the value in the normal state, much larger than the value in the s-wave case.

In high-T_c superconductors, the response function is predominantly determined by high energy quasiparticles when the measured temperature or frequency is comparable to the gap value Δ_0. As the coherence lengths of these quasiparticles are shorter than the mean free length, the system lies in the clean limit. We cannot directly use the above result to quantitatively understand or interpret the experimental data of high-T_c cuprates. On the other hand, if the measurement frequency is much lower than Δ_0, the absorption is determined by the low-lying excitations around the nodal points. In this case, the effective coherence length of nodal quasiparticles is longer than the mean free path, and the above result is still applicable.

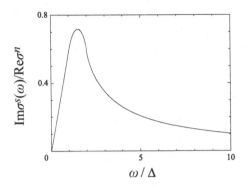

Figure 10.2 Imaginary part of the optical conductivity as a function of frequency for the d-wave superconductor in the dirty limit at zero temperature.

The imaginary part of the optical conductivity can be obtained through the Kramers–Kronig relation

$$\mathrm{Im}\sigma(\omega) = \frac{2\omega}{\pi} \int_0^\infty d\omega' \frac{\mathrm{Re}\sigma(\omega')}{\omega'^2 - \omega^2}. \tag{10.41}$$

Figure 10.2 shows the frequency dependence of the imaginary optical conductivity for the d-wave superconductor in the dirty limit. $\mathrm{Im}\sigma(\omega)$ varies linearly with ω when $\omega < \Delta$ and exhibits a peak between $\omega = \Delta$ and 2Δ. The curvature of $\mathrm{Im}\sigma(\omega)$ changes at $\omega = 2\Delta$.

From the real and imaginary parts of the optical conductivity obtained above, we can calculate the dielectric function and the reflection index using the formula

$$\epsilon = \epsilon_\infty + \frac{4\pi i}{\omega}(\mathrm{Re}\sigma + i\mathrm{Im}\sigma) = \epsilon_1 + i\epsilon_2, \tag{10.42}$$

$$n = \frac{1}{\sqrt{2}}\sqrt{\sqrt{\epsilon_1^2 + \epsilon_2^2} + \epsilon_1}, \tag{10.43}$$

$$k = \frac{1}{\sqrt{2}}\sqrt{\sqrt{\epsilon_1^2 + \epsilon_2^2} - \epsilon_1}. \tag{10.44}$$

The reflectivity $R(\omega)$ of the d-wave superconductor in the dirty limit is then found to be

$$R(\omega) = \left|\frac{n + ik - 1}{n + ik + 1}\right|^2, \tag{10.45}$$

where $R(\omega)$ can be measured directly from the infrared spectroscopy experiments. Figure 10.3 shows $R(\omega)$ as a function of frequency for the d-wave superconductor. Below the gap energy, $\omega < \Delta$, $R(\omega)$ drops linearly with ω, resulting from the linear density of states of the d-wave superconductor. The curvature of R is negative when $\Delta < \omega < 2\Delta$, and becomes positive when $\omega > 2\Delta$.

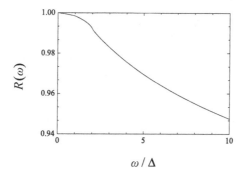

Figure 10.3 Typical frequency dependence of the reflectivity of the d-wave super-conductor in the dirty limit at zero temperature.

10.4 Effect of Elastic Impurity Scattering

The quasiparticle conductivity is determined by the imaginary part of the current–current correlation function in the clean limit. It can be expressed using the single-particle spectral function as

$$\mathrm{Re}\,\sigma_\mu(\omega) = \frac{e^2}{\pi \hbar^2 V} \sum_\mathbf{k} \left(\frac{\partial \varepsilon_\mathbf{k}}{\partial k_\mu}\right)^2 \int d\omega_1 \frac{f(\omega_1) - f(\omega_1 + \omega)}{\omega}$$
$$\mathrm{Tr}\mathrm{Im}G^R(\mathbf{k}, \omega_1)\mathrm{Im}G^R(\mathbf{k}, \omega_1 + \omega). \tag{10.46}$$

At low temperatures or low frequencies, the elastic impurity scattering is the major scattering channel, and the contribution from inelastic scattering is relatively small and negligible [261, 262].

10.4.1 Universal Conductance

At zero temperature, the conductivity is proportional to the density of states, and inversely proportional to the electron scattering rate. In normal metals, the electron density of states is energy independent and the conductivity is mainly determined by the scattering rate. However, in a d-wave superconductor, the quasiparticle density of states on the Fermi surface is proportional to the scattering rate, and these two effects cancel each other. As a consequence, the quasiparticle conductance at zero temperature is a universal constant, independent on the impurity scattering potential. P. A. Lee et al. first found this universal behavior of the d-wave superconductor. They also derived the formula of the universal conductance [263]. However, it should be pointed out that this universal property is valid only in the weak scattering limit. In the strong scattering limit, the conductance is not universal.

In the zero frequency limit at zero temperature, i.e. $\omega \to 0^+$, Eq. (10.46) becomes

$$\text{Re}\sigma_\mu(0) = \frac{e^2}{\pi \hbar^2 V} \sum_{\mathbf{k}} \left(\frac{\partial \varepsilon_{\mathbf{k}}}{\partial k_\mu} \right)^2 \text{TrIm}G^R(\mathbf{k}, 0)\text{Im}G^R(\mathbf{k}, 0). \tag{10.47}$$

The imaginary part of the zero frequency Green's function is given by

$$\text{Im}G^R(\mathbf{k}, 0) = -\frac{\Gamma_0}{\Gamma_0^2 + \xi_{\mathbf{k}}^2 + \Delta_{\mathbf{k}}^2}. \tag{10.48}$$

Substituting this into Eq. (10.47) yields the expression of the quasi-two-dimensional conductivity

$$\begin{aligned}
\text{Re}\sigma_{ab}(0) &= \frac{e^2 v_F^2}{\pi V} \sum_{\mathbf{k}} \frac{\Gamma_0^2}{\left(\Gamma_0^2 + \xi_{\mathbf{k}}^2 + \Delta_{\mathbf{k}}^2 \right)^2} \\
&= \frac{4e^2 v_F^2 N_F}{\pi^2} \int_0^{\pi/4} d\varphi \int_{-\infty}^{\infty} d\xi \frac{\Gamma_0^2}{\left(\Gamma_0^2 + \xi^2 + \Delta_0^2 \cos^2 2\varphi \right)^2} \\
&= \frac{e^2 v_F^2 N_F}{\pi \Delta_0} \int_0^{\Delta_0/\Gamma_0} \frac{dx}{\sqrt{1 - \left(\frac{\Gamma_0 x}{\Delta_0} \right)^2}} \frac{1}{\left(1 + x^2 \right)^{3/2}}. \tag{10.49}
\end{aligned}$$

For most superconductors, Δ_0 is much larger than Γ_0. The above integral can be expanded in terms of Γ_0/Δ_0. To the leading order approximation, the result is

$$\text{Re}\sigma_{ab}(0) = \sigma_0 \left[1 + \frac{\Gamma_0^2}{\Delta_0^2} \ln \frac{2\Delta_0}{\Gamma_0} + o\left(\frac{\Gamma_0^2}{\Delta_0^2} \right) \right], \tag{10.50}$$

where

$$\sigma_0 = \frac{e^2 v_F^2 N_F}{\pi \Delta_0} \tag{10.51}$$

is the universal conductance. It depends only on the intrinsic parameters of d-wave superconductors, but not on the impurity potential.

The above formula shows that the zero temperature conductivity $\text{Re}\sigma_{ab}$ equals a universal value in the limit $\Gamma_0 \ll \Delta_0$. The condition $\Gamma_0 \ll \Delta_0$ is valid in most of high-T_c superconductors. The second term in Eq. (10.50) is not negligible when the sample is not very clean and the gap is not very big. In this case, $\text{Re}\sigma_{ab}$ depends on the scattering rate and is no longer universal.

In high-T_c superconductors, the electron velocity along the c-axis is given by $v_c \approx v_\perp \cos^2(2\varphi)$. At zero temperature, to the leading order of Γ_0/Δ_0, the c-axis conductivity in the limit $\Gamma_0 \ll \Delta_0$ is determined by the formula

$$\text{Re}\sigma_c(0) = \frac{e^2 v_\perp^2}{\pi V} \sum_{\mathbf{k}} \frac{\Gamma_0^2 \cos^4 2\varphi}{\left(\Gamma_0^2 + \varepsilon_{\mathbf{k}}^2 + \Delta_{\mathbf{k}}^2 \right)^2} \approx \frac{e^2 v_\perp^2 N_F \Gamma_0^2}{\pi \Delta_0^3}. \tag{10.52}$$

Reσ_c depends on the scattering rate. The reason Reσ_c is not universal is because the c-axis velocity of electrons v_c depends strongly on the angle of the in-plane momentum. It changes the weight of conductivity contributed by the quasiparticles on different parts of the Fermi surface. In particular, the contribution of quasiparticles around the nodal points is suppressed. The cancellation between the density of states and the scattering rate does not occur.

10.4.2 Low Temperature Behavior

At low temperatures, the in-plane conductivity of quasiparticles is determined by the equation

$$\text{Re}\sigma_{ab}(T) = -\frac{e^2 v_F^2}{2\pi V} \sum_{\mathbf{k}} \int d\omega \frac{\partial f}{\partial \omega} \text{TrIm} G^R(\mathbf{k}, \omega) \text{Im} G^R(\mathbf{k}, \omega). \tag{10.53}$$

In the gapless regime, $T \ll \Gamma_0$, $\partial f/\partial \omega$ decays exponentially with ω and the integral in Eq. (10.53) can be calculated using the Sommerfeld expansion. The two leading terms are given by

$$\text{Re}\sigma_{ab}(T) \approx \frac{e^2 v_F^2}{2\pi V} \sum_{\mathbf{k}} \left[g_{\mathbf{k}}(0) + \frac{\pi^2 T^2}{6} g_{\mathbf{k}}''(0) \right], \tag{10.54}$$

with

$$g_{\mathbf{k}}(\omega) = \text{TrIm} G^R(\mathbf{k}, \omega) \text{Im} G^R(\mathbf{k}, \omega). \tag{10.55}$$

The first term in Eq. (10.54) gives the zero temperature conductivity, and the second term is the correction resulting from the finite temperature thermal fluctuation.

At low frequencies, the second order derivative of $g_{\mathbf{k}}(\omega)$ with respect to ω can be evaluated using the expression of the retarded Green's function given in Eq. (8.54). In the limit $\Gamma_0 \ll \Delta_0$, the momentum summation of $g_{\mathbf{k}}''(0)$ is approximately given by

$$\frac{1}{V} \sum_{\mathbf{k}} g_{\mathbf{k}}''(0) \approx \frac{4N_F a^2}{\Delta_0 \Gamma_0^2}. \tag{10.56}$$

The corresponding in-plane conductivity is

$$\text{Re}\sigma_{ab}(T) \approx \sigma_0 \left(1 + \frac{\pi^2 a^2 T^2}{3\Gamma_0^2} \right), \tag{10.57}$$

Reσ_{ab} at low temperatures scales as T^2. This is a consequence of the Sommerfeld expansion in the limit $T \ll \Gamma_0$, independent of the value of Γ_0. This T^2-dependence of the conductivity is also a universal property of d-wave superconductors.

In the intrinsic regime, $\Gamma(\omega) \ll T \ll \Delta_0$. Using Eq. (8.57) and the identity

$$\text{Im}G^R(\mathbf{k}, \omega)\text{Im}G^R(\mathbf{k}, \omega) = \frac{1}{2}\text{Re}G^R(\mathbf{k}, \omega)\left[G^{R*}(\mathbf{k}, \omega) - G^R(\mathbf{k}, \omega)\right], \quad (10.58)$$

it can be shown that

$$\frac{1}{V}\sum_{\mathbf{k}}\text{TrIm}G^R(\mathbf{k}, \omega)\text{Im}G^R(\mathbf{k}, \omega) \approx \frac{\pi\omega N_F}{\Delta_0\Gamma(\omega)}. \quad (10.59)$$

Thus the in-plane conductivity is proportional to the average of $|\omega|/\Gamma(\omega)$ on the Fermi surface,

$$\text{Re}\sigma_{ab}(T) = -\frac{e^2 v_F^2 N_F}{2\Delta_0}\int d\omega \frac{\omega}{\Gamma(\omega)}\frac{\partial f}{\partial \omega}, \quad (10.60)$$

consistent with the result obtained by the Boltzmann transport theory of quasiparticles.

In the Born scattering limit, substituting the value of Γ in Eq. (8.58) to Eq. (10.60), we find that

$$\text{Re}\sigma_{ab}(T) \simeq \frac{e^2 v_F^2 N_F}{\Gamma_N}. \quad (10.61)$$

It is temperature independent and equals the conductivity in the normal state. This is a special property of d-wave superconductors in the Born scattering limit. In real materials, the intrinsic region is generally very narrow and this temperature independent conductivity is valid only in a very narrow temperature range, which is difficult to detect.

In the unitary limit, $\Gamma(\omega)$ is given by Eq. (8.59). Substituting it into Eq. (10.60) and keeping the leading order term, we obtain

$$\text{Re}\sigma_{ab}(T) \simeq -\frac{4e^2 v_F^2 N_F}{\pi^2\Delta_0^2\Gamma_N}\int d\omega\omega^2 \ln^2\frac{2\Delta_0}{|\omega|}\frac{\partial f}{\partial \omega}$$

$$\approx \frac{4e^2 v_F^2 N_F k_B^2 T^2}{3\Delta_0^2\Gamma_N}\ln^2\frac{2\Delta_0}{k_B T}. \quad (10.62)$$

$\text{Re}\sigma_{ab}$ is now a function of $T^2 \ln T$, different than in the Born limit. If we neglect the weak logarithmic correction, it approximately scales as T^2, close to the temperature dependence in the gapless regime. Nevertheless, the coefficients of the T^2 terms are different in these two regions.

10.4.3 Finite Frequency Dependence

In the low frequency limit, the optical conductivity is given by

$$\text{Re}\sigma_\mu(\omega) = \frac{e^2 v_F^2}{4\pi\omega V}\sum_{\mathbf{k}} A(\mathbf{k}, \omega) \quad (10.63)$$

at zero temperature. In the above expression,

$$A(\mathbf{k}, \omega) = \int_{-\omega}^{0} d\omega_1 \operatorname{Re} \operatorname{Tr} \left[G^{R*}(\mathbf{k}, \omega_1) - G^R(\mathbf{k}, \omega_1) \right] G^R(\mathbf{k}, \omega_1 + \omega). \quad (10.64)$$

Inserting the quasiparticle Green's function, which is approximately given by Eq. (8.54), into Eq. (10.64) and taking the sum over ω_1, we obtain

$$A(\mathbf{k}, \omega) = \operatorname{Re} \frac{2}{2i a \Gamma_0 + a^2 \omega} \ln \frac{E_{\mathbf{k}}^2 + \Gamma_0^2}{E_{\mathbf{k}}^2 + (\Gamma_0 - i a \omega)^2}$$
$$- \frac{1}{a^2 \omega} \ln \frac{\left[E_{\mathbf{k}}^2 + (\Gamma_0 + i a \omega)^2\right]\left[E_{\mathbf{k}}^2 + (\Gamma_0 - i a \omega)^2\right]}{\left(E_{\mathbf{k}}^2 + \Gamma_0^2\right)^2}. \quad (10.65)$$

This expression of $A(\mathbf{k}, \omega)$ is rather complicated and difficult to handle. Nevertheless, at low frequencies, $A(\mathbf{k}, \omega)$ can be expanded in terms of ω. The leading two terms are

$$A(\mathbf{k}, \omega) = \frac{4\Gamma_0^2 \omega}{\left(E_{\mathbf{k}}^2 + \Gamma_0^2\right)^2} + \frac{2E_{\mathbf{k}}^2 - \Gamma_0^2}{\left(E_{\mathbf{k}}^2 + \Gamma_0^2\right)^4} \frac{8 a^2 \Gamma_0^2 \omega^3}{3} + o(\omega^5). \quad (10.66)$$

The first term leads to the universal conductance at zero temperature. The second term gives the finite frequency correction to the conductivity [261].

In the limit $\Gamma_0 \ll \Delta_0$, the optical conductivity is given by

$$\operatorname{Re}\sigma_{ab}(\omega) \approx \sigma_0 \left(1 - \frac{a^2 \Gamma_0^2 \omega^2}{2\Delta_0^4} + o(\omega^4) \right), \qquad (\omega \ll \Gamma_0 \ll \Delta_0). \quad (10.67)$$

The finite frequency correction to the conductivity is negative. Comparing with Eq. (10.57), we find that there is a similarity as well as a difference in the temperature and frequency dependencies of $\operatorname{Re}\sigma_{ab}$. The similarity is that $\operatorname{Re}\sigma_{ab}(T)$ depends on T^2 while $\operatorname{Re}\sigma(\omega)$ depends on ω^2. The difference is that $\operatorname{Re}\sigma_{ab}(T)$ increases with increasing temperature, while $\operatorname{Re}\sigma(\omega)$ decreases with increasing frequency.

10.5 Microwave Conductivity of Cuprate Superconductors

The low frequency conductivity of high-T_c superconductors can be determined through the measurement of surface resistance R_s. The microwave frequency is typically in the range of $1 \sim 10$ GHz. The corresponding energy scale is $10 \sim 100$ μeV. From the measurements for the YBCO and BSCCO single crystals, it was found that the surface resistance drops nearly four orders of magnitude below T_c within a temperature window which is narrower than $T_c/5$ [202, 264]. This fast decay of R_s implies that the quasiparticle lifetime increases very rapidly below T_c, unlike in a metal or alloy-based superconductor. After this fast decrease, R_s begins to grow gradually and drops again linearly with temperature after reaching a maximum at about $T_c/3$ [264].

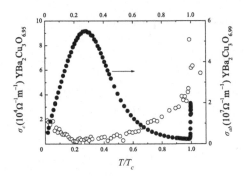

Figure 10.4 Temperature dependence of the microwave conductivities of the YBa$_2$Cu$_3$O$_x$ superconductor along both the ab-plane and the c-axis. The measured frequency is 1.14 GHz for σ_{ab} and 22 GHz for σ_c. This figure is plotted based on the data published in Ref. [264] and on the homepage of the UBC experimental group (www.physics.ubc.ca/ supercon/supercon.html).

In the superconducting state, R_s varies linearly with the microwave conductivity and cubically with the magnetic penetration depth, i.e. $R_s \propto \sigma \lambda^3$. Thus from the measurement of R_s and λ, we can find the temperature dependence of σ.

Figure 10.4 shows the typical temperature dependence of the microwave conductivity along the direction parallel to the CuO$_2$ plane as well as that along the c-axis. In the CuO$_2$ plane, Reσ_{ab} exhibits a nonmonotonic temperature dependence in accordance with the nonmonotonic behavior of the temperature dependence of R_s. Below T_c, Reσ_{ab} first shows a fast increase with decreasing temperature, and then drops linearly with temperature after reaching a maximum at about $T \sim T_c/3$.

The nonmonotonic temperature dependence of Reσ_{ab} results from the competition between the increasing quasiparticle lifetime and the decreasing quasiparticle density of states in the superconducting state. The results shown in Fig. 10.4 could be understood from the generalized Drude formula

$$\mathrm{Re}\sigma_{ab} = \frac{e^2 n_e \tau}{m^*}, \tag{10.68}$$

where n_e is the density of normal electrons (or quasiparticles) and τ is the quasiparticle lifetime. In the superconducting state, n_e decreases with decreasing temperature. Meanwhile, the quasiparticle scattering is weakened, hence the scattering lifetime is increased. Near T_c, the superfluid density is very small and the change of n_e is small, but τ increases quickly. This leads to the fast increase of Reσ_{ab} just below T_c. At low temperatures, the quasiparticle lifetime does not change too much. It increases slowly with lowering temperature, but the quasiparticle density drops quickly. This leads to the drop of Reσ_{ab} in the low temperature regime.

Equation (10.68) provides a simple picture for qualitatively understanding the temperature dependence of Reσ_{ab}. However, it does not provide a microscopic scenario for quantitatively understanding the temperature dependence of Reσ_{ab}, especially the linear temperature dependence of Reσ at low temperatures.

In fully gapped superconductors, Reσ always decays exponentially at low temperatures. Clearly, the linear conductivity cannot be attributed to the s-wave pairing. In a d-wave superconductor, Reσ contributed to by the elastic impurity scattering scales as T^2. Thus it is also difficult to use the d-wave pairing to explain this linear temperature dependence. Whether this difficulty can be resolved by further considering inelastic scatterings or electron–electron interaction needs further investigation.

The c-axis conductivity of high-T$_c$ superconductors shows a completely different temperature dependence than the in-plane conductivity. Just below T_c, as shown in Fig. 10.4, the c-axis conductivity drops with lowering temperature just below T_c. This is a common feature of high-T$_c$ superconductors. By further lowering temperature, the c-axis conductivity of the YBa$_2$Cu$_3$O$_x$ superconductor begins to increase (see Fig. 10.4). This nonmonotonic behavior of the c-axis conductivity is only observed in the YBCO superconductor, not in other high-T$_c$ superconductors. It is unknown why the c-axis conductivity of YBCO varies nonmonotonically with temperature. The increase of the c-axis conductivity at low temperatures may result from the scattering of magnetic impurities in YBCO.

Hence the microscopic conductivities along the c-axis and ab-plane exhibit completely different temperature dependence in high-T$_c$ superconductors. It is difficult to understand this difference within the conventional framework of transport theory. However, considering the anisotropy of the hopping matrix elements along the c-axis,

$$t_c = -t_\perp \cos^2 2\varphi, \tag{10.69}$$

it is not difficult to understand qualitatively this difference. φ is the azimuthal angle of the in-plane momentum. Under appropriate approximations, one can even make quantitative predictions for the temperature dependence of conductivity [265].

The conductivity is determined by the imaginary part of the current–current correlation function. If we only consider the contribution from the coherent quasiparticles by neglecting the vertex correction, the c-axis conductivity is determined by the formula

$$\text{Re}\sigma_c = -\frac{\alpha_c}{\pi} \int_{-\infty}^{\infty} d\omega \frac{\partial f(\omega)}{\partial \omega} \int_0^{2\pi} \frac{d\varphi}{2\pi} \cos^4 2\varphi M(\varphi), \tag{10.70}$$

where

$$M(\varphi) = \frac{\pi}{\Gamma_\varphi} \text{Re} \frac{(\omega + i\Gamma_\varphi)^3 - \omega\Delta_0^2 \cos^2 2\varphi}{\left[(\omega + i\Gamma_\varphi)^2 - \Delta_0^2 \cos^2 2\varphi\right]^{3/2}}, \tag{10.71}$$

and $\alpha_c = e^2 t_{\perp}^2 N_F / 4$. In deriving the above formula, the retarded single-particle Green's function is assumed to be

$$G_{\text{ret}}(\mathbf{k}, \omega) = \frac{1}{\omega - \xi_\mathbf{k} \tau_3 - \Delta_\varphi \tau_1 + i\Gamma_\varphi}, \tag{10.72}$$

where $\xi_\mathbf{k} = \varepsilon_{ab}(\mathbf{k}) - t_\perp \cos k_z \cos^2 \varphi - \mu$ is the energy dispersion of electrons. Γ_φ is the quasiparticle scattering rate.

In the superconducting state, the integral in Eq. (10.70) cannot be evaluated analytically. Nevertheless, when the temperature is less than T_c but much larger than the scattering rate Γ_φ, $T_c > T \gg \Gamma_\varphi$, we can expand the integral on the right-hand side of Eq. (10.70) in terms of Γ_φ. Up to the leading order of Γ_φ, $\text{Re}\sigma_c$ can be expressed as

$$\text{Re}\sigma_c \approx -\alpha_c \int_{-\infty}^{\infty} d\omega \frac{\partial f(\omega)}{\partial \omega} \int_0^{2\pi} \frac{d\varphi}{2\pi} \frac{\cos^4(2\varphi)}{\Gamma_\varphi} \text{Re} \frac{|\omega|}{\sqrt{\omega^2 - \Delta_\varphi^2}}. \tag{10.73}$$

In the superconducting state, the average scattering rate of quasiparticles on the Fermi surface $\tau_0^{-1} = \langle \Gamma_\varphi \rangle_{FS}$ can be estimated from the measurement data of the microwave conductivity and the superfluid density in the CuO_2 plane, using the generalized Drude formula Eq. (10.68). For the optimally doped YBCO [264], τ_0^{-1} is less than 1 K at low temperatures. Γ_0 increases with increasing temperature. At $T = 60$ K, τ_0^{-1} is approximately equal to 6 K. This estimation shows that the condition $T_c > T \gg \Gamma_\varphi$ is satisfied and the leading order approximation in Γ_φ is valid at least when the temperature is not too low.

In high-T_c superconductors, Γ_φ shows a strong φ dependence [266, 267]. The scattering rate of electrons along the gap nodal directions is much smaller than that along the antinodal ones. Ioffe and Mills [268] proposed a phenomenological "cold spot" model to describe this anisotropy. In this model, Γ_φ is assumed to have the form

$$\Gamma_\varphi = \Gamma_0 \cos^2 2\varphi + \tau^{-1}(T), \tag{10.74}$$

where Γ_0 is a temperature independent parameter. τ^{-1} is angular independent but temperature dependent. The cold spot form of Γ_φ is proposed based on the analysis of experimental data. How to derive this formula theoretically remains an open question.

Substituting Eq. (10.74) into Eq. (10.73) and performing the integral over φ, we find that σ_c is approximately given by

$$\text{Re}\sigma_c \approx \frac{9\alpha_c \zeta[3] T^3}{2\Gamma_0 \Delta_0^3} - \frac{(2\ln 2) T\alpha_c}{\tau \Gamma_0^2 \Delta_0} + \frac{\alpha_c \sigma_a}{\alpha_a \tau^2 \Gamma_0^2}, \tag{10.75}$$

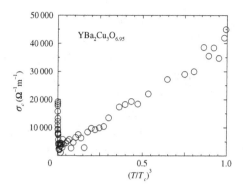

Figure 10.5 The cubic temperature dependence of the *c*-axis conductivity data previously shown in Fig. 10.4 for YBa$_2$Cu$_3$O$_{6.95}$.

where

$$\sigma_a \approx -\frac{T\tau\alpha_a}{\Delta_0} \int_{-\infty}^{\infty} dx \frac{\partial f(xT)}{\partial x} \frac{|x|}{\sqrt{1 + T^2\Gamma_0\tau x^2/\Delta_0^2}}, \tag{10.76}$$

$\alpha_a = e^2 v_F^2 N_F/4$ and $\zeta(3) = 1.202$.

In the high temperature limit, when the condition $\Gamma_0\tau T^2/\Delta_0^2 \gg 1$ is satisfied, Reσ_c is mainly determined by the first term on the right-hand side of Eq. (10.75). In this case, Reσ_c varies cubically with temperature,

$$\mathrm{Re}\sigma_c \approx \frac{9\alpha_c\zeta(3)T^3}{2\Gamma_0\Delta_0^3}. \tag{10.77}$$

This is a remarkable result. It shows that different from Reσ_{ab}, Reσ_c drops with decreasing temperature just below T_c, consistent with experimental measurements. Furthermore, Reσ_c does not depend on the scattering rate τ^{-1}. Thus the cubic temperature dependence of Reσ_c is universal.

The universal cubic temperature dependence of Reσ_c results from the interplay between the anisotropic *c*-axis hopping integral and the anisotropic scattering rate of electrons. For the YBCO superconductor, as $\tau_0^{-1} > \tau^{-1}$ at $T/\Delta_0 \gg 1/\sqrt{\Gamma_0\tau_0}$, the condition that $\Gamma_0\tau T^2/\Delta_0^2 \gg 1$ is satisfied. Based on the experimental data of the normal state resistivity, Ioffe and Mills estimated Γ_0 to be around 0.6 eV [268]. If we take $\tau_0^{-1} \sim 6$ K for YBCO at $T = 60$ K, then $1/\sqrt{\Gamma_0\tau_0}$ is estimated to be around 0.03. Hence for the optimally doped YBCO superconductor, $\Gamma_0\tau T^2/\Delta_0^2 \gg 1$ is valid at least when $T/\Delta_0 \gg 0.03$, i.e. $T/T_c \gg 0.06$.

Figure 10.5 replots the experimental data of Reσ_c shown in Fig. 10.4 as a function of $(T/T_c)^3$ for YBa$_2$Cu$_3$O$_{6.95}$ [265]. Within the measurement errors, Reσ_c agrees with the T^3-law theoretically predicted from 30 K to T_c. This universal

T^3-dependence is also consistent with the low frequency measurement data of conductivity for $Bi_2Sr_2CaCu_2O_{8+x}$ and other superconductors [269]. The agreement between the theoretical analysis and the experimental result shows again that it is crucial to include the anisotropy of the c-axis hopping matrix elements in the analysis of the c-axis transport properties.

At low temperatures, the impurity scattering becomes dominant, which ruins the universal behavior of the c-axis conductivity. The low-temperature conductivity along the c-axis is no longer universal. Different materials exhibit different temperature dependencies.

10.6 Heat Current Density Operator

In the superconducting state, the system is dissipation free, and the optical conductivity vanishes in the zero-frequency limit. However, the thermal conductivity remains finite. The thermal conductivity measures the response of the system to a temperature gradient. It reveals the property of quasiparticle excitations and can be used to extract information on the superconducting state.

In a d-wave superconductor, the thermal conductivity possesses certain peculiar properties. For example, in the zero temperature limit, the thermal conductivity is a universal quantity, which depends only on the Fermi velocity and the derivatives of the gap function at the nodal points, not on the impurity scattering and the strength of the Coulomb interaction. In §10.4.1, we showed that the low frequency electric conductivity of d-wave superconductors is universal at zero temperature. However, the universality of the electric conductivity is not as robust as the thermal conductivity. Vertex corrections induced by the Coulomb interaction as well as impurity scattering strongly affect the universal behavior of the electric conductivity, but weakly the universal behavior of the thermal conductivity. From the measurement of the universal thermal conductivity, we can determine not only the pairing symmetry, but also the ratio between the Fermi velocity and the gap slope at the nodal points and other parameters of d-wave superconductors.

Similar to the optical conductivity, the contribution of electrons to the thermal conductivity is determined by the correlation function of the heat current. The heat current represents the energy transfer in the presence of a temperature gradient. Unlike the electric current density operator, the energy flow density depends on both the kinetic energy and the interaction potential. This implies that the definition of the heat current density is different for different Hamiltonians. Nevertheless, the energy is always conserved, and the heat current density operator $\mathbf{j}_E$ should obey the continuous equation of energy

$$\partial_t h(\mathbf{r}) + \nabla \cdot \mathbf{j}_E(\mathbf{r}) = 0, \tag{10.78}$$

where $h(\mathbf{r})$ is the energy density. This is the energy flow continuity equation in analogy to the electric charge continuity equation. The expression for the heat current operator can be derived based on this equation.

In momentum space, the energy density and the heat current density operators are defined by

$$h(\mathbf{q}) = \int d\mathbf{r} e^{-i\mathbf{q}\cdot\mathbf{r}} h(\mathbf{r}), \tag{10.79}$$

$$\mathbf{j}_E(\mathbf{q}) = \int d\mathbf{r} e^{-i\mathbf{q}\cdot\mathbf{r}} \mathbf{j}_E(\mathbf{r}). \tag{10.80}$$

Substituting them into (10.78), we obtain the heat current continuity equation in momentum space

$$\partial_t h(\mathbf{q}) + i\mathbf{q} \cdot \mathbf{j}_E(\mathbf{q}) = 0. \tag{10.81}$$

Below we derive the expression of the heat current density operator for the d-wave superconductor under the mean-field approximation. There are two approaches that can be used. The first is to use the BCS mean-field Hamiltonian for the d-wave superconductor, and derive the expression for $\mathbf{j}_E$ based on the equations of motion and the continuity equation of electron operators. The second is to start from the BCS Hamiltonian and derive the heat current operator using the equation of motion of electrons before taking the mean-field approximation. The mean-field approximation is then taken to decouple the heat current operator into a quadratic form. These two methods yield the same result. Although the second approach is more commonly used in the literature [270, 271], the first approach is simpler to implement. Below we derive the expression for the heat current density $\mathbf{j}_E$ using this approach.

We start from the BCS mean-field Hamiltonian,

$$H = \sum_{\mathbf{k}\sigma} \xi_{\mathbf{k}} c_{\mathbf{k}\sigma}^\dagger c_{\mathbf{k}\sigma} + \sum_{\mathbf{k}} \Delta_{\mathbf{k}} \left(c_{\mathbf{k}\uparrow}^\dagger c_{-\mathbf{k}\downarrow}^\dagger + c_{-\mathbf{k}\downarrow} c_{\mathbf{k}\uparrow} \right). \tag{10.82}$$

For simplicity, we assume that the gap function is invariant under spatial inversion, i.e. $\Delta_{-\mathbf{k}} = \Delta_{\mathbf{k}}$. From the Fourier transformations of fermion operators

$$c_{r\sigma} = \frac{1}{\sqrt{V}} \sum_{\mathbf{k}} e^{i\mathbf{k}\cdot\mathbf{r}} c_{\mathbf{k}\sigma}, \qquad c_{\mathbf{k}\sigma} = \frac{1}{\sqrt{V}} \int d\mathbf{r} e^{-i\mathbf{k}\cdot\mathbf{r}} c_{r\sigma}, \tag{10.83}$$

it can be shown that the energy density corresponding to H is

$$h(\mathbf{r}) = \frac{1}{V} \sum_{\mathbf{k}\mathbf{k}'} e^{-i(\mathbf{k}-\mathbf{k}')\cdot\mathbf{r}} \left(\sum_\sigma \frac{\xi_{\mathbf{k}} + \xi_{\mathbf{k}'}}{2} c_{\mathbf{k}\sigma}^\dagger c_{\mathbf{k}'\sigma} + \frac{\Delta_{\mathbf{k}} + \Delta_{\mathbf{k}'}}{2} c_{-\mathbf{k}'\uparrow}^\dagger c_{\mathbf{k}\downarrow}^\dagger \right.$$
$$\left. + \frac{\Delta_{\mathbf{k}} + \Delta_{\mathbf{k}'}}{2} c_{-\mathbf{k}\downarrow} c_{\mathbf{k}'\uparrow} \right). \tag{10.84}$$

$h(\mathbf{r}) = h^{\dagger}(\mathbf{r})$ is Hermitian. In momentum space, the corresponding energy density operator is

$$h(\mathbf{q}) = \sum_{\mathbf{k}} \left(\sum_{\sigma} \frac{\xi_{\mathbf{k}} + \xi_{\mathbf{k}+\mathbf{q}}}{2} c_{\mathbf{k}\sigma}^{\dagger} c_{\mathbf{k}+\mathbf{q}\sigma} + \frac{\Delta_{-\mathbf{k}} + \Delta_{-\mathbf{k}-\mathbf{q}}}{2} c_{-\mathbf{k}-\mathbf{q}\uparrow}^{\dagger} c_{\mathbf{k}\downarrow}^{\dagger} \right.$$

$$\left. + \frac{\Delta_{\mathbf{k}} + \Delta_{\mathbf{k}+\mathbf{q}}}{2} c_{-\mathbf{k}\downarrow} c_{\mathbf{k}+\mathbf{q}\uparrow} \right). \tag{10.85}$$

From the time derivative of $h(\mathbf{q})$ and the equations of motion of electrons

$$i\partial_t c_{\mathbf{k}\sigma} = [c_{\mathbf{k}\sigma}, H] = \xi_{\mathbf{k}} c_{\mathbf{k}\sigma} + \delta_{\sigma\uparrow} \Delta_{\mathbf{k}} c_{-\mathbf{k}\downarrow}^{\dagger} - \delta_{\sigma\downarrow} \Delta_{-\mathbf{k}} c_{-\mathbf{k}\uparrow}^{\dagger}, \tag{10.86}$$

$$i\partial_t c_{\mathbf{k}\sigma}^{\dagger} = \left[c_{\mathbf{k}\sigma}^{\dagger}, H \right] = -\xi_{\mathbf{k}} c_{\mathbf{k}\sigma}^{\dagger} - \delta_{\sigma\uparrow} \Delta_{\mathbf{k}} c_{-\mathbf{k}\downarrow} + \delta_{\sigma\downarrow} \Delta_{-\mathbf{k}} c_{-\mathbf{k}\uparrow}, \tag{10.87}$$

we find that

$$\partial_t h(\mathbf{q}) = \sum_{\mathbf{k}\sigma} \frac{\xi_{\mathbf{k}+\mathbf{q}} - \xi_{\mathbf{k}}}{2} \left(c_{\mathbf{k}\sigma}^{\dagger} \dot{c}_{\mathbf{k}+\mathbf{q}\sigma} - \dot{c}_{\mathbf{k}\sigma}^{\dagger} c_{\mathbf{k}+\mathbf{q}\sigma} \right) + \sum_{\mathbf{k}} \frac{\Delta_{\mathbf{k}+\mathbf{q}} - \Delta_{\mathbf{k}}}{2} \left(\dot{c}_{-\mathbf{k}-\mathbf{q}\uparrow}^{\dagger} c_{\mathbf{k}\downarrow}^{\dagger} \right.$$

$$\left. - c_{-\mathbf{k}-\mathbf{q}\uparrow}^{\dagger} \dot{c}_{\mathbf{k}\downarrow}^{\dagger} + c_{-\mathbf{k}\downarrow} \dot{c}_{\mathbf{k}+\mathbf{q}\uparrow} - \dot{c}_{-\mathbf{k}\downarrow} c_{\mathbf{k}+\mathbf{q}\uparrow} \right). \tag{10.88}$$

In the long wavelength limit $q \to 0$, we have

$$\xi_{\mathbf{k}+\mathbf{q}} - \xi_{\mathbf{k}} = \mathbf{q} \cdot \partial_{\mathbf{k}} \xi_{\mathbf{k}}, \tag{10.89}$$

$$\Delta_{\mathbf{k}+\mathbf{q}} - \Delta_{\mathbf{k}} = \mathbf{q} \cdot \partial_{\mathbf{k}} \Delta_{\mathbf{k}}. \tag{10.90}$$

Combining this with the continuity equation, we find the heat current density operator in the long wavelength limit to be

$$\mathbf{j}_E(\mathbf{q}) = \frac{i}{2} \sum_{\mathbf{k}\sigma} \partial_{\mathbf{k}} \xi_{\mathbf{k}} \left(c_{\mathbf{k}\sigma}^{\dagger} \dot{c}_{\mathbf{k}+\mathbf{q}\sigma} - \dot{c}_{\mathbf{k}\sigma}^{\dagger} c_{\mathbf{k}+\mathbf{q}\sigma} \right) + \frac{i}{2} \sum_{\mathbf{k}} \partial_{\mathbf{k}} \Delta_{\mathbf{k}} \left(\dot{c}_{-\mathbf{k}-\mathbf{q}\uparrow}^{\dagger} c_{\mathbf{k}\downarrow}^{\dagger} \right.$$

$$\left. - c_{-\mathbf{k}-\mathbf{q}\uparrow}^{\dagger} \dot{c}_{\mathbf{k}\downarrow}^{\dagger} + c_{-\mathbf{k}\downarrow} \dot{c}_{\mathbf{k}+\mathbf{q}\uparrow} - \dot{c}_{-\mathbf{k}\downarrow} c_{\mathbf{k}+\mathbf{q}\uparrow} \right). \tag{10.91}$$

Taking the Fourier transformation with respect to time t,

$$c_{\mathbf{k}\sigma}(t) = \frac{1}{\sqrt{2\pi}} \int d\omega e^{-i\omega t} c_{\mathbf{k}\sigma}(\omega), \tag{10.92}$$

$$c_{\mathbf{k}\sigma}^{\dagger}(t) = \frac{1}{\sqrt{2\pi}} \int d\omega e^{i\omega t} c_{\mathbf{k}\sigma}^{\dagger}(\omega), \tag{10.93}$$

we find that the heat current density operator in the frequency space

$$\mathbf{j}_E(\mathbf{q}, \Omega) = \int dt e^{i\Omega t} \mathbf{j}_E(\mathbf{q}, t) \tag{10.94}$$

is given by

$$
\mathbf{j}_E(\mathbf{q}, \Omega) = \sum_{\mathbf{k}} \int d\omega \left(\omega + \frac{\Omega}{2}\right) \left\{ \partial_{\mathbf{k}}\xi_{\mathbf{k}} \sum_{\sigma} c^\dagger_{\mathbf{k}\sigma}(\omega) c_{\mathbf{k}+\mathbf{q}\sigma}(\omega + \Omega) \right.
$$
$$
- \partial_{\mathbf{k}}\Delta_{\mathbf{k}} \left[c^\dagger_{-\mathbf{k}-\mathbf{q}\uparrow}(\omega) c^\dagger_{\mathbf{k}\downarrow}(-\omega - \Omega) \right.
$$
$$
\left. \left. + c_{-\mathbf{k}\downarrow}(\omega + \Omega) c_{\mathbf{k}+\mathbf{q}\uparrow}(-\omega) \right] \right\}. \tag{10.95}
$$

This result can also be obtained directly from the BCS Hamiltonian without taking the mean-field approximation. The mean-field approximation is imposed only after the expression for the time-derivative of the heat current density operator is obtained. These two kinds of approaches are equivalent. The difference lies in that the first approach the mean-field approximation is applied to the Hamiltonian, while in the second approach this approximation is applied to the heat current operator.

10.7 Universal Thermal Conductivity

According to the linear response theory, the thermal conductivity in the absence of electric current is determined by the equation [6]

$$
\kappa = \frac{1}{k_B \hbar T^2} \mathrm{Re}\Pi_E(\omega \to 0), \tag{10.96}
$$

where Π_E is the thermal polarization function which is determined by the correlation of heat current density operators

$$
\Pi_E(i\omega_n) = \frac{i}{i\omega_n \beta} \int_0^\beta d\tau e^{i\omega_n \tau} \langle T_\tau j_E^\mu(\tau) j_E^\mu(0) \rangle, \tag{10.97}
$$

where $\mathbf{j}_E = \mathbf{j}_E(q = 0)$. In the Matsubara frequency representation, the heat current density operator (10.91) becomes

$$
\mathbf{j}_E(\tau) = \frac{1}{2\beta^2} \sum_{\mathbf{k}\omega'_m \omega_m} (\omega_m + \omega'_m) e^{i(\omega_m - \omega'_m)\tau} \left\{ \sum_{\sigma} (\partial_{\mathbf{k}}\xi_{\mathbf{k}}) c^\dagger_{\mathbf{k}\sigma}(i\omega_m) c_{\mathbf{k}\sigma}(i\omega'_m) \right.
$$
$$
\left. + \partial_{\mathbf{k}}\Delta_{\mathbf{k}} \left[c^\dagger_{\mathbf{k}\uparrow}(i\omega_m) c^\dagger_{-\mathbf{k}\downarrow}(-i\omega'_m) - c_{-\mathbf{k}\downarrow}(-i\omega'_m) c_{\mathbf{k}\uparrow}(i\omega_m) \right] \right\}. \tag{10.98}
$$

Substituting it into Eq. (10.97), we obtain

$$
\Pi_E(i\omega_n) = -\frac{i}{4i\omega_n \beta^2} \sum_{\mathbf{k}\omega_m} (2\omega_m + \omega_n)^2
$$
$$
\left[(\nabla_{\mathbf{k}}\xi_{\mathbf{k}})^2 \mathrm{Tr} G(\mathbf{k}, i\omega_m) \sigma_3 G(\mathbf{k}, i\omega_m + i\omega_n) \sigma_3 \right.
$$
$$
\left. + (\nabla_{\mathbf{k}}\Delta_{\mathbf{k}})^2 \mathrm{Tr} G(\mathbf{k}, i\omega_m) \sigma_1 G(\mathbf{k}, i\omega_m + i\omega_n) \sigma_1 \right]. \tag{10.99}
$$

Using the spectral representation of the Green's function

$$G(\mathbf{k}, i\omega_m) = -\int_{-\infty}^{\infty} \frac{d\omega}{\pi} \frac{\text{Im}G^R(\mathbf{k}, \omega)}{i\omega_m - \omega}, \tag{10.100}$$

we can rewrite Eq. (10.99) as

$$\Pi_E(i\omega_n) = \frac{i}{i\omega_n 4\pi^2 \beta} \sum_{\mathbf{k}} \int_{-\infty}^{\infty} d\omega d\omega' \frac{(2\omega + i\omega_n)^2 f(\omega) - (2\omega' - i\omega_n)^2 f(\omega')}{\omega + i\omega_n - \omega'}$$

$$\left[(\nabla_{\mathbf{k}} \xi_{\mathbf{k}})^2 \, Tr\text{Im}G^R(\mathbf{k}, \omega) \, \sigma_3 \text{Im}G^R(\mathbf{k}, \omega') \, \sigma_3 \right.$$
$$\left. + (\nabla_{\mathbf{k}} \Delta_{\mathbf{k}})^2 \, Tr\text{Im}G^R(\mathbf{k}, \omega) \, \sigma_1 \text{Im}G^R(\mathbf{k}, \omega') \, \sigma_1 \right]. \tag{10.101}$$

By performing the analytic continuation $i\omega_n \to \omega + i\delta$ and taking the limit $\omega \to 0$, this expression can be simplified. The real part is

$$\text{Re}\Pi_E(\omega \to 0)$$
$$= -\frac{1}{\pi\beta} \sum_{\mathbf{k}} \int_{-\infty}^{\infty} d\omega \omega^2 \frac{\partial f(\omega)}{\partial \omega} \left[(\nabla_{\mathbf{k}} \xi_{\mathbf{k}})^2 \, Tr\text{Im}G^R(\mathbf{k}, \omega) \, \sigma_3 \text{Im}G^R(\mathbf{k}, \omega) \, \sigma_3 \right.$$
$$\left. + (\nabla_{\mathbf{k}} \Delta_{\mathbf{k}})^2 \, Tr\text{Im}G^R(\mathbf{k}, \omega) \, \sigma_1 \text{Im}G^R(\mathbf{k}, \omega) \, \sigma_1 \right]. \tag{10.102}$$

In the limit of zero temperature, $-\partial f(\omega)/\partial\omega \to \delta(\omega)$, one can set the frequency ω in the Green's function in Eq. (10.102) to zero. Furthermore, using the integral formula

$$\int_{-\infty}^{\infty} d\omega \omega^2 \frac{\partial f(\omega)}{\partial \omega} = -\frac{\pi^2}{3\beta^2}, \tag{10.103}$$

$\text{Re}\Pi_E$ can be simplified as

$$\text{Re}\Pi_E(\omega \to 0) = \frac{\pi}{3\beta^3} \sum_{\mathbf{k}} \left[(\nabla_{\mathbf{k}} \xi_{\mathbf{k}})^2 \, Tr\text{Im}G^R(\mathbf{k}, 0) \, \sigma_3 \text{Im}G^R(\mathbf{k}, 0) \, \sigma_3 \right.$$
$$\left. + (\nabla_{\mathbf{k}} \Delta_{\mathbf{k}})^2 \, Tr\text{Im}G^R(\mathbf{k}, 0) \, \sigma_1 \text{Im}G^R(\mathbf{k}, 0) \, \sigma_1 \right]. \tag{10.104}$$

The zero frequency retarded Green's function is given by

$$G^R(\mathbf{k}, 0) = \frac{1}{-\xi_{\mathbf{k}}\sigma_3 - \Delta_{\mathbf{k}}\sigma_1 + i\Gamma}, \tag{10.105}$$

where $\Gamma = \Gamma(\omega = 0)$ is the quasiparticle scattering rate. Substituting the above result into Eq. (10.104), we obtain

$$\text{Re}\Pi_E(\omega \to 0) = \frac{\pi}{3\beta^3} \sum_{\mathbf{k}} \frac{\Gamma^2}{(\Gamma^2 + E_{\mathbf{k}}^2)^2} \left[(\nabla_{\mathbf{k}} \xi_{\mathbf{k}})^2 + (\nabla_{\mathbf{k}} \Delta_{\mathbf{k}})^2 \right]. \tag{10.106}$$

When Γ is small, the momentum summation in Eq. (10.106) contributes mainly from the nodal region. In this case,

$$\sum_{\mu} (\nabla_{\mathbf{k}} \xi_{\mathbf{k}})^2 \approx \hbar^2 v_F^2, \tag{10.107}$$

$$\sum_{\mu} (\nabla_{\mathbf{k}} \Delta_{\mathbf{k}})^2 \approx \hbar^2 v_2^2. \tag{10.108}$$

The momentum summation can be performed independently around the four gap nodes. The gradients of $\xi_{\mathbf{k}}$ and $\Delta_{\mathbf{k}}$ are perpendicular to each other at each nodal point, which gives

$$\frac{1}{V} \sum_{\mathbf{k}} \approx 4 \int \frac{dk_1 dk_2}{4\pi^2} = \frac{1}{\pi^2} \int \frac{d\xi d\Delta}{\hbar^2 v_F v_2}, \tag{10.109}$$

where k_1 and k_2 are the wave vectors along the tangential direction of $\xi_{\mathbf{k}}$ and $\Delta_{\mathbf{k}}$, respectively. Thus $\mathrm{Re}\Pi_E(\omega \to 0)$ can be further expressed as

$$\mathrm{Re}\Pi_E(\omega \to 0) \approx \frac{1}{3\pi\beta^3} \frac{v_F^2 + v_2^2}{v_F v_2} \int d\xi d\Delta \frac{\Gamma^2}{\left(\Gamma^2 + \xi^2 + \Delta^2\right)^2}. \tag{10.110}$$

Generally speaking, the integral in the above equation is a function of Γ. On the other hand, this integral is dimensionless based on the dimensional analysis. Thus we expect that this integral is Γ-independent. An explicit calculation confirms this expectation

$$\int d\xi d\Delta \frac{\Gamma^2}{\left(\Gamma^2 + \xi^2 + \Delta^2\right)^2} = 2\pi \int_0^\infty E dE \frac{\Gamma^2}{\left(\Gamma^2 + E^2\right)^2} = \pi. \tag{10.111}$$

Substituting this result into Eq. (10.96), we finally obtain the universal formula for the thermal conductivity in the limit of zero temperature

$$\frac{\kappa}{T} \approx \frac{k_B^2}{3\hbar} \frac{v_F^2 + v_2^2}{v_F v_2}. \tag{10.112}$$

It shows that κ only depends on two fundamental parameters v_F and v_2, but not on the quasiparticle scattering rate Γ. This result was first derived by Durst and P. A. Lee [271]. They also showed that this result is robust against the vertex corrections from the impurity scattering as well as the Coulomb interaction to the leading order approximation. It suggests that the thermal conductivity is more appropriate than the electric conductivity to probe the universal transport properties of d-wave superconducting quasiparticles.

Figure 10.6 shows the thermal conductivity coefficient κ_0/T in the zero temperature limit as a function of the quasiparticle scattering rate Γ_ρ for YBa$_2$Cu$_3$O$_{6.9}$ without or with partial substitution of copper atoms by zinc atoms [272]. Γ_ρ is

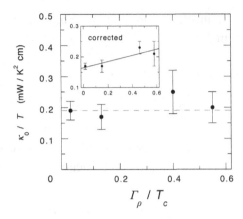

Figure 10.6 The thermal conductivity coefficient κ_0/T versus the scattering rate Γ_ρ/T_c for the YBa$_2$(Cu$_{1-x}$Zn$_x$)$_3$O$_{6.9}$ superconductor in the limit of zero temperature (from Ref.[272]). The zinc concentrations of the four samples are $x = 0$, 0.006, 0.02, 0.03. The larger x, the larger Γ_ρ is. The inset shows the results after considering the correction from the geometric effect.

determined from the measurement data of the residual resistance. The higher the zinc impurity concentration, the larger Γ_ρ is. The 3% concentration of zinc impurity suppresses T_c by 20%. Within experimental errors, the coefficient of thermal conductivity κ/T is almost a constant, independent on Γ_ρ. This indicates that, as expected, the thermal conductivity is universal in the limit of zero temperature. The similar universal behavior was observed in Bi$_2$Sr$_2$CaCu$_2$O$_8$ superconductors [273]. However, it should be pointed out that the four samples shown in Fig. 10.6 have different superconducting transition temperatures. Δ_0/v_F and v_2/v_F are also changed with the change of the zinc concentration. Thus the universality does not imply that κ/T is completely independent of doping.

The universal formula given by Eq. (10.112) indicates that v_2/v_F can be determined through the measurement of thermal conductivity. Then using the value of v_F determined by the ARPES experiments, we can further determine the value of v_2. Based on this idea, Sutherland et al. [274] measured the doping dependence of the thermal conductivities for YBCO and LSCO superconductors. They found that κ/T decreases with decreasing doping level in the limit of zero temperature, but remains finite even at very low doping. This shows unambiguously that the gap nodes exist at all levels of doping. From the measurement data, they calculated the doping dependence of Δ_0, and found that Δ_0 is not proportional to T_c, unlike what would be expected from the BCS theory. Instead, they found that Δ_0 increases as the doping level or T_c is decreased. The low temperature κ/T measures the quasiparticle excitations around the gap nodes, and the value of Δ_0 obtained from the universal relation Eq. (10.112) is actually the amplitude or the slope of the gap function around the gap nodes. Furthermore, they found that the doping dependence of Δ_0 agrees qualitatively with that of the maximal pseudogap obtained from the ARPES or other experimental measurements. It suggests that the superconducting gap is inherently connected to the pseudogap.

11

Raman Spectroscopy

11.1 Raman Response Function

Raman scattering occurs when electromagnetic radiation or light, usually from a laser, impinges on a crystal and interacts with electrons, phonons, or other elementary excitations. Unlike the light absorption discussed in the preceding chapter, which is dominantly a single photon process, the Raman scattering is usually a two-photon process. The Raman spectroscopy measures the cross section of inelastic scattering of light by quasiparticles. The spectrum of the scattered light, termed the Raman spectrum, quantifies the intensity of the scattered light as a function of its frequency difference to the incident light. Through the analysis of Raman spectroscopy, one can acquire useful information on the electronic structures, the lattice, and magnetic excitations of solids. Raman spectroscopy has served as a powerful experimental tool in the study of condensed matter physics. In the superconducting phase, the superconducting pairing strongly affects the photoelectric interaction. Through the selection rules of the Raman process, Raman spectroscopy can be used to explore the pairing symmetry and the underlying mechanism. It has played an important role in the study of both conventional and high-T_c superconductors.

The interaction between photons and electrons is described by the Hamiltonian

$$H_I = H_1 + H_2, \tag{11.1}$$

where

$$H_1 = -\frac{e}{\hbar} \sum_{\mathbf{q}} J_{\alpha,\mathbf{q}} A_{\alpha,-\mathbf{q}} \tag{11.2}$$

$$H_2 = \frac{e^2}{2\hbar^2} \sum_{\mathbf{q}_1 \mathbf{q}_2} \tau^{\alpha\beta}_{\mathbf{q}_1+\mathbf{q}_2} A_{\alpha,-\mathbf{q}_1} A_{\beta,-\mathbf{q}_2}, \tag{11.3}$$

and $J_{\alpha,\mathbf{q}}$ is the current operator of electrons

$$J_{\alpha,\mathbf{q}} = \sum_{\mathbf{k}} \frac{\partial \varepsilon_{\mathbf{k}}}{\partial k_\alpha} c^\dagger_{\mathbf{k}+\frac{\mathbf{q}}{2},\sigma} c_{\mathbf{k}-\frac{\mathbf{q}}{2},\sigma}, \tag{11.4}$$

and $\tau^{\alpha\beta}_{\mathbf{q}_1+\mathbf{q}_2}$ is a second order tensor defined by

$$\tau^{\alpha\beta}_{\mathbf{q}} = \sum_{\mathbf{k}} \frac{\partial^2 \varepsilon_{\mathbf{k}}}{\partial k_\alpha \partial k_\beta} c^\dagger_{\mathbf{k}+\frac{\mathbf{q}}{2},\sigma} c_{\mathbf{k}-\frac{\mathbf{q}}{2},\sigma}. \tag{11.5}$$

In the above equations, summation is implied over the repeated greek indices. If $\varepsilon_{\mathbf{k}} = \mathbf{k}^2/2m$ is Galilean invariant, then $\tau^{\alpha\beta}_{\mathbf{q}}$ is proportional to the single-particle density matrix. In order to study microscopically the light scattering process, we need to consider a quantized electromagnetic field $A_{\alpha,\mathbf{q}}$. We define $a_{\mathbf{q},e_\alpha}$ and $a^\dagger_{\mathbf{q},e_\alpha}$ as the creation and annihilation operators of a photon polarized along the direction e_α, then it can be shown by utilizing the Maxwell equations that

$$A_{\alpha,\mathbf{q}} = g_{\mathbf{q}} e_\alpha \left(a_{\mathbf{q},e_\alpha} + a^\dagger_{-\mathbf{q},e_\alpha} \right), \tag{11.6}$$

where

$$g_{\mathbf{q}} = \sqrt{\frac{hc^2}{\omega_{\mathbf{q}} V}}, \tag{11.7}$$

$\omega_{\mathbf{q}} = c|\mathbf{q}|$ is the photon energy, and V is the volume of the system.

Raman scattering involves the incident and scattered photons, and the scattered electrons. If we denote the energy, momentum, and polarization of the incident photon as $(\omega_i, \mathbf{k}_i, e^I)$ and the initial electron state as $|i\rangle$, then the initial wavefunction of the system can be represented as

$$|I\rangle = a^\dagger_{\mathbf{k}_i, e^I} |i\rangle. \tag{11.8}$$

Similarly, the final state of the system can be expressed as

$$|F\rangle = a^\dagger_{\mathbf{k}_f, e^S} |f\rangle, \tag{11.9}$$

where $(\omega_f, \mathbf{k}_f, e^S)$ are the energy, momentum, and polarization of the scattered photon, respectively. $|f\rangle$ is the wavefunction of the scattered electron.

The Raman scattering is a two-photon process, one photon in the initial state and another in the final state. Thus Raman scattering measures the second order response of electromagnetic field $A_{\mathbf{k}}$. It can be generated by the first order perturbation of H_2 and the second order perturbation of H_1. The latter involves an intermediate process during which a virtual photon is emitted or absorbed. When the energy of the emitted or absorbed virtual photon matches the energy difference between the initial and the intermediate states of electrons, a resonant transition happens. Thus this process is essentially resonant. By contrast, the first order perturbation process of H_2 is non-resonant. It does not need a virtual intermediate state. As long as the momentum and energy are conserved, there are no special requirements for the initial and final electron states.

An important quantity in describing the Raman scattering is the transition probability $R_{I,F}$ from the initial state $|I\rangle$ to the final state $|F\rangle$. According to the Fermi golden rule, this transition probability reads

$$R_{I,F} = \frac{2\pi}{\hbar} |\langle F|M|I\rangle|^2 \delta(E_I - E_F), \qquad (11.10)$$

where M is the effective scattering operator. We use $\langle F|M_N|I\rangle$ and $\langle F|M_R|I\rangle$ to represent the contributions from the nonresonant and resonant scattering processes, respectively. The total scattering amplitude $\langle F|M|I\rangle$ is simply a sum of these two terms

$$\langle F|M|I\rangle = \langle F|M_N|I\rangle + \langle F|M_R|I\rangle. \qquad (11.11)$$

According to the perturbation theory, it can be shown that the nonresonant transition amplitude is given by

$$\langle F|M_N|I\rangle = \langle F|H_2|I\rangle = \frac{e^2 g_{\mathbf{k}_i} g_{\mathbf{k}_f} e_\alpha^I e_\beta^S}{\hbar^2} \langle f|\tau_{\mathbf{q}}^{\alpha\beta}|i\rangle. \qquad (11.12)$$

Similarly, it can be shown that the resonant transition amplitude is given by

$$
\begin{aligned}
\langle F|M_R|I\rangle &= -\sum_J \langle F|H_1|J\rangle \frac{1}{E_J - E_I} \langle J|H_1|I\rangle \\
&= -\frac{e^2 g_{\mathbf{k}_i} g_{\mathbf{k}_f} e_\alpha^I e_\beta^S}{\hbar^2} \sum_l \left[\frac{\langle f|J_{\beta,\mathbf{k}_f}|l\rangle\langle l|J_{\alpha,-\mathbf{k}_i}|i\rangle}{\varepsilon_l - \varepsilon_i - \omega_i} \right. \\
&\qquad\qquad \left. + \frac{\langle f|J_{\alpha,-\mathbf{k}_i}|l\rangle\langle l|J_{\beta,\mathbf{k}_f}|i\rangle}{\varepsilon_l - \varepsilon_i + \omega_f} \right]. \qquad (11.13)
\end{aligned}
$$

The first term describes the process where the electron first absorbs a photon at frequency ω_i, and then emits a photon at frequency ω_f. The second term corresponds to where the electron first emits a photon at frequency ω_f, and then absorbs a photon at frequency ω_i.

Substituting the above results into Eq. (11.10), and considering the thermal equilibrium distribution for the initial electron states, the scattering probability is found to be

$$\tilde{R}(\mathbf{q}, \omega) = \frac{2\pi e^4}{\hbar^5} \sum_{i,f} \frac{e^{-\beta\xi_i}}{Z} \left| g_{\mathbf{k}_i} g_{\mathbf{k}_f} e_\alpha^I e_\beta^S \langle f|T_{\mathbf{q}}^{\alpha\beta}|i\rangle \right|^2 \delta(\varepsilon_f - \varepsilon_i - \omega), \qquad (11.14)$$

where $\xi_i = \varepsilon_i - \varepsilon_F$, $\mathbf{q} = \mathbf{k}_i - \mathbf{k}_f$, and $\omega = \omega_i - \omega_f$. Z is the partition function, and

$$\langle f|T_{\mathbf{q}}^{\alpha\beta}|i\rangle = \sum_l \left[\frac{\langle f|J_{\beta,\mathbf{k}_f}|l\rangle\langle l|J_{\alpha,-\mathbf{k}_i}|i\rangle}{\omega_i - \varepsilon_l + \varepsilon_i} - \frac{\langle f|J_{\alpha,-\mathbf{k}_i}|l\rangle\langle l|J_{\beta,\mathbf{k}_f}|i\rangle}{\omega_f + \varepsilon_l - \varepsilon_i} \right].$$
$$+ \langle f|\tau_{\mathbf{q}}^{\alpha\beta}|i\rangle \qquad (11.15)$$

Using the fluctuation-dissipation theorem, Eq. (G.16),

$$\sum_{ij} \frac{e^{-\beta\xi_i}}{Z} \left|\langle j|\rho_\gamma(\mathbf{q})|i\rangle\right|^2 \delta\left(\varepsilon_f - \varepsilon_i - \omega\right) = -\frac{1}{\pi\left(1 - e^{-\beta\omega}\right)} \mathrm{Im}R\left(\mathbf{q},\omega\right), \quad (11.16)$$

we can reexpress Eq. (11.14) as

$$\tilde{R}(\mathbf{q},\omega) = -\frac{2e^4 g_{\mathbf{k}_i}^2 g_{\mathbf{k}_f}^2}{\hbar^5 \left(1 - e^{-\beta\omega}\right)} \mathrm{Im}R\left(\mathbf{q},\omega\right). \qquad (11.17)$$

$R(q,\omega)$ is the Fourier transform of the retarded Raman response function defined by

$$R(\mathbf{q},t) = -i\theta(t)\langle[\rho_\gamma(\mathbf{q},t),\rho_\gamma(-\mathbf{q},0)]\rangle, \qquad (11.18)$$

$\rho_\gamma(\mathbf{q})$ is the Raman density operator containing both the resonant and non-resonant parts. After neglecting the contribution of surface reflection, it can be shown that $\tilde{R}(\mathbf{q},\omega)$ is proportional to the differential cross section of the Raman scattering

$$\frac{\partial^2\sigma}{\partial\Omega\partial\omega} \propto \tilde{R}(\mathbf{q},\omega). \qquad (11.19)$$

Thus what is measured by the Raman scattering is just the imaginary part of the Raman response function $R(\mathbf{q},\omega)$.

11.2 Vertex Function

In solids, because the velocity of light is much larger than the Fermi velocity of electrons, the energy transfer ω of photons after scattering could be large, but the corresponding momentum transfer $\mathbf{q}$ is small. According to the uncertainty principle, the momentum transfer $\mathbf{q}$ is roughly of the order of the inverse magnetic penetration depth, which can be approximately taken as zero. At low temperatures, when the momentum transfer is very small and the frequency of the incident light is much smaller than the band gap, there is no resonant scattering. In this case, only the nonresonant scattering needs to be considered, and the effective mass approximation can be used for $\rho_\gamma(\mathbf{q})$ [275]

$$\rho_\gamma(\mathbf{q}) = \sum_{\mathbf{k}} \gamma_{\mathbf{k}} c_{\mathbf{k}+\mathbf{q}/2}^\dagger c_{\mathbf{k}-\mathbf{q}/2}, \qquad (11.20)$$

with

$$\gamma_{\mathbf{k}} = \sum_{\alpha\beta} e^S_\beta \frac{\partial^2 \varepsilon_{\mathbf{k}}}{\partial k_\alpha \partial k_\beta} e^I_\alpha. \tag{11.21}$$

If $\gamma = 1$, ρ_γ is just the electron density operator defined by

$$\rho_1(\mathbf{q}) = \sum_{\mathbf{k}} c^\dagger_{\mathbf{k}+\mathbf{q}/2} c_{\mathbf{k}-\mathbf{q}/2}. \tag{11.22}$$

The intensity of the Raman scattering depends on the polarization of the incident light e^I and that of the scattered light e^S. By adjusting the polarization directions of the incident and scattered lights, Raman spectroscopy can measure the scattering of quasiparticles on different parts of the Fermi surface. This is useful to the analysis of the momentum dependence of the gap function over the Fermi surface. For high-T_c superconductors, the polarization vectors of the incident and scattered lights are usually chosen along the directions of (100), (010), (110), and ($1\bar{1}0$). The values of the Raman response functions for different symmetry modes can be measured and derived based on different combinations of these polarization vectors.

In the study of high-T_c superconductivity, three symmetry modes, A_{1g}, B_{1g}, and B_{2g}, are usually measured. These symbols are representations of the D_{4h} group of the crystalline symmetries. Roughly speaking, A_{1g} carries the s-wave symmetry, B_{1g} carries the $d_{x^2-y^2}$ symmetry, and B_{2g} carries the d_{xy} symmetry. The subindex g represents the even parity. Table 11.1 shows the relation between these symmetry modes and the polarization vectors of incident and scattered lights. Under the effective mass approximation, the Raman vertex functions for these three symmetry modes are

$$\gamma_{\mathbf{k}} = \begin{cases} \dfrac{1}{2}\left(\dfrac{\partial^2 \varepsilon_{\mathbf{k}}}{\partial k_x^2} + \dfrac{\partial^2 \varepsilon_{\mathbf{k}}}{\partial k_y^2} \right), & A_{1g} \\[3mm] \dfrac{1}{2}\left(\dfrac{\partial^2 \varepsilon_{\mathbf{k}}}{\partial k_x^2} - \dfrac{\partial^2 \varepsilon_{\mathbf{k}}}{\partial k_y^2} \right), & B_{1g} \\[3mm] \dfrac{\partial^2 \varepsilon_{\mathbf{k}}}{\partial k_x \partial k_y}. & B_{2g} \end{cases} \tag{11.23}$$

The vertex function $\gamma_{\mathbf{k}}$ depends on the band structure of electrons. Near the Fermi surface, $\gamma_{\mathbf{k}}$ can be expanded using the harmonic modes of the crystal [276]. To the leading order approximation, $\gamma_{\mathbf{k}}$ can be expressed as

$$\gamma_{\mathbf{k}} = \begin{cases} \gamma_0 + \gamma(A_{1g})\cos(4\phi), & A_{1g} \\ \gamma(B_{1g})\cos(2\phi), & B_{1g} \\ \gamma(B_{2g})\sin(2\phi), & B_{2g} \end{cases} \tag{11.24}$$

where $\phi = \arctan(k_y/k_x)$. These expressions for $\gamma_{\mathbf{k}}$ are more convenient to use in the analytical calculation.

Table 11.1. *The polarization directions of incident and scattered lights and the corresponding symmetry modes commonly used in the Raman scattering experiments.*

	e^I	e^S
$A_{1g} + B_{2g}$	$\frac{1}{\sqrt{2}}(\hat{x} + \hat{y})$	$\frac{1}{\sqrt{2}}(\hat{x} + \hat{y})$
$A_{1g} + B_{1g}$	$\hat{y}$	$\hat{y}$
B_{1g}	$\frac{1}{\sqrt{2}}(\hat{x} + \hat{y})$	$\frac{1}{\sqrt{2}}(\hat{x} - \hat{y})$
B_{2g}	$\hat{x}$	$\hat{y}$

From Eq. (11.24), it is clear that the A_{1g}-mode measures the average of the scattering cross section over the whole Fermi surface. The B_{1g}-mode has the same symmetry as the $d_{x^2-y^2}$-wave superconductor. It measures the Raman spectrum contributed by the quasiparticles along the antinodal directions. On the other hand, the vertex function of the B_{2g}-mode has the largest absolute value along the nodal directions. It measures mainly the contribution of quasiparticles around the gap nodes. Therefore, through the measurement of these Raman modes, especially the B_{1g}- and B_{2g}-modes, we can extract useful information on the gap function.

11.3 Vertex Correction by the Coulomb Interaction

Light radiation induces a charge fluctuation in a metal through photoelectric interaction. There is a strong screen effect resulting from the long-range Coulomb interaction between the fluctuating charges. It changes the intensity of the Raman scattering and modifies the vertex function of the Raman modes. In the long wavelength limit, the Coulomb interaction only couples to the high symmetry A_{1g}-mode and changes the corresponding Raman spectrum. But it does not affect the Raman spectra for the B_{1g}, B_{2g}, and other low symmetry modes.

The Coulomb interaction of fluctuating charges is described by the Hamiltonian

$$H_c = \frac{1}{2} \sum_q V_q \rho_1(\mathbf{q}) \rho_1(-\mathbf{q}), \qquad V_q = \frac{e^2}{\varepsilon_0 q^2}. \tag{11.25}$$

After considering the correction of this Coulomb interaction to the vertex function, whose corresponding Feymann diagram is shown in Fig. 11.1, the Raman response function becomes [277, 278]

$$R(\mathbf{q}, \omega) = R^0_{\gamma,\gamma}(\mathbf{q}, \omega) + R^0_{\gamma,1}(\mathbf{q}, \omega) \left[V_q + V_q R^0_{1,1}(\mathbf{q}, \omega) V_q + \cdots \right] R^0_{1,\gamma}(\mathbf{q}, \omega)$$

$$= R^0_{\gamma,\gamma}(\mathbf{q}, \omega) + \frac{R^0_{\gamma,1}(\mathbf{q}, \omega) V_q R^0_{1,\gamma}(\mathbf{q}, \omega)}{1 - V_q R^0_{1,1}(\mathbf{q}, \omega)}, \tag{11.26}$$

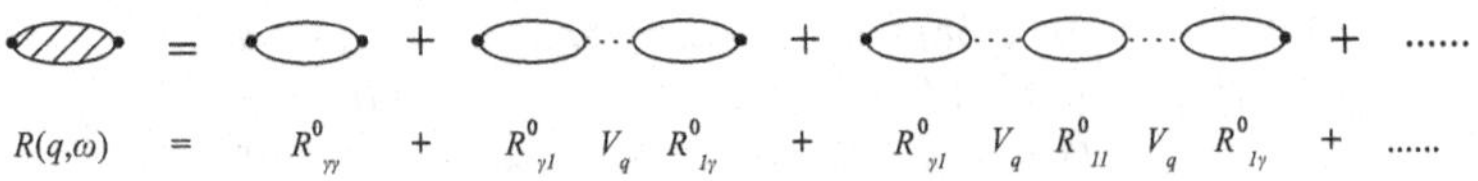

$$R(q,\omega) \quad = \quad R^0_{\gamma\gamma} \quad + \quad R^0_{\gamma l}\, V_q\, R^0_{l\gamma} \quad + \quad R^0_{\gamma l}\, V_q\, R^0_{ll}\, V_q\, R^0_{l\gamma} \quad + \quad \cdots\cdots$$

Figure 11.1 The correction to the Raman response functions due to the Coulomb interaction generated by the charge fluctuation.

where $R^0_{a,b}(\mathbf{q}, \omega)$ is the response function in the absence of the Coulomb interaction. Indices a and b represent the vertex functions. Index 1 represents the A_{1g} Raman mode.

The second term in the above equation can be reorganized to convert the response function into the form

$$R(\mathbf{q},\omega) = R^0_{\gamma,\gamma}(\mathbf{q},\omega) - \frac{R^0_{\gamma,1}(\mathbf{q},\omega)R^0_{1,\gamma}(\mathbf{q},\omega)}{R^0_{1,1}(\mathbf{q},\omega)}$$

$$+ \frac{R^0_{\gamma,1}(\mathbf{q},\omega)R^0_{1,\gamma}(\mathbf{q},\omega)}{R^0_{1,1}(\mathbf{q},\omega)\left[1 - V_q R^0_{1,1}(\mathbf{q},\omega)\right]}. \tag{11.27}$$

The third term on the right-hand side is proportional to the density–density response function of electrons

$$\chi(\mathbf{q},\omega) = \frac{R^0_{1,1}(\mathbf{q},\omega)}{1 - V_q R^0_{1,1}(\mathbf{q},\omega)}. \tag{11.28}$$

It can be shown that $\chi(\mathbf{q},\omega)$ vanishes in the limit of $q \to 0$ [278]. This is a consequence of charge conservation, since $\rho(\mathbf{q})$ is the electron density in the limit $q \to 0$ and is not alterable by an external potential. Thus in the long wavelength limit, the Raman response function is only determined by the first two terms in Eq. (11.27)

$$R(\mathbf{q},\omega) = R^0_{\gamma,\gamma}(\mathbf{q},\omega) - \frac{R^0_{\gamma,1}(\mathbf{q},\omega)R^0_{1,\gamma}(\mathbf{q},\omega)}{R^0_{1,1}(\mathbf{q},\omega)}. \tag{11.29}$$

In Eq. (11.27), $R^0_{\gamma,1}(\mathbf{q}, \omega)$ and $R^0_{1,\gamma}(\mathbf{q}, \omega)$ involve the average of the vertex function γ over the Fermi surface. For the B_{1g}, B_{2g}, or other low symmetry modes, the average of the vertex function on the whole Fermi surface is zero and the correction from the charge fluctuation to the corresponding Raman response function is also zero. Thus for all modes excluding those with the A_{1g} symmetry, only the first term on the right-hand side of Eq. (11.29) is finite.

The vertex correction to the A_{1g}-mode due to the charge fluctuation is finite. This will completely screen the ϕ-independent part in the vertex function of the A_{1g}-mode. Only the $\cos 4\phi$ or higher order harmonic components have a contribution to the A_{1g}-spectra. Furthermore, for the A_{1g}-mode, if we rewrite its vertex function as

$$\gamma_{\mathbf{k}} = \gamma_0 + \delta\gamma_{\mathbf{k}}, \tag{11.30}$$

then it is straightforward to show that the momentum independent term, i.e. the γ_0-term, has no contribution to $R(\mathbf{q}, \omega)$ and

$$R(\mathbf{q}, \omega) = R^0_{\delta\gamma, \delta\gamma}(\mathbf{q}, \omega) - \frac{R^0_{\delta\gamma, 1}(\mathbf{q}, \omega) R^0_{1, \delta\gamma}(\mathbf{q}, \omega)}{R^0_{1, 1}(\mathbf{q}, \omega)}. \tag{11.31}$$

11.4 Raman Response in a Superconducting State

To obtain the Raman response function in a superconducting state, we first evaluate the Matsubara Green's function corresponding to $R^0_{\gamma_1 \gamma_2}(\mathbf{q}, \omega)$

$$R^0_{\gamma_1 \gamma_2}(\mathbf{q}, \tau) = -\langle T_\tau \rho_{\gamma_1}(\mathbf{q}, \tau) \rho_{\gamma_2}(-\mathbf{q}, 0) \rangle, \tag{11.32}$$

and then use Eq. (11.29) to obtain the value of $R_{\gamma_1 \gamma_2}(\mathbf{q}, \omega)$ by performing the analytic continuation for the frequency. Using the definition of $\rho_\gamma(\mathbf{q})$, $R^0_{\gamma_1 \gamma_2}(\mathbf{q}, \tau)$ can be expressed as

$$R^0_{\gamma_1 \gamma_2}(\mathbf{q}, \tau) = \sum_\mathbf{k} \gamma_{1\mathbf{k}} \gamma_{2\mathbf{k}} \mathrm{Tr} G\left(\mathbf{k} + \frac{\mathbf{q}}{2}, -\tau\right) \sigma_3 G\left(\mathbf{k} - \frac{\mathbf{q}}{2}, \tau\right) \sigma_3. \tag{11.33}$$

Taking the Fourier transformation with respect to the imaginary time τ,

$$R^0_{\gamma_1 \gamma_2}(\mathbf{q}, i\omega_n) = \int_0^\beta d\tau \, e^{i\omega_n \tau} R(\mathbf{q}, \tau), \tag{11.34}$$

we obtain

$$R^0_{\gamma_1 \gamma_2}(\mathbf{q}, i\omega_n) = \frac{1}{\beta} \sum_{\mathbf{k}\omega_m} \gamma_{1\mathbf{k}} \gamma_{2\mathbf{k}} \mathrm{Tr} G\left(\mathbf{k} + \frac{\mathbf{q}}{2}, i\omega_m\right) \sigma_3 G\left(\mathbf{k} - \frac{\mathbf{q}}{2}, i\omega_m + i\omega_n\right) \sigma_3. \tag{11.35}$$

Inserting the BCS mean-field expression of the single-particle Green's function into the above equation, we obtain the expression of $R^0_{\gamma_1 \gamma_2}(q, i\omega_n)$ in the weak coupling limit. In the long wavelength limit ($q \to 0$), the result is

$$R^0_{\gamma_1 \gamma_2}(0, i\omega_n) = \sum_\mathbf{k} \frac{4\gamma_{1\mathbf{k}} \gamma_{2\mathbf{k}} \Delta^2_\mathbf{k}}{E_\mathbf{k}} \frac{1}{(i\omega_n)^2 - 4E^2_\mathbf{k}} \tanh \frac{\beta E_\mathbf{k}}{2}. \tag{11.36}$$

The corresponding retarded Green function $R^0(0, \omega)$ is

$$R^0_{\gamma_1 \gamma_2}(0, \omega) = \sum_\mathbf{k} \frac{4\gamma_{1\mathbf{k}} \gamma_{2\mathbf{k}} \Delta^2_\mathbf{k}}{E_\mathbf{k}} \frac{1}{(\omega + i0^+)^2 - 4E^2_\mathbf{k}} \tanh \frac{\beta E_\mathbf{k}}{2}. \tag{11.37}$$

Its imaginary part is

$$\mathrm{Im} R^0_{\gamma_1 \gamma_2}(0, \omega) = -\frac{4\pi}{\omega^2} \tanh \frac{\beta \omega}{4} \sum_\mathbf{k} \gamma_{1\mathbf{k}} \gamma_{2\mathbf{k}} \Delta^2_\mathbf{k} [\delta(\omega - 2E_\mathbf{k}) + \delta(\omega + 2E_\mathbf{k})]. \tag{11.38}$$

In obtaining this expression, it is assumed that $\gamma_{\mathbf{k}}$ does not include any harmonic component whose average over the Fermi surface is finite.

In the low energy limit, the momentum summation of any function in the above equations can be approximately expressed as an integral over the energy $\xi_{\mathbf{k}}$ and the average of other variables over the Fermi surface. For example, for the function $\mathcal{A}(\xi_{\mathbf{k}}, \Delta_{\mathbf{k}})$

$$\frac{1}{V}\sum_{\mathbf{k}} \mathcal{A}(\xi_{\mathbf{k}}, \Delta_{\mathbf{k}}) = \left\langle \frac{N_F}{2} \int d\xi\, \mathcal{A}(\xi, \Delta_{\mathbf{k}}) \right\rangle_{FS}. \tag{11.39}$$

Using the integral formula of the δ-function

$$\int d\xi\, \delta\,(\omega - 2E_{\mathbf{k}}) = \mathrm{Re}\frac{|\omega|}{2\sqrt{\omega^2 - 4\Delta_{\mathbf{k}}^2}}, \tag{11.40}$$

and Eq. (11.29), the imaginary part of the Raman response function $R\,(0,\omega)$ is simplified to

$$-\,\mathrm{Im}R\,(0,\omega) = \frac{\pi N_F}{\omega}\mathrm{Re}\left\langle \frac{\gamma_{\mathbf{k}}^2 \Delta_{\mathbf{k}}^2}{\sqrt{\omega^2 - 4\Delta_{\mathbf{k}}^2}} \right\rangle_{FS} \tanh\frac{\beta\omega}{4}. \tag{11.41}$$

To obtain the real part of $R_{\gamma_1\gamma_2}^0\,(0,\omega)$, we first integrate out $\xi_{\mathbf{k}}$. At zero temperature, the result is

$$\mathrm{Re}R_{\gamma_1\gamma_2}^0\,(0,\omega) = -\left\langle \frac{2\gamma_{1\mathbf{k}}\gamma_{2\mathbf{k}}N_F\Delta_{\mathbf{k}}^2}{\omega\sqrt{|4\Delta_{\mathbf{k}}^2 - \omega^2|}}f(\mathbf{k},\omega) \right\rangle_{FS}, \tag{11.42}$$

where

$$f(\mathbf{k},\omega) = \begin{cases} 2\arctan\dfrac{\omega}{\sqrt{4\Delta_{\mathbf{k}}^2 - \omega^2}}, & \omega^2 < 4\Delta_{\mathbf{k}}^2, \\[2ex] \ln\dfrac{\omega - \sqrt{\omega^2 - 4\Delta_{\mathbf{k}}^2}}{\omega + \sqrt{\omega^2 - 4\Delta_{\mathbf{k}}^2}}, & \omega^2 > 4\Delta_{\mathbf{k}}^2. \end{cases} \tag{11.43}$$

11.4.1 Isotropic s-Wave Superconductor

For the isotropic s-wave superconductor, as $\Delta_{\mathbf{k}} = \Delta$ and the average of $\gamma_{\mathbf{k}}$ (or $\delta\gamma_{\mathbf{k}}$ for the A_{1g} mode) over the Fermi surface is zero, it is simple to show that

$$R_{\gamma,1}^0(0,\omega) = 0. \tag{11.44}$$

In this case, the Raman response function is completely determined by the first term on the right-hand side of Eq. (11.29), and we have

$$-\,\mathrm{Im}R\,(0,\omega) = \theta(|\omega|-2\Delta)\frac{\pi N_F \Delta^2 \langle \gamma_{\mathbf{k}}^2 \rangle_{FS}}{\omega\sqrt{\omega^2-4\Delta^2}}\tanh\frac{\beta\omega}{4}. \qquad (11.45)$$

Hence, $\mathrm{Im}R(0,\omega)$ vanishes at low frequency $|\omega| < 2\Delta$, but exhibits a square-root divergence when ω is above and approaching 2Δ.

11.4.2 d-Wave Superconductor

For a d-wave superconductor, the Raman response behaves differently than in the s-wave case. Since $\Delta_{\mathbf{k}} = \Delta_0 \cos 2\phi$, there is no threshold for ω. However, as the quasiparticle density of states diverges at $\omega = \Delta_0$, $\mathrm{Im}R(0,\omega)$ of the B_{1g}-mode diverges at $\omega = 2\Delta_0$. This is a logarithmic divergence, weaker than the square root divergence in the s-wave superconductor.

The vertex function of the B_{2g}-mode vanishes in the antinodal direction. This cancels the divergence of the density of states. Therefore, there is no divergence in the Raman spectrum of the B_{2g}-mode at $\omega = 2\Delta_0$. The peak position of $\mathrm{Im}R(0,\omega)$ for this mode is determined by the average of the vertex function $\gamma_{\mathbf{k}}^2$ and the single-particle spectral weight over the Fermi surface. It is not located at $\omega = 2\Delta_0$, instead at a lower frequency around $\omega \sim 1.6\Delta_0$ as shown in Fig. 11.2.

As for the A_{1g}-mode, without considering the effect of Coulomb screening, the Raman spectrum of the A_{1g}-mode would also exhibit a logarithmic divergence at $\omega = 2\Delta_0$. However, this divergence is removed by the Coulomb screening. The Coulomb interaction strongly affects the Raman scattering cross section in the long wavelength limit. It completely screens the long wavelength part of the effective mass, or the ϕ-angle independent part of the vertex function γ_k, under the effective mass approximation. For the vertex function of the A_{1g}-mode given in Eq. (11.24), i.e.

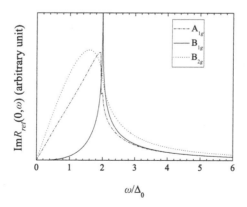

Figure 11.2 The Raman scattering spectra in the ideal d-wave superconductor at zero temperature. The vertex function is given in Eq. (11.24).

$$\gamma_{\mathbf{k}} = \gamma_0 + \gamma(A_{1g})\cos 4\phi, \tag{11.46}$$

the spectral peak is still located at $\omega = 2\Delta_0$ (Fig. 11.2). On the other hand, if the vertex function of the A_{1g}-mode contains $\cos 8\phi$ or even higher order harmonic components, the spectral peak of the A_{1g}-mode will shift toward lower frequency. Thus, in real materials, it is expected that the spectral peak of the A_{1g}-mode is always located at a frequency lower than $2\Delta_0$, i.e. $\omega < 2\Delta_0$.

The Raman spectrum of the A_{1g}-mode is not directly measured experimentally. It is usually obtained from the subtraction of the spectrum measured with $e^I = e^S = \hat{y}$ or $e^I = e^S = (\hat{x} + \hat{y})/\sqrt{2}$ and that of the B_{1g}- or B_{2g}-mode. The subtraction introduces errors in addition to the measurement error. The response function of the A_{1g}-mode is sensitive to the coefficients of higher order harmonic components. Thus it is difficult to predict quantitatively the line shape of the A_{1g}-Raman spectrum.

At low temperatures, the ω-dependence of the Raman spectrum of the B_{1g}-mode in the d-wave superconductor is different from those of the A_{1g}- and B_{2g}-modes especially at low frequencies $\omega \ll 2\Delta_0$. Since the vertex function of the B_{1g}-mode, $\gamma_{\mathbf{k}} = \gamma(B_{1g})\cos 2\phi$, has the same symmetry as the $d_{x^2-y^2}$-wave energy gap, the contribution to the Raman spectrum of this mode from the nodal quasiparticles is suppressed. As a result, the low energy Raman spectra of the B_{1g}-mode are much weaker than those of the B_{2g}- and A_{1g}-modes. For the latter two modes, the low energy spectra vary linearly with ω, resulting from the linear quasiparticle density of states. However, for the B_{1g}-mode, we find that the Raman spectrum scales as ω^3 at low frequencies simply based on the dimensional analysis.

$$-\,\mathrm{Im} R\,(0,\omega) = \begin{cases} \dfrac{\pi N_F \gamma^2(B_{2g})}{16\Delta_0}\omega\tanh\dfrac{\beta\omega}{4}, & B_{2g}, \\[3mm] \dfrac{3\pi N_F \gamma^2(B_{1g})}{128\Delta_0}\omega^3\tanh\dfrac{\beta\omega}{4}, & B_{1g}. \end{cases} \tag{11.47}$$

The Raman scattering cross section of the B_{1g}-mode exhibits a weaker frequency dependence than the A_{1g}- and B_{2g}-modes. This is due to the coincidence of the zeros of the B_{1g} vertex function and the nodes of the d-wave gap function. In general, if the zeros of the vertex function of a Raman mode coincide with the gap nodes, then the ω-dependence of the corresponding Raman response function is weaker. This property can be used to determine the nodal directions of the gap function.

11.5 Effect of Nonmagnetic Impurity Scattering

The disorder scattering has strong effects on the Raman spectra. Neglecting the vertex correction, the Raman response function can be expressed using the single-particle spectral function as

$$R(0, \omega) = \sum_{\mathbf{k}} \int \frac{d\omega_1 d\omega_2}{\pi^2} \gamma_{\mathbf{k}}^2 \text{Tr} \left[\text{Im} G^R(\mathbf{k}, \omega_1) \sigma_3 \text{Im} G^R(\mathbf{k}, \omega_2) \sigma_3 \right]$$

$$\frac{f(\omega_1) - f(\omega_2)}{\omega + \omega_1 - \omega_2 + i0^+}. \tag{11.48}$$

At zero temperature, its imaginary part is given by

$$\text{Im} R(0, \omega) = -\frac{1}{\pi} \sum_{\mathbf{k}} \gamma_{\mathbf{k}}^2 \int_0^{\omega} d\omega_1 \text{TrIm} G^R(\mathbf{k}, \omega_1 - \omega) \tau_3 \text{Im} G^R(\mathbf{k}, \omega_1) \tau_3. \tag{11.49}$$

One effect of nonmagnetic impurity scattering is to smear out the divergence of the B_{1g}-mode at $\omega = 2\Delta_0$. In addition to this, the correction of the impurity scattering to the Raman response function occurs mainly in the low energy part. In the gapless regime, $\omega \ll \Gamma_0$, i.e. the energy is smaller than the quasiparticle scattering rate, then the ω-dependence of the Raman response functions can be obtained through the series expansion. The zero temperature result is given by

$$- \text{Im} R(0, \omega) \approx \begin{cases} \dfrac{2N_F \gamma^2(B_{2g})}{\pi \Delta_0} \omega, & B_{2g}, \\[2ex] \dfrac{2N_F \gamma^2(B_{1g})}{\pi \Delta_0} \left(\dfrac{\Gamma_0}{\Delta_0} \right)^2 \left(\ln \dfrac{2\Delta_0}{\Gamma_0} \right) \omega, & B_{1g}. \end{cases} \tag{11.50}$$

Compared with the results of the intrinsic d-wave superconductor, the Raman spectrum $\text{Im} R(0, \omega)$ of the B_{2g}-mode is still a linear function of ω, but the linear coefficient changes. For the B_{1g}-mode, $\text{Im} R(0, \omega)$ changes completely. It varies linearly with ω and decays more slowly than in the intrinsic d-wave superconductor. In the weak scattering limit, the correction to the Raman spectrum is proportional to Γ_0/Δ_0, and $\text{Im} R(0, \omega)$ is not much different from that of the intrinsic d-wave superconductor.

11.6 Experimental Results of Cuprate Superconductors

The above result indicates that the Raman scattering is a powerful tool for exploring the anisotropy of the superconducting gap function. By varying the polarizations of the incident and scattered lights, the Raman spectroscopy can probe the momentum dependence of the gap function over the Fermi surface. For the isotropic s-wave superconductor, the Raman scattering is insensitive to the polarization directions. The peak positions of the Raman spectra for all symmetry modes, including A_{1g}, B_{1g}, and B_{2g}, are located at $\omega \simeq 2\Delta_0$.

However, for the d-wave superconductor, the Raman spectral peaks are located at different frequencies for different symmetry modes. The contribution to the B_{1g}-mode comes mainly from the antinodal region on the Fermi surface with the maximal energy gaps. Similarly to the s-wave superconductor, the spectral peak is located at $\omega \simeq 2\Delta_0$. On the other hand, the contribution to the B_{2g}-mode is mainly from the quasiparticle excitations around the nodal region of the $d_{x^2-y^2}$-wave energy gap, while the contribution from the antinodal region is completely suppressed. The spectral peak is located at $\omega \simeq 1.6\Delta_0$. The Raman spectrum of the A_{1g}-mode is simply an algebraic average of the contribution of quasiparticles over the entire Fermi surface if ignoring the charge fluctuation induced by the incident light. But the Coulomb screening of electrons significantly modifies the spectrum. The measured spectrum of the A_{1g}-mode depends strongly on the energy dispersion of electrons and is system dependent. Its peak can appear around $\omega \sim 2\Delta_0$ or at a frequency much lower than $2\Delta_0$.

The low frequency Raman scattering cross section varies exponentially with ω in the s-wave superconductor, but exhibits a power-law dependence of ω in the d-wave superconductor. The nonmagnetic scattering does not change the low frequency exponential behavior in the s-wave superconductor, but does change the low frequency behavior in the d-wave case. The impurity scattering broadens the peaks of Raman spectra but usually does not change the peak positions. In the intrinsic d-wave superconductor, the low frequency Raman response functions varies linearly with ω for the A_{1g}- and B_{2g}-modes, but cubically with ω for the B_{1g}-mode. However, in a disordered d-wave superconductor, the Raman response functions of the B_{1g}-mode also scales linearly with ω at low frequencies, similar to the B_{2g}-mode.

In Fig. 11.3, the measurement data of the Raman spectra of the A_{1g}-, B_{1g}-, and B_{2g}-modes for $Bi_2Sr_2CaCu_2O_8$ are shown and compared with the theoretical calculation [279]. The spectra of these Raman modes show qualitatively the same behavior in other hole-doped high-T_c superconductors [280–283]. As revealed by

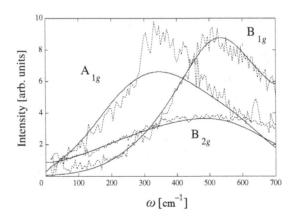

Figure 11.3 The Raman spectra of $Bi_2Sr_2CaCu_2O_8$ $(T_c = 90K)$ at $T = 20K$. The solid curves are obtained by theoretical calculation. (From Ref. [279])

the experimental data, the low frequency Raman scattering cross section decays linearly with ω for the A_{1g}- and B_{2g}-modes. For the B_{1g}-mode, the low frequency Raman spectrum decays faster and scales approximately as ω^3 within the experimental errors. In some superconductors, the low frequency cross section of the B_{1g}-mode is also a linear function of ω. This may result from the impurity scattering. All these properties are consistent with the expected scaling behaviors of low frequency Raman spectra of d-wave superconductors.

In the optimally doped as well as some overdoped high-T_c superconductors, the peak energy ω_p of the B_{1g} Raman mode is located around $(6\sim10)T_c$, which is about 1/3 higher than those of the A_{1g}- and B_{2g}-modes. The difference between the peak energies of the A_{1g}- and B_{2g}-modes is very small. The peak energy of the B_{2g}-mode is slightly higher. These differences in the peak energies of the A_{1g}-, B_{1g}-, and B_{2g}-modes agree qualitatively with the theoretical calculations by considering the vertex correction induced by the Coulomb interaction for d-wave superconductors. If we assume that the peak energy ω_p of the B_{1g} Raman mode is equal to 2Δ, then the value of $2\Delta/T_c$ thus obtained is much larger than the weak-coupling BCS s- or d-wave superconductors. In the underdoped high-T_c superconductors, the value of $2\Delta/T_c$ determined by the B_{1g} peak increases with decreasing doping and is generally around or larger than 10.

The fact that $2\Delta/T_c$ is larger than the theoretical prediction based on the weak coupling theory is closely related to the pseudogap effect observed in high-T_c cuprates. One possibility is that electrons are already paired at a temperature much higher than T_c, but the phase coherence among Cooper pairs is not established at the same temperature due to strong phase fluctuations. In the weak-coupling BCS theory, the phase fluctuation is ignored and the superconducting condensation happens immediately after electrons form Cooper pairs. Thus the superconducting transition temperature predicted in this theory is much larger than the measured value.

The Raman peaks of high-T_c superconductors are usually broad, not as sharp as those observed in the s-wave superconductors. This is likely due to the fact that the Van Hove divergence of the density of states is just logarithmic at $\omega = \Delta$ in the d-wave superconductor, much weaker than the square root divergence in the s-wave case. Elastic and inelastic scatterings can both broaden the Raman scattering peaks. The broadening due to the inelastic scattering is sensitive to temperature. It is more prominent at high temperatures. One can determine the contribution of inelastic scattering through the measurement of the variance in the temperature dependence of Raman scattering peaks.

Compared to hole-doped cuprate superconductors, the Raman spectra of electron-doped ones behave very differently. Figure 11.4 shows the Raman spectra for electron-doped high-T_c cuprates [229, 284?]. The main difference from the hole-

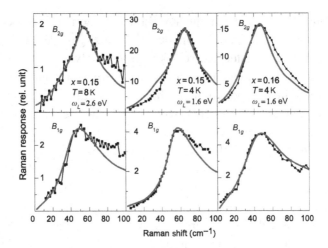

Figure 11.4 The Raman spectra of the electron-doped high-T_c superconductor Nd$_{2-x}$Ce$_x$CuO$_4$ and the comparison with the calculation based on the two-band theory. The experiment results in the first column are from Ref. [229], and those from the second and third column are from Ref. [284?]. (From Ref. [286])

doped case is that the peak of the B_{2g}-mode appears at a higher energy than that of the B_{1g}-mode. If we still use the single band model to analyze the Raman spectra in the electron-doped high-T_c cuprates, then the gap function would vary nonmonotonically with the momentum on the Fermi surface, which is inconsistent with the $d_{x^2-y^2}$-wave pairing symmetry.

However, as pointed out in §9.6, the electron-doped cuprate is not a single band system. There exist two disconnected Fermi surfaces. One of them is electron-like, located around $(\pi, 0)$ and its equivalent points. The other is hole-like, located around the nodal directions of the $d_{x^2-y^2}$-wave gap function. These two nonequivalent Fermi surfaces can be effectively described by a two-band model [226]. The weight of contribution of these two bands to a specific Raman mode is different. The vertex function of the B_{1g}-model is zero along the nodal direction of the $d_{x^2-y^2}$-gap function and reaches maximum at $(\pi, 0)$. The Raman spectrum of this mode is contributed to mainly by the quasiparticle excitations around $(\pi, 0)$ on the Fermi surfaces. On the other hand, the vertex function of the B_{2g}-model reaches the maximum along the nodal direction and becomes zero at $(\pi, 0)$. It is contributed to mainly by the quasiparticle excitations along the nodal direction. This suggests that unlike in the hole-doped case, we should not use simply the single-band model to understand the experimental results of Raman spectroscopy for electron-doped high-T_c superconductors.

Indeed, if the weak coupling two-band model introduced in §9.6 is adopted to analyze the Raman spectra, we find that there is no contradiction between the experimental results shown in Fig. 11.4 and the $d_{x^2-y^2}$-wave pairing symmetry. The Raman spectra of the B_{1g} and B_{2g}-models are mainly the contributions of the hole and electron bands, respectively. Therefore, the peak energies of these two models are determined by the energy gaps on two different bands. The peak energy of the B_{2g}-model is higher than that of the B_{1g}-model because the gap amplitude of the hole band is larger than that of the electron band. The theoretical curves shown in Fig. 11.4 are obtained from numerical calculation by considering the contribution of nonresonant Raman scattering based on the two-band model introduced in §9.6. The agreement between the calculation and the experimental results lends strong support to the assumption that the superconducting pairing in electron-doped high-T_c cuprate has $d_{x^2-y^2}$-wave symmetry.

12

Nuclear Magnetic Resonance

12.1 Spin Correlation Function

The effect of magnetic correlations plays an important role in the study of d-wave superconductivity. In particular, spin fluctuations are very strong in high-T_c cuprates, and their coupling to electrons is widely believed to be the driving force that glues electrons into Cooper pairs. A thorough investigation of spin fluctuations is crucial to the understanding of the high-T_c pairing mechanism.

Experimentally, properties of magnetic fluctuations are investigated by measuring dynamic spin correlation functions of electrons using neutron scattering and nuclear magnetic resonance (NMR) techniques. These two experimental probes are complementary, providing a comprehensive understanding of the spin dynamics of d-wave superconductors.

NMR spectroscopy relies on the measurement of a resonance phenomenon in which nuclei in a magnetic field absorb and re-emit electromagnetic radiation. It measures the Knight shift, the nuclear spin-lattice relaxation, and other physical quantities of electrons [287] and is one of the principal techniques for studying magnetic correlation in solids. The nuclear spin resonance frequency equals the Zeeman splitting energy of a nuclear spin and is proportional to the applied magnetic field. This resonance frequency is typically of the order of 10^{-7} eV, which is at the order of 100 MHz in terms of frequency, much lower than other energy scales of electrons in solids. Hence the energy window measured by NMR is very narrow. The resonance frequencies are different for different nuclei. This can be used to resolve the contributions from different nuclei. Thus the spatial resolution of NMR is very high. However, the momentum resolution of NMR is very poor according to the uncertain principle.

In a solid, the spin magnetization of an electron is defined by

$$\mathbf{M}_e(\mathbf{r}) = -\gamma_e \hbar \sum_i \mathbf{S}_i \delta(\mathbf{r} - \mathbf{r}_i), \qquad (12.1)$$

where $\gamma_e = e/m$ is the gyromagnetic ratio of the electron. $\mathbf{S}_i = c_i^\dagger(\sigma/2)c_i$ is the electron spin operator at site i and $c_i^\dagger = (c_{i\uparrow}^\dagger, c_{i\downarrow}^\dagger)$. The Zeeman energy corresponding to $\mathbf{M}_e(\mathbf{r})$ is

$$H_M = -\int d\mathbf{r}\mathbf{M}_e \cdot \mathbf{H}(\mathbf{r}, t) = \gamma_e\hbar\sum_{\mathbf{q}}\mathbf{S}(-\mathbf{q}) \cdot \mathbf{H}(\mathbf{q}, t), \tag{12.2}$$

where $\mathbf{S}(\mathbf{q})$ and $\mathbf{H}(\mathbf{q})$ are the Fourier components of the electron spin and the applied magnetic field, respectively

$$\mathbf{S}(\mathbf{q}) = \sum_i \mathbf{S}_i e^{i\mathbf{q}\cdot\mathbf{r}_i} = \sum_{\mathbf{k}} c_{\mathbf{k}+\mathbf{q}}^\dagger \frac{\sigma}{2} c_{\mathbf{k}},$$

$$\mathbf{H}(\mathbf{q}) = \frac{1}{V}\sum_i e^{i\mathbf{q}\cdot\mathbf{r}_i}\mathbf{H}(\mathbf{r}_i).$$

In an external magnetic field, the magnetic response of electrons is determined by the spin susceptibility, χ, defined by

$$\langle M_\mu(\mathbf{q}, t)\rangle = \chi_{\mu\nu}(\mathbf{q}, t)H_\nu(\mathbf{q}, t). \tag{12.3}$$

Under the linear approximation, $\chi_{\mu\nu}$ is determined by the dynamical spin–spin correlation function

$$\chi_{\mu\nu}(\mathbf{q}, t) = i\gamma_e^2\hbar^2\left\langle\left[S_\mu(\mathbf{q}, t), S_\nu(-\mathbf{q}, 0)\right]\right\rangle\theta(t), \tag{12.4}$$

where $\theta(t)$ is the step function. The corresponding Matsubara function is defined as

$$\chi_{\mu\nu}(\mathbf{q}, \tau) = \gamma_e^2\hbar^2\left\langle T_\tau S_\mu(\mathbf{q}, \tau)S_\nu(-\mathbf{q}, 0)\right\rangle, \tag{12.5}$$

where T_τ is the imaginary time-ordering operator.

In an isotropic system, if the vertex correction is negligible and $G(-\mathbf{k}, \tau) = G(\mathbf{k}, \tau)$, then the diagonal component of $\chi_{\mu\nu}$ can be expressed using the single-particle Green's function as

$$\chi_{\mu\mu}(\mathbf{q}, \tau) = \chi_{zz}(\mathbf{q}, \tau) = -\frac{\gamma_e^2\hbar^2}{4}\sum_{\mathbf{k}}\mathrm{Tr}G(\mathbf{k}, \tau)G(\mathbf{q}+\mathbf{k}, 0). \tag{12.6}$$

After Fourier transformation, it becomes

$$\chi_{zz}(\mathbf{q}, i\omega_n) = -\frac{\gamma_e^2\hbar^2}{4\beta V}\sum_{\mathbf{k}, p_m}\mathrm{Tr}G(\mathbf{k}, ip_m)G(\mathbf{q}+\mathbf{k}, i\omega_n+ip_m). \tag{12.7}$$

It can be further expressed using the single-particle spectral function as

$$\chi_{zz}(\mathbf{q}, i\omega_n) = -\frac{\gamma_e^2\hbar^2}{4\pi^2 V}\sum_{\mathbf{k}}\int d\omega_1 d\omega_2\frac{f(\omega_1) - f(\omega_2)}{i\omega_n + \omega_1 - \omega_2}$$

$$\mathrm{TrIm}G^R(\mathbf{k}, \omega_1)\mathrm{Im}G^R(\mathbf{q}+\mathbf{k}, \omega_2). \tag{12.8}$$

In an ideal d-wave superconductor, the single particle Green's function is determined by Eq. (4.4). In this case, the summation over p_m in Eq. (12.7) can be readily done. Taking analytic continuation by setting $i\omega_n \to \omega + i0^+$, we arrive at

$$\chi_{zz}(\mathbf{q}, \omega) = -\frac{\gamma_e^2 \hbar^2}{4V} \sum_{\mathbf{k}} [2f(E_\mathbf{k}) - 1] [g_{\mathbf{k},\mathbf{q}}(\omega) + g_{\mathbf{k},\mathbf{q}}(-\omega)], \qquad (12.9)$$

where

$$g_{\mathbf{k},\mathbf{q}}(\omega) = \frac{E_\mathbf{k}(\omega + E_\mathbf{k}) + \xi_\mathbf{k}\xi_{\mathbf{k}+\mathbf{q}} + \Delta_\mathbf{k}\Delta_{\mathbf{k}+\mathbf{q}}}{E_\mathbf{k}\left[(\omega + E_\mathbf{k})^2 - E_{\mathbf{k}+\mathbf{q}}^2\right]}. \qquad (12.10)$$

In the long wavelength limit, the static magnetic susceptibility reduces to

$$\lim_{q \to 0} \chi_{zz}(\mathbf{q}, 0) = -\frac{\gamma_e^2 \hbar^2}{2V} \sum_{\mathbf{k}} \frac{\partial f(E_k)}{\partial E_k} = \gamma_e^2 \hbar^2 Y(T) \qquad (12.11)$$

where $Y(T)$ is the Yoshida function defined in Eq. (9.27).

Clearly, the static susceptibility of electrons is temperature dependent. It is determined purely by the quasiparticle density of states, and scales linearly with T at low temperatures in d-wave superconductors

$$\chi_{zz}(0, 0) \approx \frac{(\ln 2)\gamma_e^2 \hbar^2 N_F k_B T}{2\Delta_0}, \qquad T \ll T_c. \qquad (12.12)$$

12.2 Hyperfine Interaction

The energy level of a nuclear spin is split under an external magnetic field. This split is determined by the Zeeman energy of the nuclear spin

$$H_{\text{Zeeman}} = -\gamma_N \hbar \mathbf{I} \cdot \mathbf{H}. \qquad (12.13)$$

The gap between two adjacent energy levels is

$$\hbar\omega_0 = -\gamma_N \hbar H, \qquad (12.14)$$

where γ_N is the gyromagnetic ratio of nuclear magnetic moment and $\mathbf{I}$ is the nuclear spin. In this nuclear spin system, if we further apply an alternating electromagnetic field to trigger the transition between two different energy levels, a resonant absorption by nuclear spins occurs when the frequency of the applied electromagnetic field equals ω_0. Such a magnetic resonance frequency typically corresponds to the radio frequency range of the electromagnetic spectrum for magnetic fields up to roughly 20 T. It is this magnetic resonant absorption which is detected in NMR.

In solid systems, in addition to the applied magnetic field, nuclear spins are also coupled to the effective magnetic field generated by electrons around the nucleus. Owing to the interaction between the nuclear spin and the magnetic moments of electrons, there is a shift in the resonance frequency. The electronic magnetic moments are contributed by both orbital angular momenta and spins. The shift induced by the orbital magnetic moment of an electron is called the chemical shift. It is also called the Van Vleck shift. The shift induced by the electron spin magnetic moment, on the other hand, is called the Knight shift, which is more pronounced in a metallic system.

The chemical shift is determined by the Hamiltonian

$$H_{\text{chem}} = -\frac{\gamma_N \hbar}{c} \mathbf{I} \cdot \int d^3 \mathbf{r} \frac{\mathbf{r} \times \mathbf{j}(\mathbf{r})}{r^3}, \tag{12.15}$$

where $\mathbf{j}$ is the gauge invariant current-density operator of electrons defined by

$$\mathbf{j} = \frac{-ie\hbar}{2m} \left(\psi^* \nabla \psi - \psi \nabla \psi^* \right) - \frac{e^2}{mc} \mathbf{A} \psi^* \psi. \tag{12.16}$$

As the electric current surrounding a nucleus is strongly screened by electromagnetic interactions, the chemical shift is also strongly affected by this effect. The electromagnetic screening, on the other hand, is affected by the chemical environment. Thus a nuclear spin may exhibit different chemical shifts under different chemical environments. The orbital angular momentum of electron in an s-orbital is zero. The contribution of this electron to the chemical shift comes mainly from the $\mathbf{A}$-term in the current operator $\mathbf{j}$. This contribution is diamagnetic, which lowers the resonance frequency. The chemical shift induced by an electron with a finite orbital angular momentum results from the contribution of both the paramagnetic and diamagnetic current terms. It can be either positive or negative, depending on the relative contribution of these two kinds of currents. Usually, the chemical shift is positive, mainly dominated by the paramagnetic contribution, and is generally very small. Since the chemical shift is determined by the electron density surrounding the nucleus, it is insensitive to temperature changes, and to the superconducting transition. Hence limited information on electron magnetic correlations and superconductivity can be extracted from the measurement of chemical shift.

The Knight shift is much larger than the chemical shift in conductors. The Knight shift results from the hyperfine interaction between an s-orbital electron and a nuclear spin, which is also called the Fermi contact interaction. It is determined by the wavefunction of electrons and governed by the Hamiltonian

$$H_{\text{hyper}} = \frac{8\pi}{3} \gamma_e \gamma_N \hbar^2 \sum_i \mathbf{S}_i \cdot \mathbf{I}_i |u(\mathbf{r}_i)|^2, \tag{12.17}$$

where $|u(\mathbf{r}_i)|^2$ is the probability of an electron at the nuclear site.

There exists a magnetic dipolar interaction between an electron with a finite orbital angular momentum (i.e. non-s-orbital) and a nuclear spin. But the energy scale of this dipolar interaction is generally very small in comparison with the hyperfine interaction. Nevertheless, non-s-orbital electrons can affect the distribution of electrons at inner shell s-orbitals, which in turn induces indirectly an effective contact interaction between these non-s-orbital electrons and nuclear spins. Furthermore, the spin moment of a magnetic ion can be transferred to an s-orbital on one of its neighboring sites via the orbital hybridization. It interacts indirectly with a neighboring nuclear spin through the Fermi-contact interaction, generating a transferred hyperfine interaction. This transferred interaction has a significant contribution to NMR in magnetic materials.

In high-T_c superconductors, the local moment of a Cu^{2+} cation interacts strongly with its neighboring nuclear spins. This interaction is even stronger than the direct hyperfine interaction on Cu. Thus in the analysis of NMR experiments of high-T_c cuprates, one should consider carefully the transferred hyperfine interactions induced by the hybridization of Cu^{2+} $3d_{x^2-y^2}$ or $4s$ orbitals with other atoms.

Assuming a magnetic field is applied along the c-axis, Mila and Rice proposed a phenomenological model to describe the transferred hyperfine interaction between the copper and oxygen atoms in the CuO_2 plane of high-T_c cuprates [288]. The model Hamiltonian is defined by

$$H_{\text{hyper}} = \sum_i \left[A\,^{63}\mathbf{I}_i \cdot \mathbf{S}_i + \sum_{\delta=\pm\hat{x},\pm\hat{y}} \left(B\,^{63}\mathbf{I}_{i+\delta} \cdot \mathbf{S}_i + C\,^{17}\mathbf{I}_{i+\delta/2} \cdot \mathbf{S}_i \right) \right], \quad (12.18)$$

where $^{63}\mathbf{I}$ and $^{17}\mathbf{I}$ are the nuclear spins of ^{63}Cu and ^{17}O, respectively. (A, B, C) are the coefficients of the hyperfine interactions. The hyperfine interaction of ^{63}Cu contains two terms. One is the interaction between the electron spin and the nuclear spin of Cu on the same site. The other is the transferred hyperfine interaction between electron and nuclear spins on two neighboring Cu sites. This transferred hyperfine interaction does not exist in conventional metals, but it is finite in high-T_c cuprates due to the strong antiferromagnetic exchange interaction between Cu^{+2} spins. The transferred hyperfine interaction between the nuclear spin of oxygen and the magnetic moment of its neighboring Cu^{2+} site is also strong. But the hyperfine interactions between the electrons and the nucleus of an O^{2-} anion are small and negligible. The effective hyperfine interactions for other atoms in high-T_c superconductors, for example, the yttrium atom in YBCO, can be similarly constructed.

In the momentum space, Eq. (12.18) can be expressed as

$$H_{\text{hyper}} = \sum_{\mathbf{q}} \left[F_{Cu}(\mathbf{q})\,^{63}\mathbf{I}(-\mathbf{q}) + F_O(\mathbf{q})\,^{17}\mathbf{I}(-\mathbf{q}) \right] \cdot \mathbf{S}(\mathbf{q}), \quad (12.19)$$

where $F_{Cu}(\mathbf{q})$ and $F_O(\mathbf{q})$ are the structural factors of the hyperfine interactions for copper and oxygen unclear spins, respectively

$$F_{Cu}(\mathbf{q}) = A + 2B(\cos q_x + \cos q_y), \tag{12.20}$$

$$F_O(\mathbf{q}) = 2C \cos \frac{q_x}{2}. \tag{12.21}$$

Generally speaking, the hyperfine interaction has the form

$$H_{\text{hyper}} = \sum_{\mathbf{q}, \mu} F(\mathbf{q})\mathbf{I}(-\mathbf{q}) \cdot \mathbf{S}(\mathbf{q}), \tag{12.22}$$

where $F(\mathbf{q})$ is the structural factor.

12.3 Knight Shift

In an external magnetic field, both the electron and nuclear spins are polarized. These polarized spins of electrons introduce an "extra" effective field at the nuclear site. As the resonance frequency is determined by the difference between the energy levels of nuclear spin in the whole magnetic field, the field induced by the electron spins leads to a shift in the resonance frequency. The relative shift in the resonance frequency is referred to as the Knight shift, which is defined by the relative change of the resonance frequency for atoms in a metal compared with the same atoms in a nonmetallic environment. This shift was first observed in a paramagnetic substance by Walter D. Knight in 1949.

The Knight shift is determined by the magnetic susceptibility of electrons and the hyperfine interaction. For the hyperfine interaction defined in Eq. (12.22), the effective magnetic field generated by the polarized electron spins on the nuclear sites equals

$$\delta H_\mu = -\frac{F(\mathbf{q})\langle S_\mu(\mathbf{q})\rangle \delta_{\mathbf{q},0}}{\gamma_N \hbar} = \frac{F(0)\chi_{\mu\mu}(0,0)H_0}{\gamma_N \gamma_e \hbar^2}, \tag{12.23}$$

where μ is the direction of the applied magnetic field. It induces a shift in the resonance frequency

$$\delta\omega = \gamma_N \delta H_\mu. \tag{12.24}$$

The Knight shift K_μ is defined by the ratio between $\delta\omega$ and the resonance frequency purely induced by the applied magnetic field

$$K_\mu = \frac{\delta\omega}{\gamma_N H_0} = \frac{F(0)\chi_{\mu\mu}(0,0)}{\gamma_N \gamma_e \hbar^2}. \tag{12.25}$$

Clearly, the Knight shift is direction-dependent. It is proportional to the electron static magnetic susceptibility.

In conventional metals, without considering the contribution of electron–electron interactions, $\chi_{\mu\mu}(0,0)$ simply equals the Pauli paramagnetic susceptibility, proportional to the electron density of states at the Fermi level

$$\chi_{zz} = \frac{\gamma_e^2 \hbar^2 N_F}{2}. \tag{12.26}$$

The corresponding Knight shift is given by

$$K = \frac{F(0)\gamma_e N_F}{2\gamma_N}. \tag{12.27}$$

In an s-wave superconductor with the spin singlet pairing, both $\chi(0,0)$ and the corresponding Knight shift drop monotonically with decreasing temperature. In particular, in an s-wave superconductor, the Knight shift decays exponentially at low temperatures, similar to other thermodynamic quantities. In a d-wave superconductor, however, the Knight shift is proportional to the quasiparticle density of states and decays just linearly at low temperatures. If the magnetic field is applied along the c-axis, the magnetic susceptibility is determined by Eq. (12.12) at low temperatures and the corresponding Knight shift is approximately given by

$$K \approx \frac{(\ln 2)F(0)\gamma_e N_F}{2\gamma_N} \frac{k_B T}{\Delta_0}. \tag{12.28}$$

12.4 Spin-Lattice Relaxation

In an applied magnetic field, the energy levels of nuclear spins are split. Upon excitations by electromagnetic radiations, a nuclear spin can occupy a higher energy level state. At a finite temperature, a nuclear spin can also be excited from a lower energy state to a higher energy one by thermal fluctuations. The nuclear spin in an excited state is unstable and tends to relax back to the ground state. The relaxation process of a nuclear spin from a nonequilibrium state to the thermal equilibrium state by exchanging energy with the environment is called the spin-lattice relaxation. Here the lattice is just the solid state environment surrounding the nuclei. During the relaxation process, the temperature of nuclear spins tends to approach the surrounding lattice temperature. At an equilibrium state, the energies gained and released by nuclear spins are balanced.

The spin-lattice relaxation rate T_1^{-1} is a characteristic quantity describing the process of a nuclear spin polarized along a particular direction relaxing to a random orientation through the scattering from surrounding electrons. The relaxation happens approximately in two steps: The first step happens with only nuclear spins, without exchanging energy with electrons. A nuclear spin tends to reach an instantaneous equilibrium state through exchanging energy with its internal degrees of freedom or with other nuclear spins. The probability of this nuclear spin at different configurations satisfies the statistical distribution of a micro-canonical ensemble described by an effective nuclear spin temperature T_N. This temperature, generally speaking, is higher than the temperature of surrounding electrons T_e. The second step is the relaxation of nuclear spins to the genuine equilibrium state by exchanging energy with electrons through the hyperfine interaction so that T_N is cooled down to the environment temperature T_e. The first step runs much faster than the second. This is the reason why we assume that the nucleus spins can reach an instantaneous micro-canonical equilibrium state. It can also be regarded as a basic assumption used in the analysis of spin-lattice relaxation. In real materials, the time scale of the first relaxation step is typically of the order of $10-100$ μs, while the time scale of the second relaxation step is typically of the order of a millisecond. Hence the assumption of two-step relaxation is valid.

Under the assumption that the intra-nucleus relaxation is much faster than the spin-lattice relaxation, the nuclear spin relaxation is just the process of the nuclear spin temperature T_N approaching the equilibrium lattice temperature through the hyperfine interaction. To describe this process, let us consider a nuclear spin system in an external magnetic field H_0, whose energy levels are given by

$$E_n = -\gamma_N \hbar n H_0, \qquad (n = -I, -I+1, \cdots, I). \tag{12.29}$$

At the instantaneous temperature T_N, the probability of the nuclear spin in the state with energy E_n is

$$p_n = \frac{\exp(-\beta_N E_n)}{Z_N}, \tag{12.30}$$

where $\beta_N = 1/k_B T_N$ and

$$Z_N = \sum_n \exp(-\beta_N E_n) \tag{12.31}$$

is the partition function of the nuclear spin. The average energy of the nuclear spin is given by

$$E_N = \sum_n p_n E_n. \tag{12.32}$$

If W_{nm} is the transition rate from the state of E_n to that of E_m, then the time-derivative of p_n is determined by the formula

$$\frac{dp_n}{dt} = \sum_m (p_m W_{mn} - p_n W_{nm}). \tag{12.33}$$

The corresponding rate of change in the energy is

$$\frac{dE_N}{dt} = \sum_n E_n \frac{dp_n}{dt} = \frac{1}{2} \sum_{mn} (p_m W_{mn} - p_n W_{nm})(E_n - E_m). \tag{12.34}$$

When the system reaches the final equilibrium state, $T = T_e$, the probability of the nuclear spin at each energy level will no longer change with time, i.e. $dp_n/dt = 0$. In this case, W_{mn} satisfies the equation

$$\frac{W_{nm}}{W_{mn}} = e^{\beta_e(E_n - E_m)}, \tag{12.35}$$

where $\beta_e = 1/k_B T_e$. Substituting this equation into Eq. (12.34), we obtain

$$\frac{dE_N}{dt} = \frac{1}{2} \sum_{mn} p_m W_{mn} \left[1 - e^{(\beta_e - \beta_N)(E_n - E_m)} \right] (E_n - E_m). \tag{12.36}$$

Typically, a nuclear magnetic moment is three to five orders of magnitudes smaller than an electron magnetic moment. Similarly, in real NMR experiments, the energy of nuclear spins $\hbar\omega_0$ is usually three to five orders of magnitude smaller than the measurement temperatures, i.e, $\hbar\omega_0 \ll k_B T_e$. The right-hand side of Eq. (12.36) can be expanded using $(\beta_N - \beta_e)\hbar\omega_0$ as a small parameter. To the first order approximation in $(\beta_N - \beta_e)\hbar\omega_0$, we find that

$$\frac{dE_N}{dt} \approx -\frac{\beta_e - \beta_N}{2} \sum_{mn} p_m W_{mn}(E_n - E_m)^2. \tag{12.37}$$

Using Eq. (12.34), we further have

$$\frac{dE_N}{dt} = \frac{d\beta_N}{dt} \sum_n E_n \frac{dp_n}{d\beta_N} = \frac{d\beta_N}{dt} \left(E_N^2 - \sum_n E_n^2 p_n \right). \tag{12.38}$$

Therefore,

$$\frac{d\beta_N}{dt} \approx \frac{\sum_{mn} p_m W_{mn}(E_n - E_m)^2}{2 \left(\sum_n E_n^2 p_n - E_N^2 \right)} (\beta_e - \beta_N). \tag{12.39}$$

By further expanding p_m up to the first order in β_N, we obtain the equation for determining the spin-lattice relaxation rate

$$\frac{d\beta_N}{dt} \approx \frac{\sum_{mn} W_{mn}(E_n - E_m)^2}{2 \sum_n E_n^2} (\beta_e - \beta_N) = \frac{\beta_e - \beta_N}{T_1}, \tag{12.40}$$

where

$$T_1^{-1} = \frac{\sum\limits_{mn} W_{mn}(E_n - E_m)^2}{2\sum\limits_n E_n^2} = \frac{\sum\limits_{mn} W_{mn}(n - m)^2}{2\sum\limits_n n^2}, \quad (12.41)$$

is the nuclear spin-lattice relaxation rate. It describes the speed of β_N approaching β_e under the electron–nucleus interaction. The transition matrix elements W_{mn} between two energy levels m and n induced by the hyperfine interaction satisfy the selection rule: $m = n, n \pm 1$. Thus the above expression can be further simplified as

$$T_1^{-1} = \frac{\sum\limits_n \left(W_{n,n+1} + W_{n+1,n} \right)}{2\sum\limits_n n^2}. \quad (12.42)$$

In a spin-lattice relaxation process, the nuclear spin undergoes a transition in which it either absorbs or releases energy. In order to conserve energy, electrons must simultaneously undergo a transition from one state to another by releasing or absorbing energy. According to the Fermi golden rule, the transition rate of a nuclear spin from the state E_n to E_m for the hyperfine interaction defined in Eq. (12.22) is given by

$$W_{mn} = \frac{2\pi}{\hbar} \sum_{\alpha,\alpha'} \left| \langle n, \alpha | \sum_{q\mu} F_\mu(\mathbf{q}) I_\mu S_\mu(\mathbf{q}) | m, \alpha' \rangle \right|^2 \delta(E_{\alpha'} - E_\alpha + \omega_{mn}), \quad (12.43)$$

where $\omega_{mn} = E_m - E_n$. $|\alpha\rangle$ and E_α are the many-body eigen–wavefunction and the corresponding eigen–energy, respectively. This formula of W_{mn} can be also expressed as

$$W_{mn} = \sum_{q\mu\nu} A_{\mu\nu}(\mathbf{q}, m, n) \sum_{\alpha,\alpha'} \langle \alpha | S_\mu(\mathbf{q}) \alpha' \rangle \langle \alpha' | S_\nu(-\mathbf{q}) \alpha \rangle \delta(E_{\alpha'} - E_\alpha + \omega_{mn}), \quad (12.44)$$

with

$$A_{\mu\nu}(\mathbf{q}, m, n) = \frac{2\pi}{\hbar} |F(\mathbf{q})|^2 \langle n | I_\mu | m \rangle \langle m | I_\nu | n \rangle. \quad (12.45)$$

From the fluctuation-dissipation theorem, it can be shown that the right-hand side of Eq. (12.44) is proportional to the imaginary part of the magnetic susceptibility

$$\sum_{\alpha,\alpha'} \langle \alpha | S_\mu(\mathbf{q}) \alpha' \rangle \langle \alpha' | S_\nu(-\mathbf{q}) \alpha \rangle \delta(E_{\alpha'} - E_\alpha + \omega) = \frac{\text{Im} \chi_{\mu\nu}(\mathbf{q}, \omega)}{\pi \gamma_e^2 \hbar^2 \left(1 - e^{-\beta\omega} \right)}. \quad (12.46)$$

Therefore, we have

$$
W_{mn} = \frac{1}{\pi \gamma_e^2 \hbar^2 \left(1 - e^{-\beta \omega_{mn}}\right)} \sum_{q\mu\nu} A_{\mu\nu}(\mathbf{q}, m, n) \mathrm{Im} \chi_{\mu\nu}(\mathbf{q}, \omega_{mn})
$$

$$
\simeq \frac{k_B T}{\pi \gamma_e^2 \hbar^2} \sum_{q\mu\nu} A_{\mu\nu}(\mathbf{q}, m, n) \lim_{\omega \to 0} \frac{\mathrm{Im} \chi_{\mu\nu}(\mathbf{q}, \omega)}{\omega}. \tag{12.47}
$$

This allows us to represent the spin-lattice relaxation rate as

$$
T_1^{-1} = k_B T \sum_{q\mu\nu} \tilde{A}_{\mu\nu}(\mathbf{q}) \lim_{\omega \to 0} \frac{\mathrm{Im} \chi_{\mu\nu}(\mathbf{q}, \omega)}{\omega}, \tag{12.48}
$$

where

$$
\tilde{A}_{\mu\nu}(\mathbf{q}) = \frac{|F(\mathbf{q})|^2}{\gamma_e^2 \hbar^3} \frac{\sum_{mn} \langle n | I_\mu | m \rangle \langle m | I_\nu | n \rangle (E_n - E_m)^2}{\sum_n E_n^2}. \tag{12.49}
$$

In an isotropic system, the spin susceptibility is diagonal and direction independent

$$
\chi_{\mu\nu} = \chi_{zz} \delta_{\mu,\nu}. \tag{12.50}
$$

In this case, T_1^{-1} becomes

$$
T_1^{-1} = \frac{2 k_B T}{\gamma_e^2 \hbar^3} \sum_q |F(\mathbf{q})|^2 \lim_{\omega \to 0} \frac{\mathrm{Im} \chi_{zz}(\mathbf{q}, \omega)}{\omega}. \tag{12.51}
$$

In the above derivation, the following identity is utilized

$$
\frac{\sum_{\mu mn} \langle n | I_\mu | m \rangle \langle m | I_\mu | n \rangle (E_n - E_m)^2}{\sum_n E_n^2} = 2. \tag{12.52}
$$

Based on the expression of structure factors, Eqs. (12.20) and (12.21), we know that antiferromagnetic fluctuations affect the NMR results at copper and oxygen sites differently. The antiferromagnetic fluctuation can enhance the spin-lattice relaxation at the copper sites, but has almost no effect on the spin-lattice relaxation at the oxygen sites. From the experimental measurement of high-T_c cuprates, it was found that the relaxation rate T_1^{-1} of Cu is significantly larger than that of O or Y. Furthermore, in the normal state, T_1^{-1} at the copper sites does not satisfy the Korringa relation that generally holds in a Landau Fermi liquid system. It implies that the antiferromagnetic fluctuation does have a strong impact on the spin-lattice relaxation rate. Nevertheless, in the discussion of low-temperature spin-lattice relaxation of d-wave superconductors, it is the nodal quasi particle excitations rather than the antiferromagnetic fluctuations that play the more important role. To the leading order approximation, one can neglect the contribution of antiferromagnetic fluctuations to the spin-lattice relaxation.

If $F(\mathbf{q}) = F(0)$ does not depend on $\mathbf{q}$, the spin-lattice relaxation rate becomes

$$T_1^{-1} = \frac{2k_B T F^2(0)}{\gamma_e^2 \hbar^3} \sum_{\mathbf{q}} \lim_{\omega \to 0} \frac{\mathrm{Im}\chi_{zz}(\mathbf{q}, \omega)}{\omega}. \tag{12.53}$$

This expression is commonly used in the analysis of experimental data. It catches the main feature of the spin-lattice relaxation and holds in most cases. For more general cases, we need to know the expression of $F(\mathbf{q})$.

From Eq. (12.8), $\mathrm{Im}\chi_{zz}$ can be written using the single-particle spectra function as

$$\lim_{\omega \to 0} \frac{\mathrm{Im}\chi_{zz}(\mathbf{q}, \omega)}{\omega} = -\frac{\gamma_e^2 \hbar^2}{4\pi V} \sum_{\mathbf{k}} \int d\omega \frac{\partial f(\omega)}{\partial \omega} Tr \mathrm{Im} G(\mathbf{q} + \mathbf{k}, \omega) \, \mathrm{Im} G(\mathbf{k}, \omega). \tag{12.54}$$

12.4.1 Isotropic s-Wave Superconductor

Using Eq. (12.54), it is simple to show that the following equation holds for the s-wave superconductor

$$T_1^{-1} = -\frac{\pi k_B T F^2(0)}{\hbar} \int d\omega \frac{\partial f(\omega)}{\partial \omega} \rho^2(\omega) \left(\frac{\omega^2 + \Delta^2}{\omega^2} \right), \tag{12.55}$$

where $\rho(\omega)$ is the quasiparticle density of states.

In the normal state, $\Delta = 0$ and $\rho(\omega) \approx N_F$,

$$T_1^{-1} = \frac{\pi k_B T F^2(0) N_F^2}{\hbar}. \tag{12.56}$$

Combining this equation with the expression of the Knight shift, Eq. (12.27), we obtain the following equation

$$k_B T K^2 T_1 = \frac{\hbar}{4\pi} \frac{\gamma_e^2}{\gamma_N^2}, \tag{12.57}$$

which is also called Korringa relation. It shows that $k_B T K^2 T_1$ is a universal quantity in the normal state, depending on the ratio of γ_e/γ_N, but not on the electronic and lattice structures.

In the superconducting state,

$$\rho(\omega) = \frac{N_F \omega}{\sqrt{\omega^2 - \Delta^2}}. \tag{12.58}$$

Substituting this into (12.55) yields

$$T_{1,s}^{-1} = -\frac{\pi k_B T F^2(0) N_F^2}{\hbar} \int d\omega \frac{\omega^2 + \Delta^2}{\omega^2 - \Delta^2} \frac{\partial f(\omega)}{\partial \omega}. \tag{12.59}$$

As $\rho(\omega)$ diverges at $\omega = \Delta$, $T_1^{-1}(T)$ should also diverge. However, in real materials, this divergence is smeared out by strong coupling, impurity scattering, and other effects, leaving only a peak at T_c as a residual character of the s-wave superconductor. This characteristic peak is often called the coherent or Hebel–Slichter coherence peak [289]. It has been observed in most metal or alloy-based superconductors. Nevertheless, there are exceptions. For example, in the strong coupling s-wave superconductor $TlMo_6Se_{7.5}$, this coherence peak is not observed.

12.4.2 d-Wave Superconductor

In a d-wave or other unconventional superconductor whose gap function is averaged to zero over the entire Fermi surface, the spin-lattice relaxation rate is determined purely by the quasiparticle density of states $\rho(\omega)$ if the form factor is momentum independent, i.e. $F(q) = F(0)$,

$$T_1^{-1} = -\frac{\pi k_B T F^2(0)}{\hbar} \int d\omega \frac{\partial f(\omega)}{\partial \omega} \rho^2(\omega). \tag{12.60}$$

In a d-wave superconductor, the density of states diverges logarithmically at the gap edge

$$\rho(\omega \to \Delta) = \frac{N_F}{\pi} \ln \frac{8}{|1 - \Delta/\omega|}, \tag{12.61}$$

much more weakly than the square-root divergence of the density of states in an s-wave superconductor. In this case, there is no divergence in the integral of Eq. (12.55). Nevertheless, T_1^{-1} still exhibits a small coherence peak just below T_c. This peak is not as robust as in the s-wave superconductor. It is rather fragile against antiferromagnetic fluctuation, strong coupling, and other effects, and difficult to observe in a d-wave superconductor.

At low temperatures, the integral in Eq. (12.55) is contributed to mainly by low-lying excitations. In this case, $\rho(\omega)$ is proportional to ω, and T_1^{-1} is approximately given by

$$T_1^{-1}(T) \approx \frac{\pi^3 N_F^2 F^2(0) k_B^3 T^3}{3\hbar \Delta_0^2}, \qquad T \ll T_c. \tag{12.62}$$

Combining this expression with Eq. (12.28), we find that the Korringa relation still holds for d-wave superconductors at low temperatures

$$k_B T K^2 T_1 = \frac{3\hbar \ln^2 2}{4\pi^3} \frac{\gamma_e^2}{\gamma_N^2}. \tag{12.63}$$

Again $k_B T K^2 T_1$ is a universal constant proportional to γ_e^2 / γ_N^2, but the coefficient is changed in comparison with the normal state.

12.5 Effect of Impurity Scattering

The spin susceptibility $\chi(0,0)$ is approximately proportional to the real part of the in-plane current–current correlation function of electrons in the low-energy long wavelength limit. The latter, on the other hand, is proportional to the paramagnetic contribution to the superfluid density

$$\chi_{zz}(0,0) = -\frac{\gamma_e^2 \hbar^2}{2e^2 v_F^2} \Pi_{ab}(0,0). \tag{12.64}$$

Because the diamagnetic contribution to the superfluid density is nearly temperature independent, the temperature dependence of $\chi(0,0)$ is therefore similar to that of the in-plane superfluid density.

In the gapless regime, it is simple to show, following the derivation for Π_{ab} presented in Chapter 8, the low temperature static uniform spin susceptibility varies quadratically with temperature in d-wave superconductors

$$\chi_{zz}(0,0) = \frac{\gamma_e^2 \hbar^2 \Gamma_0 N_F}{\pi a \Delta_0} \left[\ln \frac{2\Delta_0}{\Gamma_0} + \frac{a^2 k_B^2 T^2}{6\Gamma_0^2} + o(T^4) \right]. \tag{12.65}$$

Therefore, the Knight shift in the same temperature range, correct up to the order of T^2, is given by

$$K = \frac{\gamma_e \Gamma_0 F(0) N_F}{\pi a \gamma_N \Delta_0} \left[\ln \frac{2\Delta_0}{\Gamma_0} + \frac{a^2 k_B^2 T^2}{6\Gamma_0^2} + o(T^4) \right]. \tag{12.66}$$

The Knight shift is finite at zero temperature because the density of states of quasi-particles at the Fermi level is finite in disordered d-wave superconductors. The quadratic temperature dependence of the Knight shift is a consequence of the Sommerfeld expansion.

The spin-lattice relaxation rate T^{-1} is determined by the imaginary part of the magnetic susceptibility. In a disordered system, the quasiparticle density of states is finite at the Fermi energy. In the gapless regime, the quasiparticle density of states in the unitary and Born scattering limits is given by Eq. (8.84) and Eq. (8.85), respectively. Substituting these equations into Eq. (12.60), the spin-lattice relaxation rate is found to be

$$T_1^{-1} = \frac{\pi k_B T F^2(0) \rho^2(0)}{\hbar} \left[1 + \frac{\pi^4 \Delta_0^2 k_B^2 T^2}{3\Gamma_0^2 \Gamma_N^2} + o(T^4) \right] \tag{12.67}$$

in the Born scattering limit, and

$$T_1^{-1} = \frac{\pi k_B T F^2(0) \rho^2(0)}{\hbar} \left[1 - \frac{\pi^2 k_B^2 T^2}{6\Gamma_0^2} + o(T^4) \right] \qquad (12.68)$$

in the unitary scattering limit. In either limit, T_1^{-1} scales linearly with T at low temperatures, different from the T^3-behavior in a pure d-wave superconductor.

The linear temperature of the spin-lattice relaxation rate results from the finite density of states of electrons on the Fermi surface. The leading correction to the linear temperature dependence of T_1^{-1} is proportional to T^3. But this T^3-term behaves quite differently in these two scattering limits. The coefficient of the T^3-term is positive in the Born scattering limit, but negative in the unitary scattering limit. This difference results from the difference in the energy dependence of low-energy density of states. From this difference one can in principle determine which limit the impurity scattering potential is in.

12.6 Impurity Resonance States

Both the Knight shift and the spin-lattice relaxation are sensitive to the magnetic structure surrounding a nucleus whose NMR spectroscopy is measured. This property of NMR can be used to probe the magnetic structure in the vicinity of an impurity.

Around zinc or other nonmagnetic impurities in high-T_c superconductors, it has been found from NMR measurements that the impurity contribution to the spin susceptibility is Curie–Weiss-like [290–296]. One possible explanation for this phenomenon is that in a system with strong antiferromagnetic fluctuation, a zinc or other nonmagnetic impurity would induce certain unscreened magnetic moments around the impurity. These induced moments lead to the Curie–Weiss behavior of the spin susceptibility. However, this interpretation is not consistent with other experimental observations, which rules out the possibility that the Curie-Weiss behavior observed in the NMR experiments is truly due to the contribution of induced magnetic moments:

(1) In the zinc-doped $YBa_2(Cu_{1-x}Zn_x)O_8$ sample, the μSR measurement did not find any evidence of induced local moments [297].

(2) If the Curie–Weiss behavior is indeed induced by the unscreened local moments, then the spin-lattice relaxation rate should increase monotonically with decreasing temperature. However, in high-T_c superconductors the impurity contribution to $(T_1 T)^{-1}$ decays exponentially at low temperatures. This exponential decay indicates that the induced magnetic moment (if it exists) is frozen, although the mechanism of this frozen effect is unclear [294].

(3) In overdoped high-T_c cuprates, the magnetic correlations are strongly suppressed and the chance of creating magnetic moments by a nonmagnetic impurity is very slim if not completely impossible [298].

In the analysis of the NMR data, an important but often overlooked point is the contribution of nonmagnetic resonance states generated by nonmagnetic impurities in high-T_c superconductors. From the discussion presented in Chapter 7, we know that a zinc or other nonmagnetic impurity may create a sharp low energy resonance state in the superconducting state of d-wave superconductors [299, 300]. In the absence of an external magnetic field, this low energy resonance state is not magnetically polarized. Nevertheless, its contribution to the magnetic susceptibility is finite. In fact, as will be shown below, if the temperature is higher than the resonance energy, the contribution of the resonance state to the spin susceptibility is Curie–Weiss-like. Thus as far as the zero-field spin susceptibility is concerned, the nonmagnetic resonance state behaves like a local magnetic impurity. On the other hand, if the temperature is lower than the resonance energy, it is difficult to excite an electron to the resonance state and the contribution of this state to the spin-lattice relaxation or the Knight shift is very small, exhibiting an activated behavior. This explains naturally the exponential behavior of the $(T_1 T)^{-1}$ at low temperatures.

In a system without translation invariance, the spin-lattice relaxation rate on site $\mathbf{r}$ can be similarly derived as for Eq. (12.51). The result is given by

$$\frac{1}{T_1(\mathbf{r})T} = \frac{2k_B}{\gamma_e^2 \hbar^3} \sum_{\mathbf{j},\mathbf{l}} F_{\mathbf{j},\mathbf{r}} F_{\mathbf{l},\mathbf{r}} \lim_{\omega \to 0} \frac{\mathrm{Im}\chi_{zz}(\mathbf{j},\mathbf{l},\omega)}{\omega}, \tag{12.69}$$

where $\mathbf{j}$ or $\mathbf{l}$ is the coordinate of $\mathbf{r}$ or any of its four nearest neighboring sites. $F_{\mathbf{j},\mathbf{r}} = A$ is the structure factor of the hyperfine interaction at site $\mathbf{j} = \mathbf{r}$. $F_{\mathbf{j},\mathbf{r}} = B$ is the indirect hyperfine interaction induced by the exchange interaction of copper spins if $\mathbf{j} \neq \mathbf{r}$.

In the limit $\omega \to 0$, the magnetic susceptibility $\mathrm{Im}\chi_{zz}(\mathbf{j},\mathbf{l},\omega)$ can be expressed using the electron Green's function as

$$\lim_{\omega \to 0} \frac{\mathrm{Im}\chi_{zz}(\mathbf{j},\mathbf{l},\omega)}{\omega} = -\frac{\gamma_e^2 \hbar^2}{2\pi} \int_{-\infty}^{\infty} d\varepsilon A(\mathbf{j},\mathbf{l},\varepsilon) \frac{\partial f(\varepsilon)}{\partial \varepsilon}, \tag{12.70}$$

where

$$A(\mathbf{j},\mathbf{j}',\varepsilon) = \left[\mathrm{Im}G_{11}(\mathbf{j},\mathbf{j}';\varepsilon)\right]^2 + \left[\mathrm{Im}G_{12}(\mathbf{j},\mathbf{j}';\varepsilon)\right]^2. \tag{12.71}$$

In a single-impurity system, the electron Green's function is given by Eq. (7.34). In the unitary scattering limit, there are poles in $G(\mathbf{r},\mathbf{r}';\omega)$. These poles correspond to the impurity induced resonance state. In high-T_c superconductors, the phase shift induced by the zinc-impurity scattering potential is $\delta_0 \approx 0.48\pi$ and the scattering parameter $c \approx 0.0629$. The corresponding resonance frequency Ω'

with its imaginary part Ω'' is significantly smaller than the superconducting gap, i.e. $(\Omega', \Omega'') \ll \Delta_0$.

Substituting Eq. (12.70) into Eq. (12.69), $1/T_1 T$ can be rewritten as

$$\frac{1}{T_1(\mathbf{r})T} = -\frac{k_B}{\pi \hbar} \int d\varepsilon \frac{\partial f(\varepsilon)}{\partial \varepsilon} \sum_{\mathbf{jl}} F_{\mathbf{j,r}} F_{\mathbf{l,r}} A(\mathbf{j}, \mathbf{l}; \varepsilon). \tag{12.72}$$

In the low energy limit, the Green's function of electrons without considering the correction from the impurity scattering, $G^0(\mathbf{r}, \omega)$, is a smooth function of ω. Its imaginary part $\mathrm{Im}G^0(\mathbf{r}, \omega)$ approaches zero as $\omega \to 0$. Hence, at $r \neq 0$, the imaginary part of $G^0(\mathbf{r}, \Omega')$ is much smaller than its real part. In this case, we can neglect the imaginary part of $G^0(\mathbf{r}, \omega)$, and express the correction to the Green's function as

$$\delta G(\mathbf{r}, \mathbf{r}' \omega) \approx \mathrm{Re}G^0(\mathbf{r}, 0) T(\omega) \mathrm{Re}G^0(-\mathbf{r}', 0). \tag{12.73}$$

At low temperatures, the spin-lattice relaxation rate is mainly determined by the resonant state. If the system is particle–hole symmetric, then the contribution to the spin-lattice relaxation rate from the resonance state can be expressed using Eq. (12.73) as

$$\delta [T_1(\mathbf{r})T]^{-1} \simeq -\frac{k_B}{\pi \hbar} \int d\varepsilon \frac{\partial f(\varepsilon)}{\partial \varepsilon} Z^2(\mathbf{r}, \varepsilon), \tag{12.74}$$

with

$$Z(\mathbf{r}, \varepsilon) = \sum_{\mathbf{j}} F_{\mathbf{j,r}} \left(\left[\mathrm{Re}G^0_{11}(\mathbf{j}, 0) \right]^2 + \left[\mathrm{Re}G^0_{12}(\mathbf{j}, 0) \right]^2 \right) T''_{11}(\varepsilon). \tag{12.75}$$

If the temperature is much larger than the resonance energy but far smaller than T_c, i.e. $k_B T_c \gg k_B T \gg \Omega'$, the integral in Eq. (12.74) is contributed to mainly by the pole of $T_{11}(\varepsilon)$. In this case, $\partial f(\varepsilon)/\partial \varepsilon|_{\varepsilon = \Omega'} \sim 1/T$, and the spin-lattice relaxation

$$\delta [T_1(r)T]^{-1} \sim \frac{1}{T} \tag{12.76}$$

has the standard Curie–Weiss form, similarly to a system of magnetic impurities. This $1/T$-behavior of $(T_1 T)^{-1}$ agrees well with the experimental result in the superconducting state of zinc-doped $YBa_2Cu_4O_8$ and $YBa_2Cu_3O_{6.7}$[301].

On the other hand, if the temperature is much lower than the resonance frequency, i.e. $k_B T \ll \Omega'$, $\partial f(\varepsilon)/\partial \varepsilon$ decays exponentially with temperature, the impurity contribution to the spin-lattice relaxation also drops exponentially. As previously mentioned, an exponential decay of $(T_1 T)^{-1}$ is often regarded as a signature of spin frozen [294]. However, in this case, this exponential decay is purely due to the fact that the resonance energy is higher than the temperature and it is difficult to activate electrons to the impurity resonance state by thermal excitations.

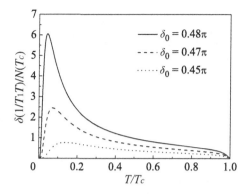

Figure 12.1 Temperature dependence of the impurity correction to the spin-lattice relaxation rate on the neighboring sites of the impurity. The normalization constant $N(T) = (T_1 T)^{-1}|_{T=T_c}$ is the spin-lattice relaxation rate at $T = T_c$.

Therefore, $\delta[T_1(r)T]^{-1}$ varies nonmonotonically with temperature. By lowering temperature $\delta[T_1(r)T]^{-1}$ increases at the beginning, and then drops exponentially after reaching a maximum at a temperature close to the resonance energy. In the limit $c \to 0$, the peak temperature of $\delta[T_1(\dot{r})T]^{-1}$ is approximately located at $k_B T_f \simeq 0.65\Omega'$. In the zinc-substituted BSCCO, the energy of the impurity resonance state is approximately $\Omega' \sim 17$ K [302]. The peak temperature of $[T_1(r)T]^{-1}$ induced by the resonance state is estimated to be $T_f \sim 11$ K, close to the experiment value of 10 K for YBCO [294].

Figure 12.1 shows the temperature dependence of the impurity contribution to the spin-lattice relaxation rate $\delta(1/T_1 T)$ on one of the nearest neighboring sites of the impurity. Clearly, the impurity contribution to $\delta(1/T_1 T)$ is very sensitive to the value of phase shift δ_0. In the unitary limit, $\delta_0 = \pi/2$, $\delta(T_1 T)^{-1}$ increases monotonically with decreasing temperature. But the peak value of $\delta(T_1 T)^{-1}$ decreases quickly with decreasing δ_0.

The Knight shift is determined by the real part of the magnetic susceptibility $\mathrm{Re}\chi_{zz}$. At site $\mathbf{r}$, it can be expressed as

$$K(\mathbf{r}) = \frac{1}{\gamma_e \gamma_n \hbar^2} \sum_{\mathbf{j}} F_{\mathbf{j},\mathbf{r}} \mathrm{Re}\chi_{zz}(\mathbf{j}), \qquad (12.77)$$

where

$$\mathrm{Re}\chi_{zz}(\mathbf{j}) = \frac{\mu_B^2}{\pi} \int d\varepsilon \frac{\partial f(\varepsilon)}{\partial \varepsilon} \mathrm{Tr}\, \mathrm{Im} G\,(\mathbf{j},\mathbf{j},\varepsilon). \qquad (12.78)$$

The contribution of the impurity resonance state to $\mathrm{Re}\chi$ is approximately given by

$$\delta\mathrm{Re}\chi_{zz}(\mathbf{j}) \approx \frac{\mu_B^2 \mathrm{Tr}\mathrm{Re}G^0(\mathbf{j},0)\mathrm{Re}G^0(\mathbf{j},0)}{\pi} \int d\varepsilon T_{11}''(\varepsilon) \frac{\partial f(\varepsilon)}{\partial \varepsilon}. \qquad (12.79)$$

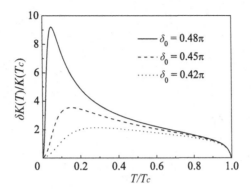

Figure 12.2 The impurity contribution to the Knight shift on the neighboring sites. $K(T_c)$ is the Knight shift at $T = T_c$.

Similar to $\operatorname{Im}\chi$, the variation of $\delta\operatorname{Re}\chi(\mathbf{j})$ with temperature is determined mainly by the resonance pole,

$$\delta\operatorname{Re}\chi_{zz}(\mathbf{j}) \sim \left.\frac{\partial f(\varepsilon)}{\partial \varepsilon}\right|_{\varepsilon=\Omega'}. \tag{12.80}$$

In the limit $k_B T_c \gg k_B T \gg \Omega'$, $\partial f(\Omega')/\partial\omega \sim 1/T$. From Eq. (12.78) we have

$$\delta K(\mathbf{r}) \sim \frac{1}{T}. \tag{12.81}$$

Thus in this temperature range the resonance contribution to the Knight shift is Curie–Weiss like. In the low temperature limit, $k_B T \ll \Omega'$, the resonance contribution is negligibly small. $K(T)$ decays to zero exponentially, again different from the contribution of free magnetic moments.

Figure 12.2 shows the impurity correction to the Knight shift as a function of temperature. It indicates that if temperature is not too low, the NMR Knight shift is approximately Curie–Weiss-like, similar to the contribution of localized magnetic moments. From this Curie–Weiss behavior of the magnetic susceptibility, one can define an effective magnetic moment μ_{eff} corresponding to this resonance state

$$\frac{\mu_{\text{eff}}^2}{3k_B T} = \sum_{\mathbf{j}} \delta\operatorname{Re}\chi_{zz}(\mathbf{j}), \tag{12.82}$$

where $\mathbf{j}$ is to sum over all four nearest neighboring sites of the impurity. Figure 12.3 shows the effective magnetic moment obtained with this equation. In obtaining the result shown in this figure, the energy–momentum dispersion relation of electron proposed by Norman et al. is used [303].

The result in Fig. 12.3 shows that in the temperature range which is neither too low, nor too close to the superconducting transition temperature, the effective

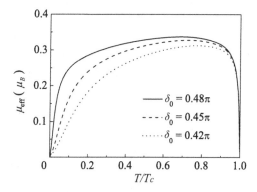

Figure 12.3 Temperature dependence of the effective magnetic moment corresponding to a resonance state induced by a single nonmagnetic impurity.

moment corresponding to the nonmagnetic impurity is approximately $0.3\mu_B$. This value is close to the effective moment estimated from the NMR data obtained in the slightly overdoped YBCO with Zn impurities. It shows that at least in this material, the Curie–Weiss behavior of the NMR spectroscopy is mainly the contribution of impurity induced resonance states.

12.7 Experimental Results of Cuprate Superconductors

NMR has an excellent frequency resolution, and can distinguish responses from different types of nuclei. It can be used to detect spatial magnetic fluctuations of electron spins around different nuclei. This is important to a comprehensive understanding of magnetic interactions, especially the property of local spin fluctuations.

The nuclear spin-lattice relaxation rate T_1^{-1} in an intrinsic d-wave superconductor shows a weak temperature dependence at low temperatures. It scales as T^3. In order to distinguish this T^3-behavior from the exponential behavior in an s-wave superconductor, the measurement resolution needs to be high. Theoretical calculations for NMR are often obtained in the limit of zero magnetic field, but experimental measurements are usually done in a strong external magnetic field.

In the normal state, the applied magnetic field affects weakly the electronic structure, and what NMR measures is the intrinsic property of electrons. This situation changes in the superconducting state. In particular, an external magnetic field may suppress the critical temperature and generate magnetic flux lines, which can alter the NMR spectra.

In fact, the NMR spectra of high-T_c cuprates are changed by varying magnetic fields. For example, the ^{63}Cu nuclear spin-relaxation rate in YBa$_2$Cu$_3$O$_7$ is four times larger in an external magnetic field of 8.31 T than in the zero field at $T = 0.2T_c$

[304]. Moreover, high-T_c cuprates are layered materials, the NMR results depend on whether the magnetic field is applied along the c-axis or parallel to the ab-plane [304]. Therefore, in the comparison of experimental results with theoretical calculations, effects of applied magnetic fields on the superconducting state need to be carefully considered. The experiment data of NMR for high-T_c cuprate superconductors were acquired mainly from copper, oxygen, and yttrium atoms.

Below T_c, the spin-lattice relaxation rate T_1^{-1} drops monotonically with lowering temperatures in almost all nuclei, including copper, oxygen, and yttrium. However, T_1^{-1} on the copper site is one order of magnitude larger than that on the oxygen site. It implies that the antiferromagnetic fluctuation is very strong even in the superconducting state [305–307]. This is strong evidence supporting the antiferromagnetic fluctuation mechanism of high-T_c superconductivity.

Just below T_c, T_1^{-1} decays quickly and the Hebel–Slichter coherent peak is not observed in all high-T_c compounds [305, 308–312]. The Hebel–Slichter peak appears in most of s-wave superconductors. Nevertheless, the absence of this coherence peak does not rule out the possibility of s-wave superconducting pairing because the strong coupling and other physical effects can smear out this peak even in s-wave superconductors. But the absence of the Hebel–Slichter peak does suggest that the density of states of quasiparticles in high-T_c superconductors is not strongly divergent or even not divergent at all at $\omega = \Delta$. In this sense, the NMR results favor more the d-wave scenario of high-T_c superconductivity, since the density of states of the d-wave superconductor diverges just logarithmically, weaker than the square-root divergence of the s-wave superconductor.

The spin-lattice relaxation rate T_1^{-1} of ^{63}Cu or ^{17}O in YBa$_2$Cu$_3$O$_{7-\delta}$ scales as T^3 in an intermediate temperature range with $T < T_c/2$. At very low temperatures, $T \ll T_c$, T_1^{-1} drops more slowly than T^3 in YBa$_2$Cu$_3$O$_{7-\delta}$ [308, 313, 314], and shows a linear temperature dependence in Bi$_2$Sr$_2$CaCu$_2$O$_8$ [309, 312], overdoped La$_{2-x}$Sr$_x$CuO$_{4-x}$ [310], and Tl$_2$Ba$_2$Ca$_2$Cu$_3$O$_{10}$ [307]. Both T^3 at relatively high temperatures and the linear low temperature dependence of T_1^{-1} agree with the behavior of disordered d-wave superconductors. In the underdoped La$_{2-x}$Sr$_x$CuO$_{4-x}$ [310], T_1^{-1} of ^{63}Cu tends to saturate at very low temperature. This saturation cannot be explained by the effect of nonmagnetic impurity scattering in d-wave superconductors. Whether it is due to the experimental background or other physical effects, such as antiferromagnetic fluctuations, needs further clarification.

The Knight shift of nuclear spins K decreases monotonically as temperature is decreased in the superconducting phase. Its value at zero temperature, $K(0)$, is small but finite [52, 308, 309, 312, 315–317]. The monotonic decay of the Knight shift in the superconducting phase is a strong indication of singlet pairing. If superconducting electrons are spin-triplet paired, the Knight shift does not change much across the superconducting critical point, in particular, it will not drop steeply with temperature in the superconducting state. The shift in the NMR resonance frequency resulting from the coupling of nuclear spins with the orbital moments of electrons, K_{orb}, is usually temperature independent. It is also insensitive to the electron or hole doping level. In normal metals, K_{orb} can be obtained by measuring the Knight shift in the corresponding diamagnetic insulator in which electron spins are completely quenched. However, high-T_c superconductors are doped antiferromagnetic insulators and electron spins have a large contribution to the Knight shift, so careful analysis of experimental data should be done in order to determine K_{orb}. For the s-wave or intrinsic d-wave superconductor, the spin susceptibility vanishes at zero temperature, and $K(0) = K_{orb}$ contributes completely from the orbital angular momentum. However, in disordered d-wave superconductors, the quasiparticle density of states is finite at the Fermi surface, and its contribution to $K(0)$ is also finite.

13

Neutron Scattering Spectroscopy

13.1 Neutron Scattering and Magnetic Susceptibility

Neutron scattering is an important spectroscopic method of measuring bulk properties of solids, since neutrons are charge neutral and interact weakly with charged particles in a solid. The scattering of neutrons by nuclei in a solid, for example, is widely used to measure phonon excitation spectra. Furthermore, a neutron carries a magnetic moment whose interaction with the magnetic moments of electrons provides an ideal tool for probing magnetic properties of solids. Elastic neutron scattering is capable of measuring static magnetic long-range orders. The inelastic neutron scattering, on the other hand, is able to probe dynamic spin excitations. In particular, the cross section of the inelastic neutron scattering is determined by the spin susceptibility [318, 319].

Neutron scattering by electrons results from the interaction between their magnetic moments. Each electron carries a magnetic moment,

$$\mathbf{m}(\mathbf{r}) = g_e \mu_B \psi^\dagger(\mathbf{r}) \frac{\sigma}{2} \psi(\mathbf{r}), \tag{13.1}$$

where $g_e = -2$ is the Lande factor of an electron. An electron moment at $\mathbf{r}'$ generates an effective magnetic field described by the vector potential

$$\mathbf{A}(\mathbf{r}) = \frac{\mu_0}{4\pi} \nabla_\mathbf{r} \times \frac{\mathbf{m}(\mathbf{r}')}{|\mathbf{r} - \mathbf{r}'|}. \tag{13.2}$$

Integrating over all the electrons yields the total magnetic field in the system

$$\mathbf{B}(\mathbf{r}) = \frac{\mu_0}{4\pi} \int d\mathbf{r}' \, \nabla_\mathbf{r} \times \left[\nabla_\mathbf{r} \times \frac{\mathbf{m}(\mathbf{r}')}{|\mathbf{r} - \mathbf{r}'|} \right]. \tag{13.3}$$

Its Fourier transform is

$$\mathbf{B}(\mathbf{q}) = g_e \mu_0 \mu_B \hat{\mathbf{q}} \times \left[\hat{\mathbf{q}} \times \mathbf{S}(\mathbf{q}) \right], \tag{13.4}$$

where $\mathbf{S}(\mathbf{q})$ is the Fourier component of the electron spin density.

306

The electron–neutron interaction is given by the Hamiltonian

$$H_I = -\sum_i \frac{\gamma \mu_N}{2} \int d\mathbf{r} \psi_N^\dagger(\mathbf{r}) \sigma_i \psi_N(\mathbf{r}) B_i(\mathbf{r})$$

$$= -g \sum_{ij} \int \frac{d^3 q}{(2\pi)^3} \psi_N^\dagger(\mathbf{k} - \mathbf{q}) \sigma_i \psi_N(\mathbf{k}) P_{ij}(\mathbf{q}) S_j(\mathbf{q}), \tag{13.5}$$

where

$$P_{ij}(\mathbf{q}) = \delta_{ij} - \hat{q}_i \hat{q}_j \tag{13.6}$$

is the transverse projection operator at momentum $\mathbf{q}$. $g = -\gamma \mu_0 \mu_N \mu_B$, ψ_N is the neutron field operator, $\gamma = -1.9$ is the gyromagnetic ratio of neutron, and $\mu_N = e\hbar/(2m_N)$ is the neutron Bohr magneton [319].

Using the Fermi golden rule, the scattering rate $1/\tau$ of a neutron with a momentum $\mathbf{k}$ is found to be

$$\frac{1}{\tau} = \frac{2\pi}{\hbar} g^2 \sum_{ij} \sum_{\mathbf{k}', \alpha\beta} \left| \langle \mathbf{k}'\alpha | \psi^\dagger(\mathbf{k}') \sigma_{i,\alpha\beta} \psi_N(\mathbf{k}) | \mathbf{k}\beta \rangle \right|^2 \delta(\mathbf{k}' - \mathbf{k} - \mathbf{q})$$

$$\times \sum_{mn} \frac{e^{-\beta E_n}}{Z} \left| \langle m | P_{ij}(\mathbf{q}) S_j(\mathbf{q}) | n \rangle \right|^2 \delta(\hbar\omega - E_m + E_n), \tag{13.7}$$

where τ is the lifetime, and $|\mathbf{k}, \alpha\rangle$ is the neutron state with momentum $\mathbf{k}$ and spin eigenvalue α. $|n\rangle$ is a many-body eigenstate of electrons with Z the corresponding partition function. $\mathbf{q}$ and $\hbar\omega$ are respectively the transferred momentum and energy between electrons and neutrons during the inelastic scattering.

In Eq. (13.7), the contribution from electrons can be simplified as

$$\sum_{ij} \sum_{mn} \frac{e^{-\beta E_n}}{Z} \left| \langle m | P_{ij}(\mathbf{q}) S_j(\mathbf{q}) | n \rangle \right|^2 \delta(\hbar\omega - E_m + E_n)$$

$$= \sum_{ij} P_{ij}(\mathbf{q}) \sum_{mn} \frac{e^{-\beta E_n}}{Z} \langle n | S_i(-\mathbf{q}) | m \rangle \langle m | S_j(\mathbf{q}) | n \rangle \delta(\hbar\omega - E_m + E_n)$$

$$= \sum_{ij} P_{ij} S_{ij}(\mathbf{q}, \omega)$$

$$= S_\perp(\mathbf{q}, \omega), \tag{13.8}$$

where S_{ij} is the dynamic spin structure factor defined by

$$S_{ij}(\mathbf{q}, \omega) = Z^{-1} \sum_n e^{-\beta E_n} \int dt e^{i\omega t} \langle n | S_i(-\mathbf{q}, t) S_j(\mathbf{q}) | n \rangle, \tag{13.9}$$

and $S_\perp(\mathbf{q}, \omega)$ is the transverse dynamic spin structure factor.

The integral over $\mathbf{k}'$ can be approximately represented [319] as

$$\int \frac{k'^2 dk'}{(2\pi)^3} d\Omega = \int dE' d\Omega' \left(\frac{m_N k_f}{8\pi^3 \hbar^2} \right). \tag{13.10}$$

k_f is the wave vector of the neutron in the final state. Furthermore, if the neutron spin is not polarized, then

$$\sum_{\alpha\beta} \left| \langle \mathbf{k}'\alpha | \psi_N^\dagger(\mathbf{k}') \sigma_{i,\alpha\beta} \psi_N(\mathbf{k}) | \mathbf{k}\beta \rangle \right|^2 = \mathrm{Tr}(\sigma_i^2) = 2. \tag{13.11}$$

In this case, we find that the mean free path $l = v_N \tau$, which is determined by the product of the scattering time τ and the neutron velocity v_N, is given by

$$\frac{1}{l} = \frac{m_N}{\hbar k_i \tau} = c \int d\Omega_{\mathbf{k}'} dE_{\mathbf{k}'} \frac{k_f}{k_i} \left(\frac{g m_N}{2\pi \hbar^2} \right)^2 S_\perp(\mathbf{q}, \omega), \tag{13.12}$$

where k_i is the initial wave vector of the neutron and c is a constant of order 1.

The cross section per unit volume $\bar{\sigma}$ is defined as

$$\bar{\sigma} = n_e \sigma = \frac{1}{l}, \tag{13.13}$$

where n_e is the electron density. Hence, the differential scattering cross section per unit volume is [319]

$$\frac{d^2\bar{\sigma}}{d\Omega d\omega} = c r_0^2 \frac{k_f}{k_i} S_\perp(\mathbf{q}, \omega), \tag{13.14}$$

where r_0 is a unit length scale

$$r_0 = \frac{|g| m_N}{2\pi \hbar^2} = \frac{|\gamma|}{2} \frac{\mu_0}{4\pi} \frac{e^2}{m} = \frac{|\gamma|}{2} \frac{1}{4\pi \epsilon_0} \frac{e^2}{mc^2} = \frac{|\gamma|}{2} r_{e,cl}. \tag{13.15}$$

$r_{e,cl} = 0.28 \times 10^{-12}$ cm is the classic electron radius, which is roughly equal to the nuclear radius.

The spin structure factor is related to the magnetic susceptibility by the fluctuation-dissipation theorem, Eq. (G.16),

$$S_{ij}(\mathbf{q}, \omega) = \frac{1}{\pi} [1 + n_B(\omega)] \, \mathrm{Im} \chi_{ij}(\mathbf{q}, \omega), \tag{13.16}$$

where

$$n_B(\omega) = \frac{1}{e^{\beta\omega} - 1} \tag{13.17}$$

is the Bose occupation number, and $\mathrm{Im} \chi_{ij}(\mathbf{q}, \omega)$ is the Fourier transform of the retarded spin susceptibility defined in Eq. (12.4).

The differential scattering cross section therefore is

$$\frac{d^2\bar{\sigma}}{d\Omega d\omega} = cr_0^2 \frac{k_f}{k_i} \left[1 + n_B(\omega)\right] P_{ij}(\mathbf{q}) \text{Im}\chi_{ij}(\mathbf{q}, \omega). \qquad (13.18)$$

In a magnetic disordered state, there is no preferred spatial direction,

$$P_{ij}(\mathbf{q}) \text{Im}\chi_{ij}(\mathbf{q}, \omega) = \sum_{ij} (\delta_{ij} - \hat{q}_i \hat{q}_j) \text{Im}\chi_{ij}(\mathbf{q}, \omega) = 2\text{Im}\chi(\mathbf{q}, \omega), \qquad (13.19)$$

where

$$\text{Im}\chi = \text{Im}\chi_{ii} = \sum_i \text{Tr}\frac{1}{3}\text{Im}\chi_{ii}. \qquad (13.20)$$

The differential scattering cross section now reduces to

$$\frac{d^2\bar{\sigma}}{d\Omega d\omega} \sim r_0^2 \frac{k_f}{k_i} \left[1 + n_B(\omega)\right] \text{Im}\chi(\mathbf{q}, \omega). \qquad (13.21)$$

At zero temperature, it becomes

$$\frac{d^2\bar{\sigma}}{d\Omega d\omega} \sim r_0^2 \frac{k_f}{k_i} \text{Im}\chi(\mathbf{q}, \omega). \qquad (13.22)$$

13.2 Magnetic Resonances in High-T$_c$ Superconductors

A striking phenomenon observed by inelastic neutron scattering measurement is the appearance of a resonance peak in the superconducting state of high-T$_c$ cuprates. This phenomenon reveals an intimate connection between antiferromagnetic correlations and high-T$_c$ superconductivity. The resonance peak was first observed in the optimally doped high-T$_c$ cuprate YBa$_2$Cu$_3$O$_7$ (T_c = 93 K) [320–322]. It is a spin triplet collective mode. As the response function is strongly enhanced near the antiferromagnetic ordering wave vector $\mathbf{Q} = (\pi, \pi)$, the peak is due to the contribution of magnetic excitations, rather than the phonon excitations [115, 323]. More specifically, as shown in Fig. 13.1, the resonance energy is centered around $\omega_0 = 41$ meV at the wave vector $\mathbf{Q}$ in the optimally doped YBa$_2$Cu$_3$O$_7$. Moreover, this resonance energy is found to be nearly temperature independent in the superconducting state [323]. By lowering the temperature, the width of the resonance is sharpened and becomes narrower than the instrumental resolution, which is typically at the order of 10 meV, at a temperature far lower than the critical temperature. The resonance intensity is weakened with increasing temperature and vanishes at the superconducting critical temperature. In other words, the resonance peak appears only in the superconducting state.

Similar spin resonances are also observed in the underdoped YBa$_2$Cu$_3$O$_{6+x}$ samples [324, 325]. The resonance is as sharp as in the optimally doped case, but the

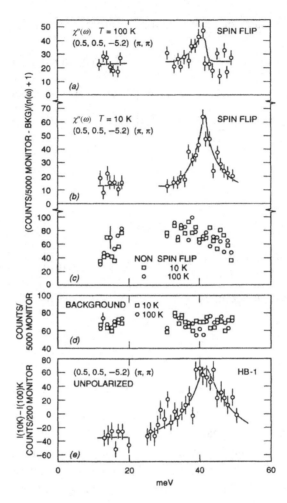

Figure 13.1 The magnetic resonances observed in the inelastic neutron scattering experiment on the optimally doped $YBa_2Cu_3O_7$. The constant-$\mathbf{q}$ scan for spin-flip scattering of polarized neutrons at $\mathbf{q} = (\pi, \pi)$ and (a) $T = 100$ K and (b) $T = 10$ K. (c) The phonon contribution to the non-spin-flip scattering. (d) The analyzer-turned background. (e) The difference between the unpolarized neutron scatterings at 10 K and 100 K. (From Mook et al. [115])

peak energy is shifted downward to $\sim$35 meV for the $x = 0.6$ case ($T_c = 62.7$ K). Furthermore, as shown in Fig. 13.2, a precursor of the resonance is observed in the normal state, whose intensity grows gradually as the temperature approaches to T_c. The resonance becomes more pronounced after crossing the critical temperature, and the intensity exhibits a cusp at T_c. In contrast, the spectral intensity at an off-resonance frequency lower than the superconducting energy gap drops with

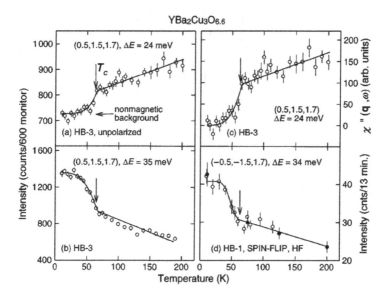

Figure 13.2 Temperature dependences of the inelastic neutron scatterings in the underdoped YBa$_2$Cu$_3$O$_{6.6}$ at an off-resonance energy 24 meV shown in (a) and (c), and at the resonance energy 35 meV shown in (b) and (d). T_c is indicated as arrows. (From Dai et al. [324])

decreasing temperature. This suggests that the spectral weight is redistributed and moves toward the resonance energy in the superconducting state.

A more systematic investigation on the doping dependence of the resonance energy and intensity was performed by Fong et al. [326]. The resonance energy, as shown in Fig. 13.3(c), increases with doping. However, the resonance intensity decreases with doping, due to the weakening effect of antiferromagnetic correlations by doping. On the other hand, the resonance width in the momentum space, $\Delta q \approx 0.25 A^{-1}$, is found to be nearly doping independent. This width corresponds to a finite magnetic correlation length of a few lattice constants.

The magnetic resonances are also observed in both the optimally doped and overdoped Bi$_2$Sr$_2$CaCu$_2$O$_{8+\delta}$ [327, 328]. For the optimally doped case ($T_c = 91$ K), the resonance is similar to YBa$_2$Cu$_3$O$_7$: The resonance occurs at (π, π) in the superconducting state at ~ 43 meV [327]. In the overdoped case ($T_c = 83$ K), the resonance frequency drops to ~ 38 meV. Unlike in YBa$_2$Cu$_3$O$_7$, the resonance peaks in both optimally doped and overdoped Bi$_2$Sr$_2$CaCu$_2$O$_{8+\delta}$ are much broadened. The broadening energy is of order of 20 meV, which may arise from the disorder effect. Combined with the data from the optimally and underdoped YBa$_2$Cu$_3$O$_{7-x}$, it seems that the resonance energy scales almost linearly with T_c [328].

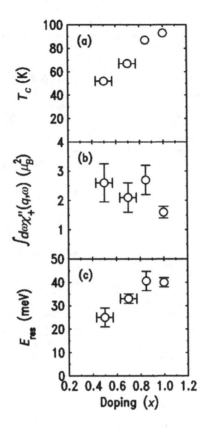

Figure 13.3 (a) Doping dependences of T_c, (b) the resonance intensity measured in the inelastic neutron scattering experiments, and (c) the resonance energy on the underdoped and optimally doped $YBa_2Cu_3O_{6+x}$. (From Fong et al. [326])

Magnetic resonances are also observed in the optimally doped monolayer cuprate $Tl_2Ba_2CuO_{6+\delta}$ ($T_c \approx 90$ K) [329]. As in $YBa_2Cu_3O_7$, this resonance peak is observed only in the superconducting state. Moreover, the resonances are very sharp and their widths are resolution-limited at low temperatures. However, the resonance energy, ~47 meV, is higher than $YBa_2Cu_3O_7$ and $Bi_2Sr_2CaCu_2O_{8+\delta}$. This is likely due to the structural difference between the monolayer and bilayer cuprates.

However, the resonance peak is not observed in another monolayer material $La_{2-x}Sr_xCuO_{4+\delta}$ [330]. In fact, the magnetic spectrum of $La_{2-x}Sr_xCuO_4$ is more complicated. Instead of showing a peak at (π, π), it exhibits four incommensurate magnetic peaks at $\mathbf{Q}_\delta = (1 \pm \delta, 1)\pi$ and $(1, 1 \pm \delta)\pi$. This four-peak structure is believed to result from the stripe instability in this class of material. In the optimally doped case, $x = 0.16$, the width of each incommensurate peak is about $\delta \approx 0.2$ at 10 meV. With the increase of the energy, the peak momentum shifts toward $\mathbf{Q} = (\pi, \pi)$ [331].

An applied magnetic field can strongly affect the magnetic resonance peak. It further demonstrates the close relation between the magnetic resonance and the superconducting coherence. For example, a modest magnetic field at 6.8 T along the *c*-axis can significantly suppress the resonance intensity in the underdoped $YBa_2Cu_3O_{6.6}$ [332]. However, the resonance energy is not changed by this magnetic field, because the Zeeman energy induced by this field, ~ 0.8 meV, is much less than the resonance energy. On the other hand, the field effect is highly anisotropic. The suppression to the resonance peak is weaker but visible when the magnetic field is applied parallel to in the *ab*-plane.

13.3 Implications of the Magnetic Resonances

The magnetic resonance observed in high-T_c superconductors can be used to estimate the condensation energy transferred from the magnetic interaction, namely how much magnetic exchange energy is saved in the superconducting state [333]. According to Scalapino and White [334], the difference in the expectation values of the Heisenberg interaction

$$H_J = J \sum_{\langle ij \rangle} \mathbf{S}_i \cdot \mathbf{S}_j \tag{13.23}$$

in the superconducting and normal states per unit cell is related to the imaginary part of dynamic spin susceptibilities by the formula

$$\Delta E_J = \frac{3\hbar}{2} J \left(\frac{a}{2\pi} \right)^2 \int_{\pi/a}^{\pi/a} d^2q \int_0^\infty \frac{d\omega}{\pi} X(\mathbf{q}, \omega), \tag{13.24}$$

where a is the lattice constant and

$$X(\mathbf{q}, \omega) = [\mathrm{Im}\chi_N(\mathbf{q}, \omega) - \mathrm{Im}\chi_S(\mathbf{q}, \omega)] \left[\cos(q_x a) + \cos(q_y a) \right]. \tag{13.25}$$

$\mathrm{Im}\chi_N(q, \omega)$ is the extrapolated value of the imaginary normal state susceptibility in the zero temperature limit. For $YBa_2Cu_3O_7$ or other bilayer cuprate superconductors, $X(\mathbf{q}, \omega)$ should be replaced by the expression

$$X(\mathbf{q}, \omega) = \frac{1}{2} \left[X^+(\mathbf{q}, \omega) + X^-(\mathbf{q}, \omega) \right], \tag{13.26}$$

in which

$$X^{\pm}(\mathbf{q}, \omega) = \left[\text{Im}\chi_N^{\pm}(\mathbf{q}, \omega) - \text{Im}\chi_S^{\pm}(\mathbf{q}, \omega) \right]$$
$$\left[J \left(\cos(q_x a) + \cos(q_y a) \right) \pm \frac{1}{2} J_{\perp} \right], \qquad (13.27)$$

and $J_{\perp}$ is the inter-bilayer magnetic exchange constant. Superscripts "+" and "−" refer to the in-phase and out-of-phase spin fluctuations between the two layers.

For the optimally doped $YBa_2Cu_3O_7$, the contribution from the interlayer coupling is negligible because $J_{\perp} \ll J$. Furthermore, the major difference between $\text{Im}\chi_N(q, \omega)$ and $\text{Im}\chi_S(q, \omega)$ is the 41 meV resonance mode in the out-of-phase channel. From the data published by Fong et al. [335], it is estimated that

$$\hbar \int_0^{\infty} d\omega \text{Im}\chi_S^-(\mathbf{Q}, \omega) \approx 0.51 \qquad (13.28)$$

at $T = 10$ K and $\mathbf{Q} = (\pi, \pi)$. The resonance width in the momentum space is about $\delta q \approx 0.23$ Å^{-1}. This leads to a rough estimation for ΔE_J,

$$\Delta E_J = \frac{3}{2} \left(\frac{a}{2\pi} \right)^2 \pi (\delta q)^2 \frac{0.51}{\pi} \frac{1}{2} 2J \approx 0.016 J. \qquad (13.29)$$

ΔE_J is about 20 K if we take $J \sim 100$ meV.

On the other hand, the condensation energy per unit cell can be estimated from the thermodynamic critical field H_c,

$$E_c = \frac{1}{2} \mu_0 H_c^2, \qquad H_c^2 = \frac{\Phi_0}{8\pi \xi_0 \lambda}, \qquad (13.30)$$

where $\Phi_0 = h/(2e)$ is the flux quantum, ξ_0 is the correlation length, and λ is the penetration depth. For the optimally doped $YBa_2Cu_3O_7$, the lattice constants are $a = 3.9$ Å and $c = 11.6$ Å, $\xi_0 \approx 12-20$Å, and $\lambda \approx 1\,300-1\,500$ Å. Substituting this set of parameters into Eq. (13.30), it is estimated that $E_c \approx 4-12$ K [333], which is of the same order as ΔE_J. This rough estimation suggests that the antiferromagnetic exchange interaction may have a substantial contribution to the superconducting condensation energy.

13.4 Origin of the Magnetic Resonance

After the discovery of the neutron resonance peak in the superconducting state of high-T_c cuprates, a number of scenarios have been proposed to unveil its microscopic driving force. It is widely believed that the resonance peak results from the interplay between antiferromagnetism and superconductivity and is a feedback from the opening of a d-wave pairing gap in the fermionic spectrum. The simplest picture put forward by various groups is that the neutron resonance is a spin excitation mode emerged due to an attractive residual spin interaction between d-wave super-conducting quasiparticles [336–338]. Alternatively, it was interpreted as a collective mode, also known as the π-mode, in the particle–particle channel [339, 340]. Below we give a brief introduction to these pictures. More detailed discussion could be found, for example, from Ref. [337].

13.4.1 Spin Excitonic Resonance Mode

A spin-1 collective excitation mode in the particle–hole channel (or a spin exciton) is highly damped and difficult to see in the normal state due to its interaction with other low-lying excitations. In the superconducting state, this damping effect is greatly suppressed by the opening of the pairing energy gap. This would sharpen the width of the particle–hole spin excitation mode and allow it to be observed in the neutron scattering measurement. To understand this clearly, let us consider how the dynamic spin susceptibility $\chi(\mathbf{q}, \omega)$ is renormalized by an effective Hubbard interaction U.

We start from the spin susceptibility in the absence of interaction, i.e. $\chi_0(\mathbf{q}, \omega)$. In the presence of interaction, it becomes

$$\chi(\mathbf{q}, \omega) = \frac{\chi_0(\mathbf{q}, \omega)}{1 - U\chi_0(\mathbf{q}, \omega)} \tag{13.31}$$

under the random-phase approximation. The imaginary part of the susceptibility, which is proportional to the differential cross section measured by the inelastic neutron scattering spectroscopy, is

$$\mathrm{Im}\,\chi(\mathbf{q}, \omega) = \frac{\mathrm{Im}\,\chi_0(\mathbf{q}, \omega)}{[1 - U\mathrm{Re}\,\chi_0(\mathbf{q}, \omega)]^2 + [U\mathrm{Im}\,\chi_0(\mathbf{q}, \omega)]^2}. \tag{13.32}$$

In the normal state, the unperturbed susceptibility, $\chi_0(\mathbf{q}, \omega)$ is described by the Lindhardt-type response function

$$\chi_0(\mathbf{q}, \omega) = \int \frac{d^2\mathbf{k}}{(2\pi)^2} \frac{f(\xi_{\mathbf{k}+\mathbf{q}}) - f(\xi_{\mathbf{k}})}{\omega - (\xi_{\mathbf{k}+\mathbf{q}} - \xi_{\mathbf{k}}) + i0^+}. \tag{13.33}$$

If the kinetic energy is dominated by the nearest-neighbor hopping, the energy dispersion of electrons is

$$\xi_{\mathbf{k}} = -2t(\cos k_x + \cos k_y) - \mu. \tag{13.34}$$

It contains a nesting vector $\mathbf{Q} = (\pi, \pi)$ and satisfies the equation

$$\xi_{\mathbf{k}+\mathbf{Q}} + \mu = -(\xi_{\mathbf{k}} + \mu). \tag{13.35}$$

For a state on the Fermi surface, i.e. $\xi_{\mathbf{k}_F} = 0$, the following nesting condition holds

$$\xi_{\mathbf{k}_F+\mathbf{Q}} - \xi_{\mathbf{k}_F} = -2\mu. \tag{13.36}$$

At half-filling, $\mu = 0$, the Fermi surface itself is nested. This gives rise to a logarithmic divergence in the antiferromagnetic spin susceptibility, leading to the antiferromagnetic long-range order in the ground state. Upon hole doping, $\mu < 0$, the nesting becomes dynamic at a finite energy $\Omega = 2|\mu|$, which renders a sharp peak (or more precisely a logarithmic divergence), which is broadened by the Landau damping effect, in $\text{Im}\chi_0(\mathbf{Q}, \omega)$ at $\omega = \Omega$.

In the superconducting state, the unperturbed dynamic spin susceptibility is given by Eq. (12.7) [341]. Up to a constant prefactor, the diagonal component of the spin susceptibility is

$$\chi_0(\mathbf{q}, \omega) \propto \sum_{\mathbf{k}, p_m} \text{Tr} G^{(0)}(\mathbf{k}, ip_m) G^{(0)}(\mathbf{q} + \mathbf{k}, i\omega_n + ip_m). \tag{13.37}$$

Inserting (4.4) into the above expression and taking the Matsubara frequency summation yields

$$\chi_0(\mathbf{q}, \omega) \propto \frac{1}{2} [I_1(\mathbf{q}, \omega) - I_2(\mathbf{q}, \omega)], \tag{13.38}$$

where I_1 results from the particle–hole excitations of Bogoliubov quasiparticles

$$I_1(\mathbf{q}, \omega) = \int \frac{d^2\mathbf{k}}{(2\pi)^2} \left(1 + \frac{\xi_{\mathbf{k}}\xi_{\mathbf{k}+\mathbf{q}}}{E_{\mathbf{k}}E_{\mathbf{k}+\mathbf{q}}}\right) \frac{f(E_{\mathbf{k}+\mathbf{q}}) - f(E_{\mathbf{k}})}{\omega - (E_{\mathbf{k}+\mathbf{q}} - E_{\mathbf{k}}) + i0^+} \tag{13.39}$$

and I_2 contributes from the pair-creation and annihilation process

$$I_2(\mathbf{q}, \omega) = \int \frac{d^2\mathbf{k}}{(2\pi)^2} \left(1 - \frac{\xi_{\mathbf{k}}\xi_{\mathbf{k}+\mathbf{q}} + \Delta_{\mathbf{k}}\Delta_{\mathbf{k}+\mathbf{q}}}{E_{\mathbf{k}}E_{\mathbf{k}+\mathbf{q}}}\right) [1 - f(E_{\mathbf{k}+\mathbf{q}}) - f(E_{\mathbf{k}})]$$
$$\frac{E_{\mathbf{k}+\mathbf{q}} + E_{\mathbf{k}}}{(\omega + i0^+)^2 - (E_{\mathbf{k}+\mathbf{q}} + E_{\mathbf{k}})^2}. \tag{13.40}$$

Exactly at zero temperature, as there is no thermal excitation of superconducting quasiparticles, the Fermi distribution function $f(E_{\mathbf{k}})$ vanishes so that $I_1 = 0$. However, I_2 is finite even at zero temperature

$$I_2(\mathbf{q}, \omega) = \int \frac{d^2\mathbf{k}}{(2\pi)^2} \left(1 - \frac{\xi_{\mathbf{k}}\xi_{\mathbf{k}+\mathbf{q}} + \Delta_{\mathbf{k}}\Delta_{\mathbf{k}+\mathbf{q}}}{E_{\mathbf{k}}E_{\mathbf{k}+\mathbf{q}}}\right)$$
$$\frac{E_{\mathbf{k}+\mathbf{q}} + E_{\mathbf{k}}}{(\omega + i0^+)^2 - (E_{\mathbf{k}+\mathbf{q}} + E_{\mathbf{k}})^2}, \quad (T = 0). \tag{13.41}$$

This implies that the susceptibility contributes mainly by the I_2 term at low temperatures.

The integrand in I_2 diverges when the frequency equals the pair energy of quasiparticles, or the negative pair energy of quasiholes,

$$\omega = E_{\mathbf{k}+\mathbf{q}} + E_{\mathbf{k}}. \tag{13.42}$$

The integral of I_2, on the other hand, is dominated by the contribution where the density of states of a pair of quasiparticle excitations becomes singular. This happens when the Fermi surface or a constant energy surface is nested. Without doping, the Fermi surface, at which $\xi_k = 0$, is perfectly nested and the nested wave vector is $\mathbf{Q} = (\pi, \pi)$. This nested surface also exists at finite doping, but its energy is shifted to $\xi_k = -\mu$. This nested constant energy surface has the largest contribution to the integral in I_2, and the corresponding energy is

$$\omega = \pm 2\sqrt{\mu^2 + \Delta_k^2}. \tag{13.43}$$

As the absolute value of μ is generally much larger than the superconducting energy gap, i.e. $|\mu| \gg |\Delta_k|$ in the doped system where the magnetic resonance is observed, the singular energy is approximately given by

$$\omega \approx \omega_0 = \pm 2|\mu| \tag{13.44}$$

independent on $\mathbf{k}$. If the integral in I_2 becomes sufficient large so that $1 - U\mathrm{Re}\chi_0(\mathbf{Q}, \omega_0)$ approaches zero, a resonance peak would appear in the neutron scattering spectrum.

In general, the resonance peak emerges when the condition

$$|1 - U\mathrm{Re}\chi_0(\mathbf{Q}, \omega_0)| \ll U\mathrm{Im}\chi_0(\mathbf{Q}, \omega_0) \tag{13.45}$$

is satisfied. As the number of excitation states that satisfy the momentum and energy conservations are dramatically suppressed by the superconducting energy gap, $\mathrm{Im}\chi_0(\mathbf{Q}, \omega_0)$ is also significantly reduced in comparison with the normal state. Furthermore, since the intensity of the resonance scales as $1/\mathrm{Im}\chi_0(\mathbf{Q}, \omega_0)$, the resonance peak is sharpened in the superconducting state.

Besides the frequency dependent terms, there is also a coherence form factor in I_2. When the nesting condition is satisfied, $\mathbf{q} = \mathbf{Q}$, the coherence factor is

$$1 - \frac{\xi_k \xi_{k+Q} + \Delta_k \Delta_{k+Q}}{E_k E_{k+Q}} = 1 - \frac{\xi_k \xi_{k+Q}}{E_k E_{k+Q}} - \frac{\Delta_k \Delta_{k+Q}}{E_k E_{k+Q}}. \tag{13.46}$$

In a d-wave superconducting pairing state, since the gap function Δ_k and Δ_{k+Q} always have opposite signs, it shows that the coherence factor is enhanced. On the other hand, in an s-wave pairing state, the product $\Delta_k \Delta_{k+Q}$ is always positive and the coherence factor is reduced. This suggests that the resonance peak should be more pronounced in the d-wave superconducting state than in the s-wave one.

The above analysis relies strongly on the assumption that there is a strong nesting effect, induced by a nested energy surface connected by the wave vector $\mathbf{Q}$, that can dramatically increase the integral in I_2 in high-T_c copper oxides. This nesting effect leads to a quasisingular behavior in the spin spectral function when the condition (13.45) is satisfied. In the superconducting state, the d-wave pairing gap enhances the coherence factor, but suppresses the spectral weight of unperturbed spin excitation spectrum $\text{Im}\,\chi_0(\mathbf{q}, \omega)$. This yields the resonance as observed by the experiments.

13.4.2 π-*Resonance Mode*

Unlike the excitonic mode in the particle–hole channel, the π-mode is a collective mode in the particle–particle channel [339, 340]. This mode involves the change of the charge number in addition to the change of spin. It does not couple to the particle–hole excitation that is detected by neutron scattering in the normal state. However, in the superconducting state, particles are mixed with holes by the pairing order parameter, allowing the π-mode to be probed in the neutron scattering measurement. A natural consequence of this scenario is that the resonance energy is nearly independent of temperature since it already exists in the normal state, but the intensity of resonance is proportional to $|\Delta_0|^2$ resulting from the particle–hole mixing.

The idea of π-resonance mode is motivated by the η-pairing picture first introduced for the Hubbard model. The η-pairing is a well-defined collective mode in the particle–particle channel [94]. The η-operators are defined for the single-band Hubbard model Eq. (2.35) on a bipartite lattice. On the two-dimensional square lattice, the η-operators carry the momentum $\mathbf{Q} = (\pi, \pi)$,

$$\eta^\dagger = \frac{1}{N} \sum_\mathbf{k} c^\dagger_{\mathbf{k}+\mathbf{Q}\uparrow} c^\dagger_{-\mathbf{k}\downarrow}. \tag{13.47}$$

Their commutator equals the particle number operator

$$[\eta^\dagger, \eta] = \sum_\mathbf{k} \left(c^\dagger_{\mathbf{k},\uparrow} c_{\mathbf{k},\uparrow} + c^\dagger_{\mathbf{k},\downarrow} c_{\mathbf{k},\downarrow} - 1 \right) = \hat{N} - \frac{N}{2}. \tag{13.48}$$

Hence, $\eta^\dagger$, η, and $\hat{N} - N/2$ form an SU(2) algebra, which is often referred to as the pseudospin algebra. It is simple to show that the single-band Hubbard model (2.35) satisfies

$$[H, \eta^\dagger] = \left(\frac{U}{2} - \mu \right) \eta^\dagger, \qquad [H, \eta] = - \left(\frac{U}{2} - \mu \right) \eta, \tag{13.49}$$

where μ is the chemical potential.

At half-filling, the Hubbard model is particle–hole symmetric and the chemical potential $\mu = U/2$. In this case, $\eta^\dagger$ and η commute with H, and the pseudo-spin SU(2) symmetry is exact.

Away from the half-filling, for example, if $n < 1$, or equivalently $\mu < U/2$, applying $\eta^\dagger$ on the many-body ground state $|0\rangle$ generates an excited eigen-mode $\eta^\dagger |\Omega\rangle$,

$$H\eta^{\dagger}|\Omega\rangle = [H, \eta^{\dagger}]|0\rangle = \left(\frac{U}{2} - \mu\right)\eta^{\dagger}|0\rangle, \tag{13.50}$$

which carries momentum $\mathbf{Q}$ and energy $\omega = (U/2) - \mu$.

The η-mode could be understood from the negative-U Hubbard model in which the superconductivity coexists with the charge-density wave order. The superconducting and charge-density wave order parameters are defined by

$$\Delta^{\dagger} = \sum_{\mathbf{k}} c^{\dagger}_{\mathbf{k}\uparrow} c^{\dagger}_{-\mathbf{k}\downarrow}, \qquad O_{cdw} = \sum_{\mathbf{k},\sigma} c^{\dagger}_{\mathbf{k}+\mathbf{Q},\sigma} c_{\mathbf{k},\sigma}. \tag{13.51}$$

The superconducting order parameter $\Delta^{\dagger}$ is complex. $(\text{Re}\Delta, \text{Im}\Delta, O_{cdw})$ form a three-vector representation of the pseudospin SU(2) algebra. It is simple to show that

$$[\eta^{\dagger}, \Delta] = O_{cdw}. \tag{13.52}$$

At half-filling, the superconducting order is degenerate with the charge-density wave one, as a manifestation of the pseudospin SU(2) symmetry. $\eta^{\dagger}|\Omega\rangle$ is simply the Goldstone mode. Hole doping away from the half-filling, the pseudospin SU(2) symmetry is broken, and the ground state remains superconducting. Nevertheless, $\eta^{\dagger}|\Omega\rangle$ remains an eigenstate but gains an finite excitation energy.

The above pseudospin symmetry of the spin-$\frac{1}{2}$ Hubbard model also exists in the four-component spin-3/2 Hubbard model defined on a bipartite lattice [342, 343]. Depending on the coupling parameters, this model system could exhibit either a pseudospin SU(2) or an SO(7) symmetry at half-filling. Either of these symmetries correlates the superconducting order in the particle–particle channel with the charge density wave order or the spin-quadruple density wave order in the particle–hole channel.

The π-mode is a generalization of the above η-pairing picture to a system with repulsive interaction in a d-wave superconducting state with strong antiferromagnetic fluctuations. To describe quantitatively this mode, let us consider an extended Hubbard model that includes explicitly a Heisenberg antiferromagnetic interaction term

$$H_J = J \sum_{\langle ij \rangle} \mathbf{S}_i \cdot \mathbf{S}_j. \tag{13.53}$$

No constraint is imposed to the electron number on one site. For simplicity, we decouple the Hubbard interaction at the mean-field level. This yields a non-interacting Hamiltonian

$$H_0 = \sum_{\mathbf{k},\sigma} \xi_{\mathbf{k}} c^{\dagger}_{\mathbf{k}\sigma} c_{\mathbf{k}\sigma}, \tag{13.54}$$

with

$$\xi_{\mathbf{k}} = -2t \left(\cos k_x + \cos k_y \right) + \frac{U}{2} \langle n \rangle - \mu. \tag{13.55}$$

The term with the average occupation number of electrons per site, $\langle n \rangle$, results from the mean-field approximation of the Hubbard interaction.

The π-mode is a triplet spin excitation state

$$|\pi\rangle = \pi_{\mathbf{Q}}^{\dagger}|0\rangle \tag{13.56}$$

created by the operator

$$\pi_{\mathbf{Q}}^{\dagger} = \sum_{\mathbf{k}} \varphi_{\mathbf{k}} c_{\mathbf{k}+\mathbf{Q}\uparrow}^{\dagger} c_{-\mathbf{k}\uparrow}^{\dagger}, \tag{13.57}$$

where $\varphi_{\mathbf{k}}$ is the wavefunction of this π-mode. This π-mode is a single mode approximation to the lowest triplet excitation in the particle–particle channel. $|0\rangle$ is the ground state of the total Hamiltonian

$$H = H_0 + H_J - E_0. \tag{13.58}$$

Here we add a constant E_0 to the Hamiltonian so that its ground state energy is zero.

We determine the wavefunction of the π-mode by variationally minimizing its energy

$$E_{\pi} = \frac{\langle \pi | H | \pi \rangle}{\langle \pi | \pi \rangle}. \tag{13.59}$$

This expectation value can be evaluated directly using the above expressions. Assuming the Wick theorem holds for operators acting on the many-body ground state $|0\rangle$ of H, and after a lengthy calculation, we find that

$$\langle \pi | H | \pi \rangle = \frac{J}{2N} \sum_{\mathbf{k}\mathbf{k}'} g_{\mathbf{k},\mathbf{k}'} \left(1 - n_{-\mathbf{k}'\uparrow} - n_{\mathbf{k}'+\mathbf{Q}\uparrow} \right) \chi_{\mathbf{k},\mathbf{k}'} + \sum_{\mathbf{k}} \omega_{\mathbf{k}} \chi_{\mathbf{k},\mathbf{k}}, \tag{13.60}$$

and

$$\langle \pi | \pi \rangle = \sum_{\mathbf{k}} \chi_{\mathbf{k},\mathbf{k}}, \tag{13.61}$$

where

$$g_{\mathbf{k},\mathbf{k}'} = \cos k_x \cos k_x' + \cos k_y \cos k_y', \tag{13.62}$$

is a formal factor resulting from the Heisenberg interaction and

$$\chi_{\mathbf{k},\mathbf{k}'} = \left(\varphi_{\mathbf{k}'}^{*} - \varphi_{-\mathbf{k}'-\mathbf{Q}}^{*} \right) \varphi_{\mathbf{k}} \left(1 - n_{-\mathbf{k}\uparrow} \right) \left(1 - n_{\mathbf{k}+\mathbf{Q}\uparrow} \right). \tag{13.63}$$

$n_{k\sigma}$ is the particle occupation number of momentum $\mathbf{k}$ and spin σ in the ground state

$$n_{\mathbf{k}\sigma} = \langle 0 | c_{\mathbf{k}\sigma}^\dagger c_{\mathbf{k}\sigma} | 0 \rangle. \tag{13.64}$$

$\omega_{\mathbf{k}}$ is a momentum dependent energy

$$\omega_{\mathbf{k}} = \frac{J}{2N} \sum_{\mathbf{q}} \left[4 \left(n_{\mathbf{q}\uparrow} - n_{\mathbf{q}\downarrow} \right) - \gamma_{\mathbf{k}+\mathbf{q}} \left(n_{\mathbf{q}\uparrow} + 2 n_{\mathbf{q}\downarrow} - n_{-\mathbf{q}\uparrow} - 2 n_{-\mathbf{q}\downarrow} \right) \right]$$
$$+ \xi_{\mathbf{k}+\mathbf{Q}} + \xi_{-\mathbf{k}}, \tag{13.65}$$

with

$$\gamma_{\mathbf{k}} = \cos k_x + \cos k_y. \tag{13.66}$$

If the ground state is spin polarization free, then the occupation number of up-spin electrons equals that of down-spin electrons and the above expression becomes

$$\omega_{\mathbf{k}} = \xi_{\mathbf{k}+\mathbf{Q}} + \xi_{-\mathbf{k}} = U \langle n \rangle - 2\mu, \tag{13.67}$$

independent of momentum $\mathbf{k}$ in a system with only nearest neighbor hopping terms whose single-particle energy dispersion $\xi_{\mathbf{k}}$ is defined by Eq. (13.55).

Taking the derivative of the energy expectation value of the π-mode, Eq. (13.59), with respect to $(\varphi_{\mathbf{k}'}^* - \varphi_{-\mathbf{k}'-\mathbf{Q}}^*)$, we obtain the self-consistent equation that determines the wave function $\varphi_{\mathbf{k}}$

$$\frac{J}{2N} \left(1 - n_{-\mathbf{q}\uparrow} - n_{\mathbf{q}+\mathbf{Q}\uparrow} \right) \sum_{\mathbf{k}} g_{\mathbf{k},\mathbf{q}} f_{\mathbf{k}} = \left(E_\pi - \omega_{\mathbf{q}} \right) f_{\mathbf{q}}, \tag{13.68}$$

where

$$f_{\mathbf{k}} = \varphi_{\mathbf{k}} \left(1 - n_{-\mathbf{k}\uparrow} \right) \left(1 - n_{\mathbf{k}+\mathbf{Q}\uparrow} \right). \tag{13.69}$$

To simplify the equation, we factorize $g_{\mathbf{k},\mathbf{k}'}$ using the symmetric functions in the extended s- and d-wave channels

$$g_{\mathbf{k},\mathbf{k}'} = \frac{1}{2} \gamma_{\mathbf{k}} \gamma_{\mathbf{k}'} + \frac{1}{2} d_{\mathbf{k}} d_{\mathbf{k}'} \tag{13.70}$$

where

$$d_{\mathbf{k}} = \cos k_x - \cos k_y. \tag{13.71}$$

Close to half-filling,

$$\gamma_{\mathbf{k}} = \cos k_x + \cos k_y \approx 0, \tag{13.72}$$

we therefore have

$$g_{\mathbf{k},\mathbf{k}'} \approx \frac{1}{2} d_{\mathbf{k}} d_{\mathbf{k}'}. \tag{13.73}$$

Inserting it into Eq. (13.68), we find the equation that determines the energy

$$\frac{J}{4N} \sum_{\mathbf{q}} \frac{d_{\mathbf{q}}^2 \left(1 - n_{-\mathbf{q}\uparrow} - n_{\mathbf{q}+\mathbf{Q}\uparrow}\right)}{E_\pi - \omega_{\mathbf{q}}} = 1. \tag{13.74}$$

It is simple to show that the corresponding solution of the wavefunction is

$$f_{\mathbf{q}} = \frac{d_{\mathbf{q}} \left(1 - n_{-\mathbf{q}\uparrow} - n_{\mathbf{q}+\mathbf{Q}\uparrow}\right)}{E_\pi - \omega_{\mathbf{q}}}, \tag{13.75}$$

hence

$$\varphi_{\mathbf{q}} = \frac{d_{\mathbf{q}} \left(1 - n_{-\mathbf{q}\uparrow} - n_{\mathbf{q}+\mathbf{Q}\uparrow}\right)}{\left(E_\pi - \omega_{\mathbf{q}}\right) \left(1 - n_{-\mathbf{q}\uparrow}\right) \left(1 - n_{\mathbf{q}+\mathbf{Q}\uparrow}\right)}. \tag{13.76}$$

In case $\omega_{\mathbf{q}} = U \langle n \rangle - 2\mu = \omega_0$, the energy of the π-mode is simply given by

$$E_\pi = \omega_0 + \frac{J}{4N} \sum_{\mathbf{q}} d_{\mathbf{q}}^2 \left(1 - n_{-\mathbf{q}\uparrow} - n_{\mathbf{q}+\mathbf{Q}\uparrow}\right)$$

$$= \omega_0 + \frac{J}{4N} \sum_{\mathbf{q}} d_{\mathbf{q}}^2 \left(1 - n_{\mathbf{q}}\right). \tag{13.77}$$

The corresponding wavefunction is

$$\varphi_{\mathbf{q}} \propto \frac{d_{\mathbf{q}} \left(1 - n_{-\mathbf{q}\uparrow} - n_{\mathbf{q}+\mathbf{Q}\uparrow}\right)}{\left(1 - n_{-\mathbf{q}\uparrow}\right) \left(1 - n_{\mathbf{q}+\mathbf{Q}\uparrow}\right)} \approx d_{\mathbf{q}} \tag{13.78}$$

up to a normalization constant. The corresponding π-mode creation operator is

$$\pi_{\mathbf{Q}}^\dagger = \frac{A}{\sqrt{N}} \sum_{\mathbf{k}} d_{\mathbf{k}} c_{\mathbf{Q}+\mathbf{k}\uparrow}^\dagger c_{-\mathbf{k}\uparrow}^\dagger, \tag{13.79}$$

where N is the lattice size. A is a normalization constant determined by the equation

$$\langle \pi | \pi \rangle = \frac{2A^2}{N} \sum_{\mathbf{k}} d_{\mathbf{k}}^2 (1 - n_{-\mathbf{k}\uparrow})(1 - n_{\mathbf{k}+\mathbf{Q}\uparrow}) = 1, \tag{13.80}$$

so that

$$A = \left[\frac{2}{N} \sum_{\mathbf{k}} d_{\mathbf{k}}^2 (1 - n_{-\mathbf{k}\uparrow})(1 - n_{\mathbf{k}+\mathbf{Q}\uparrow})\right]^{-1/2}. \tag{13.81}$$

The π-mode is a spin triplet excitation in the particle–particle channel, which is not the spin excitation directly probed by neutron scattering spectroscopy in the particle–hole channel. This mode could be detected by the neutron scattering, however, in a superconducting state thanks to the particle–hole mixing of Cooper pairs.

The transverse dynamic spin structure factor at zero temperature could be represented as

$$\text{Im}\chi(\mathbf{Q},\omega) = \sum_n |\langle n|S_{\mathbf{Q}}^+|0\rangle|^2 \delta(\omega - \omega_n), \tag{13.82}$$

where $|n\rangle$ is an excitation state of H with ω_n the corresponding excitation energy.

$$S_{\mathbf{Q}}^+ = \frac{1}{\sqrt{N}} \sum_{\mathbf{q}} c_{\mathbf{q}+\mathbf{Q}\uparrow}^\dagger c_{\mathbf{q}\downarrow} \tag{13.83}$$

is the spin flip operator of momentum $\mathbf{Q}$. The contribution from the π-mode under the single-mode approximation to the dynamic spin structure factor is

$$\text{Im}\chi_\pi(\mathbf{Q},\omega) = |\langle \pi|S_{\mathbf{Q}}^+|0\rangle|^2 \delta(\omega - E_\pi). \tag{13.84}$$

Using the property $\pi_{\mathbf{Q}}|0\rangle = 0$, the above expression can be also written as

$$\text{Im}\chi_\pi(\mathbf{Q},\omega) = \left|\langle 0|[\pi_{\mathbf{Q}}, S_{\mathbf{Q}}^\dagger]|0\rangle\right|^2 \delta(\omega - E_\pi). \tag{13.85}$$

The commutator $[\pi_{\mathbf{Q}}, S_{\mathbf{Q}}^\dagger]$ is nothing but the d-wave pairing operator

$$[\pi_{\mathbf{Q}}, S_{\mathbf{Q}}^\dagger] = \frac{2A}{N} \sum_{\mathbf{k}} d_{\mathbf{k}} c_{\mathbf{k}\uparrow} c_{-\mathbf{k}\downarrow}. \tag{13.86}$$

In the normal state, the expectation value of this operator vanishes. However, in the superconducting state, its expectation is proportional to the gap order parameter Δ_0 in the ground state

$$\langle 0|[\pi_{\mathbf{Q}}, S_{\mathbf{Q}}^+]|0\rangle = \frac{2A\Delta_0}{g}, \tag{13.87}$$

where g is the coupling constant of the pairing interaction. Thus the contribution of the π-mode is

$$\text{Im}\chi_\pi(\mathbf{Q},\omega) = \frac{4A^2\Delta_0^2}{g^2}\delta(\omega - \omega_\pi). \tag{13.88}$$

This expression holds under the single-mode approximation. In real systems, the delta-function will be broadened by the interaction of this π-mode with other excitation states as well as the disorder effect, and become a broad resonance peak. The intensity of the resonance peak is proportional to Δ_0^2. As both A and g are nearly temperature independent, the spectral weight of the resonance should follow the temperature dependence of Δ_0^2 which decreases with increasing temperature and vanishes in the normal state. This is an intrinsic property of the π-mode. It could be used to reveal the microscopic origin of the resonance peak observed in the neutron scattering measurement.

14

Mixed State

14.1 Caroli–de Gennes–Matricon Vortex Core State

A type-II superconductor is in a mixed state with quantized magnetic flux lines, when exposed to an external magnetic field that is higher than the lower critical field but lower than the upper critical field. The flux lines are also the vortex lines of the superconducting order parameter. The presence of vortices changes the quasiparticle excitation spectra, leading to the change of thermodynamic, as well as transport, properties of the superconducting state. Quasiparticle excitations around vortex cores behave differently in the s- and d-wave superconductors. This leads to the difference in the field dependence of the specific heat and other physical quantities in these two kinds of superconductor.

The vortex state is a good example of inhomogeneity that can be studied with the BdG equation. It turns out that in each vortex, not all of the excitation states are pushed out of the superconducting energy gap. Instead, there are a number of discretized energy states within the energy gap. These vortex excitation states are called Caroli–de Gennes–Matricon states. They are localized around the vortex core. For better understanding the physics of the mixed state in a d-wave state, let us first consider the Caroli–de Gennes–Matricon vortex core states in an s-wave superconductor.

14.1.1 BdG Equation in the Extreme Type-II Limit

We start from the BdG equation for the s-wave superconductor

$$\begin{pmatrix} H_0(\mathbf{r}) & \Delta(\mathbf{r}) \\ \Delta^*(\mathbf{r}) & -H_0(\mathbf{r}) \end{pmatrix} \begin{pmatrix} u(\mathbf{r}) \\ v(\mathbf{r}) \end{pmatrix} = E \begin{pmatrix} u(\mathbf{r}) \\ v(\mathbf{r}) \end{pmatrix}, \qquad (14.1)$$

where

$$H_0(\mathbf{r}) = \frac{1}{2m} (-i\hbar\nabla - e\mathbf{A})^2 - \mu, \qquad (14.2)$$

and E is the eigenenergy. We seek the solution of this equation in the extreme type-II limit $\xi \ll \lambda$.

In a single-vortex state, the magnetic field is roughly confined within an area of $\pi \lambda^2$ and approximately given by

$$B \sim \frac{\Phi_0}{\pi \lambda^2}, \tag{14.3}$$

which is nearly zero in the large λ limit. This suggests that we can ignore the gauge vector term in H_0 in solving the vortex core state in the limit $\lambda \gg \xi$, so that

$$H_0(\mathbf{r}) = -\frac{\hbar^2}{2m}\nabla^2 - \mu. \tag{14.4}$$

The gap parameter $\Delta(\mathbf{r})$ is now a complex. It acquires a phase when $\mathbf{r}$ swirls around the core center. In the polar coordinates, $\mathbf{r} = (r, \varphi, z)$ (φ is the azimuthal angle), the gap function could be represented as

$$\Delta(\mathbf{r}) = \Delta(r)e^{i\varphi}. \tag{14.5}$$

On the basis of Ginzburg–Landau theory, we expect that $\Delta(r) \approx r$ as $r \to 0$ and $\Delta(r) = \Delta_\infty$ as $r \to \infty$ with Δ_∞ the equilibrium gap.

It is simple to show that the angular momentum along the z-axis

$$L_z = -i\hbar\frac{\partial}{\partial\varphi} - \frac{\hbar}{2}\sigma_z \tag{14.6}$$

is conserved. This motivates us to seek the solution of the form

$$\begin{pmatrix} u(\mathbf{r}) \\ v(\mathbf{r}) \end{pmatrix} = e^{ik_z z}e^{in\varphi}\begin{pmatrix} f(r)e^{i\varphi} \\ g(r) \end{pmatrix}, \tag{14.7}$$

which is an eigenstate of L_z

$$L_z\begin{pmatrix} u(\mathbf{r}) \\ v(\mathbf{r}) \end{pmatrix} = \left(n + \frac{1}{2}\right)\hbar\begin{pmatrix} u(\mathbf{r}) \\ v(\mathbf{r}) \end{pmatrix}, \tag{14.8}$$

where n is an integer. Substituting (14.7) into (14.1), we find the equation that determines f and g

$$\begin{pmatrix} h_{n+1}(r) & \Delta(r) \\ \Delta(r) & -h_n(r) \end{pmatrix}\begin{pmatrix} f(r) \\ g(r) \end{pmatrix} = E\begin{pmatrix} f(r) \\ g(r) \end{pmatrix}, \tag{14.9}$$

where

$$h_n(r) = -\frac{\hbar^2}{2m}\left(\frac{d^2}{dr^2} + \frac{1}{r}\frac{d}{dr} - \frac{n^2}{r^2} + k_\parallel^2\right), \tag{14.10}$$

and

$$k_{\parallel} = \sqrt{k_f^2 - k_z^2} \tag{14.11}$$

is the momentum in the xy-plane. $k_{\parallel}$ is assumed to be of the same order as $k_f = \sqrt{2m\mu}/\hbar$. Furthermore, it is straightforward to show that the vector

$$\begin{pmatrix} u(\mathbf{r}) \\ v(\mathbf{r}) \end{pmatrix} = e^{ik_z z} e^{-in\varphi} \begin{pmatrix} g(r) \\ -f(r)e^{-i\varphi} \end{pmatrix} \tag{14.12}$$

is also a solution of Eq. (14.9) with an eigenenergy $(-E)$, corresponding to the antiquasiparticle eigenstate of the former solution. Thus without loss of generality, we can always assume that n is nonnegative definite.

14.1.2 Envelope Functions

Deep inside the vortex core, $r \to 0$, the gap parameter $\Delta(r) \to 0$, the BdG equation becomes the simple Bessel's differential equations

$$\left[\frac{d^2}{dr^2} + \frac{1}{r} \frac{d}{dr} - \frac{(n+1)^2}{r^2} + k_+^2 \right] f(r) = 0, \tag{14.13}$$

$$\left(\frac{d^2}{dr^2} + \frac{1}{r} \frac{d}{dr} - \frac{n^2}{r^2} + k_-^2 \right) g(r) = 0, \tag{14.14}$$

where

$$k_{\pm} = \sqrt{k_{\parallel}^2 \pm \frac{2mE}{\hbar^2}} \approx k_{\parallel} \pm k_0, \tag{14.15}$$

with

$$k_0 = \frac{Em}{\hbar^2 k_{\parallel}}. \tag{14.16}$$

The solutions of $f(r)$ and $g(r)$ are given by

$$f = A_+ J_{n+1}(k_+ r), \qquad g = A_- J_n(k_- r), \tag{14.17}$$

where $A_{\pm}$ are normalization constants. J_n is the nth order Bessel function of the first kind, which is finite at the origin ($r = 0$) for positive n.

In general, $f(r)$ and $g(r)$ are mixed by $\Delta(r)$. To find the solution, we rewrite Eq. (14.9) as

$$[h_l(r)\sigma_3 + \Delta(r)\sigma_1] \begin{pmatrix} f(r) \\ g(r) \end{pmatrix} = \left(E - \frac{\hbar^2}{4m} \frac{2n+1}{r^2} \right) \begin{pmatrix} f(r) \\ g(r) \end{pmatrix}, \tag{14.18}$$

where l is the square average of n and $n + 1$

$$l = \sqrt{\frac{n^2 + (n+1)^2}{2}} = \sqrt{n^2 + n + \frac{1}{2}}. \tag{14.19}$$

Now we use the Henkel functions of the first and second kinds to expand $f(r)$ and $g(r)$ as

$$\begin{pmatrix} f(r) \\ g(r) \end{pmatrix} = H_l^{(1)}(k_\parallel r) \begin{pmatrix} \tilde{f}(r) \\ \tilde{g}(r) \end{pmatrix} + h.c. \tag{14.20}$$

$H_l^{(1)}$ is the Hankel functions of the first kind. It is a linear combination of the Bessel functions of the first and second kinds, corresponding to the propagating wave solution of the Bessel equation. The Hankel function contains the rapidly oscillating part of the radial wavefunction while $(\tilde{f}, \tilde{g})$ are envelope functions which account for slow variations in the amplitude and phase caused by the slowly varying gap function $\Delta(r)$.

$H_l^{(1)}$ is the solutions of the Bessel equation, namely

$$h_l(r) H_l^{(1)}(k_\parallel r) = 0. \tag{14.21}$$

For real $l > 0$, the Bessel functions of the first and second kinds are all real. In particular, the Bessel function of the first kind is the real part of the Hankel function, and the Bessel function of the second kind is the imaginary part of the first Hankel function. In the limit $x = k_\parallel r \to +\infty$,

$$H_l^{(1)}(x) \sim \sqrt{\frac{2}{\pi x}} \exp\left[i \left(x - \frac{l\pi}{2} + \frac{l^2}{2x} - \frac{\pi}{4} \right) \right]. \tag{14.22}$$

From (14.18), it is simple to show that the envelope functions are governed by the equation

$$\left[-\frac{\hbar^2}{2m} \left(\frac{d^2}{dr^2} + \frac{1}{r}\frac{d}{dr} + 2\frac{d \ln H_l^{(1)}}{dr}\frac{d}{dr} \right) \sigma_3 + \Delta(r)\sigma_1 \right] \begin{pmatrix} \tilde{f}(r) \\ \tilde{g}(r) \end{pmatrix}$$

$$= \left(E - \frac{\hbar^2}{4m}\frac{2n+1}{r^2} \right) \begin{pmatrix} \tilde{f}(r) \\ \tilde{g}(r) \end{pmatrix}. \tag{14.23}$$

14.1.3 Perturbative Expansion

It is difficult to solve Eq. (14.23) for an arbitrarily given $\Delta(r)$. Nevertheless, it can be accurately solved in the limit that the vortex core energy is far below the superconducting energy gap, i.e. $E \ll \Delta_\infty$, or equivalently, $|n| \ll k_f \xi$.

In the limit $k_\parallel r \gg 1$, we may neglect the first two derivative terms in the bracket before σ_3 in (14.23). Furthermore,

$$\frac{d \ln H^{(1)}(k_\parallel r)}{dr} \approx i k_\parallel, \qquad (k_\parallel r \gg 1). \qquad (14.24)$$

Thus in that limit, Eq. (14.23) becomes

$$\left[-i\hbar v_\parallel \sigma_3 \frac{d}{dr} + \Delta(r)\sigma_1 \right] \left(\begin{array}{c} \tilde{f}(r) \\ \tilde{g}(r) \end{array} \right) = \left[E - \frac{\hbar^2 (n + \frac{1}{2})}{2mr^2} \right] \left(\begin{array}{c} \tilde{f}(r) \\ \tilde{g}(r) \end{array} \right), \qquad (14.25)$$

where $v_\parallel = \hbar k_\parallel / m$. The centrifugal term on the right-hand side is a small quantity in comparison with the energy gap Δ_∞,

$$\frac{1}{\Delta_\infty} \frac{\hbar^2 \left(n + \frac{1}{2} \right)}{2mr^2} \sim \frac{\xi}{\hbar v_f} \frac{\hbar^2 \left(n + \frac{1}{2} \right)}{2mr^2} \sim \frac{\xi}{r} \frac{n}{k_f r} \ll 1. \qquad (14.26)$$

Hence, both terms on the right-hand side of Eq. (14.25) can be treated as perturbations in the low energy limit $E \ll \Delta_\infty$.

To solve the above equation, we take the ansatz

$$\left(\begin{array}{c} \tilde{f}(r) \\ \tilde{g}(r) \end{array} \right) = \left(\begin{array}{c} e^{i\psi(r)/2} \\ e^{-i[\psi(r)+\pi]/2} \end{array} \right) e^{-K(r)}, \qquad (14.27)$$

and assume $|\psi(r)| \ll 1$ to account for the effect of perturbation. $K(r)$ is an integral of the gap function

$$K(r) = \frac{1}{\hbar v_\parallel} \int_0^r \Delta(r') dr'. \qquad (14.28)$$

Inserting (14.27) into (14.25), and keeping the leading terms in $\psi(r)$ on both sides of the equation, we obtain

$$\frac{d}{dr} \psi(r) - \frac{2\Delta(r)}{\hbar v_\parallel} \psi(r) = 2k_0 - \frac{2n + 1}{2k_\parallel r^2}. \qquad (14.29)$$

The solution is

$$\psi(r) = - \int_r^{+\infty} dr_1 e^{2[K(r) - K(r_1)]} \left(2k_0 - \frac{2n + 1}{2k_\parallel r_1^2} \right). \qquad (14.30)$$

Since

$$K(r) - K(r_1) = -\frac{1}{\hbar v_\parallel} \int_r^{r_1} dr' \Delta(r'), \qquad (14.31)$$

and the integrand in (14.30) decays exponentially, $|\psi(r)| \ll 1$ is justified.

14.1.4 Vortex Core Bound State Energies

To find the bound state energies, we divide r into two regions, separated by a radius r_c such that the gap function $\Delta(r)$ is roughly zero in the region $r < r_c$. Then r_c does not need to be precisely determined, provided the following condition is satisfied

$$n/k_\parallel \ll r_c < \xi. \tag{14.32}$$

Roughly speaking, both $f(r)$ and $g(r)$ are standing waves convoluted by slow varying envelope functions. In the region $r > r_c$, the radial eigenfunction is given by

$$\begin{pmatrix} f(r) \\ g(r) \end{pmatrix} = H_l^{(1)}(k_\parallel r)e^{-K(r)} \begin{pmatrix} e^{i\psi(r)/2} \\ e^{-i[\psi(r)+\pi]/2} \end{pmatrix} + h.c. \tag{14.33}$$

The phase difference between $f(r)$ and $g(r)$ is

$$\delta\phi(r) = \psi(r) + \frac{\pi}{2}, \qquad (r > r_c). \tag{14.34}$$

In the region $r < r_c$, the solution is given by Eq. (14.17). As $k_\parallel r_c \gg 1$, we may use the asymptotic form (14.22) of the first Hankel function (its real part is the Bessel function of the first kind), which yields

$$f(r_c^-) \sim \cos\left[(k_\parallel + k_0)r_c + \frac{(n+1)^2}{2(k_\parallel + k_0)r_c} - \frac{n+1}{2}\pi - \frac{\pi}{4}\right], \tag{14.35}$$

$$g(r_c^-) \sim \cos\left[(k_\parallel r_c - k_0)r_c + \frac{n^2}{2(k_\parallel - k_0)r_c} - \frac{n}{2}\pi - \frac{\pi}{4}\right]. \tag{14.36}$$

Hence, the phase difference, up to the leading order in $k_0 r_c$, is

$$\delta\phi(r_c^-) = 2k_0 r_c + \frac{2n+1}{2k_\parallel r_c} - \frac{\pi}{2} + \pi. \tag{14.37}$$

It should be noted that there is an ambiguity in defining the phase of a wavefunction even if it is real since two wavefunctions that differ from each other just by a minus sign or a phase π describe the same quantum state. We add a phase π in the above phase difference so that it can match the phase difference at $r = r_c^+$ in the limit $E \ll \Delta_\infty$.

For consistency, we need to match the phase difference from the two sides of $r = r_c$. This leads to the equation

$$\psi(r_c) = 2k_0 r_c + \frac{2n+1}{2k_\parallel r_c}. \tag{14.38}$$

As $\Delta(r)$ is nearly zero in the region $r < r_c$, $K(r_c)$ is small and can be approximately treated as zero. So we have

$$\psi(r_c) \approx -\int_{r_c}^{+\infty} dr e^{-2K(r)} \left(2k_0 - \frac{2n+1}{2k_{\parallel} r^2}\right)$$

$$\approx 2k_0 r_c + \frac{2n+1}{2k_{\parallel} r_c} - 2k_0 \int_0^{\infty} dr e^{-2K(r)}$$

$$+ \int_{r_c}^{+\infty} dr e^{-2K(r)} \left(1 - e^{2K(r)}\right) \frac{2n+1}{2k_{\parallel} r^2}. \tag{14.39}$$

In the last term, $e^{-2K(r)}$ decays exponentially with r, thus the integration is mainly contributed by the small r part. In that case

$$1 - e^{2K(r)} \approx -\frac{\Delta(r)r}{\hbar v_{\parallel}} \tag{14.40}$$

is approximately given by the average of $\Delta(r)$ from $r = 0$ to r. The last term becomes

$$\int_{r_c}^{+\infty} dr e^{-2K(r)} \left(1 - e^{2K(r)}\right) \frac{2n+1}{2k_{\parallel} r^2}$$

$$\approx -\int_{r_c}^{+\infty} dr e^{-2K(r)} \frac{(2n+1)\Delta(r)}{2\hbar v_{\parallel} k_{\parallel} r}$$

$$\approx -\int_0^{+\infty} dr e^{-2K(r)} \frac{(2n+1)\Delta(r)}{2\hbar v_{\parallel} k_{\parallel} r}. \tag{14.41}$$

Here the lower limit of the integral r_c is set to 0 because $\left(1 - e^{2K(r)}\right)$ removes the singularity of the integral at $r \to 0$. The matching equation now becomes

$$2k_0 \int_0^{\infty} dr e^{-2K(r)} = -\int_0^{+\infty} dr e^{-2K(r)} \frac{(2n+1)\Delta(r)}{2\hbar v_{\parallel} k_{\parallel} r}. \tag{14.42}$$

Thus the vortex core bound energy is given by

$$E = -\frac{n + \frac{1}{2}}{2k_{\parallel}} \frac{\int_0^{\infty} dr e^{-2K(r)} \frac{\Delta(r)}{r}}{\int_0^{\infty} dr e^{-2K(r)}}. \tag{14.43}$$

The gap function $\Delta(r)$ is unknown, but it should be a universal function of r/ξ

$$\Delta(r) = \Delta_{\infty} \alpha\left(\frac{r}{\xi}\right). \tag{14.44}$$

$\xi = \hbar v_f/(\pi \Delta_\infty)$ is the coherence length. The ratio of the two integrals now reduces to

$$\frac{\int_0^\infty dr\, e^{-2K(r)} \frac{1}{r} \Delta(r)}{\int_0^\infty dr\, e^{-2K(r)}} = \frac{\Delta_\infty}{\xi} \eta \left(\frac{2\Delta_\infty \xi}{\hbar v_\parallel} \right) = \frac{4\Delta_\infty^2}{\hbar v_f} \eta \left(\frac{2k_f}{\pi k_\parallel} \right), \tag{14.45}$$

where

$$\eta(a) = \frac{\pi \int_0^\infty dx \exp\left[-a \int_0^x dx'\, \alpha(x') \right] \frac{\alpha(x)}{x}}{4 \int_0^\infty dx \exp\left[-a \int_0^x dx'\, \alpha(x') \right]} \tag{14.46}$$

is a function of the order of 1. Thus we have

$$E \approx - \left(n + \frac{1}{2} \right) \frac{\Delta_\infty^2}{\varepsilon_f} \eta \left(\frac{2k_f}{\pi k_\parallel} \right). \tag{14.47}$$

ε_F is the Fermi energy. As mentioned previously, for each given E, there is another vortex core state with energy $-E$. Thus the energy levels of the deep vortex core states are evenly distributed. The level distance is roughly proportional to $\Delta_\infty^2/\varepsilon_F$.

14.2 Semiclassical Approximation

Similarly to the s-wave superconductor, the vortex structure of the d-wave supercon-ductor is determined by the superconducting coherence length ξ and the magnetic penetration depth λ. The radius of the vortex core roughly equals the coherence length ξ. The pairing order parameter is suppressed inside the vortex core and van-ishes right at the core center. The external magnetic field passes through the vortex core as well as its vicinity within the characteristic length scale of the penetration depth λ.

High-T_c cuprates are typical type-II superconductors. The lower critical field is of the order of a few hundred gauss, but the upper critical field is generally above 50 T. Along the CuO_2 planes, the penetration depth is about 10^3 Å, but the coherence length of Cooper pairs is just around $15-20$ Å. The ratio of the magnetic penetration depth to the coherence length is about $\lambda/\xi_0 \sim 10^2$. Thus the size of a vortex core of high-T_c superconductors is very small, and the spatial distribution of the magnetic field is very broad. The magnetic field can be approximately regarded as uniformly distributed in the whole system excluding the vortex cores if the average intervortex distance R lies between the above two length scales, i.e. $\xi \ll R \ll \lambda$.

In the mixed state, there exist two kinds of low energy excitation. The first is the fermionic core states localized mainly inside the vortex cores. In 1964, Caroli, de Gennes, and Matricon [344] showed that there are quasi–particle bound states inside a vortex core by solving the Bogoliubov-de Gennes equation for the isotropic s-wave superconductor. In 1989, Hess et al. observed, for the first time, this kind of vortex bound state in the superconducting state of $NbSe_2$ through STM experiments [345, 346], verifying the theoretical prediction. In the d-wave superconductor, there are no vortex bound states because the gap function vanishes along the nodal directions. But there are sharp resonance states around each vortex. These resonance states behave similarly to the s-wave core bound states. It is difficult to distinguish a resonance state from a bound state if the energy resolution is not sufficiently high.

The second is the quasiparticle excitations induced by the applied magnetic field outside the cores. In the isotropic s-wave superconductor, the numbers of this kind of excitation are suppressed by the superconducting energy gap, and its contribution is very small at low temperatures. However, it is different in the d-wave superconductor. Owing to the existence of gap nodes, quasiparticles are relatively easy to excite outside vortex cores. Since the volume in which these quasiparticle excitations are populated is significantly larger than the size of the vortex cores, their contribution to thermodynamic quantities is also much larger than that of the vortex core states. Thus in the mixed state of the d-wave superconductor, as first pointed out by Volovik [347], the low-energy physics is predominately governed by the quasiparticle excitations induced by the applied field outside the vortex cores. In fact, this is also a basic assumption made in the analysis of physical properties of the mixed state of d-wave superconductors.

In addition, the vortex lines always form a lattice, either regular or irregular, in the mixed state. This vortex lattice can scatter quasiparticles and affect thermodynamic and dynamic properties of superconductors. This scattering effect is weak and negligible if the applied field is not very high and the intervortex distance is large. However, if the applied field is very high, namely close to the upper critical field, the scattering becomes strong and this effect should be more seriously considered.

It is difficult to study comprehensively properties of quasiparticle excitations in the mixed state of d-wave superconductors. There are two reasons for this. First, the vortex line in a d-wave superconductor is not rotationally invariant, unlike in its s-wave counterpart. In particular, the effective coherence length of Cooper pairs diverges along the nodal direction. This implies that the quasiparticle excitation inside the core is not perfectly confined, hence not forming a bound state in the d-wave superconductor, and the wavefunction of a vortex core state can escape to infinity along the gap nodal directions. It is impossible to find a rigorous solution for the vortex core states by solving the Bogoliubov–de Gennes equation for the d-wave superconductor. Second, the scattering process of superconducting quasiparticles by the vortex cores is complicated and lacks any systematic study.

Certain approximations have to be used in order to calculate the microscopic structures of vortex lines and related quasiparticle excitation spectra from quantum theory.

In the discussions below, we use the d_{xy}-wave superconductor as an example to discuss the properties of quasiparticle excitations in the mixed state. The results can be generalized to the $d_{x^2-y^2}$-wave superconductor straightforwardly.

In the mixed state, the dynamics of quasiparticles is governed by the Hamiltonian defined in Eq. (3.71). For the d_{xy} superconductor, Eq. (3.71) becomes

$$
\hat{H} = \begin{pmatrix} \frac{1}{2m}\left(\boldsymbol{p} - \frac{e}{c}\boldsymbol{A}\right)^2 + U(\boldsymbol{r}) - \varepsilon_F & \frac{1}{4p_F^2}\{p_x, \{p_y, \Delta_0(\boldsymbol{r})\}\} \\ \frac{1}{4p_F^2}\{p_x, \{p_y, \Delta_0(\boldsymbol{r})\}\} & -\frac{1}{2m}\left(\boldsymbol{p} + \frac{e}{c}\boldsymbol{A}\right)^2 - U(\boldsymbol{r}) + \varepsilon_F \end{pmatrix}.
$$
(14.48)

In the limit $\xi \ll R \ll \lambda$, the magnetic field is uniformly distributed except inside or in the close proximity to the vortex cores. The amplitude of the gap function $\Delta_0(\boldsymbol{r})$ is also approximately coordinate independent. But the phase of $\Delta_0(\boldsymbol{r})$ varies in space. It winds by 2π around a close loop enclosing a vortex flux. Hence, outside the vortex, we can take the approximation

$$
\Delta_0(\boldsymbol{r}) \approx \Delta_0 e^{i\phi(\boldsymbol{r})},
$$

and assume Δ_0 to be $\boldsymbol{r}$ independent.

The phase of $\Delta_0(\boldsymbol{r})$ can be gauged out by a unitary transformation and replaced by an effective vector potential acting on quasiparticles. The corresponding gauge transformation is defined by

$$
\hat{H} \to U^{-1}\hat{H}U, \quad U = \begin{pmatrix} e^{i\phi_e(\boldsymbol{r})} & 0 \\ 0 & e^{-i[\phi(\boldsymbol{r})-\phi_e(\boldsymbol{r})]} \end{pmatrix},
$$
(14.49)

where $\phi_e(\boldsymbol{r})$ is an arbitrary phase function. To ensure that the transformation matrix U is single-valued by winding a vortex line, we usually set $\phi_e(\boldsymbol{r}) = 0$ or $\phi_e(\boldsymbol{r}) = \phi(\boldsymbol{r})$. For a single vortex line, these are the only two values $\phi_e(\boldsymbol{r})$ can take. However, in a system with many vortex lines, $\phi_e(\boldsymbol{r})$ can also take other expressions.

It is usually convenient to use a single-valued U to solve the Hamiltonian. However, to understand qualitatively the physical property of the mixed state, sometimes it is more convenient to use a non-single-valued transformation. For example, if we take $\phi_e(\boldsymbol{r}) = \phi(\boldsymbol{r})/2$, then the Hamiltonian Eq. (14.48) can be greatly simplified [348]. The resulting Hamiltonian reads

$$
\hat{H} = \begin{pmatrix} \frac{1}{2m}(\boldsymbol{p} + m\boldsymbol{v}_s)^2 + U(\boldsymbol{r}) - \varepsilon_F & \frac{\Delta_0}{p_F^2}\left(p_x p_y + \frac{i\hbar^2}{2}\phi''_{xy}\right) \\ \frac{\Delta_0}{p_F^2}\left(p_x p_y - \frac{i\hbar^2}{2}\phi''_{xy}\right) & -\frac{1}{2m}(\boldsymbol{p} - m\boldsymbol{v}_s)^2 - U(\boldsymbol{r}) + \varepsilon_F \end{pmatrix},
$$
(14.50)

where

$$v_s = \frac{1}{m} \left(\frac{\hbar}{2} \nabla \phi - \frac{e}{c} A \right) \tag{14.51}$$

is the velocity of supercurrent.

In the study of quasiparticle properties in the mixed state, a frequently used approximation is to treat the velocity of supercurrent as a static field rather than a dynamic variable. This is a semiclassical approximation. Under this approximation, one can solve the Hamiltonian defined by Eq. (14.50) to obtain an analytic solution.

In Eq. (14.50), if v_s is coordinate independent, and both the disorder potential $U(r)$ and the spatial variation of the phase ϕ vanish, the above Hamiltonian can be readily diagonalized. The quasiparticle excitation spectrum such obtained is

$$E_k = \sqrt{\left(\frac{\hbar^2}{2m} k^2 + \frac{1}{2} m v_s^2 - \varepsilon_F \right)^2 + \Delta_k^2} + \hbar k \cdot v_s,$$

where $\Delta_k = \Delta_0 \hat{k}_x \hat{k}_y$. The role of the v_s^2 term is to change the Fermi energy. This term can be absorbed in the redefinition of ε_F. Then the above expression becomes

$$E_k = \sqrt{\xi_k^2 + \Delta_k^2} + \hbar k \cdot v_s, \tag{14.52}$$

in which

$$\xi_k = \frac{\hbar^2}{2m} k^2 - \varepsilon_F.$$

The first term on the right-hand side of Eq. (14.52) is the quasiparticle dispersion in the absence of the supercurrent. The second term is the correction from the supercurrent which is usually dubbed as the Doppler shift. If the supercurrent velocity does not vary in space, the expression for the Doppler shift $\delta \varepsilon(k) = \hbar k \cdot v$ is exact. This shift results from the finite center-of-mass momentum of Cooper pairs in the presence of supercurrent.

In the s-wave superconductor, the correction to the quasiparticle spectrum from the Doppler shift is too small to qualitatively alter the gap structure. Hence the Doppler shift does not greatly affect the low energy properties of the s-wave superconductor. However, for the d-wave superconductor, the Doppler shift can change significantly the gap structure of quasiparticles near the nodal lines by lifting the chemical potential. The Fermi surface is no longer just a point. Instead, the volume of the Fermi surface and the corresponding zero-energy density of states becomes finite, proportional to the energy scale of the Doppler shift.

In real superconductors, the supercurrent velocity v_s is spatially dependent. In particular, it varies significantly around the vortex cores. If the variance is small in comparison with the coherence length of Cooper pairs, i.e. $|\nabla v_s| \xi_0 \ll |v_s|$,

Eq. (14.52) holds approximately. $E_\mathbf{k}$ defined by Eq. (14.52) can be approximately taken as the quasiparticle energy dispersion at the point where $\boldsymbol{v}_s$ is defined.

Under the semiclassical approximation, the supercurrent velocity $\boldsymbol{v}_s$ in the mixed state can be determined from the supercurrent density and the classical equations of electromagnetic fields

$$\frac{4\pi\lambda^2}{c}\nabla \times \boldsymbol{j}_s + \boldsymbol{H} = \hat{z}\Phi_0 \sum_i \delta(\boldsymbol{r} - \boldsymbol{R}_i), \tag{14.53}$$

$$\nabla \times \boldsymbol{H} = \frac{4\pi}{c}\boldsymbol{j}_s. \tag{14.54}$$

Equation (14.53) is the London equation in the presence of vortex lines. It holds approximately in the limit $\xi \ll R \ll \lambda$ and $\boldsymbol{R}_i$ is the coordinate of the magnetic vortex core center. The solution to the above equations is

$$\boldsymbol{j}_s = \frac{c\Phi_0}{4\pi}\int \frac{d^2k}{4\pi^2}\frac{i\boldsymbol{k}\cdot\hat{z}}{1+\lambda^2k^2}e^{i\boldsymbol{k}\cdot\boldsymbol{r}}. \tag{14.55}$$

By further using the definition $\boldsymbol{j}_s = en_s\boldsymbol{v}_s$, $\Phi_0 = hc/(2e)$, and the relation between superfluid density n_s and λ, i.e. $n_s = mc^2/(4\pi e^2\lambda^2)$, the supercurrent velocity is found to be

$$\boldsymbol{v}_s = \frac{\hbar}{4\pi m}\sum_i \int d^2k\frac{i\boldsymbol{k}\times\hat{z}}{k^2+\lambda^{-2}}e^{i\boldsymbol{k}\cdot(\boldsymbol{r}-\boldsymbol{R}_i)}. \tag{14.56}$$

In the limit $\lambda \to \infty$, the above integral can be solved analytically. It gives

$$\boldsymbol{v}_s = \sum_i \frac{\hbar}{2m|\boldsymbol{r}_i|}\hat{z}\times\hat{\boldsymbol{r}}_i, \tag{14.57}$$

where $\boldsymbol{r}_i \equiv \boldsymbol{r} - \boldsymbol{R}_i$. It shows that the change of the supercurrent velocity $\boldsymbol{v}_s$ is indeed small in comparison with the coherence length and the semiclassical approximation is valid away from the vortex cores. Therefore, the semiclassical approximation can be used in the calculation of physical quantities that are predominately determined by the quasiparticle excitations outside the vortex cores.

14.3 Low-Energy Density of States

In the mixed state, there exist two types of quasiparticle excitations, residing inside and outside the vortex core, respectively. Below we discuss their contributions to the low energy density of states.

Inside the vortex core, the superconducting order parameter is suppressed, but there are circulating screening currents. In order to evaluate the contribution of quasiparticle excitations inside the cores, each vortex core can be regarded as a potential well with height Δ_0^2/ε_F and radius ξ_0. In the s-wave superconductor, the scattering potential is isotropic and the quasiparticles can form a few bound states inside the core. In the d-wave superconductor, there are no bound states due to the existence of gap nodes. Instead, the low-lying states of quasiparticles are resonance states. The eigenfunctions or energies of these resonant or bound states are determined purely by the intrinsic parameters of the superconductor in the limit $\lambda \gg \xi_0$, independent of the strength of applied magnetic field. Thus the contribution of these resonant or bound states to the low energy density of states by each vortex core is also independent of the external magnetic field. Since the density of vortex lines is proportional to the external magnetic field H, this means that the vortex contribution to the low-energy density of states is also proportional to H, i.e.

$$\rho_{core} \sim H \tag{14.58}$$

irrespective of the pairing symmetry.

Away from the vortex cores, the correction to the quasiparticle spectrum by the Doppler shift behaves differently in the s- and d-wave superconductors. In the s-wave superconductor, the correction is negligibly small compared to the quasiparticle excitation gap. Thus the low-energy quasiparticle density of states is contributed to mainly by the core excitations at low temperatures, and is proportional to H as given by Eq. (14.58). For the d-wave superconductor, the Doppler shift correction to the excitation spectrum is larger than the gap value near the nodal lines. The contribution from the quasiparticle excitations outside the core cannot be neglected. In fact, its contribution is larger than the core excitations.

Under the semiclassical approximation, both the supercurrent velocity and the quasiparticle density of states are functions of coordinates. On average, the quasiparticle density of states contributed by each magnetic flux line is determined by the formula

$$\rho_{out}(\omega, H) = \int d\varepsilon \rho_0(\omega + \varepsilon) P(\varepsilon, H), \tag{14.59}$$

where

$$P(\varepsilon, H) = \frac{1}{A} \int d^2r \delta \left(\varepsilon - \hbar v_s(\mathbf{r}) \cdot \mathbf{k} \right), \tag{14.60}$$

is the Doppler distribution function. It measures the average distribution of the Doppler shift in space. The domain of integration in Eq. (14.60) is the region of one flux line, and A is the corresponding area.

When the magnetic field is changed, the coordinates of the vortex core centers R_i are also changed. The density of vortices increases with the external magnetic field. The field dependence of R_i is determined, on average, by the following formula

$$R_i(H) = x^{-1} R_i(H_0),$$

where $x = \sqrt{H/H_0}$ and H_0 is a reference magnetic field. Here it is implicitly assumed that the system behaves similarly in the two fields, H and H_0. Using Eq. (14.56), it can be shown that the supercurrent velocity satisfies the following scaling relation

$$v_s(\mathbf{r}, \lambda, H) = x v_s(x\mathbf{r}, x\lambda, H_0). \qquad (14.61)$$

v_s depends on the penetration depth λ, which determines the characteristic length scale of v_s at long distance. For the system with $R \ll \lambda$, the effect of λ on v_s is very small on the length scale discussed here. Thus λ can be approximately taken as infinity. In this case,

$$v_s(\mathbf{r}, H) \approx x v_s(x\mathbf{r}, H_0). \qquad (14.62)$$

Substituting this equation into Eq. (14.60), and considering the fact that the area per flux line scales as $A \to x^2 A$ under the change of magnetic field from H to H_0, we find the following expression for the distribution function $P(\varepsilon, H)$

$$P(\varepsilon, H) = x^{-1} P(x^{-1}\varepsilon, H_0). \qquad (14.63)$$

By further substituting it into Eq. (14.59), we then obtain the following equation

$$\rho_{\text{out}}(\omega, H) = \int d\varepsilon \rho_0(\omega + x\varepsilon) P(\varepsilon, H_0). \qquad (14.64)$$

If $\omega = 0$ and the correction of the Doppler shift to the energy is small compared to the maximal gap, $\rho_0(x\varepsilon)$ is approximately a linear function of ε, $\rho_0(x\varepsilon) \approx x N_F \varepsilon / \Delta_0$. Thus at zero frequency, the average density of states contributed by the quasiparticle excitations outside the vortex core is approximately given by

$$\rho_{\text{out}}(0, H) = \alpha \sqrt{H}, \qquad (14.65)$$

where the coefficient α is determined purely by the system parameters, independent of H

$$\alpha = \frac{N_F}{\Delta_0 \sqrt{H_0}} \int d\varepsilon \varepsilon P(\varepsilon, H_0). \qquad (14.66)$$

Comparing Eq. (14.65) with Eq. (14.58), it is clear that the quasiparticle excitations outside the vortex cores contribute more to the low-energy density of states than the vortex core states at low magnetic fields, i.e. $\rho_{\text{core}}/\rho_{\text{out}} \ll 1$. Thus to the leading order approximation, the density of states at the Fermi surface scales as

$$\rho(H) \sim \sqrt{H}. \qquad (14.67)$$

This is an important result for the d-wave superconductor, which was first obtained by Volovik [347]. It implies that one can neglect the vortex core excitations in the study of the thermodynamic and dynamic properties of d-wave superconductors in the mixed state. As only the quasiparticles outside the vortex cores need to be considered, this greatly simplifies the calculation for the excitation spectra. The coefficient of the $\sqrt{H}$ term in the density of states depends on the distribution function $P(\varepsilon, H)$. $P(\varepsilon, H)$, on the other hand, depends on the distribution of vortices. A detailed discussion of the expressions of $P(\varepsilon, H)$ at different vortex distributions is given in Ref. [349]. In general, $P(\varepsilon, H)$ is obtained by numerical calculations.

At low temperatures, the contribution of quasiparticle excitations to the specific heat coefficient C_v/T is proportional to the low-energy density of states. Thus the low temperature specific heat of electrons in the mixed state is proportional to $\sqrt{H}$ [347]

$$C_v \sim T\sqrt{H}. \tag{14.68}$$

This square-root field dependence of the specific heat is a characteristic property of d-wave superconductors. In the s-wave superconductor, the low-energy density of states is dominated by the quasiparticle excitations inside the cores. It is proportional to the magnetic field H, so is the low temperature specific heat coefficient C_v/T.

The specific heat contains the contribution from both electrons and phonons. Generally it is difficult to separate the electron contribution from the phonon one. This is the major obstacle to the analysis of experimental data of specific heat. However, phonons do not couple to the magnetic field. Their contribution to the specific heat does not depend on the applied field. This means that the difference of the specific heat at different magnetic fields is purely the contribution of electrons. This property can be used to test the $\sqrt{H}$ scaling behavior of d-wave superconductors in a finite magnetic field.

The $\sqrt{H}$ scaling behavior of low temperature specific heat in the d-wave superconductor was first verified experimentally by Moler et al. [350] for YBCO superconductors. Figure 14.1 shows the field dependence of the specific heat coefficient C_v/T that they obtained from the measurement data in the low temperature limit. The experimental results agree with the theoretical prediction. It shows that the low energy excitations in the d-wave mixed states is indeed contributed to mainly by the quasiparticle excitations outside the vortex cores. Later on, the $\sqrt{H}$ behavior of the specific heat coefficient was further confirmed in YBCO [351, 352], $Y_{0.8}Ca_{0.2}Ba_2Cu_3O_{6+x}$ [353] and $La_{2-x}Sr_xCuO_4$ [354]. Wen et al. [354] also found that the doping dependence of the maximal gap they obtained from the specific heat experiment is consistent with that obtained by the thermal conductance measurement [274]. However, in the underdoped $Y_{0.8}Ca_{0.2}Ba_2Cu_3O_{6+x}$[353], the specific heat varies almost linearly with H. This linear field dependence of the specific heat may result from the impurity scattering. It may also arise from the fact that the measurement temperature is still not low enough and the specific heat contains a significant contribution from the quasiparticles far away from the gap nodes.

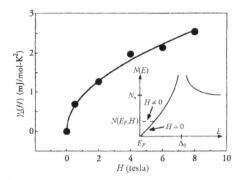

Figure 14.1 The specific heat coefficient $\gamma_\perp(H) = C_v(H)/T$ as a function of H in the low temperature limit for $YBa_2Cu_3O_{6.95}$. The solid line is the fitting curve for the experimental results with the formula $\gamma_\perp(H) = A\sqrt{H}$, and $A = 0.91 \, mJ/ml \, K^2T^{1/2}$. The inset shows the density of states both at zero field and at finite field. (From Ref. [350])

In the dirty scattering limit, Kubert and Hirschfeld found that the quasiparticle contribution to the specific heat scales as $H \ln H$, significantly different from the $\sqrt{H}$ behavior [355].

14.4 Universal Scaling Laws

Around the gap nodes, the quasiparticles of d-wave superconductors are Dirac-like. Their energy varies linearly with momentum. Thus the energy and momentum have the same scaling dimension under the scaling transformation. As a result of this, various thermodynamic quantities exhibit strong scaling behaviors [356]. Below, we take the d_{xy}-superconductor as an example to derive the scaling laws for several different thermodynamic quantities in the mixed states under the linear approximation of the energy dispersion.

There are four nodes in the quasiparticle spectra of the d_{xy}-superconductor. If the scattering among these four nodes is neglected, the contributions to the thermodynamic quantities from these four nodes are independent. The total contribution is simply the sum of the contribution from each node.

Around the gap node $k = (k_F, 0)$, the Hamiltonian of quasiparticles, Eq. (14.50), can be linearized according to the method introduced in §3.6. The linearized Hamiltonian corresponding to Eq. (3.90) is given by

$$\hat{H}_0(r, H) = \begin{pmatrix} v_F(p_x + mv_{s,x}) + U(r) & \dfrac{\Delta_0}{p_F}p_y \\[2mm] \dfrac{\Delta_0}{p_F}p_y & -v_F(p_x - mv_{s,x}) - U(r) \end{pmatrix}, \quad (14.69)$$

which is valid at $T \ll \Delta_0^2/\varepsilon_F$. This is the full Hamiltonian for describing the quasiparticle excitations outside the vortex cores.

Under the scaling transformation, $r \to xr$, if we assume that both the number of vortices and the ratio between the average volume of each vortex and the sample size are invariant, and the disorder potentials are uncorrelated, then it can be shown that the above Hamiltonian satisfies the following scaling equation

$$\hat{H}_0(r, H) = x\hat{H}_0(xr, H_0). \tag{14.70}$$

Of course, in order to keep the total number of vortices unchanged under the scaling transformation, the total area of the system should scale with the magnetic field as

$$S_A(H) = x^{-2}S_A(H_0), \quad H = x^2 H_0.$$

Equation (14.70) can be verified by analyzing the scaling behavior of each individual term in $\hat{H}_0$ under the transformation $r \to xr$. The variation of the momentum operator under the scaling transformation is simple. From the definition of the momentum operator, we have

$$-i\hbar\partial_r = x\left(-i\hbar\partial_{(xr)}\right). \tag{14.71}$$

Generally the random potential is uncorrelated and its average is zero, $\langle U(r)\rangle = 0$. The spatial correlation of the random potential is a δ-function

$$\langle U(r)U(r')\rangle = U_0\delta(r - r').$$

This short-ranged random potential has no characteristic length scale. In a two-dimensional system, $\delta(xr) = x^{-2}\delta(r)$. Thus $U(r)$ should satisfy the following relation under the scaling transformation

$$U(r) = xU(xr). \tag{14.72}$$

Substituting the above equations and the scaling formula of the supercurrent velocity, Eq. (14.62), into (14.69), we then obtain Eq. (14.70).

The scaling relation revealed by Eq. (14.70) results from the linear approximation. This approximation is not valid in the strong impurity scattering limit, because the low-energy density of states is changed by the random impurity potential and is no longer zero at the Fermi level.

From Eq. (14.70), it can be shown that the eigenvalue E_n and the corresponding eigenfunction $\tilde{\psi}_n$ of $\hat{H}_0$ transform under the scaling transformation as

$$\tilde{\psi}_n(r, H) = \tilde{\psi}_n(xr, H_0), \tag{14.73}$$

$$E_n(H) = xE_n(H_0). \tag{14.74}$$

This gives the equation that the internal energy satisfies at different magnetic fields

$$U(T, H) = \sum_n E_n(H) f\left(\frac{E_n(H)}{T}\right) = xU(x^{-1}T, H_0), \qquad (14.75)$$

where f is the Fermi distribution function. From the scaling behavior of the system size $S_A(H) = x^{-2}S_A(H_0)$, we then obtain the scaling law of the internal energy density

$$u(T, H) = H^{3/2} F_U(T/\sqrt{H}), \qquad (14.76)$$

in which F_U is an unknown scaling function. The specific heat per unit area is determined by the derivative of $u(T, H)$ with respect to temperature

$$C_v(T, H) = H F_U'(T/\sqrt{H}) = T\sqrt{H} F_C(T/\sqrt{H}), \qquad (14.77)$$

where $F_C(T/\sqrt{H})$ is a universal scaling function of $T/\sqrt{H}$.

By integrating the specific heat with respect to temperature, we obtain the scaling formula of the entropy

$$S = \int dT\, T C_v(T, H) = \sqrt{H} \int dT\, T^2 F_C(T/\sqrt{H}) = H^2 F_S(T/\sqrt{H}), \quad (14.78)$$

where F_S is a universal scaling function of the entropy.

The free energy is defined by $F = U - TS$. Its scaling law is given by the formula

$$F(H) = H^{3/2} F_F(T/\sqrt{H}). \qquad (14.79)$$

The magnetic susceptibility is proportional to the second order derivative of the free energy with respect to the magnetic field. It satisfies the following scaling law

$$\chi(T, H) = \frac{\partial^2 F}{\partial H^2} = \frac{T^2}{4H^{3/2}} F_F''(T/\sqrt{H}) = T^{-1} F_\chi(T/\sqrt{H}). \qquad (14.80)$$

F_F and F_χ are the scaling functions for the free energy and the magnetic susceptibility, respectively.

These scaling laws of thermodynamic properties are obtained under the linear approximation. They were verified through specific heat measurements in high-T_c superconductors. For YBCO superconductors, it was found that the specific heat indeed scales as $T/\sqrt{H}$, consistent with the theoretical prediction (Fig. 14.2) [251, 351, 352, 357]. Similar scaling behavior of the specific heat with $T/\sqrt{H}$ was also found in LSCO superconductors [354]. These experimental results gave a strong support to the scaling theory of thermodynamic quantities in the mixed states of d-wave superconductors.

In addition to these thermodynamic quantities, Simon and Lee [356] found that the optical and thermal conductivity tensor determined by quasiparticle excitations

Mixed State

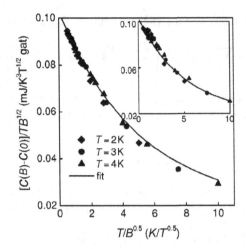

Figure 14.2 The scaling behavior of the specific heat versus the magnetic field B in the mixed state of high quality $YBa_2Cu_3O_7$ superconductor. The scaling variable is $TB^{1/2}$. The inset is the scaling behavior before subtracting the Schottky impurity term. (From Ref. [251])

also exhibits approximate scaling behavior as a function of $T/\sqrt{H}$. In particular, they found that up to the leading order approximation, the thermal Hall conductance κ_{xy} satisfies the following scaling law,

$$\kappa \sim T^2 F_{xy}(T/\sqrt{H}). \tag{14.81}$$

This result agrees with the experimental results for YBCO superconductors.

Appendix A

Bogoliubov Transformation

The Bogoliubov transformation is used to diagonalize a bilinear Hamiltonian of fermions or bosons. The simplest bilinear Hamiltonian that can be diagonalized by the Bogoliubov transformation has the form

$$H = \lambda(a^\dagger a + b^\dagger b) + \left(\gamma a^\dagger b^\dagger + h.c.\right).$$ (A.1)

where (a, b) is a pair of fermion or boson operators. This kind of Hamiltonian does not conserve the particle number. It is widely used in the mean-field study of many-body physics.

The Bogoliubov transformation is canonical. It maintains the commutation rules of the creation and annihilation operators. In Fermi systems, the Bogoliubov transformation is a unitary transformation because the fermion creation and annihilation operators can be transformed to each other by taking a particle-hole transformation. However, in Bose systems, Bogoliubov transformation is no longer unitary. Instead, it is symplectic.

Below we discuss the Bogoliubov transformation for the fermion and boson systems separately. For simplicity, we assume that γ is real. It is straightforward to generalize the results to the case with complex γ.

A.1 Fermi Systems

For fermions, Eq. (A.1) can be written in the matrix form as

$$H = \left(\begin{array}{cc} a^\dagger & b \end{array}\right) \left(\begin{array}{cc} \lambda & \gamma \\ \gamma & -\lambda \end{array}\right) \left(\begin{array}{c} a \\ b^\dagger \end{array}\right) + \lambda.$$ (A.2)

The corresponding Bogoliubov transformation is defined by

$$\left(\begin{array}{c} a \\ b^\dagger \end{array}\right) = \left(\begin{array}{cc} u & v \\ -v & u \end{array}\right) \left(\begin{array}{c} \alpha \\ \beta^\dagger \end{array}\right).$$ (A.3)

The inverse transformation is

$$\left(\begin{array}{c} \alpha \\ \beta^\dagger \end{array}\right) = \left(\begin{array}{cc} u & -v \\ v & u \end{array}\right) \left(\begin{array}{c} a \\ b^\dagger \end{array}\right),$$ (A.4)

where α and β are fermion operators, satisfying the anticommutation relations. In order to maintain the Fermi–Dirac statistics of these operators, the transformation matrix must satisfy the equation

$$u^2 + v^2 = 1. \tag{A.5}$$

This is also the condition that the transformation matrix is unitary. Both u and v are real if γ is real.

After the transformation, the Hamiltonian becomes

$$H = \begin{pmatrix} \alpha^\dagger & \beta \end{pmatrix} \begin{pmatrix} \lambda\left(u^2 - v^2\right) - 2\gamma uv & 2\lambda uv + \gamma\left(u^2 - v^2\right) \\ 2\lambda uv + \gamma\left(u^2 - v^2\right) & -\lambda\left(u^2 + v^2\right) + 2\gamma uv \end{pmatrix} \begin{pmatrix} \alpha \\ \beta^\dagger \end{pmatrix} + \lambda. \tag{A.6}$$

This Hamiltonian is diagonalized if u and v also satisfy the following equation

$$\gamma(u^2 - v^2) + 2uv\lambda = 0. \tag{A.7}$$

By solving Eqs. (A.5) and (A.7), we find that

$$u = \sqrt{\frac{1}{2} + \frac{\lambda}{2\omega}}, \tag{A.8}$$

$$v = -\text{sgn}(\gamma)\sqrt{\frac{1}{2} - \frac{\lambda}{2\omega}}, \tag{A.9}$$

where $\text{sgn}(\gamma) = 1$ if $\gamma \geqslant 0$ or -1 otherwise, and

$$\omega = \sqrt{\lambda^2 + \gamma^2}. \tag{A.10}$$

After the diagonalization, the Hamiltonian becomes

$$H = \begin{pmatrix} \alpha^\dagger & \beta \end{pmatrix} \begin{pmatrix} \omega & 0 \\ 0 & -\omega \end{pmatrix} \begin{pmatrix} \alpha \\ \beta^\dagger \end{pmatrix} + \lambda$$

$$= \omega(\alpha^\dagger \alpha + \beta^\dagger \beta) - \omega + \lambda, \tag{A.11}$$

A.2 Bose Systems

Again we rewrite Eq. (A.1) in the matrix form

$$H = \begin{pmatrix} a^\dagger & b \end{pmatrix} \begin{pmatrix} \lambda & \gamma \\ \gamma & \lambda \end{pmatrix} \begin{pmatrix} a \\ b^\dagger \end{pmatrix} - \lambda. \tag{A.12}$$

The Bogoliubov transformation is now defined as

$$\begin{pmatrix} a \\ b^\dagger \end{pmatrix} = \begin{pmatrix} u & v \\ v & u \end{pmatrix} \begin{pmatrix} \alpha \\ \beta^\dagger \end{pmatrix}. \tag{A.13}$$

Similarly to the fermion systems, u and v are not independent. They satisfy the following equation

$$u^2 - v^2 = 1, \tag{A.14}$$

if α and β are boson operators.

Substituting Eq. (A.13) into Eq. (A.12), we have

$$H = \begin{pmatrix} \alpha^\dagger & \beta \end{pmatrix} \begin{pmatrix} \lambda\left(u^2 + v^2\right) + 2\gamma uv & 2\lambda uv + \gamma\left(u^2 + v^2\right) \\ 2\lambda uv + \gamma\left(u^2 + v^2\right) & \lambda\left(u^2 + v^2\right) + 2\gamma uv \end{pmatrix} \begin{pmatrix} \alpha \\ \beta^\dagger \end{pmatrix} - \lambda. \quad (A.15)$$

To set the off-diagonal terms to zero, we obtain another equation that u and v satisfy

$$2uv\lambda + \gamma(u^2 + v^2) = 0. \quad (A.16)$$

By solving Eqs. (A.14) and (A.16), we find that

$$u = \sqrt{\frac{1}{2} + \frac{\lambda}{2\omega}}, \quad (A.17)$$

$$v = -\text{sgn}(\gamma)\sqrt{-\frac{1}{2} + \frac{\lambda}{2\omega}}, \quad (A.18)$$

where

$$\omega = \sqrt{\lambda^2 - \gamma^2}. \quad (A.19)$$

This solution is valid when $\lambda \geqslant |\gamma|$. Otherwise, the system described by Hamiltonian Eq. (A.13) is unstable. The diagonalized Hamiltonian then becomes

$$H = \omega(\alpha^\dagger \alpha + \beta^\dagger \beta) + \omega - \lambda. \quad (A.20)$$

The inverse transformation of Eq. (A.13) is

$$\begin{pmatrix} \alpha \\ \beta^\dagger \end{pmatrix} = \begin{pmatrix} u & -v \\ -v & u \end{pmatrix} \begin{pmatrix} a \\ b^\dagger \end{pmatrix}. \quad (A.21)$$

Appendix B

Hohenberg Theorem

In 1967, Hohenberg proved an important theorem on the superfluid or superconducting orders [25]. It states that in both one and two dimensions, there is no superfluid long-range order in bosonic systems and no superconducting long-range order in electronic systems at any finite temperature. This theorem shows that there is no BCS-type superconductor in pure one- or two-dimensional materials. But it does not rule out the possibility of non-BCS-type superconducting phase transition, for example the Kosterlitz–Thouless (KT) transition, in low dimensions. The Hohenberg theorem puts a constraint on the pairing mechanism and serves as an important guiding principle in the study of superconductivity. Below we give an introduction to the key steps and formulas used in the proof of this theorem.

B.1 Bogoliubov Inequality

The proof of the Hohenberg theorem uses the Bogoliubov inequality defined below. Given a Hamiltonian H, the Bogoliubov inequality reads

$$\langle \{A, A^\dagger\} \rangle \left\langle \left[[C, H], C^\dagger \right] \right\rangle \geqslant 2k_B T \left| \langle [C, A] \rangle \right|^2, \tag{B.1}$$

where A and C are two arbitrary operators, and $\langle X \rangle$ represents the thermodynamic average of X defined by

$$\langle X \rangle = \frac{\mathrm{Tr} X \exp(-\beta H)}{\mathrm{Tr} \exp(-\beta H)}, \tag{B.2}$$

where $\beta = 1/k_B T$. This inequality was also used by Mermin and Wagner [358] to prove the absence of ferromagnetic and antiferromagnetic long-range orders in one- or two-dimensional Heisenberg spin models at any finite temperatures. There are many ways to prove this inequality. A relatively simple one was given by Mermin and Wagner in Ref. [358]. Below we briefly introduce their method.

We first define the inner product of A and B as

$$(A, B) = P \sum_{ij} A_{ij}^* B_{ij} \frac{W_i - W_j}{E_j - E_i + i0^\dagger}, \tag{B.3}$$

where $A_{ij} = \langle i\,|A|\,j\rangle$. P is to take the principal value for the expression behind it.

$$W_i = \frac{\exp(-\beta E_i)}{\text{Tr}\exp(-\beta H)} \tag{B.4}$$

is the Boltzmann weight of the ith eigenstate of H. Using the inequality

$$|\tanh x| \leqslant |x|, \tag{B.5}$$

it is simple to show that the following inequality is valid

$$0 < \frac{W_i - W_j}{E_j - E_i} \leqslant \frac{1}{2}\beta(W_i + W_j). \tag{B.6}$$

We then obtain the inequality

$$(A, A) \leqslant \frac{1}{2}\beta\langle\left\{A, A^\dagger\right\}\rangle. \tag{B.7}$$

Similarly, using

$$A_{ij}^* A_{ij} B_{kl}^* B_{kl} + A_{kl}^* A_{kl} B_{ij}^* B_{ij} \geqslant A_{ij}^* B_{ij} B_{kl}^* A_{kl} + B_{ij}^* A_{ij} A_{kl}^* B_{kl}, \tag{B.8}$$

it can be shown that (A, B) satisfies the Schwartz inequality:

$$(A, A)(B, B) \geqslant |(A, B)|^2. \tag{B.9}$$

Taking $B = [C^\dagger, H]$, from the definition we find that

$$(A, B) = \langle[C^\dagger, A^\dagger]\rangle, \tag{B.10}$$

$$(B, B) = \langle[C^\dagger, [H, C]]\rangle. \tag{B.11}$$

Substituting these expressions into (B.9) and using the inequality Eq. (B.7), we then obtain the Bogoliubov inequality (B.1).

B.2 Physical Meaning of the Bogoliubov Inequality

For a comprehensive understanding of the Bogoliubov inequality, let us analyze the physical meaning of each term in Eq. (B.1). We start by introducing the time-dependent correlation function of operators A and B

$$\tau_{AB}(t - t') = \langle[A^\dagger(t), B(t')]\rangle, \tag{B.12}$$

and the corresponding spectral function

$$\tau_{AB}(\omega) = \int dt\,e^{i\omega t}\tau_{AB}(t) = \sum_{ij} 2\pi\delta(\omega + E_i - E_j)A_{ji}^* B_{ji}(W_i - W_j). \tag{B.13}$$

The response function of A and B is defined by

$$\chi_{AB}(\omega) = \int \frac{d\omega'}{2\pi}\frac{\tau_{AB}(\omega')}{\omega' - \omega} = \sum_{ij} A_{ij}^* B_{ij}\frac{W_j - W_i}{E_i - E_j - \omega}. \tag{B.14}$$

Comparing this expression with Eq. (B.3), we find that the inner product of A and B is just the zero-frequency response function

$$(A, B) = P\chi_{AB}(\omega = 0) \equiv \chi^s_{AB}. \tag{B.15}$$

Therefore, the Schwartz inequality (B.9) can be also expressed as

$$\chi^s_{AA}\chi^s_{BB} \geqslant |\chi^s_{AB}|^2. \tag{B.16}$$

This inequality reveals the relation between different response functions.

From the definition, it can also be shown that the expectation value of the anticommutator of $A^\dagger$ and B and the spectral function of A and B satisfy the following equation

$$\langle\{A^\dagger, B\}\rangle = \int \frac{d\omega}{2\pi} \tau_{AB}(\omega) \coth \frac{\beta\omega}{2}. \tag{B.17}$$

This equation associates the fluctuation (left-hand side) with the dissipation (right-hand side), and is commonly known as the fluctuation-dissipation theorem. Hence the Bogoliubov inequality is just a constraint between fluctuations and correlations.

B.3 Bose System

Now we use the Bogoliubov inequality to prove that there is no superfluid long-range order in one- or two-dimensional bosonic systems at any finite temperature by *reductio ad absurdum*. We first assume that the system has a superfluid long-range order with the order parameter defined by

$$\langle a_{\mathbf{k}}\rangle = \sqrt{Vn_0}\delta(\mathbf{k}), \tag{B.18}$$

where $a_{\mathbf{k}}$ is the boson operator, and V is the volume of the system.

From the previous discussion, we know that $\langle\{A, A^\dagger\}\rangle$ in the Bogoliubov inequality (B.1) describes the fluctuation of the system. In the superfluid state, the fluctuation arises from the bosonic excitations at finite momenta, which have destructive effect on superfluidity. To describe this effect, it is natural to set

$$A = a_{\mathbf{k}}, \tag{B.19}$$

$$C = \rho_{\mathbf{k}} = \sum_{\mathbf{q}} a^\dagger_{\mathbf{q}+\mathbf{k}} a_{\mathbf{q}}. \tag{B.20}$$

We then have

$$\langle[C, A]\rangle = -\langle a_{\mathbf{q}=0}\rangle = -\sqrt{n_0}, \tag{B.21}$$

$$\langle\{A, A^\dagger\}\rangle = 2\langle a^\dagger_{\mathbf{k}} a_{\mathbf{k}}\rangle + 1. \tag{B.22}$$

To evaluate the commutator between C and H, we assume

$$H = \sum_{\mathbf{q}} \varepsilon_{\mathbf{q}} a^\dagger_{\mathbf{q}} a_{\mathbf{q}} + H_I, \tag{B.23}$$

and the density operator C commutes with the interaction term H_I. Under this assumption, the continuity equation of electric charges holds

$$\frac{\partial \rho}{\partial t} + \nabla \cdot \mathbf{j} = 0, \tag{B.24}$$

and

$$[[C, H], C^\dagger] = \sum_q \left(\varepsilon_{\mathbf{k}+\mathbf{q}} + \varepsilon_{\mathbf{q}-\mathbf{k}} - 2\varepsilon_{\mathbf{q}} \right) a_q^\dagger a_q. \tag{B.25}$$

If we further assume that the dispersion relation of free bosons is given by

$$\varepsilon_{\mathbf{k}} = \frac{\hbar^2 \mathbf{k}^2}{2m}, \tag{B.26}$$

then

$$[[C, H], C^\dagger] = \frac{\hbar^2 \mathbf{k}^2}{m} \sum_q a_q^\dagger a_q. \tag{B.27}$$

Substituting the above results into the inequality Eq. (B.3), we obtain

$$\langle a_{\mathbf{k}}^\dagger a_{\mathbf{k}} \rangle \geqslant -\frac{1}{2} + \frac{k_B T m}{\hbar^2 \mathbf{k}^2} \frac{n_0}{n}, \tag{B.28}$$

where n is the density of bosons. The right-hand side diverges quadratically as $\mathbf{k} \to 0$, and its integral over momentum also diverges in both one and two dimensions. Clearly, this infrared divergence will invalidate the following sum rule

$$\frac{1}{V} \sum_{\mathbf{k} \neq 0} \langle a_{\mathbf{k}}^\dagger a_{\mathbf{k}} \rangle = n - n_0 \tag{B.29}$$

at any finite temperature ($T \neq 0$). It indicates that the assumption made in Eq. (B.18) is invalid. Therefore, there is no superfluid long-range order in one- and two-dimensional Bose systems at finite temperatures.

B.4 Fermi Systems

Similar to the proof for the Bose system, we assume that there is a superconducting long-range order in a Fermi system. The order parameter is defined by

$$\Delta = \frac{1}{V} \sum_q \gamma_q \langle c_{q\uparrow} c_{-q\downarrow} \rangle, \tag{B.30}$$

which is assumed to be finite and the pairing function γ_q is nonsingular. Similar to Eqs. (B.19) and (B.20), we define

$$A = \frac{1}{V} \sum_q \gamma_q c_{\mathbf{k}+q\uparrow} c_{-q\downarrow}, \tag{B.31}$$

$$C = \rho_{\mathbf{k}} = \sum_{q\sigma} c_{q+\mathbf{k}\sigma}^\dagger c_{q\sigma}, \tag{B.32}$$

where A and C are the Fourier components of the pairing and density operators at momentum $\mathbf{k}$, respectively. The commutator between the above two operators is

$$\langle [A, C] \rangle = \Delta + \eta_{\mathbf{k}}, \tag{B.33}$$

where

$$\eta_{\mathbf{k}} = \frac{1}{V} \sum_{\mathbf{q}} \gamma_{\mathbf{q}-\mathbf{k}} \langle c_{\mathbf{q}\uparrow} c_{-\mathbf{q}\downarrow} \rangle. \tag{B.34}$$

$\eta_{\mathbf{k}}$ is a function of $\mathbf{k}$. In the limit $\mathbf{k} \to 0$,

$$\lim_{\mathbf{k}\to 0} \eta_{\mathbf{k}} = \Delta. \tag{B.35}$$

Similarly, we define the Hamiltonian as

$$H = \sum_{\mathbf{q}\sigma} \varepsilon_{\mathbf{q}} c_{\mathbf{q}\sigma}^{\dagger} c_{\mathbf{q}\sigma} + H_I. \tag{B.36}$$

If the electric charge is conserved and C commutes with the interaction H_I, we have

$$\langle [[C, H], C^{\dagger}] \rangle = \sum_{\mathbf{q}\sigma} \left(\varepsilon_{\mathbf{k}+\mathbf{q}} + \varepsilon_{\mathbf{q}-\mathbf{k}} - 2\varepsilon_{\mathbf{q}} \right) \langle c_{\mathbf{q}\sigma}^{\dagger} c_{\mathbf{q}\sigma} \rangle = \frac{\hbar^2 \mathbf{k}^2 n V}{m}. \tag{B.37}$$

In obtaining this equation, the energy dispersion of fermions is assumed to have the form

$$\varepsilon_{\mathbf{k}} = \frac{\hbar^2 \mathbf{k}^2}{2m}. \tag{B.38}$$

The average value of the anticommutator of A and $A^{\dagger}$ is given by

$$\langle \{A, A^{\dagger}\} \rangle = \frac{1}{V} [F(\mathbf{k}) + R(\mathbf{k})], \tag{B.39}$$

where

$$F(\mathbf{k}) = \frac{1}{V} \sum_{\mathbf{q}\mathbf{q}'} \gamma_{\mathbf{q}} \gamma_{\mathbf{q}'}^{*} \langle c_{-\mathbf{q}'\downarrow}^{\dagger} c_{\mathbf{k}+\mathbf{q}'\uparrow}^{\dagger} c_{\mathbf{k}+\mathbf{q}\uparrow} c_{-\mathbf{q}\downarrow} \rangle, \tag{B.40}$$

$$R(\mathbf{k}) = \frac{1}{V} \sum_{\mathbf{q}} |\gamma_{\mathbf{q}}|^2 \left(1 - \langle c_{\mathbf{q}\downarrow}^{\dagger} c_{\mathbf{q}\downarrow} \rangle - \langle c_{\mathbf{q}+\mathbf{k}\uparrow}^{\dagger} c_{\mathbf{q}+\mathbf{k}\uparrow} \rangle \right). \tag{B.41}$$

Since $\gamma_{\mathbf{q}}$ is nonsingular and $0 \leqslant \langle c_{\mathbf{q}\sigma}^{\dagger} c_{\mathbf{q}\sigma} \rangle \leqslant 1$, $R(\mathbf{k})$ is always finite. The integral of $F(\mathbf{k})$ with respect to $\mathbf{k}$ equals

$$\frac{1}{V} \sum_{\mathbf{k}} F(\mathbf{k}) = \int d\mathbf{r}_1 d\mathbf{r}_2 \gamma(\mathbf{r} - \mathbf{r}_2) \gamma^{*}(\mathbf{r} - \mathbf{r}_1) \langle c_{\mathbf{r}_1\downarrow}^{\dagger} c_{\mathbf{r}\uparrow}^{\dagger} c_{\mathbf{r}\uparrow} c_{\mathbf{r}_2\downarrow} \rangle, \tag{B.42}$$

where

$$\gamma(\mathbf{r}) = \frac{1}{V} \sum_{\mathbf{q}} \gamma_{\mathbf{q}} e^{i\mathbf{q}\cdot\mathbf{r}}. \tag{B.43}$$

Physically, $\langle a^{\dagger} b \rangle$ can be considered as an inner product between operators a and b. It is simple to show that they satisfy the definition of inner products as well as the Schwartz inequality

$$|\langle a^{\dagger} b \rangle|^2 \leqslant \langle a^{\dagger} a \rangle \langle b^{\dagger} b \rangle. \tag{B.44}$$

To apply this expression to $\langle c_{\mathbf{r}_1\downarrow}^\dagger c_{\mathbf{r}\uparrow}^\dagger c_{\mathbf{r}\uparrow} c_{\mathbf{r}_2\downarrow} \rangle$, we obtain the following inequality

$$\frac{1}{V} \sum_{\mathbf{k}} F(\mathbf{k}) < \left| \int d\mathbf{r}' |\gamma(\mathbf{r} - \mathbf{r}')| \langle \rho_\downarrow(\mathbf{r}')\rho_\uparrow(\mathbf{r})\rangle^{1/2} \right|^2 \equiv f_0, \qquad (B.45)$$

where f_0 is finite because the density–density correlation function is nonsingular. Since $F(\mathbf{k} = 0)$ is positive-definite, we obtain the following inequality for $F(\mathbf{k})$

$$\frac{1}{V} \sum_{\mathbf{k}\neq 0} F(\mathbf{k}) = \frac{1}{V} \sum_{\mathbf{k}} F(\mathbf{k}) - \frac{F(\mathbf{k} = 0)}{V} < f_0. \qquad (B.46)$$

In addition, according to the Bogoliubov inequality, we find that $F(\mathbf{k})$ also satisfies the inequality

$$F(\mathbf{k}) \geqslant \frac{2k_B T m |\Delta + \eta_{\mathbf{k}}|^2}{\hbar^2 k^2 n} - R(\mathbf{k}). \qquad (B.47)$$

The momentum integration of the right-hand side is infrared divergent in both one and two dimensions. This clearly conflicts with Eq. (B.46), which implies that there is no superconducting long-range order in one- and two-dimensional Fermi systems at any finite temperature.

Appendix C

Degenerate Perturbation Theory

The degenerate perturbation theory is a useful tool for studying low energy physics in strongly correlated systems. It is widely used to derive low energy effective models of strongly correlated systems. The theory starts by assuming that the Hamiltonian H of a quantum system is a sum of two terms,

$$H = H_0 + H_I, \tag{C.1}$$

with H_0 the unperturbed Hamiltonian whose ground states are degenerate and can be diagonalized analytically, and H_I the perturbation which is small compared to H_0. The goal of the theory is to find systematically the corrections of H_I to the eigenvalues and eigenstates of H_0 by perturbation expansions. It is particularly useful when the energy scale of the problem is much smaller than the energy difference between the degenerate ground states and the first-excited states. In this case, the perturbation can be done to transform H into an effective Hamiltonian H_{eff} which acts only on the ground state subspace of H_0. It is sufficient to use this effective Hamiltonian to investigate low energy physics of the system. Both the t–J model and the Kondo lattice model are these kinds of effective Hamiltonians. The former is the effective low energy model of the single-band or the three-band Hubbard model. The latter is the effective low energy model of the periodic Anderson model.

Let us consider the Schrödinger equation of eigenstates,

$$(H_0 + H_I)|\Psi\rangle = E|\Psi\rangle. \tag{C.2}$$

After a simple transformation, this equation can be reexpressed as

$$
\begin{aligned}
|\Psi\rangle &= \frac{1}{E - H_0} H_I |\Psi\rangle \\
&= \frac{1}{E - H_0} P H_I |\Psi\rangle + \frac{1}{E - H_0}(1 - P)H_I |\Psi\rangle \\
&= \sum_{\alpha} a_\alpha |\alpha\rangle + \frac{1}{E - H_0}(1 - P)H_I |\Psi\rangle,
\end{aligned}
\tag{C.3}
$$

where $\{|\alpha\rangle\}$ are the degenerate ground states of H_0, and P is the corresponding projection operator,

$$a_\alpha = \frac{\langle \alpha | H_I | \Psi \rangle}{E - E_0}, \qquad P = \sum_\alpha |\alpha\rangle\langle\alpha|. \tag{C.4}$$

Solving Eq. (C.3) iteratively yields

$$|\Psi\rangle = \frac{1}{1-A}\sum_\alpha a_\alpha |\alpha\rangle = \left(1 + \frac{1}{1-A}A\right)\sum_\alpha a_\alpha |\alpha\rangle, \qquad (C.5)$$

where

$$A \equiv \frac{1}{E - H_0}(1 - P)H_I, \qquad (C.6)$$

whose projection onto the ground states of H_0 is zero, i.e. $PAP = 0$.

Using Eqs. (C.3) and (C.5), we find that

$$(E - E_0)\sum_\alpha a_\alpha |\alpha\rangle = \left[H_I \frac{1}{1-A} - (E - H_0)\frac{1}{1-A}A\right]\sum_\alpha a_\alpha |\alpha\rangle. \qquad (C.7)$$

This indicates that the eigenvalues and eigenstates of H are determined by the following effective Hamiltonian

$$H_{\text{eff}}(E) = \left[H_I \frac{1}{1-A} - (E - H_0)\frac{1}{1-A}A\right]P. \qquad (C.8)$$

If only the correction to the ground states of H_0 is considered, the effective Hamiltonian can be simplified as

$$H_{\text{eff}} = P\left[H_I \frac{1}{1-A} - (E - H_0)\frac{1}{1-A}A\right]P = PH_I \frac{1}{1-A}P. \qquad (C.9)$$

By expansion in the order of A, the above Hamiltonian becomes

$$H_{\text{eff}} = PH_I \frac{1}{1-A}P = PH_I \sum_{n=0} A^n P. \qquad (C.10)$$

This is the formula that is commonly used in practical calculations.

Appendix D

Anderson Theorem

In conventional s-wave superconductors, nonmagnetic impurity scattering has very little effect on the superconducting transition temperature and other physical quantities. This phenomenon was first noticed by Anderson. He gave an insightful explanation of this phenomenon based on the self-consistent mean-field theory in 1959 [163], which is commonly referred to as the Anderson theorem.

The Anderson theorem results from the fact that nonmagnetic impurity scattering does not break the time-reversal invariance of s-wave superconductors. Rigorously speaking, it is valid only when the superconducting correlation length ξ is much larger than the scattering mean free path l, i.e. $\xi \gg l$. In the opposite limit $l \gg \xi$, the time-reversal symmetry is still conserved, but the electronic band structures and the pairing interactions are strongly renormalized by the scattering potentials. This can significantly affect the superconducting properties of s-wave superconductors and break the Anderson theorem.

In d-wave superconductors, as the gap function changes sign in momentum space, even nonmagnetic impurities can strongly affect superconducting properties no matter whether the impurity potential is in the weak or strong scattering limit. They interfere with the pairing phase and serve as pair-breakers. In particular, impurities can significantly change the low energy or low temperature properties of d-wave superconductors. This is an important factor of d-wave superconductors that needs to be considered in the comparison of theoretical calculations with experimental results.

Two approximations are assumed in the proof of the Anderson theorem. First, the variation of the pairing gap function $\Delta(r)$ is small in space so that it can be replaced by its average value, $\Delta(r) = \Delta$. This approximation implies that the self-consistent mean-field equation of the energy gap is just a result of spatial averaging. Second, the scattering potential does not change the density of states around the Fermi surface of normal electrons. These two approximations are generally valid if the disorder scattering potential is not very strong. But the first approximation holds only when the correlation length ξ is much larger than the scattering mean free path l. Under these approximations, the Bogoliubov–de-Gennes self-consistent field equation is given by

$$
\begin{pmatrix} H_0(\mathbf{r}) & \Delta \\ \Delta & -H_0(\mathbf{r}) \end{pmatrix} \begin{pmatrix} u_n(\mathbf{r}) \\ v_n(\mathbf{r}) \end{pmatrix} = E_n \begin{pmatrix} u_n(\mathbf{r}) \\ v_n(\mathbf{r}) \end{pmatrix},
\tag{D.1}
$$

in which

$$
H_0(\mathbf{r}) = -\frac{\hbar^2}{2m}\nabla^2 + U(\mathbf{r}) - \mu,
\tag{D.2}
$$

and $U(\mathbf{r})$ is the impurity scattering potential.

The gap function Δ does not depend on $\mathbf{r}$. This greatly simplifies the calculation of the self-consistent gap equation. If $w_n(\mathbf{r})$ is the eigenstate of normal electrons

$$H_0 w_n(\mathbf{r}) = \xi_n w_n(\mathbf{r}),$$ (D.3)

then $u_n(\mathbf{r})$ and $v_n(\mathbf{r})$ can be expressed using $w_n(\mathbf{r})$ as

$$u_n(\mathbf{r}) = u_n w_n(\mathbf{r}), \qquad v_n(\mathbf{r}) = v_n w_n(\mathbf{r}).$$ (D.4)

Substituting these expressions into Eq. (D.1), we obtain the equation for determining the coefficients u_n and v_n

$$\begin{pmatrix} \xi_n & \Delta \\ \Delta & -\xi_n \end{pmatrix} \begin{pmatrix} u_n \\ v_n \end{pmatrix} = E_n \begin{pmatrix} u_n \\ v_n \end{pmatrix}.$$ (D.5)

This equation has exactly the same form as the standard BCS mean-field equation for a translation invariant system. The difference is that the momentum is not conserved and the basis states are now characterized by the quantum number n of H_0, instead of the momentum. By diagonalizing Eq. (D.5), we obtain the quasiparticle eigenenergy

$$E_n = \sqrt{\xi_n^2 + \Delta^2},$$ (D.6)

and the corresponding eigenfunction

$$u_n = \sqrt{\frac{1}{2}\left(1 + \frac{\xi_n}{E_n}\right)}, \qquad v_n = -\sqrt{\frac{1}{2}\left(1 - \frac{\xi_n}{E_n}\right)}.$$ (D.7)

The energy gap is determined by the self-consistent equation

$$\Delta = -g \sum_n u_n(\mathbf{r}) v_n(\mathbf{r}) \tanh \frac{\beta E_n}{2}.$$ (D.8)

Substituting the above solutions into Eq. (D.8), we obtain

$$\begin{aligned}
\Delta &= g \sum_n \langle w_n^2(\mathbf{r}) \rangle \frac{\Delta}{2\sqrt{\xi_n^2 + \Delta^2}} \tanh \frac{\beta \sqrt{\xi_n^2 + \Delta^2}}{2} \\
&= g \int_{\omega_0}^{\omega_0} d\xi \rho(\omega) \frac{\Delta}{2\sqrt{\xi^2 + \Delta^2}} \tanh \frac{\beta \sqrt{\xi^2 + \Delta^2}}{2},
\end{aligned}$$ (D.9)

where $\langle A \rangle$ is the spatial average of A, and

$$\rho(\xi) = \sum_n \delta(\xi - \xi_n) \langle w_n^2(\mathbf{r}) \rangle$$ (D.10)

is the spatial average of electron density of states in the normal state. Since the impurity scattering does not change the density of states of normal electrons around the Fermi surface according to the previous assumption, Eq. (D.9) has exactly the same form as the gap equation for the impurity-free system with $U(\mathbf{r}) = 0$. Thus the impurity scattering does not change the transition temperature T_c of the s-wave superconductor. This is the proof first given by Anderson. It is consistent with experimental observations for conventional superconductors.

Appendix E

Sommerfeld Expansion

In the calculation of thermodynamic or dynamic quantities of electronic systems, we often encounter the following integral

$$I = \int_{-\infty}^{\infty} d\varepsilon\, g(\varepsilon)\, f(\varepsilon), \tag{E.1}$$

where $g(\varepsilon)$ is an arbitrary function of ε and $f(\varepsilon)$ is the Fermi distribution function

$$f(\varepsilon) = \frac{1}{e^{(\varepsilon-\mu)/k_B T} + 1}. \tag{E.2}$$

To ensure that Eq. (E.1) integrable, $g(\varepsilon)$ is assumed to be at most exponentially divergent as $\varepsilon \to \infty$, and approach 0 as $\varepsilon \to -\infty$. It is impossible to rigorously solve this integral in most cases. But if we only want to know its low-temperature behavior, the Sommerfeld expansion could be used to obtain an approximate expression for this integral [359].

We first define a function

$$K(\varepsilon) = \int_{-\infty}^{\varepsilon} d\varepsilon\, g(\varepsilon), \tag{E.3}$$

whose derivative with respect to ε is $g(\varepsilon)$. Integrating (E.1) by parts leads to the following expression

$$I = -\int_{-\infty}^{\infty} d\varepsilon\, K(\varepsilon)\, \frac{df(\varepsilon)}{d\varepsilon}. \tag{E.4}$$

If the deviation of the energy from the chemical potential is much larger than the temperature, $|\varepsilon - \mu| \gg k_B T$, $df(\varepsilon)/d\varepsilon$ decays exponentially with ε. Hence the integral in Eq. (E.4) is important only in the vicinity of the Fermi level. This implies that I can be evaluated by performing the Taylor expansion for $K(\varepsilon)$ at $\varepsilon = \mu$ at low temperatures,

The Taylor expansion of $K(\varepsilon)$ is given by

$$K(\varepsilon) = K(\mu) + \sum_{n=1} \frac{(\varepsilon - \mu)^n}{n!} \left(\frac{dK(\varepsilon)}{d\varepsilon}\right)_{\varepsilon=\mu}. \tag{E.5}$$

Substituting this into Eq. (E.4) and integrating over ε, we then obtain the expression for the Sommerfeld expansion

$$I = \int_{-\infty}^{\mu} d\varepsilon g\left(\varepsilon\right) + \sum_{n=1} a_n (k_B T)^{2n} \left[\frac{d^{2n-1}}{d\varepsilon^{2n-1}} g(\varepsilon)\right]_{\varepsilon=\mu}, \qquad (E.6)$$

where

$$a_n = -\frac{1}{(2n)!} \int dx x^{2n} \frac{d}{dx} \frac{1}{e^x + 1} = \frac{\left(2^{2n} - 2\right) \pi^{2n}}{(2n)!} B_n, \qquad (E.7)$$

and B_n is the Bernoulli number. The first five Bernoulli numbers are

$$B_1 = \frac{1}{6}, \quad B_2 = \frac{1}{30}, \quad B_3 = \frac{1}{42}, \quad B_4 = \frac{1}{30}, \quad B_5 = \frac{5}{66}. \qquad (E.8)$$

Appendix F

Single-Particle Green's Function

F.1 Retarded Green's Function

In a many-body system, the retarded single-particle Green's function of electrons is defined by

$$G_{\text{ret},\sigma}(\mathbf{k}, t - t') = -i\theta(t - t')\left\langle\left\{c_{\mathbf{k}\sigma}(t), c_{\mathbf{k}\sigma}^{\dagger}(t')\right\}\right\rangle, \tag{F.1}$$

where $\sigma = (\uparrow, \downarrow)$ is the spin index, $\{\ \}$ is the anticommutator operation, and $\langle\ \rangle$ represents the thermal average.

$$c_{\mathbf{k}\sigma}(t) = e^{-iHt}c_{\mathbf{k}\sigma}e^{iHt} \tag{F.2}$$

is the electron operator in the Heisenberg representation.

In the Lehmann representation, G_{ret} is expressed in terms of the matrix elements of many-body eigenstates as

$$G_{\text{ret},\sigma}(\mathbf{k}, t - t') = -i\theta(t - t')e^{\beta\Omega}\sum_{nm}\langle n|c_{\mathbf{k}\sigma}|m\rangle\langle m|c_{\mathbf{k}\sigma}^{\dagger}|n\rangle e^{i(E_n - E_m)(t - t')}$$

$$\left(e^{-\beta E_n} - e^{-\beta E_m}\right). \tag{F.3}$$

From its Fourier transform

$$G_{\text{ret},\sigma}(\mathbf{k}, \omega) = \int_{-\infty}^{+\infty} dt\, e^{i(\omega + i0^+)t} G_{\text{ret},\sigma}(\mathbf{k}, t), \tag{F.4}$$

we obtain the expression of the Green's function in the frequency space

$$G_{\text{ret},\sigma}(\mathbf{k}, \omega) = e^{\beta\Omega}\sum_{nm}\frac{\langle n|c_{\mathbf{k}\sigma}|m\rangle\langle m|c_{\mathbf{k}\sigma}^{\dagger}|n\rangle(e^{-\beta E_n} + e^{-\beta E_m})}{\omega + E_n - E_m + i0^+}, \tag{F.5}$$

where Ω is the thermodynamic potential of the grand canonical ensemble

$$\Omega = -\frac{1}{\beta}\ln \text{Tr}\, e^{-\beta H}. \tag{F.6}$$

The fermion spectra function $A(\mathbf{k}, \omega)$ is determined by the imaginary part of the retarded Green's function

$$A_\sigma(\mathbf{k}, \omega) = -\frac{1}{\pi} \mathrm{Im} G_{\mathrm{ret}, \sigma}(\mathbf{k}, \omega). \tag{F.7}$$

Using the eigenstates of the Hamiltonian, it can be represented as

$$A_\sigma(\mathbf{k}, \omega) = e^{\beta\Omega} \sum_{nm} \langle n|c_{\mathbf{k}\sigma}|m\rangle\langle m|c_{\mathbf{k}\sigma}^\dagger|n\rangle \left(e^{-\beta E_n} + e^{-\beta E_m}\right) \delta\left(\omega + E_n - E_m\right). \tag{F.8}$$

$A_\sigma(\mathbf{k}, \omega)$ is a sum of the following two terms

$$A_\sigma(\mathbf{k}, \omega) = A_{\sigma,+}(\omega) + A_{\sigma,-}(\omega), \tag{F.9}$$

where

$$A_{\sigma,+}(\omega) = \sum_{nm} e^{\beta\Omega} \langle n|c_{\mathbf{k}\sigma}|m\rangle\langle m|c_{\mathbf{k}\sigma}^\dagger|n\rangle e^{-\beta E_n} \delta\left(\omega + E_n - E_m\right), \tag{F.10}$$

$$A_{\sigma,-}(\omega) = \sum_{nm} e^{\beta\Omega} \langle n|c_{\mathbf{k}\sigma}|m\rangle\langle m|c_{\mathbf{k}\sigma}^\dagger|n\rangle e^{-\beta E_m} \delta\left(\omega + E_n - E_m\right). \tag{F.11}$$

They are related by the formula

$$A_{\sigma,+}(\omega) = e^{\beta\omega} A_{\sigma,-}(\omega). \tag{F.12}$$

The retarded Green's function can also be obtained from the integral of the spectral function

$$G_{\mathrm{ret}, \sigma}(\mathbf{k}, \omega) = \int d\varepsilon \frac{A_\sigma(\mathbf{k}, \varepsilon)}{\omega - \varepsilon + i0^+}. \tag{F.13}$$

This is the Lehmann representation of the Green's function.

The integral of $A_\sigma(\mathbf{k}, \omega)$ over the momentum $\mathbf{k}$ equals the electron density of states

$$\rho(\omega) = \frac{1}{V} \sum_{\mathbf{k}\sigma} A_\sigma(\mathbf{k}, \omega). \tag{F.14}$$

On the other hand, the integral of $A(\mathbf{k}, \omega)$ over the frequency satisfies the following sum rules

$$\int_{-\infty}^{\infty} d\omega A_\sigma(\mathbf{k}, \omega) = 1, \tag{F.15}$$

$$\sum_\sigma \int_{-\infty}^{\infty} d\omega f(\omega) A_\sigma(\mathbf{k}, \omega) = n(\mathbf{k}) = \sum_\sigma \langle c_{\mathbf{k}\sigma}^\dagger c_{\mathbf{k}\sigma}\rangle, \tag{F.16}$$

where $n(\mathbf{k})$ is the momentum distribution function of electrons.

F.2 Matsubara Green's Function

In an interacting system, it is difficult to directly evaluate the retarded Green's function. An approach that is often used is to first evaluate the imaginary time Green's function, or the

Matsubara Green function, using the pertubation theory, and then find the retarded Green's function through analytic continuation.

The single-particle Matsubara Green's function is defined by

$$G_\sigma(\mathbf{k}, \tau - \tau') = -\left\langle T_\tau c_\sigma(\mathbf{k}, \tau) c_\sigma^\dagger(\mathbf{k}, \tau') \right\rangle,$$ (F.17)

where T_τ is the imaginary time ordering operator,

$$c_{\mathbf{k}\sigma}(\tau) = e^{-H\tau} c_{\mathbf{k}\sigma} e^{H\tau}, \qquad c_{\mathbf{k}\sigma}^\dagger(\tau) = e^{-H\tau} c_{\mathbf{k}\sigma}^\dagger e^{H\tau},$$ (F.18)

and $0 \leqslant \tau \leqslant \beta$. Since $c_\mathbf{k}$ and $c_\mathbf{k}^\dagger$ are fermion operators, it is simple to show that

$$G_\sigma(\mathbf{k}, \tau) = -G_\sigma(\mathbf{k}, \tau + \beta),$$ (F.19)

hence G_σ is an antiperiodic function of τ from $\tau = 0$ to β. Correspondingly, the Fourier transform of $G_\sigma(\mathbf{k}, \tau)$

$$G_\sigma(\mathbf{k}, i\omega_n) = \int_0^\beta G_\sigma(\mathbf{k}, \tau) e^{i\omega_n \tau}$$ (F.20)

contains only the odd frequency terms with

$$\omega_n = \frac{(2n + 1)\pi}{\beta},$$ (F.21)

where n an integer.

The Lehmann representation of $G_\sigma(\mathbf{k}, i\omega_n)$ is given by

$$G_\sigma(\mathbf{k}, i\omega_n) = e^{\beta\Omega} \sum_{nm} \frac{\langle n|c_{\mathbf{k}\sigma}|m\rangle \langle m|c_{\mathbf{k}\sigma}^\dagger|n\rangle (e^{-\beta E_n} + e^{-\beta E_m})}{i\omega_n + E_n - E_m}.$$ (F.22)

The Matsubara Green's function is related to the retarded Green's function by the analytic continuation. Comparing (F.22) with (F.5), we obtain

$$G_{\text{ret},\sigma}(\mathbf{k}, \omega) = G_\sigma(\mathbf{k}, i\omega_n)|_{i\omega_n \to \omega + i0^+}.$$ (F.23)

Similarly, the Matsubara Green's function can be represented using the spectral function as

$$G_\sigma(\mathbf{k}, i\omega_n) = \int d\varepsilon \frac{A_\sigma(\mathbf{k}, \varepsilon)}{i\omega_n - \varepsilon} = -\frac{1}{\pi} \int d\varepsilon \frac{\text{Im} G_{\text{ret},\sigma}(\mathbf{k}, \varepsilon)}{i\omega_n - \varepsilon}.$$ (F.24)

In a non-interacting system of electrons described by the Hamiltonian H_0, the Mastubara Green's function is the solution of the equation

$$(\partial_\tau - H_0) G_\sigma^{(0)}(\mathbf{k}, \tau) = \delta(\tau),$$ (F.25)

where

$$H_0 = \sum_{\mathbf{k}\sigma} \xi_k c_{\mathbf{k}\sigma}^\dagger c_{\mathbf{k}\sigma}.$$ (F.26)

In the imaginary frequency space, it is simple to show that the solution is

$$G_\sigma^{(0)}(k, i\omega_n) = \frac{1}{i\omega_n - \xi_k}.$$ (F.27)

In an interacting system, the Green's function is renormalized by the coupling between electrons. The Green's function is now determined by the Dyson equation

$$G_\sigma(\mathbf{k}, i\omega_n) = \frac{1}{i\omega_n - \xi_\mathbf{k} - \Sigma_\sigma(\mathbf{k}, i\omega_n)}, \tag{F.28}$$

where $\Sigma_\sigma(\mathbf{k}, i\omega_n)$ is the self-energy. The corresponding retarded Green's function is

$$G_{\sigma,\text{ret}}(\mathbf{k}, \omega) = \frac{1}{\omega + i0^+ - \xi_\mathbf{k} - \Sigma_\sigma(\mathbf{k}, \omega + i0^+)}. \tag{F.29}$$

The spectral function can now be expressed as

$$A_\sigma(\mathbf{k}, \omega) = -\frac{1}{\pi} \frac{\text{Im}\Sigma_\sigma(\mathbf{k}, \omega)}{[\omega - \xi_\mathbf{k} - \text{Re}\Sigma_\sigma(\mathbf{k}, \omega)]^2 + [\text{Im}\Sigma_\sigma(\mathbf{k}, \omega)]^2}. \tag{F.30}$$

In the noninteracting case, $\Sigma_\sigma(\mathbf{k}, \omega) = 0$, the spectral function is simply a delta-function

$$A_\sigma^{(0)}(\mathbf{k}, \omega) = \delta(\omega - \xi_\mathbf{k}). \tag{F.31}$$

F.3 Frequency Summation

In the evaluation of a Feynman diagram with Matsubara Green's functions, we often meet the problem of imaginary frequency summation

$$S = \frac{1}{\beta} \sum_{\omega_n} K(i\omega_n), \tag{F.32}$$

where $K(z)$ is a complex meromorphic function. For fermions, ω_n takes the value given in Eq. (F.21). For bosons, the Matsubara Green function is a periodic function of τ from $\tau = 0$ to β, and

$$\omega_n = \frac{2\pi n}{\beta}. \tag{F.33}$$

In order to calculate the above summation, we first consider the following contour integral,

$$I = \lim_{R \to +\infty} \oint \frac{dz}{2\pi i} \frac{K(z)}{e^{\beta z} \pm 1}, \tag{F.34}$$

where the contour is along the circular path of radius R centered at the origin. The $\pm$ corresponds to the boson and fermion systems, respectively. In case $\lim_{z \to \infty} |zK(z)| \to 0$ uniformly, we have $I \to 0$.

On the other hand, such an integral can also be evaluated by summing over all the residues at all the poles. The poles of the integrand include those of $K(z)$ and $z = \omega_n$ for the above defined ω_n.

By applying the residue theorem, it is straightforward to show that

$$S = \sum_i n_f(z_i) \text{Res} K(z_i), \qquad \text{(fermion)} \tag{F.35}$$

for the fermion case, and

$$S = -\sum_i n_b(z_i) \mathrm{Res}\, K(z_i), \qquad \text{(boson)} \qquad \text{(F.36)}$$

for the boson case. In the above equations, z_i is the ith pole of $K(z)$ and $\mathrm{Res}\, K(z_i)$ is the residue at that pole.

$$n_{f,b}(z) = \frac{1}{e^{\beta z} \pm 1} \qquad \text{(F.37)}$$

are the Fermi and Bose distribution functions, respectively.

As an example, let us use the above results to evaluate the following summation

$$S = \frac{1}{\beta} \sum_{i\omega_n} \frac{1}{i\omega_n - \xi_1} \frac{1}{i\omega_n + ip_m - \xi_2}, \qquad \text{(F.38)}$$

where $\omega_n = (2n+1)\pi/\beta$ and $p_m = 2m\pi/\beta$. $K(z)$ is now given by

$$K(z) = \frac{1}{z - \xi_1} \frac{1}{z + ip_m - \xi_2}. \qquad \text{(F.39)}$$

It has two poles, at $z = \xi_1$ and $z = \xi_2 - ip_m$, respectively. From Eq. (F.35), we have

$$S = \frac{n_f(\xi_1)}{\xi_1 + ip_m - \xi_2} + \frac{n_f(\xi_2 - ip_m)}{\xi_2 - ip_m - \xi_1} = \frac{n_f(\xi_1) - n_f(\xi_2)}{ip_m + \xi_1 - \xi_2}. \qquad \text{(F.40)}$$

The second equality holds because $n_f(\xi_2 - ip_m) = n_f(\xi_2)$.

Appendix G
Linear Response Theory

Here we derive the response function of a system described by the Hamiltonian H to a perturbative Hamiltonian H'. In particular, we consider how the expectation value of an operator J_α evolves under the perturbation, starting from the ground state of H, $|0\rangle$

$$j_\alpha(t) = \langle \Psi(t) | J_\alpha | \Psi(t) \rangle , \tag{G.1}$$

where $\Psi(t)$ is the wavefunction in the Schrödinger representation

$$|\Psi(t)\rangle = T e^{-i \int_{-\infty}^t dt' (H+H')} |0\rangle. \tag{G.2}$$

T is the time-ordering operator.

To perform the perturbation calculation, we turn to the interaction representation, where H' is treated as a perturbation. The expectation value of operator J_α now becomes

$$j_\alpha(t) = \left\langle 0 \left| U^\dagger(t) J_\alpha(t) U(t) \right| 0 \right\rangle, \tag{G.3}$$

where $U(t)$ is the unitary evolution operator, defined by

$$U(t) = T e^{-i \int_{-\infty}^t dt' H'(t')}. \tag{G.4}$$

$J_\alpha(t)$ is the operator of J in the interaction representation

$$J_\alpha(t) = e^{iHt} J_\alpha e^{-iHt}. \tag{G.5}$$

$H'(t)$ is similarly defined.

For linear response, it is sufficient to keep the linear term of H' in the following Taylor expansion

$$U(t) = 1 - i \int_{-\infty}^t dt' H'(t') + O\left(H'\right)^2. \tag{G.6}$$

This yields the approximate formula for the expectation value

$$j_\alpha(t) \approx \left\langle 0 \left| \left[1 + i \int_{-\infty}^t dt' H'(t') \right] J_\alpha(t) \left[1 - i \int_{-\infty}^t dt' H'(t') \right] \right| 0 \right\rangle$$

$$\approx \langle 0 | J_\alpha(t) | 0 \rangle - i \int_{-\infty}^t dt' \left\langle 0 \left| \left[J_\alpha(t), H'(t') \right] \right| 0 \right\rangle. \tag{G.7}$$

To the leading order of H', the change in the expectation value $j_\alpha(t)$ induced by the perturbation is

$$\delta j_\alpha(t) = j_\alpha(t) - \langle 0 | J_\alpha(t) | 0 \rangle \tag{G.8}$$

$$= -i \int_{-\infty}^{t} dt' \, \langle 0 | [J_\alpha(t), H'(t')] | 0 \rangle. \tag{G.9}$$

If H' describes an interaction of J_α with an external field $f_\alpha(t)$

$$H' = \sum_\alpha J_\alpha f_\alpha(t), \tag{G.10}$$

then

$$\delta j_\alpha(t) = -i \sum_\beta \int_{-\infty}^{t} dt' \, \langle 0 | [J_\alpha(t), J_\beta(t')] | 0 \rangle f_\beta(t')$$

$$= -i \sum_\beta \int_{-\infty}^{t} dt' \, D_{\alpha\beta}(t - t') f_\beta(t'), \tag{G.11}$$

where $D_{\alpha\beta}$ is a retarded Green's function

$$D_{\alpha\beta}(t - t') = -i\theta(t - t') \langle 0 | [J_\alpha(t), J_\beta(t')] | 0 \rangle, \tag{G.12}$$

which is also the linear response function of J_α to an external field $f_\beta(t)$.

The above derivation can be readily extended to finite temperatures. In that case, the average over the ground state $|0\rangle$ should simply be replaced by the thermal average, i.e.

$$D_{\alpha\beta}(t - t') = -i\theta(t - t') \langle [J_\alpha(t), J_\beta(t')] \rangle. \tag{G.13}$$

Its Fourier transform is defined as

$$D_{\alpha\beta}(\omega) = i \int_{-\infty}^{+\infty} D_{\alpha\beta}(t) e^{i\omega t}. \tag{G.14}$$

The imaginary part of the retarded Green's function $\mathrm{Im} D_{\alpha\beta}(\omega)$ corresponds to the dissipation under the external perturbation. It is in fact related to the fluctuation spectrum of the system, namely the Fourier component of the dynamic structure factor

$$S_{\alpha\beta}(\omega) = \int_{-\infty}^{+\infty} e^{i\omega t} \langle [J_\alpha(t), J_\beta(0)] \rangle. \tag{G.15}$$

More precisely, these two quantities are related by the so-called fluctuation-dissipation theorem:

$$-\frac{\mathrm{Im} D_{ret}(\omega)}{\pi(1 - e^{-\beta\omega})} = S_{\alpha\beta}(\omega). \tag{G.16}$$

As an example, let us consider the response of the magnetic moment m_α to an applied magnetic field

$$H' = -\sum_\alpha m_\alpha B_\alpha e^{-i\omega t}, \tag{G.17}$$

where $B_\alpha \exp(-i\omega t)$ is the α-component of the AC magnetic field. In this case, $J_\alpha = m_\alpha$, and the expectation value of m_α is just the magnetization

$$m_\alpha(t) = i \sum_\beta \int_{-\infty}^{t} dt' \langle [m_\alpha(t), m_\beta(t')] \rangle B_\beta e^{-i\omega t'}. \tag{G.18}$$

This is the Kubo formula for the magnetization and the response function is just the magnetic susceptibility

$$\chi_{\alpha\beta}(t - t') = i\theta(t - t') \langle [m_\alpha(t), m_\beta(t')] \rangle. \tag{G.19}$$

The Fourier transform of this retarded Green's function is

$$\chi_{\alpha\beta}(\omega) = i \int_{-\infty}^{t} dt' \langle [m_\alpha(t), m_\beta(t')] \rangle e^{i\omega(t-t')}. \tag{G.20}$$

References

[1] J. G. Bednorz and K. A. Müller. Possible high-T_c superconductivity in the Ba-La-Cu-O system. *Z. Phys. B*, 64:189–193, 1986.

[2] M. Tinkham. *Introduction to Superconductivity*. McGraw-Hill, 2nd ed., New York, 1996.

[3] J. R. Schrieffer. *Theory of Superconductivity*. Benjamin/Cummings, New York, 1964.

[4] J. B. Ketterson and S. N. Song. *Superconductivity*. Cambridge University Press, Cambridge, 1999.

[5] Y. Kamihara, T. Watanabe, M. Hirano, and H. Hosono. Iron-based layered superconductor La($O_{1-x} F_x$)FeAs (x = 0.05–0.12) with T_c = 26 K. *J. Am. Chem. Soc.*, 130:3296–3297, 2008.

[6] G. D. Mahan. *Many-Particle Physics*. Plenum Press, New York, 2nd ed., 1981.

[7] T. Xiang. *D-wave Superconductivity*. Science Press, Beijing, 1st ed., 2007.

[8] H. K. Onnes. The resistance of pure mercury at helium temperatures. *Commun. Phys. Lab. Univ. Leiden*, 120b, 1911.

[9] H. K. Onnes. The disappearance of the resistivity of mercury. *Commun. Phys. Lab. Univ. Leiden*, 122b, 1911.

[10] H. K. Onnes. On the sudden change in the rate at which the resistance of mercury disappears. *Commun. Phys. Lab. Univ. Leiden*, 124c, 1911.

[11] W. Meissner and R. Ochsenfeld. Ein neuer Effekt bei Eintritt der Supraleitfähigkeit. *Naturwissenschaften*, 21:787–788, 1933.

[12] F. C. Hsu, J. Y. Luo, K. W. Yeh, et al. Superconductivity in the PbO-type structure-FeSe. *Proc. Nat. Acad. Sci.*, 105:14262–14264, 2008.

[13] N. W. Ashcroft. Metallic hydrogen: A high-temperature superconductor? *Phys. Rev. Lett.*, 21:1748–1749, 1968.

[14] N. W. Ashcroft. Hydrogen dominant metallic alloys: High temperature superconductors? *Phys. Rev. Lett.*, 92:187002, 2004.

[15] A. P. Drozdov, M. I. Eremets, I. A. Troyan, V. Ksenofontov, and S. I. Shilin. Conventional superconductivity at 203 kelvin at high pressures in the sulfur hydride system. *Nature*, 569:528, 2015.

[16] J. Nagamatsu, N. Nakagawa, T. Muranaka, Y. Zenitani, and J. Akimitsu. Superconductivity at 39K in magnesium diboride. *Nature*, 410:63–64, 2001.

[17] A. Schilling, M. Cantoni, J. Guo, and H. Ott. Superconductivity above 130 K in the Hg-Ba-Ca-Cu-O system. *Nature*, 363:56–58, 1993.

[18] E. Snider, N. Dasenbrock-Gammon, R. McBride, et al. Room-temperature superconductivity in a carbonaceous sulfur hydride. *Nature*, 586:373–377, Dec 2020.

[19] C. N. Yang. Concept of off-diagonal long-range order and the quantum phases of liquid He and of superconductors. *Rev. Mod. Phys.*, 34:694–704, Oct 1962.

[20] C. J. Gorter and H. B. G. Casimir. On superconductivity I. *Physica*, 1:30–320, 1934.

[21] F. London and H. London. The electromagnetic equations of the supraconductor. *Proc. Royal Soc. London. Series A: Math. Phys. Sci.*, 149:71–88, 1935.

[22] L. N. Cooper. Bound electron pairs in a degenerate Fermi gas. *Phys. Rev.*, 104:1189–1190, 1956.

[23] J. Bardeen, L. N. Cooper, and J. R. Schrieffer. Theory of superconductivity. *Phys. Rev.*, 108:1175–1204, 1957.

[24] P. G. De Gennes. *Superconductivity of Metals and Alloys* (Advanced Book Classics). Addison-Wesley Publ. Company Inc, Boston, 1999.

[25] P. C. Hohenberg. Existence of long-range order in one and two dimensions. *Phys. Rev.*, 158:383–386, 1967.

[26] P. W. Anderson. Coherent excited states in the theory of superconductivity: Gauge invariance and the Meissner effect. *Phys. Rev.*, 110:827–835, 1958.

[27] V. L. Ginzburg and L. D. Landau. On the theory of superconductivity. *Zh. Eksperim. i. Teor. Fiz.*, 20:1064, 1950.

[28] L. P. Gorkov. Microscopic derivation of the Ginzburg–Landau equations in the theory of superconductivity. *Sov. Phys. JETP*, 9:1364–1367, 1959.

[29] B. S. Deaver and W. M. Fairbank. Experimental evidence for quantized flux in superconducting cylinders. *Phys. Rev. Lett.*, 7:43–46, 1961.

[30] R. Doll and M. Näbauer. Experimental proof of magnetic flux quantization in a superconducting ring. *Phys. Rev. Lett.*, 7:51–52, Jul. 1961.

[31] A. B. Pippard and W. L. Bragg. An experimental and theoretical study of the relation between magnetic field and current in a superconductor. *Proc. Royal Soc. London. Series A: Math. Phys. Sci.*, 216:547–568, 1953.

[32] V. J. Emery and S. A. Kivelson. Importance of phase fluctuations in superconductors with small superfluid density. *Nature*, 374:434–437, 1995.

[33] Y. J. Uemura, G. M. Luke, B. J. Sternlieb, et al. Universal correlations between T_c and n_s/m^* (carrier density over effective mass) in high-T_c cuprate superconductors. *Phys. Rev. Lett.*, 62:2317–2320, 1989.

[34] W. L. McMillan. Transition temperature of strong-coupled superconductors. *Phys. Rev.*, 167:331–344, 1968.

[35] D. J. Scalapino. Antiferromagnetic fluctuations and $d_{x^2-y^2}$-pairing in the cuprates. *Phys. C: Supercond.*, 235–240:107–112, 1994.

[36] C. de la Cruz, Q. Huang, J. W. Lynn, et al. Magnetic order close to superconductivity in the iron-based layered $LaO_{1-x}F_xFeAs$ systems. *Nature*, 453:899, 2008.

[37] P. W. Anderson. The resonating valence bond state in La_2CuO_4 and superconductivity. *Science*, 235:1196–1198, 1987.

[38] A. P. Mackenzie and Y. Maeno. The superconductivity of Sr_2RuO_4 and the physics of spin-triplet pairing. *Rev. Mod. Phys.*, 75:657–712, 2003.

[39] G. Kotliar and J. L. Liu. Superexchange mechanism and d-wave superconductivity. *Phys. Rev. B*, 38:5142–5145, 1988.

[40] N. E. Bickers, D. J. Scalapino, and S. R. White. Conserving approximations for strongly correlated electron systems: Bethe-Salpeter equation and dynamics for the two-dimensional Hubbard model. *Phys. Rev. Lett.*, 62:961–964, 1989.

[41] T. Moriya, Y. Takahashi, and K. Ueda. Antiferromagnetic spin fluctuations and superconductivity in two-dimensional metals – A possible model for high T_c oxides. *J. Phys. Soc. Jpn.*, 59:2905–2915, 1990.

[42] P. Monthoux, A. V. Balatsky, and D. Pines. Weak-coupling theory of high-temperature superconductivity in the antiferromagnetically correlated copper oxides. *Phys. Rev. B*, 46:14803–14817, 1992.

[43] T. Yamashita, A. Kawakami, T. Nishihara, Y. Hirotsu, and M. Takata. AC Josephson effect in point-contacts of Ba-Y-Cu-O ceramics. *Jpn. J. Appl. Phys.*, 26:L635, 1987.

[44] T. Yamashita, A. Kawakami, T. Nishihara, M. Takata, and K. Kishio. RF power dependence of AC Josephson current in point contacts of BaY (Tm) CuO ceramics. *Jpn. J. Appl. Phys.*, 26:L671, 1987.

[45] T. J. Witt. Accurate determination of $\frac{2e}{h}$ in Y-Ba-Cu-O Josephson junctions. *Phys. Rev. Lett.*, 61:1423–1426, 1988.

[46] H. F. C. Hoevers, P. J. M. Van Bentum, L. E. C. Van De Leemput, et al. Determination of the energy gap in a thin $YBa_2Cu_3O_{7-x}$ film by Andreev reflection and by tunneling. *Phys. C: Supercond.*, 152:105–110, 1988.

[47] P. J. M. Van Bentum, H. F. C. Hoevers, H. Van Kempen, et al. Determination of the energy gap in $YBa_2Cu_3O_{7-\delta}$ by tunneling, far infrared reflection and Andreev reflection. *Phys. C: Supercond.*, 153:1718–1723, 1988.

[48] C. E. Gough, M. S. Colclough, E. M. Forgan, R. G. Jordan, and M. Keene. Flux quantization in a high-T_c superconductor. *Nature*, 326:855, 1987.

[49] R. H. Koch, C. P. Umbach, G. J. Clark, P. Chaudhari, and R. B. Laibowitz. Quantum interference devices made from superconducting oxide thin films. *Appl. Phys. Lett.*, 51:200–202, 1987.

[50] P. L. Gammel, P. A. Polakos, C. E. Rice, L. R. Harriott, and D. J. Bishop. Little-Parks oscillations of T_c in patterned microstructures of the oxide superconductor $YBa_2Cu_3O_7$: Experimental limits on fractional-statistics-particle theories. *Phys. Rev. B*, 41:2593–2596, 1990.

[51] J. C. Campuzano, H. Ding, M. R. Norman, et al. Direct observation of particle-hole mixing in the superconducting state by angle-resolved photoemission. *Phys. Rev. B*, 53:R14737–R14740, 1996.

[52] M. Takigawa, P. C. Hammel, R. H. Heffner, and Z. Fisk. Spin susceptibility in superconducting $YBa_2Cu_3O_7$ from ^{63}Cu Knight shift. *Phys. Rev. B*, 39:7371–7374, Apr 1989.

[53] S. E. Barrett, D. J. Durand, C. H. Pennington, et al. ^{63}Cu Knight shifts in the superconducting state of $YBa_2Cu_3O_{7-\delta}$ ($T_c = 90K$). *Phys. Rev. B*, 41:6283–6296, 1990.

[54] T. Xiang. Physical properties of d-wave superconductors and pairing symmetry of high temperature superconducting electrons, in *Fundamental Research in High-T_c superconductivity*, ed. W. Z. Zhou and W. Y. Liang. Shanghai Science and Technology Press, Shanghai, 1999.

[55] F. Ma, W. Ji, J. Hu, Z. Y. Lu, and T. Xiang. First-principles calculations of the electronic structure of tetragonal alpha-FeTe and alpha-FeSe crystals: Evidence for a bicollinear antiferromagnetic order. *Phys. Rev. Lett.*, 102, 2009.

[56] W. Bao, Q. Z. Huang, G. F. Chen, et al. A novel large moment antiferromagnetic order in $K_{0.8}Fe_{1.6}Se_2$ superconductor. *Chin. Phys. Lett.*, 28:86104, 2011.

[57] F. Ma, Z. Y. Lu, and T. Xiang. Arsenic-bridged antiferromagnetic superexchange interactions in LaFeAsO. *Phys. Rev. B.*, 78:224517, 2008.

[58] X. W. Yan, M. Gao, Z. Y. Lu, and T. Xiang. Electronic structures and magnetic order of ordered-Fe-vacancy ternary iron selenides $TlFe_{1.5}Se_2$ and $AFe_{1.5}Se_2$ (A = K, Rb, or Cs). *Phys. Rev. Lett.*, 106:087005, 2011.

[59] T. Yildirim. Origin of the 150-K anomaly in LaFeAsO: Competing antiferromagnetic interactions, frustration, and a structural phase transition. *Phys. Rev. Lett.*, 101:057010, 2008.

[60] M. Yi, D. H. Lu, R. Yu, et al. Observation of temperature-induced crossover to an orbital-selective Mott phase in $A_x Fe_{2-y}Se_2$ (A=K, Rb) superconductors. *Phys. Rev. Lett.*, 110:067003, 2013.

[61] H.-J. Grafe, D. Paar, G. Lang, et al. ^{75}As NMR studies of superconducting $LaFeAsO_{0.9}F_{0.1}$. *Phys. Rev. Lett.*, 101:047003, 2008.

[62] F. Ning, K. Ahilan, T. Imai, et al. ^{59}Co and ^{75}As NMR investigation of electron-doped high-T_c superconductor $BaFe_{1.8}Co_{0.2}As_2$ (T_c = 22 K). *J. Phys. Soc. Jpn.*, 77:103705, 2008.

[63] D. Liu, C. Li, J. Huang, et al. Orbital origin of extremely anisotropic superconducting gap in nematic phase of FeSe superconductor. *Phys. Rev. X*, 8:031033, 2018.

[64] I. I. Mazin, D. J. Singh, M. D. Johannes, and M. H. Du. Unconventional superconductivity with a sign reversal in the order parameter of $LaFeAsO_{1-x}F_x$. *Phys. Rev. Lett.*, 101:057003, 2008.

[65] K. Seo, B. A. Bernevig, and J. Hu. Pairing symmetry in a two-orbital exchange coupling model of oxypnictides. *Phys. Rev. Lett.*, 101:206404, 2008.

[66] F. Wang, H. Zhai, Y. Ran, A. Vishwanath, and D. H. Lee. Functional renormalization-group study of the pairing symmetry and pairing mechanism of the FeAs-based high-temperature superconductor. *Phys. Rev. Lett.*, 102:047005, 2009.

[67] K. Kuroki, H. Usui, S. Onari, R. Arita, and H. Aoki. Pnictogen height as a possible switch between high-T_c nodeless and low-T_c nodal pairings in the iron-based superconductors. *Phys. Rev. B*, 79:224511, 2009.

[68] S. Onari and H. Kontani. Violation of Anderson's theorem for the sign-reversing s-wave state of iron-pnictide superconductors. *Phys. Rev. Lett.*, 103:177001, 2009.

[69] T. A. Maier and D. J. Scalapino. Theory of neutron scattering as a probe of the superconducting gap in the iron pnictides. *Phys. Rev. B*, 78:020514, 2008.

[70] M. M. Korshunov and I. Eremin. Theory of magnetic excitations in iron-based layered superconductors. *Phys. Rev. B*, 78:140509, 2008.

[71] A. D. Christianson, E. A. Goremychkin, R. Osborn, et al. Unconventional superconductivity in $Ba_{0.6}K_{0.4}Fe_2As_2$ from inelastic neutron scattering. *Nature*, 456:930, 2008.

[72] D. S. Inosov, J. T. Park, P. Bourges, et al. Normal-state spin dynamics and temperature-dependent spin-resonance energy in optimally doped $BaFe_{1.85}Co_{0.15}As_2$. *Nat. Phys.*, 6:178–181, 2010.

[73] S. Li, C. Zhang, M. Wang, et al. Normal-state hourglass dispersion of the spin excitations in $FeSe_xTe_{1-x}$. *Phys. Rev. Lett.*, 105:157002, 2010.

[74] T. Hanaguri, S. Niitaka, K. Kuroki, and H. Takagi. Unconventional s-wave superconductivity in Fe(Se,Te). *Science*, 328:474–476, 2010.

[75] S. Grothe, S. Chi, P. Dosanjh, et al. Bound states of defects in superconducting LiFeAs studied by scanning tunneling spectroscopy. *Phys. Rev. B*, 86:174503, 2012.

[76] C. L. Song, Y. L. Wang, Y. P. Jiang, et al. Suppression of superconductivity by twin boundaries in FeSe. *Phys. Rev. Lett.*, 109:137004, 2012.

[77] H. Yang, Z. Wang, D. L. Fang, Q. Deng, et al. In-gap quasiparticle excitations induced by non-magnetic Cu impurities in $Na(Fe_{0.96}Co_{0.03}Cu_{0.01})As$ revealed by scanning tunnelling spectroscopy. *Nat. Commun.*, 4:2749, 2013.

[78] J. X. Yin, Z. Wu, J. H. Wang, et al. Observation of a robust zero-energy bound state in iron-based superconductor Fe(Te,Se). *Nat. Phys.*, 11:543, 2015.

[79] R. M. Fernandes and J. Schmalian. Competing order and nature of the pairing state in the iron pnictides. *Phys. Rev. B*, 82:014521, 2010.

[80] Y. Laplace, J. Bobroff, F. Rullier-Albenque, D. Colson, and A. Forget. Atomic coexistence of superconductivity and incommensurate magnetic order in the pnictide $Ba(Fe_{1-x}Co_x)_2As_2$. *Phys. Rev. B*, 80:140501, 2009.

[81] S. Nandi, M. G. Kim, A. Kreyssig, et al. Anomalous suppression of the orthorhombic lattice distortion in superconducting $Ba(Fe_{1-x}Co_x)_2As_2$ single crystals. *Phys. Rev. Lett.*, 104:057006, 2010.

[82] J.-Ph. Reid, M. A. Tanatar, A. Juneau-Fecteau, et al. Universal heat conduction in the iron arsenide superconductor KFe_2As_2: Evidence of a d-wave state. *Phys. Rev. Lett.*, 109:087001, 2012.

[83] C. L. Song, Y. L. Wang, P. Cheng, et al. Direct observation of nodes and twofold symmetry in FeSe superconductor. *Science*, 332:1410–1413, 2011.

[84] F. F. Tafti, A. Juneau-Fecteau, M-É Delage, et al. Sudden reversal in the pressure dependence of T_c in the iron-based superconductor KFe_2As_2. *Nat. Phys.*, 9:349, 2013.

[85] A. Damascelli, Z. Hussain, and Z. X. Shen. Angle-resolved photoemission studies of the cuprate superconductors. *Rev. Mod. Phys.*, 75:473–541, Apr 2003.

[86] P. W. Anderson. Hall effect in the two-dimensional Luttinger liquid. *Phys. Rev. Lett.*, 67:2092–2094, 1991.

[87] T. R. Chien, Z. Z. Wang, and N. P. Ong. Effect of Zn impurities on the normal-state Hall angle in single-crystal $YBa_2Cu_{3-x}Zn_xO_{7-\delta}$. *Phys. Rev. Lett.*, 67:2088–2091, 1991.

[88] T. Timusk and B. Statt. The pseudogap in high-temperature superconductors: An experimental survey. *Rep. Prog. Phys.*, 62:61, 1999.

[89] F. Zhou, P. H. Hor, X. L. Dong, and Z. X. Zhao. Anomalies at magic charge densities in under-doped $La_{2-x}Sr_xCuO_4$ superconductor crystals prepared by floating-zone method. *Sci. Tech. Adv. Mater.*, 6:873, 2005.

[90] Z. A. Xu, N. P. Ong, Y. Y. Wang, T. Kakeshita, and S. Uchida. Vortex-like excitations and the onset of superconducting phase fluctuation in underdoped $La_{2-x}Sr_xCuO_4$. *Nature*, 406:486–488, 2000.

[91] J. M. Tranquada, B. J. Sternlieb, J. D. Axe, Y. Nakamura, and S. Uchida. Evidence for stripe correlations of spins and holes in copper oxide superconductors. *Nature*, 375:561–563, 1995.

[92] J. L. Tallon and J. W. Loram. The doping dependence of T^* – what is the real high-T_c phase diagram? *Phys. C*, 349:53–68, 2001.

[93] C. Panagopoulos, J. L. Tallon, B. D. Rainford, et al. Evidence for a generic quantum transition in high-T_c cuprates. *Phys. Rev. B*, 66:064501, 2002.

[94] C. N. Yang and S. C. Zhang. SO_4 symmetry in a Hubbard model. *Mod. Phys. Lett.*, 04:759–766, 1990.

[95] J. Zaanen, G. A. Sawatzky, and J. W. Allen. Band gaps and electronic structure of transition-metal compounds. *Phys. Rev. Lett.*, 55:418–421, 1985.

[96] V. J. Emery. Theory of high-T_c superconductivity in oxides. *Phys. Rev. Lett.*, 58:2794–2797, 1987.

[97] F. C. Zhang and T. M. Rice. Effective Hamiltonian for the superconducting Cu oxides. *Phys. Rev. B*, 37:3759–3761, 1988.

[98] E. H. Lieb and F. Y. Wu. Absence of Mott transition in an exact solution of the short-range, one-band model in one dimension. *Phys. Rev. Lett.*, 20:1445–1448, 1968.

[99] T. Xiang and J. M. Wheatley. c-axis superfluid response of copper oxide superconductors. *Phys. Rev. Lett.*, 77:4632–4635, 1996.

[100] T. Xiang, C. Panagopoulos, and J. R. Cooper. Low temperature superfluid response of high-T_c superconductors. *Int. J. Mod. Phys. B*, 12:1007–1032, 1998.

[101] D. L. Feng, N. P. Armitage, D. H. Lu, et al. Bilayer splitting in the electronic structure of heavily overdoped $Bi_2Sr_2CaCu_2O_{8+\delta}$. *Phys. Rev. Lett.*, 86:5550–5553, 2001.

[102] N. E. Hussey, M. Abdel-Jawad, A. Carrington, A. P. Mackenzie, and L. Balicas. A coherent three-dimensional Fermi surface in a high-transition-temperature superconductor. *Nature*, 425:814–817, 2003.

[103] T. Xiang, Y. H. Su, C. Panagopoulos, Z. B. Su, and L. Yu. Microscopic Hamiltonian for Zn- or Ni-substituted high-temperature cuprate superconductors. *Phys. Rev. B*, 66:174504, 2002.

[104] J. Wheatley and T. Xiang. Stability of d-wave pairing in two dimensions. *Solid State Commun.*, 88:593–595, 1993.

[105] W. Y. Liang, J. W. Loram, K. A. Mirza, N. Athanassopoulou, and J. R. Cooper. Specific heat and susceptibility determination of the pseudogap in $YBCO_{7-\delta}$. *Phys. C: Supercond.*, 263:277–281, 1996. Proceedings of the International Symposium on Frontiers of High-T_c Superconductivity.

[106] N. Momono and M. Ido. Evidence for nodes in the superconducting gap of $La_{2-x}Sr_xCuO_4$. T^2 dependence of electronic specific heat and impurity effects. *Phys. C*, 264:311–318, 1996.

[107] G. D. Mahan. Theory of photoemission in simple metals. *Phys. Rev. B*, 2:4334–4350, 1970.

[108] W. L. Schaich and N. W. Ashcroft. Model calculations in the theory of photoemission. *Phys. Rev. B*, 3:2452–2465, 1971.

[109] C. N. Berglund and W. E. Spicer. Photoemission studies of copper and silver: Theory. *Phys. Rev.*, 136:A1030–A1044, 1964.

[110] P. J. Feibelman and D. E. Eastman. Photoemission spectroscopy – correspondence between quantum theory and experimental phenomenology. *Phys. Rev. B*, 10: 4932–4947, 1974.

[111] D. S. Dessau, B. O. Wells, Z.-X. Shen, et al. Anomalous spectral weight transfer at the superconducting transition of $Bi_2Sr_2CaCu_2O_{8+\delta}$. *Phys. Rev. Lett.*, 66: 2160–2163, 1991.

[112] Ch. Renner and O. Fischer. Vacuum tunneling spectroscopy and asymmetric density of states of $Bi_2Sr_2CaCu_2O_{8+\delta}$. *Phys. Rev. B*, 51:9208–9218, Apr 1995.

[113] A. A. Kordyuk, S. V. Borisenko, T. K. Kim, et al. Origin of the peak-dip-hump line shape in the superconducting-state $(\pi, 0)$ photoemission spectra of $Bi_2Sr_2CaCu_2O_8$. *Phys. Rev. Lett.*, 89:077003, 2002.

[114] M. R. Norman, H. Ding, J. C. Campuzano, et al. Unusual dispersion and line shape of the superconducting state spectra of $Bi_2Sr_2CaCu_2O_{8+\delta}$. *Phys. Rev. Lett.*, 79: 3506–3509, 1997.

[115] H. A. Mook, M. Yethiraj, G. Aeppli, T. E. Mason, and T. Armstrong. Polarized neutron determination of the magnetic excitations in $YBa_2Cu_3O_7$. *Phys. Rev. Lett.*, 70:3490–3493, 1993.

[116] J. C. Campuzano, M. R. Norman, and M. Randeira. *Physics of Superconductors*, volume II. ed. Bennemann, K. H. and Ketterson, J. B., Springer, Berlin 2003.

[117] T. Cuk, D. H. Lu, X. J. Zhou, et al. A review of electron-phonon coupling seen in the high-T_c superconductors by angle-resolved photoemission studies (ARPES). *Phys. Stat. Sol. B*, 242:11–29, 2005.

[118] J. M. Luttinger and J. C. Ward. Ground-state energy of a many-fermion system. II. *Phys. Rev.*, 118:1417–1427, 1960.

[119] A. A. Abrikosov, L. P. Gorkov, and I. E. Dzyaloshinski. *Quantum Field Theoretical Methods in Statistical Physics*, volume 4. Pergamon, Oxford, 1965.

[120] D. S. Marshall, D. S. Dessau, A. G. Loeser, et al. Unconventional electronic structure evolution with hole doping in $Bi_2Sr_2CaCu_2O_{8+\delta}$: Angle-resolved photoemission results. *Phys. Rev. Lett.*, 76:4841–4844, 1996.

[121] M. R. Norman, H. Ding, M. Randeria, et al. Destruction of the Fermi surface in underdoped high-T_c superconductors. *Nature*, 392:157–160, 1998.

[122] H. Ding, M. R. Norman, J. C. Campuzano, et al. Angle-resolved photoemission spectroscopy study of the superconducting gap anisotropy in $Bi_2Sr_2CaCu_2O_{8+x}$. *Phys. Rev. B*, 54:R9678–R9681, 1996.

[123] Z. X. Shen, D. S. Dessau, B. O. Wells, et al. Anomalously large gap anisotropy in the a-b plane of $Bi_2Sr_2CaCu_2O_{8+\delta}$. *Phys. Rev. Lett.*, 70:1553–1556, 1993.

[124] M. L. Titov, A. G. Yashenkin, and D. N. Aristov. Quasiparticle damping in two-dimensional superconductors with unconventional pairing. *Phys. Rev. B*, 52:10626–10632, 1995.

[125] M. B. Walker and M. F. Smith. Quasiparticle-quasiparticle scattering in high-T_c superconductors. *Phys. Rev. B*, 61:11285–11288, 2000.

[126] T. Valla, T. E. Kidd, J. D. Rameau, et al. Fine details of the nodal electronic excitations in $Bi_2Sr_2CaCu_2O_{8+\delta}$. *Phys. Rev. B*, 73:184518, 2006.

[127] D. Duffy, P. J. Hirschfeld, and D. J. Scalapino. Quasiparticle lifetimes in a $d_{x^2-y^2}$ superconductor. *Phys. Rev. B*, 64:224522, 2001.

[128] T. Dahm, P. J. Hirschfeld, D. J. Scalapino, and L. Zhu. Nodal quasiparticle lifetimes in cuprate superconductors. *Phys. Rev. B*, 72:214512, 2005.

[129] A. Hosseini, R. Harris, S. Kamal, et al. Microwave spectroscopy of thermally excited quasiparticles in $YBa_2Cu_3O_{6.99}$. *Phys. Rev. B*, 60:1349–1359, 1999.

[130] A. F. Andreev. The thermal conductivity of the intermediate state in superconductors. *Sov. Phys. JETP*, 19:1228, 1964.

[131] C. R. Hu. Midgap surface states as a novel signature for $d_{a^2-b^2}$-wave superconductivity. *Phys. Rev. Lett.*, 72:1526–1529, 1994.

[132] Y. Tanaka and S. Kashiwaya. Theory of tunneling spectroscopy of d-wave superconductors. *Phys. Rev. Lett.*, 74:3451–3454, 1995.

[133] S. Kashiwaya, Y. Tanaka, M. Koyanagi, and K. Kajimura. Theory for tunneling spectroscopy of anisotropic superconductors. *Phys. Rev. B*, 53:2667–2676, 1996.

[134] G. E. Blonder, M. Tinkham, and T. M. Klapwijk. Transition from metallic to tunneling regimes in superconducting microconstrictions: Excess current, charge imbalance, and supercurrent conversion. *Phys. Rev. B*, 25:4515–4532, 1982.

[135] S. Sinha and K. W. Ng. Zero bias conductance peak enhancement in $Bi_2Sr_2CaCu_2O_8$/Pb tunneling junctions. *Phys. Rev. Lett.*, 80:1296–1299, 1998.

[136] H. Aubin, L. H. Greene, S. Jian, and D. G. Hinks. Andreev bound states at the onset of phase coherence in $Bi_2Sr_2CaCu_2O_8$. *Phys. Rev. Lett.*, 89:177001, 2002.

[137] M. Covington, M. Aprili, E. Paraoanu, et al. Observation of surface-induced broken time-reversal symmetry in $YBa_2Cu_3O_7$ tunnel junctions. *Phys. Rev. Lett.*, 79:277–280, 1997.

[138] J. Y. T. Wei, N. C. Yeh, D. F. Garrigus, and M. Strasik. Directional tunneling and Andreev reflection on $YBa_2Cu_3O_{7-\delta}$ single crystals: Predominance of d-wave pairing symmetry verified with the generalized Blonder, Tinkham, and Klapwijk theory. *Phys. Rev. Lett.*, 81:2542–2545, 1998.

[139] M. Aprili, E. Badica, and L. H. Greene. Doppler shift of the Andreev bound states at the YBCO surface. *Phys. Rev. Lett.*, 83:4630–4633, 1999.

[140] R. Krupke and G. Deutscher. Anisotropic magnetic field dependence of the zero-bias anomaly on in-plane oriented [100] $Y_1Ba_2Cu_3O_{7-x}$/In tunnel junctions. *Phys. Rev. Lett.*, 83:4634–4637, 1999.

[141] J. Bardeen. Tunnelling from a many-particle point of view. *Phys. Rev. Lett.*, 6:57–59, 1961.

[142] W. A. Harrison. Tunneling from an independent-particle point of view. *Phys. Rev.*, 123:85–89, 1961.

[143] M. Franz and A. J. Millis. Phase fluctuations and spectral properties of underdoped cuprates. *Phys. Rev. B*, 58:14572–14580, 1998.

[144] M. B. Walker and J. Luettmer-Strathmann. Josephson tunneling in high-T_c superconductors. *Phys. Rev. B*, 54:588–601, 1996.

[145] M. Sigrist and T. M. Rice. Paramagnetic effect in high T_c superconductors-a hint for d-wave superconductivity. *J. Phys. Soc. Jpn.*, 61:4283–4286, 1992.

[146] C. C. Tsuei, J. R. Kirtley, C. C. Chi, et al. Pairing symmetry and flux quantization in a tricrystal superconducting ring of $YBa_2Cu_3O_{7-\delta}$. *Phys. Rev. Lett.*, 73:593–596, 1994.

[147] D. J. Van Harlingen. Phase-sensitive tests of the symmetry of the pairing state in the high-temperature superconductors – evidence for $d_{x^2-y^2}$ symmetry. *Rev. Mod. Phys.*, 67:515–535, 1995.

[148] C. C. Tsuei and J. R. Kirtley. Pairing symmetry in cuprate superconductors. *Rev. Mod. Phys.*, 72:969–1016, 2000.

[149] D. A. Wollman, D. J. Van Harlingen, W. C. Lee, D. M. Ginsberg, and A. J. Leggett. Experimental determination of the superconducting pairing state in YBCO from the phase coherence of YBCO-Pb dc SQUIDS. *Phys. Rev. Lett.*, 71:2134–2137, 1993.

[150] D. A. Wollman, D. J. Van Harlingen, J. Giapintzakis, and D. M. Ginsberg. Evidence for $d_{x^2-y^2}$ pairing from the magnetic field modulation of $YBa_2Cu_3O_7$-Pb Josephson junctions. *Phys. Rev. Lett.*, 74:797–800, 1995.

[151] C. C. Tsuei and J. R. Kirtley. d-wave pairing symmetry in cuprate superconductors. *Phys. C*, 341:1625–1628, 2000.

[152] C. C. Tsuei, J. R. Kirtley, M. Rupp, et al. Pairing symmetry in single-layer tetragonal Tl_2Ba_2CuO superconductors. *Science*, 271:329, 1996.

[153] C. C. Tsuei, J. R. Kirtley, Z. F. Ren, et al. Pure $d_{x^2-y^2}$ order-parameter symmetry in the tetragonal superconductor $Tl_2Ba_2CuO_{6+\delta}$. *Nature*, 387:481–483, 1997.

[154] C. C. Tsuei and J. R. Kirtley. Pure d-wave pairing symmetry in high-T_c cuprate superconductors. *J. Phys. Chem. Solids*, 59:2045, 1998.

[155] C. C. Tsuei and J. R. Kirtley. Phase-sensitive evidence for d-wave pairing symmetry in electron-doped cuprate superconductors. *Phys. Rev. Lett.*, 85:182–185, 2000.

[156] C. C. Tsuei, J. R. Kirtley, G. Hammerl, et al. Robust $d_{x^2-y^2}$ pairing symmetry in hole-doped cuprate superconductors. *Phys. Rev. Lett.*, 93:187004, 2004.

[157] M. Sigrist and T. M. Rice. Unusual paramagnetic phenomena in granular high-temperature superconductors a consequence of d-wave pairing? *Rev. Mod. Phys.*, 67:503–513, 1995.

[158] W. Braunisch, N. Knauf, V. Kataev, S. Neuhausen, A. Grütz, A. Kock, B. Roden, D. Khomskii, and D. Wohlleben. Paramagnetic Meissner effect in Bi high-temperature superconductors. *Phys. Rev. Lett.*, 68:1908–1911, 1992.

[159] P. Svedlindh, K. Niskanen, P. Norling, P. Nordblad, L. Lundgren, B. Lönnberg, and T. Lundström. Anti-Meissner effect in the BiSrCaCuO-system. *Phys. C*, 162:1365–1366, 1989.

[160] D. J. Thompson, M. S. M. Minhaj, L. E. Wenger, and J. T. Chen. Observation of paramagnetic Meissner effect in niobium disks. *Phys. Rev. Lett.*, 75:529–532, 1995.

[161] A. K. Geim, S. V. Dubonos, J. G. S. Lok, M. Henini, and J. C. Maan. Paramagnetic Meissner effect in small superconductors. *Nature*, 396:144, 1998.

[162] T. M. Rice and M. Sigrist. Comment on "Paramagnetic Meissner effect in Nb". *Phys. Rev. B*, 55:14647–14648, 1997.

[163] P. W. Anderson. Theory of dirty superconductors. *J. Phys. Chem. Solids*, 11:26–30, 1959.

[164] T. Xiang and J. M. Wheatley. Nonmagnetic impurities in two-dimensional supercon-ductors. *Phys. Rev. B*, 51:11721–11727, 1995.

[165] S. H. Pan, E. W. Hudson, K. M. Lang, et al. Imaging the effects of individual zinc impurity atoms on superconductivity in $Bi_2Sr_2CaCu_2O_{8+\delta}$. *Nature*, 403:746–750, 2000.

[166] J. W. Loram, J. Luo, J. R. Cooper, W. Y. Liang, and J. L. Tallon. Evidence on the pseudogap and condensate from the electronic specific heat. *J. Phys. Chem. Solids*, 62:59–64, 2001.

[167] C. J. Wu, T. Xiang, and Z. B. Su. Absence of the zero bias peak in vortex tunneling spectra of high-temperature superconductors. *Phys. Rev. B*, 62:14427–14430, 2000.

[168] I. Martin, A. V. Balatsky, and J. Zaanen. Impurity states and interlayer tunneling in high temperature superconductors. *Phys. Rev. Lett.*, 88:097003, 2002.

[169] L. Yu. Bound state in superconductors with paramagnetic impurities. *Acta Phys. Sin.*, 21:75–91, 2005.

[170] H. Shiba. Classical spins in superconductors. *Prog. Theor. Phys.*, 40:435–451, 1968.

[171] J. Kondo. Resistance minimum in dilute magnetic alloys. *Prog. Theor. Phys.*, 32: 37–49, 1964.

[172] P. W. Anderson, G. Yuval, and D. R. Hamann. Exact results in the Kondo problem. II. scaling theory, qualitatively correct solution, and some new results on one-dimensional classical statistical models. *Phys. Rev. B*, 1:4464–4473, 1970.

[173] K. G. Wilson. The renormalization group: Critical phenomena and the Kondo problem. *Rev. Mod. Phys.*, 47:773–840, 1975.

[174] A. A. Abrikosov and L. P. Gorkov. Contribution to the theory of superconducting alloys with paramagnetic impurities. *Sov. Phys. JETP*, 12:1243, 1961.

[175] L. S. Borkowski and P. J. Hirschfeld. Kondo effect in gapless superconductors. *Phys. Rev. B*, 46:9274–9277, 1992.

[176] C. R. Cassanello and E. Fradkin. Overscreening of magnetic impurities in $d_{x^2-y^2}$-wave superconductors. *Phys. Rev. B*, 56:11246–11261, 1997.

[177] G. M. Zhang, H. Hu, and L. Yu. Marginal Fermi liquid resonance induced by a quantum magnetic impurity in d-wave superconductors. *Phys. Rev. Lett.*, 86: 704–707, 2001.

[178] A. Polkovnikov, S. Sachdev, and M. Vojta. Impurity in a d-wave superconductor: Kondo effect and STM spectra. *Phys. Rev. Lett.*, 86:296–299, 2001.

[179] A. V. Balatsky, I. Vekhter, and J.-X. Zhu. Impurity-induced states in conventional and unconventional superconductors. *Rev. Mod. Phys.*, 78:373–433, 2006.

[180] Ø. Fischer, M. Kugler, I. Maggio-Aprile, C. Berthod, and C. Renner. Scanning tun-neling spectroscopy of high-temperature superconductors. *Rev. Mod. Phys.*, 79:353–419, 2007.

[181] J. E. Hoffman. Imaging quasiparticle interference in $Bi_2Sr_2CaCu_2O_{8+\delta}$. *Science*, 297:1148–1151, 2002.

[182] K. McElroy, R. W. Simmonds, J. E. Hoffman, et al. Relating atomic-scale electronic phenomena to wave-like quasiparticle states in superconducting $Bi_2Sr_2CaCu_2O_{8+\delta}$. *Nature*, 422:592, 2003.

[183] C. Howald, H. Eisaki, N. Kaneko, M. Greven, and A. Kapitulnik. Periodic density-of-states modulations in superconducting $Bi_2Sr_2CaCu_2O_{8+\delta}$. *Phys. Rev. B*, 67:014533, 2003.

[184] T. Hanaguri, Y. Kohsaka, M. Ono, et al. Coherence factors in a high-T_c cuprate probed by quasi-particle scattering off vortices. *Science*, 323:923, 2009.

[185] Q. H. Wang and D. H. Lee. Quasiparticle scattering interference in high-temperature superconductors. *Phys. Rev. B*, 67:020511, 2003.

[186] R. S. Markiewicz. Bridging k and q space in the cuprates: Comparing angle-resolved photoemission and STM results. *Phys. Rev. B*, 69:214517, 2004.

[187] J. Friedel. Metallic alloys. *Nuovo Cimento*, 7:287, 1958.

[188] M. F. Crommie, C. P. Lutz, and D. M. Eigler. Imaging standing waves in a two-dimensional electron gas. *Nature*, 363:524, 1993.

[189] P. J. Hirschfeld, P. Wölfle, and D. Einzel. Consequences of resonant impurity scattering in anisotropic superconductors: Thermal and spin relaxation properties. *Phys. Rev. B*, 37:83–97, 1988.

[190] Y. Itoh, S. Adachi, T. Machi, Y. Ohashi, and N. Koshizuka. Ni-substituted sites and the effect on Cu electron spin dynamics of $YBa_2Cu_{3-x}Ni_xO_{7-\delta}$. *Phys. Rev. B*, 66:134511, 2002.

[191] N. Schopohl and O. V. Dolgov. T dependence of the magnetic penetration depth in unconventional superconductors at low temperatures: Can it be linear? *Phys. Rev. Lett.*, 80:4761–4762, 1998.

[192] G. E. Volovik. Comment on "T dependence of the magnetic penetration depth in unconventional superconductors at low temperatures: Can it be linear?". *Phys. Rev. Lett.*, 81:4023–4023, 1998.

[193] D. L. Novikov and A. J. Freeman. Electronic structure and Fermi surface of the $HgBa_2CuO_{4+\delta}$ superconductor: Apparent importance of the role of Van Hove singularities on high T_c. *Physica C*, 212:233–238, 1993.

[194] O. K. Andersen, A. I. Liechtenstein, O. Jepsen, and F. Paulsen. LDA energy bands, low-energy Hamiltonians, t', t'', $t_\perp(k)$, and $J_\perp$. *J. Phys. Chem. Solids*, 56:1573–1591, 1995.

[195] W. N. Hardy, D. A. Bonn, D. C. Morgan, R. X. Liang, and K. Zhang. Precision measurements of the temperature dependence of λ in $YBa_2Cu_3O_{6.95}$: Strong evidence for nodes in the gap function. *Phys. Rev. Lett.*, 70:3999–4002, 1993.

[196] K. Zhang, D. A. Bonn, S. Kamal, et al. Measurement of the ab plane anisotropy of microwave surface impedance of untwinned $YBa_2Cu_3O_{6.95}$ single crystals. *Phys. Rev. Lett.*, 73:2484–2487, 1994.

[197] D. A. Bonn, S. Kamal, K. Zhang, R. X. Liang, and W. N. Hardy. The microwave surface impedance of $YBa_2Cu_3O_{7-\delta}$. *J. Chem Phys. Solids*, 56:1941–1943, 1995.

[198] J. E. Sonier, R. F. Kiefl, J. H. Brewer, et al. New muon-spin-rotation measurement of the temperature dependence of the magnetic penetration depth in $YBa_2Cu_3O_{6.95}$. *Phys. Rev. Lett.*, 72:744–747, 1994.

[199] J. Mao, D. H. Wu, J. L. Peng, R. L. Greene, and S. M. Anlage. Anisotropic surface impedance of $YBa_2Cu_3O_{7-\delta}$ single crystals. *Phys. Rev. B*, 51:3316–3319, 1995.

[200] L. A. deVaulchier, J. P. Vieren, Y. Guldner, et al. Linear temperature variation of the penetration depth in $YBa_2Cu_3O_{7-\delta}$ thin films. *Europhys. Lett.*, 33:153, 1996.

[201] T. Jacobs, S. Sridhar, Q. Li, G. D. Gu, and N. Koshizuka. In-plane and c-axis microwave penetration depth of $Bi_2Sr_2Ca_1Cu_2O_{8+\delta}$ crystals. *Phys. Rev. Lett.*, 75:4516–4519, 1995.

[202] S. F. Lee, D. C. Morgan, R. J. Ormeno, et al. a-b plane microwave surface impedance of a high-quality $Bi_2Sr_2CaCu_2O_8$ single crystal. *Phys. Rev. Lett.*, 77:735–738, 1996.

[203] O. Waldmann, F. Steinmeyer, P. Müller, et al. Temperature and doping dependence of the penetration depth in $Bi_2Sr_2CaCu_2O_{8+\delta}$. *Phys. Rev. B*, 53:11825–11830, 1996.

[204] C. Panagopoulos, J. R. Cooper, G. B. Peacock, et al. Anisotropic magnetic penetration depth of grain-aligned $HgBa_2Ca_2Cu_3O_{8+\delta}$. *Phys. Rev. B*, 53:R2999–R3002, 1996.

[205] C. Panagopoulos, J. R. Cooper, T. Xiang, et al. Probing the order parameter and the c-axis coupling of high-T_c cuprates by penetration depth measurements. *Phys. Rev. Lett.*, 79:2320–2323, 1997.

[206] C. Panagopoulos, J. R. Cooper, N. Athanassopoulou, and J. Chrosch. Effects of Zn doping on the anisotropic penetration depth of $YBa_2Cu_3O_7$. *Phys. Rev. B*, 54:R12721–R12724, 1996.

[207] C. Panagopoulos, J. R. Cooper, T. Xiang, et al. Anisotropic penetration depth measurements of high-T_c superconductors. *Phys. C*, 282:145–148, 1997.

[208] C. Panagopoulos, J. R. Cooper, and W. Lo. *unpublished*.

[209] C. Panagopoulos and T. Xiang. Relationship between the superconducting energy gap and the critical temperature in high-T_c superconductors. *Phys. Rev. Lett.*, 81:2336–2339, 1998.

[210] D. M. Broun, D. C. Morgan, R. J. Ormeno, et al. In-plane microwave conductivity of the single-layer cuprate $Tl_2Ba_2CuO_{6+\delta}$. *Phys. Rev. B*, 56:R11443–R11446, 1997.

[211] V. J. Emery and S. A. Kivelson. Superconductivity in bad metals. *Phys. Rev. Lett.*, 74:3253–3256, 1995.

[212] D. N. Basov, R. X. Liang, D. A. Bonn, et al. In-plane anisotropy of the penetration depth in $YBa_2Cu_3O_{7-x}$ and $YBa_2Cu_4O_8$ superconductors. *Phys. Rev. Lett.*, 74: 598–601, 1995.

[213] Z. Schlesinger, R. T. Collins, F. Holtzberg, et al. Superconducting energy gap and normal-state conductivity of a single-domain $YBa_2Cu_3O_7$ crystal. *Phys. Rev. Lett.*, 65:801–804, 1990.

[214] T. A. Friedmann, M. W. Rabin, J. Giapintzakis, J. P. Rice, and D. M. Ginsberg. Direct measurement of the anisotropy of the resistivity in the a-b plane of twin-free, single-crystal, superconducting $YBa_2Cu_3O_{7-\delta}$. *Phys. Rev. B*, 42:6217–6221, 1990.

[215] R. C. Yu, M. B. Salamon, J. P. Lu, and W. C. Lee. Thermal conductivity of an untwinned $YBa_2Cu_3O_{7-\delta}$ single crystal and a new interpretation of the superconducting state thermal transport. *Phys. Rev. Lett.*, 69:1431–1434, 1992.

[216] M. B. Gaifullin, Y. Matsuda, N. Chikumoto, et al. c-axis superfluid response and quasiparticle damping of underdoped Bi:2212 and Bi:2201. *Phys. Rev. Lett.*, 83:3928–3931, 1999.

[217] P. J. Hirschfeld and N. Goldenfeld. Effect of strong scattering on the low-temperature penetration depth of a d-wave superconductor. *Phys. Rev. B*, 48:4219–4222, 1993.

[218] D. A. Bonn, S. Kamal, K. Zhang, et al. Comparison of the influence of Ni and Zn impurities on the electromagnetic properties of $YBa_2Cu_3O_{6.95}$. *Phys. Rev. B*, 50:4051–4063, 1994.

[219] J. Annett, N. Goldenfeld, and S. R. Renn. Interpretation of the temperature dependence of the electromagnetic penetration depth in $YBa_2Cu_3O_{7-\delta}$. *Phys. Rev. B*, 43:2778–2782, 1991.

[220] J. Y. Lee, K. M. Paget, T. R. Lemberger, S. R. Foltyn, and X. D. Wu. Crossover in temperature dependence of penetration depth $\lambda(t)$ in superconducting $YBa_2Cu_3O_{7-\delta}$ films. *Phys. Rev. B*, 50:3337–3341, 1994.

[221] J. Nagamatsu, N. Nakagawa, T. Muranaka, Y. Zenitani, and J. Akimitsu. Superconductivity at 39K in magnesium diboride. *Nature*, 410:63–64, 2001.

[222] N. Nakai, M. Ichioka, and K. Machida. Field dependence of electronic specific heat in two-band superconductors. *J. Phys. Soc. Jpn.*, 71:23–26, 2002.

[223] I. I. Mazin, O. K. Andersen, O. Jepsen, et al. Superconductivity in MgB_2: Clean or dirty? *Phys. Rev. Lett.*, 89:107002, 2002.

[224] T. Xiang and J. M. Wheatley. Superfluid anisotropy in YBCO: Evidence for pair tunneling superconductivity. *Phys. Rev. Lett.*, 76:134–137, 1996.

[225] C. Panagopoulos, J. L. Tallon, and T. Xiang. Effects of the CuO chains on the anisotropic penetration depth of $YBa_2Cu_4O_8$. *Phys. Rev. B*, 59:R6635–R6638, 1999.

[226] H. G. Luo and T. Xiang. Superfluid response in electron-doped cuprate superconductors. *Phys. Rev. Lett.*, 94:027001, 2005.

[227] T. Sato, T. Kamiyama, T. Takahashi, K. Kurahashi, and K. Yamada. Observation of $d_{x^2-y^2}$-like superconducting gap in an electron-doped high-temperature superconductor. *Science*, 291:1517–1519, 2001.

[228] N. P. Armitage, D. H. Lu, D. L. Feng, et al. Superconducting gap anisotropy in $Nd_{1.85}Ce_{0.15}CuO_4$: Results from photoemission. *Phys. Rev. Lett.*, 86:1126–1129, 2001.

[229] G. Blumberg, A. Koitzsch, A. Gozar, et al. Nonmonotonic $d_{x^2-y^2}$ superconducting order parameter in $Nd_{2-x}Ce_xCuO_4$. *Phys. Rev. Lett.*, 88:107002, 2002.

[230] B. Chesca, K. Ehrhardt, M. Mößle, et al. Magnetic-field dependence of the maximum supercurrent of $La_{2-x}Ce_xCuO_{4-y}$ interferometers: Evidence for a predominant $d_{x^2-y^2}$ superconducting order parameter. *Phys. Rev. Lett.*, 90:057004, 2003.

[231] L. Alff, S. Meyer, S. Kleefisch, et al. Anomalous low temperature behavior of superconducting $Nd_{1.85}Ce_{0.15}CuO_{4-y}$. *Phys. Rev. Lett.*, 83:2644–2647, 1999.

[232] J. D. Kokales, P. Fournier, L. V. Mercaldo, et al. Microwave electrodynamics of electron-doped cuprate superconductors. *Phys. Rev. Lett.*, 85:3696–3699, 2000.

[233] R. Prozorov, R. W. Giannetta, P. Fournier, and R. L. Greene. Evidence for nodal quasiparticles in electron-doped cuprates from penetration depth measurements. *Phys. Rev. Lett.*, 85:3700–3703, 2000.

[234] A. Snezhko, R. Prozorov, D. D. Lawrie, et al. Nodal order parameter in electron-doped $Pr_{2-x}Ce_xCuO_{4-\delta}$ superconducting films. *Phys. Rev. Lett.*, 92:157005, 2004.

[235] M. S. Kim, J. A. Skinta, T. R. Lemberger, A. Tsukada, and M. Naito. Magnetic penetration depth measurements of $Pr_{2-x}Ce_xCuO_{4-\delta}$ films on buffered substrates: Evidence for a nodeless gap. *Phys. Rev. Lett.*, 91:087001, 2003.

[236] J. A. Skinta, T. R. Lemberger, T. Greibe, and M. Naito. Evidence for a nodeless gap from the superfluid density of optimally doped $Pr_{1.855}Ce_{0.145}CuO_{4-y}$ films. *Phys. Rev. Lett.*, 88:207003, 2002.

[237] J. A. Skinta, M. S. Kim, T. R. Lemberger, T. Greibe, and M. Naito. Evidence for a transition in the pairing symmetry of the electron-doped cuprates $La_{2-x}Ce_xCuO_{4-y}$ and $Pr_{2-x}Ce_xCuO_{4-y}$. *Phys. Rev. Lett.*, 88:207005, 2002.

[238] A. V. Pronin, A. Pimenov, A. Loidl, A. Tsukada, and M. Naito. Doping dependence of the gap anisotropy in $La_{2-x}Ce_xCuO_4$ studied by millimeter-wave spectroscopy. *Phys. Rev. B*, 68:054511, 2003.

[239] N. P. Armitage, F. Ronning, D. H. Lu, et al. Doping dependence of an n-type cuprate superconductor investigated by angle-resolved photoemission spectroscopy. *Phys. Rev. Lett.*, 88:257001, 2002.

[240] T. Yoshida, X. J. Zhou, T. Sasagawa, et al. Metallic behavior of lightly doped $La_{2-x}Sr_xCuO_4$ with a Fermi surface forming an arc. *Phys. Rev. Lett.*, 91:027001, 2003.

[241] T. Xiang and J. M. Wheatley. Quasiparticle energy dispersion in doped two-dimensional quantum antiferromagnets. *Phys. Rev. B*, 54:R12653–R12656, 1996.

[242] T. Xiang, H. G. Luo, D. H. Lu, K. M. Shen, and Z. X. Shen. Intrinsic electron and hole bands in electron-doped cuprate superconductors. *Phys. Rev. B*, 79:014524, 2009.

[243] Z. Z. Wang, T. R. Chien, N. P. Ong, J. M. Tarascon, and E. Wang. Positive Hall coefficient observed in single-crystal $Nd_{2-x}Ce_xCuO_{4-\delta}$ at low temperatures. *Phys. Rev. B*, 43:3020–3025, 1991.

[244] W. Jiang, S. N. Mao, X. X. Xi, et al. Anomalous transport properties in superconducting $Nd_{1.85}Ce_{0.15}CuO_{4-\delta}$. *Phys. Rev. Lett.*, 73:1291–1294, 1994.

[245] P. Fournier, X. Jiang, W. Jiang, et al. Thermomagnetic transport properties of $Nd_{1.85}Ce_{0.15}CuO_{4+\delta}$ films: Evidence for two types of charge carriers. *Phys. Rev. B*, 56:14149–14156, 1997.

[246] C. Kusko, R. S. Markiewicz, M. Lindroos, and A. Bansil. Fermi surface evolution and collapse of the Mott pseudogap in $Nd_{2-x}Ce_xCuO_{4-\delta}$. *Phys. Rev. B*, 66:140513, 2002.

[247] H. Matsui, K. Terashima, T. Sato, et al. Angle-resolved photoemission spectroscopy of the antiferromagnetic superconductor $Nd_{1.87}Ce_{0.13}CuO_4$: Anisotropic spin-correlation gap, pseudogap, and the induced quasiparticle mass enhancement. *Phys. Rev. Lett.*, 94:047005, 2005.

[248] Q. S. Yuan, Y. Chen, T. K. Lee, and C. S. Ting. Fermi surface evolution in the antiferromagnetic state for the electron-doped $t - t' - t'' - J$ model. *Phys. Rev. B*, 69:214523, 2004.

[249] S. K. Yip and J. A. Sauls. Nonlinear Meissner effect in CuO superconductors. *Phys. Rev. Lett.*, 69:2264–2267, 1992.

[250] K. A. Moler, D. L. Sisson, J. S. Urbach, et al. Specific heat of $YBa_2Cu_3O_{7-\delta}$. *Phys. Rev. B*, 55:3954–3965, 1997.

[251] Y. X. Wang, B. Revaz, A. Erb, and A. Junod. Direct observation and anisotropy of the contribution of gap nodes in the low-temperature specific heat of $YBa_2Cu_3O_7$. *Phys. Rev. B*, 63:094508, 2001.

[252] I. Kosztin and A. J. Leggett. Nonlocal effects on the magnetic penetration depth in d-wave superconductors. *Phys. Rev. Lett.*, 79:135–138, 1997.

[253] M. R. Li, P. J. Hirschfeld, and P. Wölfle. Is the nonlinear Meissner effect unobservable? *Phys. Rev. Lett.*, 81:5640–5643, 1998.

[254] D. C. Mattis and J. Bardeen. Theory of the anomalous skin effect in normal and superconducting metals. *Phys. Rev.*, 111:412–417, 1958.

[255] R. A. Ferrell and R. E. Glover. Conductivity of superconducting films: A sum rule. *Phys. Rev.*, 109:1398–1399, 1958.

[256] M. Tinkham and R. A. Ferrell. Determination of the superconducting skin depth from the energy gap and sum rule. *Phys. Rev. Lett.*, 2:331–333, 1959.

[257] C. C. Homes, S. V. Dordevic, D. A. Bonn, R. X. Liang, and W. N. Hardy. Sum rules and energy scales in the high-temperature superconductor $YBa_2Cu_3O_{6+x}$. *Phys. Rev. B*, 69:024514, 2004.

[258] D. N. Basov, S. I. Woods, A. S. Katz, et al. Sum rules and interlayer conductivity of high-T_c cuprates. *Science*, 283:49–52, 1999.

[259] H. J. A. Molegraaf, C. Presura, D. Van Der Marel, P. H. Kes, and M. Li. Superconductivity-induced transfer of in-plane spectral weight in $Bi_2Sr_2CaCu_2O_{8+\delta}$. *Science*, 295:2239–2241, 2002.

[260] L. B. Ioffe and A. J. Millis. Superconductivity and the c-axis spectral weight of high-T_c superconductors. *Science*, 285:1241–1244, 1999.

[261] P. J. Hirschfeld, W. O. Putikka, and D. J. Scalapino. Microwave conductivity of d-wave superconductors. *Phys. Rev. Lett.*, 71:3705–3708, 1993.

[262] P. J. Hirschfeld, W. O. Putikka, and D. J. Scalapino. *d*-wave model for microwave response of high-T_c superconductors. *Phys. Rev. B*, 50:10250–10264, 1994.

[263] P. A. Lee. Localized states in a *d*-wave superconductor. *Phys. Rev. Lett.*, 71: 1887–1890, 1993.

[264] A. Hosseini, S. Kamal, D. A. Bonn, R. X. Liang, and W. N. Hardy. c-axis electrodynamics of YBa$_2$Cu$_3$O$_{7-\delta}$. *Phys. Rev. Lett.*, 81:1298–1301, 1998.

[265] T. Xiang and W. N. Hardy. Universal c-axis conductivity of high-T_c oxides in the superconducting state. *Phys. Rev. B*, 63:024506, 2000.

[266] T. Valla, A. V. Fedorov, P. D. Johnson, et al. Evidence for quantum critical behavior in the optimally doped cuprate Bi$_2$Sr$_2$CaCu$_2$O$_{8+\delta}$. *Science*, 285:2110–2113, 1999.

[267] T. Valla, A. V. Fedorov, P. D. Johnson, et al. Temperature dependent scattering rates at the Fermi surface of optimally doped Bi$_2$Sr$_2$CaCu$_2$O$_{8+\delta}$. *Phys. Rev. Lett.*, 85: 828–831, 2000.

[268] L. B. Ioffe and A. J. Millis. Zone-diagonal-dominated transport in high-T_c cuprates. *Phys. Rev. B*, 58:11631–11637, 1998.

[269] Y. I. Latyshev, T. Yamashita, L. N. Bulaevskii, et al. Interlayer transport of quasiparticles and cooper pairs in Bi$_2$Sr$_2$CaCu$_2$O$_{8+\delta}$ superconductors. *Phys. Rev. Lett.*, 82:5345–5348, 1999.

[270] V. Ambegaokar and A. Griffin. Theory of the thermal conductivity of superconducting alloys with paramagnetic impurities. *Phys. Rev.*, 137:A1151–A1167, 1965.

[271] A. C. Durst and P. A. Lee. Impurity-induced quasiparticle transport and universal-limit Wiedemann-Franz violation in *d*-wave superconductors. *Phys. Rev. B*, 62: 1270–1290, 2000.

[272] L. Taillefer, B. Lussier, R. Gagnon, K. Behnia, and H. Aubin. Universal heat conduction in YBa$_2$Cu$_3$O$_{6.9}$. *Phys. Rev. Lett.*, 79:483–486, 1997.

[273] S. Nakamae, K. Behnia, L. Balicas, et al. Effect of controlled disorder on quasiparticle thermal transport in Bi$_2$Sr$_2$CaCu$_2$O$_8$. *Phys. Rev. B*, 63:184509, 2001.

[274] M. Sutherland, D. G. Hawthorn, R. W. Hill, et al. Thermal conductivity across the phase diagram of cuprates: Low-energy quasiparticles and doping dependence of the superconducting gap. *Phys. Rev. B*, 67:174520, 2003.

[275] A. Abrikosov and V. M. Genkin. On the theory of Raman scattering of light in superconductors. *Zh. Eksp. Teor. Fiz.[Sov. Phys. JETP* **38**, *417 (1974)]*, 65:842, 1973.

[276] P. B. Allen. Fermi-surface harmonics: A general method for nonspherical problems. application to Boltzmann and Eliashberg equations. *Phys. Rev. B*, 13:1416–1427, 1976.

[277] M. V. Klein and S. B. Dierker. Theory of Raman scattering in superconductors. *Phys. Rev. B*, 29:4976–4991, 1984.

[278] H. Monien and A. Zawadowski. Theory of Raman scattering with final-state interaction in high-T_c BCS superconductors: Collective modes. *Phys. Rev. B*, 41:8798–8810, 1990.

[279] T. P. Devereaux and D. Einzel. Erratum: Electronic Raman scattering in superconductors as a probe of anisotropic electron pairing. *Phys. Rev. B*, 54:15547–15547, 1996.

[280] T. Staufer, R. Nemetschek, R. Hackl, P. Müller, and H. Veith. Investigation of the superconducting order parameter in Bi$_2$Sr$_2$CaCu$_2$O$_8$ single crystals. *Phys. Rev. Lett.*, 68:1069–1072, 1992.

[281] X. K. Chen, J. C. Irwin, H. J. Trodahl, T. Kimura, and K. Kishio. Investigation of the superconducting gap in La$_{2-x}$Sr$_x$CuO$_4$ by Raman spectroscopy. *Phys. Rev. Lett.*, 73:3290–3293, 1994.

[282] T. P. Devereaux, D. Einzel, B. Stadlober, et al. Electronic Raman scattering in high-T_c superconductors: A probe of $d_{x^2-y^2}$ pairing. *Phys. Rev. Lett.*, 72:396–399, 1994.

[283] C. Kendziora, R. J. Kelley, and M. Onellion. Superconducting gap anisotropy vs doping level in high-T_c cuprates. *Phys. Rev. Lett.*, 77:727–730, 1996.

[284] M. M. Qazilbash, A. Koitzsch, B. S. Dennis, et al. Evolution of superconductivity in electron-doped cuprates: Magneto-Raman spectroscopy. *Phys. Rev. B*, 72:214510, 2005.

[285] G. Blumberg, M. M. Qazilbash, B. S. Dennis, and R. L. Greene. Evolution of coherence and superconductivity in electron-doped cuprates. AIP Conference Proceedings, edited by Y. Takano, S. P. G. Blumberg, M. M. Qazilbash, B. S. Dennis, and R. L. Greene. Evolution of coherence and superconductivity in electron-doped cuprates. AIP Conference Proceedings, edited by Y. Takano, S. P. Hershfield, P. J. Hirschfeld, and A. M. Goldman, 24th International Conference on Low Temperature Physics (LT24). *Low Temperature Physics, Pts. A and B*, 850:525, 2006.

[286] C. S. Liu, H. G. Luo, W. C. Wu, and T. Xiang. Two-band model of Raman scattering on electron-doped high-T_c superconductors. *Phys. Rev. B*, 73:174517, 2006.

[287] C. P. Slichter. *Principles of Magnetic Resonance*. Springer, Berlin, 1996.

[288] F. Mila and T. M. Rice. Spin dynamics of $YBa_2Cu_3O_{6+x}$ as revealed by NMR. *Phys. Rev. B*, 40:11382–11385, 1989.

[289] L. C. Hebel and C. P. Slichter. Nuclear spin relaxation in normal and superconducting aluminum. *Phys. Rev.*, 113:1504–1519, 1959.

[290] H. Alloul, P. Mendels, H. Casalta, J. F. Marucco, and J. Arabski. Correlations between magnetic and superconducting properties of Zn-substituted $YBa_2Cu_3O_{6+x}$. *Phys. Rev. Lett.*, 67:3140–3143, 1991.

[291] P. Mendels, H. Alloul, G. Collin, et al. Macroscopic magnetic properties of Ni and Zn substituted $YBa_2Cu_3O_x$. *Phys. C: Supercond.*, 235:1595–1596, 1994.

[292] A. V. Mahajan, H. Alloul, G. Collin, and J. F. Marucco. ^{89}Y NMR probe of Zn induced local moments in $YBa_2(Cu_{1-y}Zn_y)_3O_{6+x}$. *Phys. Rev. Lett.*, 72:3100–3103, 1994.

[293] A. V. Mahajan, H. Alloul, G. Collin, and J. F. F. Marucco. ^{89}Y NMR probe of Zn induced local magnetism in $YBa_2(Cu_{1-y}Zn_y)_3O_{6+x}$. *Eur. Phys. J. B: Cond. Mat. Comp. Sys.*, 13:457–475, 2000.

[294] M. H. Julien, T. Fehér, M. Horvatić, et al. ^{63}Cu NMR evidence for enhanced antiferromagnetic correlations around Zn impurities in $YBa_2Cu_3O_{6.7}$. *Phys. Rev. Lett.*, 84:3422–3425, 2000.

[295] K. Ishida, Y. Kitaoka, K. Yamazoe, K. Asayama, and Y. Yamada. Al NMR probe of local moments induced by an Al impurity in high-T_c cuprate $La_{1.85}Sr_{0.15}CuO_4$. *Phys. Rev. Lett.*, 76:531–534, 1996.

[296] J. Bobroff, W. A. MacFarlane, H. Alloul, et al. Spinless impurities in high-T_c cuprates: Kondo-like behavior. *Phys. Rev. Lett.*, 83:4381–4384, 1999.

[297] C. Bernhard, C. Niedermayer, T. Blasius, et al. Muon-spin-rotation study of Zn-induced magnetic moments in cuprate high-T_c superconductors. *Phys. Rev. B*, 58:R8937–R8940, 1998.

[298] J. L. Tallon, J. W. Loram, and G. V. M. Williams. Comment on "spinless impurities in high-T_c cuprates: Kondo-like behavior". *Phys. Rev. Lett.*, 88:059701, 2002.

[299] A. V. Balatsky, M. I. Salkola, and A. Rosengren. Impurity-induced virtual bound states in d-wave superconductors. *Phys. Rev. B*, 51:15547–15551, 1995.

[300] M. I. Salkola, A. V. Balatsky, and D. J. Scalapino. Theory of scanning tunneling microscopy probe of impurity states in a d-wave superconductor. *Phys. Rev. Lett.*, 77:1841–1844, 1996.

[301] G. V. M. Williams and S. Krämer. Localized behavior near the Zn impurity in $YBa_2Cu_4O_8$ as measured by nuclear quadrupole resonance. *Phys. Rev. B*, 64:104506, 2001.

[302] S. H. Pan, J. P. O'neal, R. L. Badzey, et al. Microscopic electronic inhomogeneity in the high-T_c superconductor $Bi_2Sr_2CaCu_2O_{8+x}$. *Nature*, 413:282–285, 2001.

[303] M. R. Norman, M. Randeria, H. Ding, and J. C. Campuzano. Phenomenological models for the gap anisotropy of $Bi_2Sr_2CaCu_2O_8$ as measured by angle-resolved photoemission spectroscopy. *Phys. Rev. B*, 52:615–622, 1995.

[304] J. A. Martindale, S. E. Barrett, C. A. Klug, et al. Anisotropy and magnetic field dependence of the planar copper NMR spin-lattice relaxation rate in the superconducting state of $YBa_2Cu_3O_7$. *Phys. Rev. Lett.*, 68:702–705, 1992.

[305] P. C. Hammel, M. Takigawa, R. H. Heffner, Z. Fisk, and K. C. Ott. Spin dynamics at oxygen sites in $YBa_2Cu_3O_7$. *Phys. Rev. Lett.*, 63:1992–1995, 1989.

[306] S. E. Barrett, J. A. Martindale, D. J. Durand, et al. Anomalous behavior of nuclear spin-lattice relaxation rates in $YBa_2Cu_3O_7$ below T_c. *Phys. Rev. Lett.*, 66:108–111, 1991.

[307] G. Q. Zheng, Y. Kitaoka, K. Asayama, et al. Characteristics of the spin fluctuation in $Tl_2Ba_2Ca_2Cu_3O_{10}$. *J. Phys. Soc. Jpn*, 64:3184–3187, 1995.

[308] T. Imai, T. Shimizu, H. Yasuoka, Y. Ueda, and K. Kosuge. Anomalous temperature dependence of Cu nuclear spin-lattice relaxation in $YBa_2Cu_3O_{6.91}$. *J. Phys. Soc. Jpn.*, 57:2280–2283, 1988.

[309] K. Ishida, Y. Kitaoka, K. Asayama, K. Kadowaki, and T. Mochiku. Cu NMR study in single crystal $Bi_2Sr_2CaCu_2O_8$ – observation of gapless superconductivity. *J. Phys. Soc. Jpn.*, 63:1104–1113, 1994.

[310] S. Ohsugi, Y. Kitaoka, K. Ishida, G. Q. Zheng, and K. Asayama. Cu NMR and NQR studies of high-T_c superconductor $La_{2-x}Sr_xCuO_4$. *J. Phys. Soc. Jpn.*, 63:700–715, 1994.

[311] K. Magishi, Y. Kitaoka, G. Q. Zheng, et al. Spin correlation in high-T_c cuprate $HgBa_2Ca_2Cu_3O_{8+\delta}$ with T_c=133 K – an origin of T_c-enhancement evidenced by ^{63}Cu-NMR study. *J. Phys. Soc. Jpn.*, 64:4561–4565, 1995.

[312] M. Takigawa and D. B. Mitzi. NMR studies of spin excitations in superconducting $Bi_2Sr_2CaCu_2O_{8+\delta}$ single crystals. *Phys. Rev. Lett.*, 73:1287–1290, 1994.

[313] Y. Kitaoka, K. Ishida, G. Zheng, S. Ohsugi, K. Yamazoe, and K. Asayama. NMR study of symmetry of the superconducting order parameter in high-T_c cuprate superconductor. *Phys. C*, 235–240:1881, 1994.

[314] J. A. Martindale, S. E. Barrett, K. E. O'Hara, C. P. Slichter, W. C. Lee, and D. M. Ginsberg. Magnetic-field dependence of planar copper and oxygen spin-lattice relaxation rates in the superconducting state of $YBa_2Cu_3O_7$. *Phys. Rev. B*, 47: 9155–9157, 1993.

[315] M. Horvatić, T. Auler, C. Berthier, Y. Berthier, P. Butaud, W. G. Clark, J. A. Gillet, P. Ségransan, and J. Y. Henry. NMR investigation of single-crystal $YBa_2Cu_3O_{6+x}$ from the underdoped to the overdoped regime. *Phys. Rev. B*, 47:3461–3464, 1993.

[316] Y. Q. Song, M. A. Kennard, K. R. Poeppelmeier, and W. P. Halperin. Spin susceptibility in the $La_{2-x}Sr_xCuO_4$ system from underdoped to overdoped regimes. *Phys. Rev. Lett.*, 70:3131–3134, 1993.

[317] Y. Kitaoka, K. Fujiwara, K. Ishida, et al. Spin dynamics in heavily-doped high-T_c superconductors $Tl_2Ba_2CuO_{6+y}$ with a single CuO_2 layer studied by ^{63}Cu and ^{205}Tl NMR. *Phy. C: Supercond.*, 179:107–118, 1991.

[318] S. W. Lovesey. *Theory of Neutron Scattering from Condensed Matter, Vol. II.* Clarendon Press, Oxford, 1984.

[319] P. Coleman. *Introduction to Many-body Physics*. Cambridge University Press, Cambridge, 2015.

[320] J. Rossat-Mignod, L. P. Regnault, C. Vettier, et al. Neutron scattering study of the $YBa_2Cu_3O_{6+x}$ system. *Phys. C: Supercond.*, 185-189:86–92, 1991.

[321] J. Rossat-Mignod, L. P. Regnault, C. Vettier, et al. Spin dynamics in the high-T_c system $YBa_2Cu_3O_{6+x}$. *Phys. B: Condens. Matter*, 180-181:383–388, 1992.

[322] J. Rossat-Mignod, L. P. Regnault, C. Vettier, et al. Investigation of the spin dynamics in $YBa_2Cu_3O_{6+x}$ by inelastic neutron scattering. *Phys. B: Conden. Matter*, 169:58–65, 1991.

[323] H. F. Fong, B. Keimer, P. W. Anderson, et al. Phonon and magnetic neutron scattering at 41 meV in $YBa_2Cu_3O_7$. *Phys. Rev. Lett.*, 75:316–319, 1995.

[324] P. Dai, M. Yethiraj, H. A. Mook, T. B. Lindemer, and F. Doğan. Magnetic dynamics in underdoped $YBa_2Cu_3O_{7-x}$: Direct observation of a superconducting gap. *Phys. Rev. Lett.*, 77:5425–5428, 1996.

[325] H. A. Mook, P. C. Dai, S. M. Hayden, et al. Spin fluctuations in $YBa_2Cu_3O_{6.6}$. *Nature*, 395:580, 1998.

[326] H. F. Fong, P. Bourges, Y. Sidis, et al. Spin susceptibility in underdoped $YBa_2Cu_3O_{6+x}$. *Phys. Rev. B*, 61:14773–14786, 2000.

[327] H. F. Fong, P. Bourges, Y. Sidis, et al. Neutron scattering from magnetic excitations in $Bi_2Sr_2CaCu_2O_{8+\delta}$. *Nature*, 398:588–591, 1999.

[328] H. He, Y. Sidis, P. Bourges, et al. Resonant spin excitation in an overdoped high temperature superconductor. *Phys. Rev. Lett.*, 86:1610–1613, 2001.

[329] H. He, P. Bourges, Y. Sidis, et al. Magnetic resonant mode in the single-layer high-temperature superconductor $Tl_2Ba_2CuO_{6+\delta}$. *Science*, 295:1045–1048, 2002.

[330] M. A. Kastner, R. J. Birgeneau, G. Shirane, and Y. Endoh. Magnetic, transport, and optical properties of monolayer copper oxides. *Rev. Mod. Phys.*, 70:897–928, 1998.

[331] N. B. Christensen, D. F. McMorrow, H. M. Rønnow, et al. Dispersive excitations in the high-temperature superconductor $La_{2-x}Sr_xCuO_4$. *Phys. Rev. Lett.*, 93:147002, 2004.

[332] P. C. Dai, H. A. Mook, G. Aeppli, S. M. Hayden, and F. Dogan. Resonance as a measure of pairing correlations in the high-T_c superconductor $YBa_2Cu_3O_{6.6}$. *Nature*, 406:965–968, 2000.

[333] E. Demler and S. C. Zhang. Quantitative test of a microscopic mechanism of high-temperature superconductivity. *Nature*, 396:733–735, 1998.

[334] D. J. Scalapino and S. R. White. Superconducting condensation energy and an antiferromagnetic exchange-based pairing mechanism. *Phys. Rev. B*, 58:8222–8224, 1998.

[335] H. F. Fong, B. Keimer, D. Reznik, D. L. Milius, and I. A. Aksay. Polarized and unpolarized neutron-scattering study of the dynamical spin susceptibility of $YBa_2Cu_3O_7$. *Phys. Rev. B*, 54:6708–6720, 1996.

[336] D. K. Morr and D. Pines. The resonance peak in cuprate superconductors. *Phys. Rev. Lett.*, 81:1086–1089, 1998.

[337] M. Eschrig. The effect of collective spin-1 excitations on electronic spectra in high-T_c superconductors. *Adv. Phys.*, 55:47–183, 2006.

[338] P. A. Lee, N. Nagaosa, and X-G. Wen. Doping a Mott insulator: Physics of high-temperature superconductivity. *Rev. Mod. Phys.*, 78:17–85, 2006.

[339] E. Demler and S. C. Zhang. Theory of the resonant neutron scattering of high-T_c superconductors. *Phys. Rev. Lett.*, 75:4126–4129, 1995.

[340] E. Demler, W. Hanke, and S. C. Zhang. $SO(5)$ theory of antiferromagnetism and superconductivity. *Rev. Mod. Phys.*, 76:909–974, 2004.

[341] N. Bulut and D. J. Scalapino. Weak-coupling model of spin fluctuations in the superconducting state of the layered cuprates. *Phys. Rev. B*, 45:2371–2384, 1992.

[342] C. J. Wu, J. P. Hu, and S. C. Zhang. Exact SO(5) symmetry in the spin-3/2 fermionic system. *Phys. Rev. Lett.*, 91:186402, 2003.

[343] C. Wu. Hidden symmetry and quantum phases in spin-3/2 cold atomic systems. *Mod. Phys. Lett. B*, 20:1707–1738, 2006.

[344] C. Caroli, P. G. De Gennes, and J. Matricon. Bound fermion states on a vortex line in a type II superconductor. *Phys. Lett.*, 9:307–309, 1964.

[345] H. F. Hess, R. B. Robinson, R. C. Dynes, J. M. Valles, and J. V. Waszczak. Scanning-tunneling-microscope observation of the Abrikosov flux lattice and the density of states near and inside a fluxoid. *Phys. Rev. Lett.*, 62:214–216, 1989.

[346] H. F. Hess, R. B. Robinson, and J. V. Waszczak. Vortex-core structure observed with a scanning tunneling microscope. *Phys. Rev. Lett.*, 64:2711–2714, 1990.

[347] G. E. Volovik. Superconductivity with lines of gap nodes: Density of states in the vortex. *JETP Lett.*, 58:469–469, 1993.

[348] M. Franz and Z. Tešanović. Quasiparticles in the vortex lattice of unconventional superconductors: Bloch waves or Landau levels? *Phys. Rev. Lett.*, 84:554–557, 2000.

[349] I. Vekhter, P. J. Hirschfeld, and E. J. Nicol. Thermodynamics of d-wave superconductors in a magnetic field. *Phys. Rev. B*, 64:064513, 2001.

[350] K. A. Moler, D. J. Baar, J. S. Urbach, et al. Magnetic field dependence of the density of states of $YBa_2Cu_3O_{6.95}$ as determined from the specific heat. *Phys. Rev. Lett.*, 73:2744–2747, 1994.

[351] B. Revaz, J. Y. Genoud, A. Junod, et al. d-wave scaling relations in the mixed-state specific heat of $YBa_2Cu_3O_7$. *Phys. Rev. Lett.*, 80:3364–3367, 1998.

[352] D. A. Wright, J. P. Emerson, B. F. Woodfield, et al. Low-temperature specific heat of $YBa_2Cu_3O_{7-\delta}$, $0 \leqslant \delta \leqslant 0.2$: Evidence for d-wave pairing. *Phys. Rev. Lett.*, 82:1550–1553, 1999.

[353] J. L. Luo, J. W. Loram, T. Xiang, T. R. Cooper, and J. L. Tallon. The magnetic field dependence of the electronic specific heat of $Y_{0.8}Ca_{0.2}Ba_2Cu_3O_{6+x}$. *arXiv:cond-mat/0112065*, 2001.

[354] H. H. Wen, L. Shan, X. G. Wen, et al. Pseudogap, superconducting energy scale, and Fermi arcs of underdoped cuprate superconductors. *Phys. Rev. B*, 72:134507, 2005.

[355] C. Kübert and P. J. Hirschfeld. Vortex contribution to specific heat of dirty d-wave superconductors: Breakdown of scaling. *Solid State Commun.*, 105:459–463, 1998.

[356] S. H. Simon and P. A. Lee. Scaling of the quasiparticle spectrum for d-wave superconductors. *Phys. Rev. Lett.*, 78:1548–1551, 1997.

[357] A. Junod, M. Roulin, B. Revaz, et al. Specific heat of high temperature superconductors in high magnetic fields. *Phys. C: Supercond.*, 282:1399–1400, 1997.

[358] N. D. Mermin and H. Wagner. Absence of ferromagnetism or antiferromagnetism in one or two-dimensional isotropic Heisenberg models. *Phys. Rev. Lett.*, 17: 1133–1136, 1966.

[359] N. W. Ashcroft and N. D. Mermin. *Solid State Phys.* Holt, Rinehart and Winston, New York, 1976.

Index

Printed in the United States
by Baker & Taylor Publisher Services